How Much is Clean Air Worth?

How Much is Clean Air Worth? offers a comprehensive overview of the core methodologies and tools used to quantify the impacts and damage costs of pollution. The book begins by reviewing the tools used for environmental assessments and shows that a rational approach requires an impact pathway analysis (IPA) for each of the possible impacts of a pollutant, i.e. an analysis of the chain emission → dispersion → exposure-response functions → monetary valuation. The IPA methodology is explained in full and illustrated with worked examples, difficulties are discussed and uncertainties analyzed. In addition to detailed computer models, a very simple model (the "uniform world model") is presented, enabling readers to make estimates for cases where limited input data are available. Published results for electricity, waste treatment and transport are reviewed, with a thorough discussion of policy implications. This book will appeal to a broad mix of academics, graduate students and practitioners in government and industry working on cost–benefit analysis, environmental impact analysis and environmental policy.

Ari Rabl obtained his PhD in physics at Berkeley and worked for many years as research scientist for energy technologies at Argonne, NREL and Princeton University. He moved to the Ecole des Mines de Paris as Senior Scientist at the Centre Energétique et Procédés until his retirement in 2007, and now continues to work as a consultant. For the last 20 years, his work has focused on environmental impacts and costs of pollution. He is one of the principal participants of the ExternE Project series (External Costs of Energy) of the European Commission DG Research, for which he has co-ordinated several research projects.

Joseph V. Spadaro is Research Professor at the Basque Centre for Climate Change, Bilbao, Spain where he works on low carbon planning strategies, and climate change and health vulnerability. He is also Environmental Systems Engineer at Argonne National Labs, Decision and Information Sciences Division, USA, working on environmental impact assessment of energy technologies. He has been a member of the core team of principal investigators in the ExternE Project series (European Commission, DG Research) since the mid 1990s, and has acted as expert consultant for international organizations, national governments and private industry in projects related to impacts and costs of pollution.

Mike Holland is a freelance consultant based in the UK, and an Honorary Research fellow at Imperial College London. He has worked on assessment of pollution effects on ecosystems and health since 1986, and, following his work on the ExternE project, has performed a large number of cost–benefit analyses for national government and the European Commission, particularly in relation to measures to improve air quality.

How Much is Clean Air Worth?

Calculating the Benefits of Pollution Control

Ari Rabl

Joseph V. Spadaro

Mike Holland

CAMBRIDGE
UNIVERSITY PRESS

CAMBRIDGE
UNIVERSITY PRESS

University Printing House, Cambridge CB2 8BS, United Kingdom

Cambridge University Press is part of the University of Cambridge.

It furthers the University's mission by disseminating knowledge in the pursuit of education, learning and research at the highest international levels of excellence.

www.cambridge.org
Information on this title: www.cambridge.org/9781107043138

© Ari Rabl, Joseph V. Spadaro and Mike Holland 2014

First published 2014

Printed in the United Kingdom by Clays, St Ives plc

A catalogue record for this publication is available from the British Library

ISBN 978-1-107-04313-8 Hardback

Additional resources for this publication at www.cambridge.org/9781107043138

Contents

Contents

Contents

Figures

Tables

Foreword

In the Western world, it has been a long journey to achieve the current ecological and social transition. The US National Environmental Policy Act of 1969 and the European Community Environmental Action Programme of 1973 opened the way to establishing cost–benefit analysis and, later, the polluter-pays principle. The 1981 US Presidential Executive Order 12191 required Impact Assessment studies to be carried out for all major policies presented by the Federal Government, while the 1986 Single European Act stipulated that, when developing environmental policies, the European Community will take account of available scientific and technical data, and of benefits and costs of actions and lack of action.

Within this context, at the end of the eighties, the EU and USA (European Commission's DG XII – Science and Research and US Department of Energy) launched a fruitful collaboration on a joint study on fuel cycle cost that gave birth to the so-called ExternE – Externalities of Energy – project series, funded since then and with different names by the European Commission through its successive Research Framework Programmes.

Among the pioneer actors of this interdisciplinary research work, one can mention David Pearce, Ari Rabl, Anil Markandya, Olav Hohmeyer, Robert Shelton, Russell Lee, Alan Krupnick, Nick Eyre and Richard Ottinger. More recently, key researchers in the field of external costs quantification, tackling the issues of energy, environment and transport have been Mike Holland, Jacquie Berry, Rainer Friedrich, Andrea Ricci, Joseph V. Spadaro, Stale Navrud, Stefan Hirschberg and Milan Scasny.

The concept of external costs entered the European political jargon with the 1993 Jacques Delors White Paper on growth, competitiveness and employment, which stated that "energy can no longer be seen as an unlimited resource, particularly if the external costs associated with climate change, acidification and health are not taken into account (. . .). The way the Community uses its labour and environmental resources highlights some fundamental weaknesses in the incentive structure of the

Community economy attributable to public intervention (e.g. tax treatment of labour costs, transport infrastructure) as well as to market forces (environmental externalities). It is open to question whether, to an increasing extent, the economic growth figures do not reflect illusionary instead of real economic progress and whether many traditional economic concepts (e.g. GDP as traditionally conceived) may be losing their relevance to policy formulation in the future."

Since then a lot of progress has been made on external costs quantification, both from a research perspective and on EU policy implementation.

Regarding the latter, and to give some concrete examples of European policies in the last decade, it is worth mentioning references to "internalization of damage caused to the environment" in the 2000 Commission Green Paper on security of energy supply, or "the principle of recovery of the costs of water services, including environmental and resource costs, having regard to economic analysis (...) and in accordance with the polluter-pays principle," in the EU Water Framework Directive, adopted in 2000.

The Community guidelines on State aid for environmental protection published by the Commission in 2001 mentioned that "Member States may grant operating aid to new plants producing renewable energy that will be calculated on the basis of the external costs avoided (...) that must not exceed 5 euro-cents per kilowatt hour."

The 2004 European Commission Environmental Technologies Action Plan stated that getting the prices right requires the systematic internalization of costs through market-based instruments (e.g. taxes, tax breaks, subsidies, tradeable permits, deposit-refund schemes) that make producers and consumers bear the real costs of their actions or change their behavior in a cost-effective way.

The European Environmental Liability Directive, that entered into force on 1 May 2008, forces industrial polluters to pay the costs of preventing and remedying environmental damage, including where it affects biodiversity, water and human health.

The 2007 Commission Green Paper on market-based instruments adds that these "give firms an incentive, in the longer term, to pursue technological innovation to further reduce adverse impacts on the environment ('dynamic efficiency') and support employment when used in the context of environmental tax or fiscal reform."

The 2005 cost–benefit analysis of the European thematic strategy on air pollution was essentially based on ExternE research work, including the health impacts of particles. The 2008 EU Directive on ambient air quality requires that "Member States reduce exposure levels in urban areas to PM2.5 by an average of 20% by 2020 based on predicted 2010 exposure levels. Member States will be obliged to bring exposure levels below 20 μg m^{-3} by 2015 in these areas."

From an academic and research perspective, this book provides an extremely useful update and comprehensive overview of the most recent scientific developments. It clearly presents impact pathways analysis, one of the most original methodologies of ExternE: the dispersion of traditional pollutants like PM, NOx, SO$_2$ and VOC, as well as greenhouse gas emissions and toxic metals; the exposure–response functions and the monetary valuation, i.e. translation of the health and environmental impacts into easily understandable figures like euro-cents per kilowatt hour of coal-based electricity or euros per 100 passenger kilometres of a diesel car.

The relation between impact pathways analysis and life cycle assessment (LCA) is also a very good piece of reading. The exposure–response functions from health impacts have been cleverly thought out. They deal with mortality (loss of life expectancy), mortality due to PM and O$_3$, and toxic metals like arsenic, cadmium, chrome and lead.

The dispersion of pollutants is not only addressed for the air but also for the soil and water with its impacts from human ingestion, on agriculture, on ecosystems and on damage to materials.

Very concrete applications of this research work are provided for the electricity sectors covering fossil fuels, nuclear and renewable energy sources and technologies for both the European Union and the USA. External costs are also estimated for the transport sector, including an innovative comparison between conventional vehicles (gasoline and diesel) and hybrid electric vehicles.

A chapter on the change in exposure for individuals who switch from car to bicycle or to walking provides a fresh approach to addressing sustainable lifestyles and illustrating the impact of behavioral changes. The pedagogic part of this book, on economics – from the discount rate to valuation of mortality and morbidity – and on uncertainties of damage costs, demonstrates the crucial role of ethical choices.

Briefly, this Rabl–Spadaro–Holland book enriches the scientific debate on external costs, updates and completes previous research, and offers practical examples that have been or could be implemented through policy measures.

For a decade now, in upstream EU policy-making, there has been an obligation to prepare an *impact assessment* for any legislative or programmatic proposals. For such an exercise, quantification in monetary terms of the environmental impact is extremely useful. Furthermore, thanks to this research work, economic models can also be improved, particularly in their environmental dimension. As such, the Rabl–Spadaro–Holland book serves both policy-makers and academics.

Over forty years, combined European and American efforts on external costs quantification have transformed the concepts of "internalization of externalities" and "getting the prices right" from a purely theoretical exercise to a daily and genuine political message in the Western world.

Since the 2008 financial and economic crises some commentators have spoken about the "End of the West," the strength of "State Capitalism" or the "Decline of Europe." We are convinced that also thanks to the hand-in-hand work of researchers and decision-makers, of rigorous scientists calculating the benefits of pollution control and enlightened policy-makers implementing the right measures, a *renaissance* is possible.

This *renaissance* takes place through sustainable economic growth, a more efficient use of natural resources and the right value awarded to common goods like air, water and soil.

Domenico Rossetti di Valdalbero,
Principal Administrator at the European Commission,
Directorate General for Research and Innovation
Pierre Valette, *Acting Director,*
Directorate Environment – DG Research,
European Commission

Preface

The book is addressed to
- Researchers interested in the calculation of environmental impacts;
- Policy-makers and their advisors, in energy and environmental policy;
- Graduate students and advanced undergraduates in environmental science.

In the past, decisions about environmental policy were made without quantifying the benefits. Pollution had become so bad, for instance with the Great London Smog of 1952 and rivers like the Rhine becoming too poisoned for fish to survive, that the demand for cleanup became overwhelming and environmental regulations were imposed in the absence of a cost–benefit analysis (CBA). The main sources of pollution and their impacts were obvious, and the regulations were clearly beneficial.

Nowadays, the remaining environmental problems tend to be more complex and so is the task of finding suitable solutions. For example, what should we do with our waste? Should what remains after recycling be incinerated or put into landfill, either method having some harmful impacts? Fortunately, environmental science has progressed to the point where the problems can be analyzed with a fair degree of confidence and CBA can help us to identify the best solutions. When cost-effective measures are proposed, CBA is a powerful tool for convincing concerned stakeholders that such measures should indeed be implemented.

Calculation of the damage costs of pollution ("external costs") is multi-disciplinary to the extreme, requiring expertise in engineering, environmental modeling, epidemiology, ecology, economics, statistics, life cycle assessment, and so on. This presents quite a challenge for the writing of a book on the subject. We do have a broad expertise in most of these fields, demonstrated by our publications in fields as diverse as economics, dispersion modeling, epidemiology, risk analysis, life cycle assessment, energy policy, waste treatment, and transport policy. We have been very active in all phases of the ExternE (External Costs of Energy) project series of the European Commission (EC), DG Research. That is why this book shows much more on methods and results of ExternE than on equivalent work elsewhere; even though two of us (AR and JVS) live and work both in the EU and the USA.

Our goal has been to provide an introduction to the subject that is sufficiently thorough to enable readers to understand the methodologies and even to carry out their own approximate calculations. We believe that, particularly in the age of powerful computers and sophisticated software, it is crucial to understand the principles of the calculations and to have a sense of whether the results are plausible. This is because most policies or other options that are proposed will be contested by people who would be worse off, or believe that they would be worse off. Decision makers, or at least their advisors, need to have a sufficient understanding of the issues to be able to decide whether or not such critiques are justified. Therefore we present models that are simple, transparent and suitable for first approximations. Our guiding principle has been "better approximately right than precisely wrong."

While our focus is on air pollution, we also touch on related burdens, such as noise, congestion, accidents and resource use, in order to enrich the presentation of applications to power production, waste treatment and transportation. The research is rapidly evolving, but our capacities are limited and we cannot provide perfect up-to-date coverage of each topic. Experts in the respective fields may well find items to criticize, although few would have the combination of breadth and depth to offer a better comprehensive treatment of the entire subject. In fact, we have found several instances where experts of one field use the results obtained in other fields, without understanding the underlying assumptions and limitations. For example, the valuation of air pollution mortality has frequently been done by economists who do not understand what has been measured by epidemiologists and who use a number of deaths that is inappropriate (as we have explained in several publications and also in this book). Another example of trans-disciplinary misunderstanding is the use of site-specific numbers by policy makers, when they really need typical results for an entire region.

A criticism frequently lobbed towards environmental CBA is that the uncertainties are too large to inform policy assessment. In response, we emphasize that it is not the uncertainties themselves that matter, but the consequences for the decisions. The relevant question is "how large is the cost penalty for making a wrong decision because of an over- or underestimation of the damage cost, compared with the optimal decision if one had known the true cost?" As we have shown, the information is very valuable in spite of seemingly large uncertainties, because near the optimum the total social cost does not vary much as a function of the abatement level: even a sizable error entails only a small increase in the total social cost. And what is the alternative to CBA? Is it better to guess, based on vague intuitions or preconceived ideas? Without a careful CBA the error could be so large as to entail very costly decisions. To prove these points, the book contains a substantial chapter on the analysis of uncertainties.

One of the challenges of writing a book on this subject is that a flood of new studies is continually adding new elements that should ideally be used to update all the affected results. And all the updates should account for the effect of inflation on the costs ... of course, even items with large uncertainty, such as the value of a life year, should be adjusted (and we cannot help thinking of that story of the museum director who happens to overhear one of the guides saying, "this statue is 3007 years old." He asks the guide how he comes to know that and the guide replies. "when you hired me seven years ago you told me ...."). Rather than trying to keep all the numbers up to date, for the purpose of illustrating the methodology, we think that sometimes older results can be cited. For presentation of the main damage cost results of ExternE in Chapters 12 to 15, we have chosen the numbers of the NEEDS and CASES projects (ExternE, 2008) and we explain how they can be adjusted, in particular for changes in the cost of mortality and of greenhouse gases that we believe to be appropriate. We also present results from analogous studies in the USA.

For the organization of this book, two alternatives seem logical: by impact category, or in the order of the steps of the impact pathways analysis (IPA: analysis of the chain emission → dispersion → exposure–response function → monetary valuation). Nevertheless, we found it preferable to use a mixed approach because the exposure–response functions (ERF) determine what kind of dispersion modeling is required. Therefore, we present the chapters with ERFs before those on atmospheric models and multimedia models. For some burdens we treat the entire IPA in a single section. Climate change is such a complex subject that we cannot do it justice in our book and we have only one chapter where we focus on issues related to estimation of the damage costs of greenhouse gases. Development of the IPA methodology closes with a chapter on the uncertainties, where we present both Monte Carlo analysis and a simple shorthand method that we have developed for typical calculations of IPA and risk analysis.

The last part of the book presents the results of IPA analysis, with applications for electricity generation, waste treatment and transport. To close the book, we review how the results have been used by policy makers.

Our motivation has been the desire to help to bring a greater rationality to environmental decision making. All too often, decisions have been the outcome of noisy arguments and angry power play, without evaluating costs and benefits. As a result, society ends up with gross and costly inconsistencies, paying billions to avoid a death due to one particular risk, while refusing to pay a few thousand when a similar death is due to a different risk.[1] Even if a well-documented analysis of costs and benefits may not remove disagreements, at least it can provide a basis for proper and detailed discussion.

[1] Documented by Tengs and Graham (1996).

Acknowledgements

This book is the result of our participation in the ExternE project series of the European Commission, initiated by Pierre Valette and, after Pierre moved to a different area, continued by Domenico Rossetti di Valdalbero. The enthusiastic support and effective management they have both provided for ExternE has greatly contributed to the success of the research and its use by policy makers. We have had the benefit of countless interesting and informative discussions with many colleagues working on ExternE projects and on similar work elsewhere. It is not possible to list them all, or even to remember whom we met when and where and what we talked about, but they all helped to shape our thinking and we would like to thank them. At the risk of forgetting some names, we mention Anil Markandya of BC3, Nick Eyre of the University of Oxford, Fintan Hurley of IOM, Rainer Friedrich of the University of Stuttgart, Ståle Navrud of the Norwegian University of Life Sciences, Bob van der Zwaan of the Energy Research Centre, Netherlands, Milan Scasny of Charles University Prague, Andrea Ricci of ISIS, Paul Watkiss of Paul Watkiss Associates, Till Bachmann of EIFER Karlsruhe, Richard Tol of University of Sussex, Jérôme Adnot, Denis Clodic and Assaad Zoughaib of Ecole des Mines de Paris, Vincent Nedellec of Nedellec Consultants, Olivier Chanel of Université de Marseille, Bernard de Caevel and Elisabeth van Overbeke of RDC Environment, Mona Dreicer of Lawrence Livermore Laboratory, Nino Künzli of Universität Basel, Stefan Hirschberg of the Paul Scherrer Institute, Mark Delucchi of University of California, Davis, Bart Ostro of CAL EPA, Alan Krupnick of Resources for the Future, Russell Lee of Oak Ridge National Laboratory, Bob Rowe and Laurie Chestnut of Stratus Consulting, Veronika Rabl of Vision & Results, Peter Curtiss of Curtiss Engineering, and Dick Wilson of Harvard.

Our special thanks go to Jim Hammitt of Harvard with whom we have enjoyed many stimulating discussions over the years. His thorough review and constructive criticisms of an early draft of this book have been extremely helpful. Phil Graves of the University of Colorado gave us valuable feedback on several chapters, especially Chapters 1 and 9, in addition to ideas during numerous conversations.

... and some individual acknowledgments:

Ari Rabl: I want to add a personal tribute to Brigitte Desaigues, my wife and colleague for many years until a cancer stole her. She was Professor of Environmental Economics at the Sorbonne and one of the participants in the ExternE projects. Many publications and reports are testimony to our collaboration.

This is also an opportunity to thank individuals who have been most important in the development of my career, thus enabling me to contribute to this book. Marty Halpern gave me a good start as my PhD advisor at Berkeley. Roland Winston and Frank Kreith helped me in my transition from theoretical physics to solar energy, Roland by recruiting me to work with him on solar collectors at Argonne National Laboratory and the University of Chicago, Frank by bringing me to the National Renewable Energy Laboratory in Colorado. Rob Socolow invited me to join the Center for Energy and Environmental Studies of Princeton University to work on renewable energy and energy efficiency; it was the most stimulating group of colleagues and I thought I would stay for the rest of my life ... until, during a leave of absence at the Ecole des Mines de Paris, I fell in love with Brigitte and decided to move to Paris. At the Ecole des Mines I am deeply grateful to my directors, Jérôme Adnot and Denis Clodic, for their encouragement and support of my switch from energy to environment.

Joseph V. Spadaro: I am grateful to several people who have had a profound impact during my formative years and those individuals who have since enriched my professional career. I would like to remember Maria Geraci, who selflessly gave of her time and started me on my journey to discovery. Yitzhak Sharon was my physics mentor, and later on became one of my closest friends and supporters. As a teacher he taught me the basic principles of science and maths, but it was his nurturing of and belief in my abilities that finally led me to pursue a career in science and engineering. He also helped me to secure my first job as an intern in the Solid State Division of the Oak Ridge National Laboratory. I am thankful to Robert Socolow from the Princeton Environment Institute of Princeton University, who hired me to work on energy efficiency in buildings. I am thankful to Yogesh Jaluria of the Mechanical and Aerospace Engineering Dept. of Rutgers University, for his guidance and friendship during an especially difficult transition time in my personal life. Hans Holger Rogner was my director at the International Atomic Energy Agency in Vienna. He continues to be a great source of inspiration, and I have greatly benefited from our time spent together. I am indebted to Jérôme Adnot and Denis Clodic of the Ecole des Mines de Paris, for their friendship and enthusiastic support of my PhD work. The night before my defense, Jérôme spent several hours coaching me on my French presentation. His efforts and mine were rewarded the next day when the jury acknowledged how much my French had improved since moving to Paris. Most importantly, I want to acknowledge here the many years of friendship and collaboration with my PhD advisor and co-author of this book, Ari Rabl. A lifetime to say thank you is not enough! On a personal note, I want to mention my wife Santuzza Spadaro, for her constant support and love over the years.

Mike Holland: I have been extremely fortunate to work with a series of teams over the years who have provided extremely stimulating and pleasurable working environments – the Department of Forestry and Natural Resources at the University of Edinburgh, the Pollution Effects Group at Imperial College, Silwood Park, the ExternE Team and the Strategic Studies Department at what was the UK's Energy Technology Support Unit (ETSU). The friendships forged in each made life far more pleasurable (and productive) than it would otherwise have been. Of course, I must also thank my family: my parents, Ray and Audrey, not just for their love but also the value that they placed in my education over a good many years; and Amanda, Amy and Evie for the happiness and joy that they bring.

1 Introduction

*The responsibility of those who exercise power in a democratic society is not to
reflect inflamed public feeling but to help form its understanding*
*Felix Frankfurter (former Supreme Court Justice) (1928) Carved in stone on the
wall of the Federal Court House, Boston*

Summary

In this chapter we explain why one needs to evaluate environmental costs
and benefits. Cost–benefit analysis (CBA) is necessary for many choices
relating to public policy, especially in the field of environmental protection,
to avoid costly mistakes. Even when other, non-monetary criteria must also
be taken into account, a CBA should be carried out whenever appropriate.
Without a monetary evaluation of damage costs one can only do a cost-
effectiveness analysis, as illustrated in Section 1.3. In Section 1.4 we explain
how to determine the optimal level of pollution abatement, as a simple
example of the use of a CBA. Impact pathway analysis (IPA), the methodology for quantifying damage costs or environmental benefits, is sketched in
Section 1.5. The internalization of external costs is addressed in Section 1.6.

1.1   Why quantify environmental benefits?

The answer emerges through asking another question: "how else can we
decide how much to spend to protect the environment?" The simple
demand for "zero pollution" sometimes made by well-meaning but naïve
environmentalists is totally unrealistic: our economy would be paralyzed
because the technologies for perfectly clean production do not exist.[1]

---

[1] If you don't believe it, try to think of an economic activity that does not involve pollution.
Maybe growing tomatoes in your yard and selling them at the local farmer's market? But
few people live within walking distance of the local farmers' market, and even if you do,
most of your customers have to drive or at least use a bicycle – and don't forget the
pollution emitted during the production of a bicycle. Chemical and physical processes

In the past, most decisions about environmental policy were made without quantifying the benefits. During the 1960s and 1970s increasing pollution and growing prosperity led to increased demand for cleaner air, and at the same time there was sufficient technological progress in the development of equipment such as flue gas desulfurization to allow cleanup without prohibitive costs. The demand for cleanup became overwhelming and environmental regulations were imposed with no cost–benefit analysis.

Not only would the scientific basis for environmental cost–benefit analysis (CBA) have been inadequate in the past, but much simpler criteria seemed satisfactory for making decisions. Conditions in cities like London were initially so bad, and effects on health so serious, that major action was taken on the assumption that this was for the common good (an assumption borne out by subsequent analysis). The classic example is the Great London Smog of 1952, during which 4000 additional deaths were recorded over what would ordinarily have been expected. Another criterion was based on the idea that a toxic substance has no effect below a certain threshold dose. If that is the case, it is sufficient to reduce the emission of a pollutant to below the level where even the highest doses remain below the threshold. Based on this idea standards for ambient air quality were developed, for example, by the World Health Organization, and governments imposed regulations that forced industry to reduce emissions to reach these standards.

However, the situation is changing. Epidemiologists have not been able to find no-effect thresholds for air pollutants, in any case not at the level of an entire population. For example, the most recent guidelines from the World Health Organization say that there seems to be no such threshold for PM (particulate matter). The available evidence suggests that exposure–response functions (ERF) are linear at low dose for PM, and probably for other air pollutants as well.[2] In particular, for the neurotoxic impacts of Pb the ERF is at least linear without threshold, and it may even be above the straight line at low dose. Linearity is already generally assumed for substances that initiate cancers.

At the same time, the incremental cost of reducing the emission of pollutants (called the "abatement cost")[3] increases sharply as lower

---

are by their very nature not "pure," i.e. they always produce at least some byproducts in addition to what we want. We can and should reduce pollution as much as is practical (in a sense to be defined below) but zero pollution is not feasible in an industrialized society.

[2] One implication of linear no-threshold ERFs is that the traditional preoccupation with pollution peaks is not really relevant: it is the long-term average exposure that matters. Of course, since on average the peaks are proportional to the long-term average exposure, reducing the peaks also reduces the average exposure.

[3] Some authors, for instance IPCC, use "mitigation" instead of "abatement."

emission levels are demanded. Thus the question "how much to spend?" acquires an increasing urgency, while the natural criterion for answering it by reference to a threshold has vanished. Even worse, we are facing a proliferation of small risks as ever more sensitive scientific methods identify ever greater numbers of substances that can have harmful effects, possibly (or probably, in the case of many carcinogens) without a safe threshold. Thus we have to deal with a new paradigm, and monetary evaluation of environmental benefits is required, to make our decisions consistent with our preferences. Optimal decisions are not obvious because of the complexity of the links.

Many people have objected to environmental CBA, feeling that one cannot assign monetary values to items such as a beautiful landscape, an endangered species or human life. This objection is based on a fundamental misunderstanding of what is involved in the monetary evaluation. In reality it is not the intrinsic value of the item in question, but society's collective willingness to pay (WTP) to avoid losing the item. For instance our WTP (even our ability to pay) to avoid an anonymous premature death is limited, even if we feel that the value of life is infinite. In any case, we have to make decisions, and a judgment about monetary values is implicit in our decisions. For example, if a city refuses to replace a railway level crossing by a bridge, at a cost x to avoid y traffic deaths, its "value of life" is less than x/y; if the bridge is built, the implied value is at least x/y.[4] Environmental CBA makes implicit judgments explicit. Stakeholders are free to disagree with the analysis, but they need to justify why they disagree.

Ultimately, the objective of a CBA of proposed policies or regulations is to render the decisions more consistent, in particular to avoid inconsistencies of the type where a billion is spent in one sector to avoid the loss of a life year, while refusing to spend a thousand for the same thing in another sector – inconsistencies amply documented by Tengs and Graham (1996), Lutter *et al.* (1999) and Morrall (2003). Tengs and Graham calculate that in the USA a more consistent allocation of resources could save about 0.5% of GDP, without any reduction in the protection

---

[4] The unfortunate term "value of statistical life" (VSL) used by economists often evokes hostility, "You cannot put a price on life" being a typical reaction. But that is a total misunderstanding of the problem. It is not "and how much for your grandmother?" The objective is not to determine the intrinsic value of life, of a beautiful landscape, of a cultural monument or a species threatened by extinction; rather it is the WTP to avoid the loss of the good in question. The WTP (including ability to pay) is limited, even for people who say that the good is priceless. Really VSL is the "willingness to pay for avoiding the risk of an anonymous premature death," and we prefer the term "value of prevented fatality" (VPF) which is more appropriate and less likely to evoke negative reactions. Another good term is "value of (mortality) risk reduction," which has been proposed in the USA.

of the population. Inconsistent valuations are especially likely when regulations are imposed in response to the latest "risk of the day" that happens to attract intense attention in the media (dioxins, mad cow disease, asbestos, electromagnetic fields etc.).[5]

Without monetary valuation of environmental goods one can do a cost-effectiveness analysis (CEA), i.e. rank choices with comparable outcomes, for instance options for reducing PM_{10} emission, or one can compare years of life saved by reducing air pollution versus years of life saved by reducing water pollution. But how can one compare choices with incommensurate outcomes, such as closing a factory to avoid pollution or keeping it going to avoid unemployment? Paying for putting particle filters on diesel buses to reduce particulate emissions or continuing to suffer the health impacts of the particles in our cities? Raising the price of cars to make them cleaner? General principles such as sustainable development or the precautionary principle provide no guidance (except in their most extreme and totally impractical interpretation of demanding zero pollution) because the difficulties lie in the specifics of each situation.

The extra cost of a cleaner environment must be paid, ultimately by the tax payer or consumer. For example, if a factory is forced to spend more for environmental protection (or pay a tax on pollution), this cost is passed on to the consumer – and if the cost is too high to be charged to the consumer because of competition from countries with less strict regulations, the owner of the factory has less incentive to continue the investments necessary to keep the factory up to date and will choose other more profitable investments instead; eventually the factory becomes unprofitable and is closed. Even if immediate tradeoffs do not cross budget categories, ultimately the money we spend on reducing pollution is not available for other good causes, such as the education of our children.

Links can be subtle and unexpected. When evaluating a decision, one should not forget the consequences of the alternatives and the induced effects. For example, lowering the limit for the allowable emission of dioxins from waste incinerators will avoid some cancer deaths, but it

---

[5] Usually there is little correlation between the magnitude of a risk and the amount of media attention, or it may even be negative, since small risks are more difficult to quantify: the greater the uncertainty, the more disagreement there is between the experts – and the greater is the entertainment value for the media. Much of the apparent irrationality of risk perception arises from binary thinking, a simple shortcut to help us to deal with the complexities of most decisions: A is safe, B is not, … Most of the time such shortcuts are better than getting paralyzed by an attempt to weigh quantitative criteria. But trouble comes when we are finding more and more small risks. Most potential new risks are nonzero and we, or the media, tend to assign them to the "not safe" category. So we have the cognitive illusion of living in an ever more dangerous world.

will raise the cost of waste disposal. And poverty also kills, as demonstrated by numerous studies. The reasons are poor education, unhealthy housing, inability to pay for medical services, etc. For example, in the USA, Keeney (1994) has found that for each \$5 to 10 million of cost imposed by a regulation there will be on average one additional premature death due to this cost. Lutter *et al.* (1999) estimate that a \$15 million decrease in income is associated with the loss of an additional statistical life, and therefore, regulations that cost more than \$15 million per expected life saved are likely to cause a net increase in mortality.

The value of clean air is not infinite. With excessive pollution control the costs are not worth the benefits. There is a socially optimal level of pollution – which decreases with increasing prosperity and technological progress. The link to prosperity is complex. People with different levels of income and in different situations will rightly have different priorities for its allocation: access to food, clean water and shelter will, for example, come ahead of most people's priority for clean air. Problems arise when the actions of one group impact unequally on others. Of particular relevance here are the risks linked to climate change, where those living on a dollar a day will be at higher risk than those who are richer and contribute far more to greenhouse gas emissions. Likewise, the costs and benefits of air pollution abatement are different for different income groups.

1.2   Cost is not the only criterion for decisions

A cost–benefit analysis (CBA) should not be a simple automatic criterion for environmental decisions. The results have to be used with care because the uncertainties can be large, and it can be too easy to manipulate assumptions to get the result that one particular group might want to see.

Furthermore, other considerations may be as important as cost, particularly equity (who pays and who benefits?). For many options for reducing air pollution the distribution of costs and benefits among the population is sufficiently uniform so as not to raise serious equity issues. But there are exceptions, for instance, if low emission zones are created in cities by prohibiting the use of older, more polluting vehicles: the benefits accrue to all while the costs fall on the owners of the excluded vehicles. If suitable compensation schemes cannot be devised, a policy option may be problematic even if the total benefits are clearly larger than the total costs.

For the choice of energy systems, one should also take into account issues such as supply security, the right to impose risk on future generations (nuclear waste or global warming), the dangers of proliferation and the acceptability of a large accident. Such issues involve societal value judgments, beyond the costs one can quantify. That is not an argument against monetary valuation. Rather, the appropriate approach is to quantify the costs as much as possible for input into the decision process. For example, one may be able to evaluate equity implications explicitly in monetary terms, as is being done for the implications of a carbon tax, by evaluating how the tax burden would be distributed among different socio-economic groups. Equity issues tend to weigh heavily in the decisions of elected officials, because they want to be re-elected.

The question of acceptability complicates any quantification of mortality risk. Mortality risk can differ in the nature of the death (e.g. by accident, by cancer or by other illness) as well as in attributes that influence the perception of a risk, for instance:

• is the risk voluntary or involuntary?
• is the risk natural or manmade?
• to what extent is it associated with an activity that is considered to be socially desirable?
• how much control does an individual have over the exposure or consequences?

Such attributes affect the importance that people place on avoiding a particular risk. This can confound even a direct comparison of risks in physical units (number of deaths or years of life lost), quite apart from any controversies surrounding the value in monetary terms.

However, such limitations do not render a CBA useless. If the analysis has been carried out with care, clearly stating the underlying assumptions, it brings hidden consequences into the open and helps focus the debate onto the facts. In particular, it can indicate whether a proposed decision reflects true preferences or merely a cognitive illusion.

1.3   Cost-effectiveness analysis

To show what can and what cannot be done without evaluation of damage costs, let us look at cost-effectiveness analysis (CEA). This quantifies the costs and the effectiveness of methods for achieving a goal, for instance, reducing the emission of a pollutant, and then ranks them according to their improvement/cost ratio. As an example, we show in Fig. 1.1 the marginal abatement costs for greenhouse gas emissions for various technology options in the EU, as calculated by IIASA (the International Institute for Applied Systems Analysis) (2010). For each option, IIASA determined the cost and the potential reduction if the option were implemented in the entire EU. Of course, one should implement the less expensive options before the more expensive ones. Therefore, they are presented in order of increasing cost (from left to right), and the x-axis shows the cumulative emission reduction if the various technology

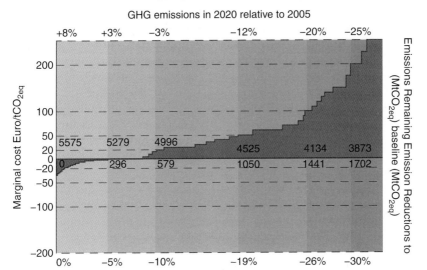

Fig. 1.1 Example of cost-effectiveness analysis: marginal abatement costs for various technology options in the EU. From IIASA (2010), reproduced with permission from the International Institute for Applied Systems Analysis (IIASA).

choices are implemented in the order of increasing cost. In this particular example of a CEA, the x-axis labels are a little complicated because the emission reduction is shown in several different ways (as % reduction relative to 1990 and 2005, as cumulative reduction in $Mt_{CO_{2eq}}$/yr,[6] and as remaining emissions).[7]

According to these data, up to 296 $Mt_{CO_{2eq}}$/yr can be avoided at negative abatement cost, because there are options that bring net savings. If the damage cost were equal to 50 €/t_{CO_2} (100 €/t_{CO_2}) the emissions could be reduced by as much as 1050 $Mt_{CO_{2eq}}$/yr (1441 $Mt_{CO_{2eq}}$/yr).

---

[6] The notation CO_{2eq} indicates that all greenhouse gases are included and expressed as equivalent CO_2 emissions.

[7] Note that the abatement costs are uncertain because for the most part they involve new technologies. Over time, as these technologies are applied more and more, their costs will decrease thanks to learning. For that reason, abatement cost curves for the near term, such as Fig. 1.1, show much higher costs than cost curves that are projected to the more distant future, with assumptions about learning rates, such as the marginal abatement cost curve in Fig. 1.2b.

That kind of information is valuable because it indicates how much can be achieved and at what cost. But it does not indicate how much should be done. Cost-effectiveness analysis is necessary but is not sufficient in itself. To determine the optimal level, one also needs to know the damage costs.

A further problem is that CEA typically focuses on the cost of achieving one particular goal, without reference to other issues. For example, in the case given, no account is taken of other impacts of the measures introduced to reduce carbon emissions. Some of these will be beneficial, such as the reduction in regional air pollutant emissions (SO_2 etc.) associated with reduced fossil fuel use. Some of them will involve a tradeoff of the carbon abatement benefit with other risks, for example, those associated with nuclear power generation. Cost–benefit analysis therefore provides a mechanism for a much more comprehensive assessment of the impacts of policies.

1.4   The optimal level of pollution abatement

Taken as a whole, society must pay the cost of pollution abatement and suffer the damage cost of the remaining pollution. To find the optimum, one has to minimize the sum of the abatement cost $C_{ab}(E)$ and the damage cost $C_{dam}(E)$,

$$C_{tot}(E) \; = \; C_{dam}(E) \; + \; C_{ab}(E) \qquad (1.1)$$

as a function of the emission level E of the pollutant. Setting the derivative of C_{tot} equal to zero one finds that the optimal emission level E_{opt} of the pollutant corresponds to the point where

$$\frac{dC_{dam}}{dE} + \frac{dC_{ab}}{dE} = 0 \;\; at \;\; E = E_{opt}. \qquad (1.2)$$

Economists call the derivative of the cost a marginal cost, and say that the optimum is where the marginal damage cost is equal to the marginal abatement cost.

For the important case of pollutants with linear or near-linear dose–response functions (the case for the most important impacts of the classical air pollutants), the marginal damage cost is independent of E. For greenhouse gases the marginal damage cost increases with E. Figure 1.2 illustrates the optimization with an example, the abatement of worldwide CO_2 emissions. Of course, this is an extraordinarily complex problem (even without politics interfering with the search for the truth), and Fig. 1.2 presents an

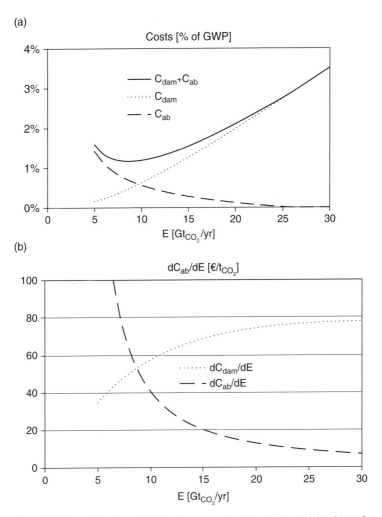

Fig. 1.2 Example of optimizing the emission level E: optimization of global CO_2 emissions, based on a simple steady-state analysis; (a) total costs, (b) marginal costs.

extreme simplification, based on a simple steady-state analysis (van der Zwaan and Rabl, 2008).[8] Part (a) shows the damage cost C_{dam} and the abatement cost C_{ab} as well as their sum and Part (b) shows their derivatives. The current emissions are 27 Gt_{CO_2}/yr. The optimal emission level is the one

---

[8] A more realistic dynamic analysis has been published by Rabl and van der Zwaan (2009), and some results can be found in Section 11.5.3.

that minimizes the total cost. It is the point where the marginal damage and the abatement costs are equal: 8.7 $GtCO_2$/yr (about a third of current emissions) under steady-state conditions. Of course in reality, the emissions and costs change over time and a more rigorous dynamic analysis is needed. Nonetheless, the key conclusion is quite robust, namely, that the emissions should be reduced by a factor of about three relative to the business-as-usual scenario; that conclusion has been confirmed by a more rigorous dynamic analysis by Rabl and van der Zwaan (2009).

Strictly speaking there should be a third term in the optimization: the cost of defensive or adaptive measures. For example, the damage caused by sea level rise due to CO_2 can be reduced by building dykes, and the damage to some materials due to SO_2 or O_3 can be reduced by appropriate surface treatments. If the damage, for a given emission level, can be reduced by defensive or adaptive measures, their cost, C_{def}, should be included in Eqs. (1.1) and (1.2) by considering C_{dam} as the net damage cost. For the classical air pollutants the potential for defensive measures is often so limited that it can be neglected in the optimization, but for greenhouse gases these are very important.

For another example, consider the National Emission Ceilings Directive of the EU. This directive fixed limits for the annual emissions of SO_2, NO_x, VOC and NH_3 for each member country, to be attained by 2010. In preparation for the negotiations leading to these limits, IIASA was asked to assemble abatement cost data for each country. The data for France are shown in Fig. 1.3; the points are from IIASA (1998), the line is

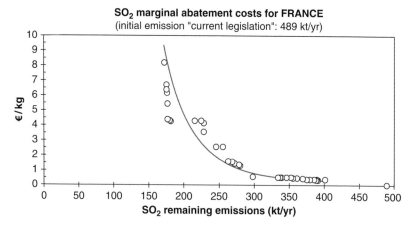

Fig. 1.3 Marginal cost of SO_2 abatement as a function of the total emissions of France. The solid line shows our curve fit to the data points of IIASA (1998).

our curve fit (Rabl *et al.*, 2005). The rightmost point shows the emission level at the time, 489 kt/yr, corresponding to zero spending for additional abatement.

Example 1.1 The national emission ceiling for France was 375 kt_{SO_2}/yr in 2010, and the actual emissions were much lower, 303 kt_{SO_2}/yr in 2009. But how low should they be if the marginal damage cost for emissions in France is equal to the ExternE (2008) estimate of 6.8 €/Kg_{SO_2}?

Solution Drawing a horizontal line at 6800 €/t_{SO_2} in Fig. 1.3, we see that the social optimum is much lower, at about 175 kt_{SO_2}/yr, less than half the current ceiling. Needless to say, the actual values of the directive are the result of political considerations, where the social optimum has not been the primary criterion.

1.5 How to quantify environmental benefits

1.5.1 Impact pathway analysis

Policy decisions must act on the sources of pollution. Therefore it is not sufficient to calculate the damage per exposure; one needs to know which source of a pollutant causes how much damage, i.e. what benefit one can expect from a reduction in specific emissions. This requires an impact pathway analysis (IPA), tracing the passage of the pollutant from where it is emitted to the affected receptors (population, crops, forests, buildings, etc.). The principal steps of this analysis can be grouped as follows:

- Emission: specification of the relevant technologies and pollutants (e.g. kg of NO_x emitted by a power plant per GWh_e);
- Dispersion: calculation of increased pollutant concentrations in all affected regions (e.g. incremental concentration of ozone, using models of atmospheric dispersion and chemistry for ozone formation due to this NO_x);
- Impact: calculation of the dose from the increased concentration and calculation of impacts (damage in physical units) from this dose, using an exposure–response function (e.g. cases of asthma due to the increase in ozone);
- Cost: economic evaluation of these impacts (e.g. multiplication by the cost incurred by a case of asthma).

The impacts and costs are summed over all receptors of concern. The work involves a multidisciplinary systems analysis, with input from engineers, dispersion modelers, epidemiologists, ecologists and economists. By now the methodology is well developed, thanks to the ExternE project series which the European Commission, DG Research, has funded since 1991, and by analogous research in the USA (ORNL/RFF 1994, Abt 2004, NRC 2010).

Of course, the work is fraught with difficulties. Much of the required information is uncertain and in some cases even missing completely. However, the alternative of not doing any analysis is worse. Remember the basic rule of precaution: "think before you act!" Before taking a decision, evaluate its consequences as much as can reasonably be achieved. If a complete IPA is not feasible, for instance when no monetary values are available, carry the analysis as far as possible and then use other methods for choosing the most desirable action, such as risk–risk comparisons or cost-effectiveness analysis (i.e. a ranking of abatement costs to find which option has the lowest cost per avoided pollution).

Damage cost estimates are notorious for their large uncertainties (see Chapter 11), and many people have questioned the usefulness of all this work. The first reply to this critique is that even an uncertainty by a factor of three (the typical uncertainty of air pollution damage costs) is better than infinite uncertainty. Secondly, in many cases the benefits are either so much larger or so much smaller than the costs that the implication for a decision is clear even in the face of uncertainty. Thirdly, if policy decisions are made without a significant bias in favor of either costs or benefits, some of the resulting decisions will err on the side of cost, others on the side of benefit. Rabl *et al.* (2005) have examined the consequences of such unbiased errors for the abatement of NO_x, SO_2 and CO_2, and found a very reassuring result: compared with the minimal social cost that one would incur with perfect knowledge, the extra social cost due to errors is remarkably small, amounting to less than 10 to 20% in most cases, even if the damage costs are in error by a factor of three times. But without any knowledge of the damage costs, the extra social cost could be very large.

Of course, the results of an IPA depend on the current state of scientific knowledge. Estimates of impacts and damage costs may grow or shrink. New impacts may turn out to be important. A little historical perspective may be good to keep us from being overconfident. For that reason we include, in Chapter 13 on electric power, a review of older fuel chain assessments.

1.5.2.   *Why not simply ask people how much they value each impact?*

Instead of all the complications of an IPA, why not simply ask people, by means of a questionnaire, how much they value each impact? It is really a matter of the appropriate expertise. The general public is of course the expert on general preferences and values. But not on dispersion modeling and exposure–response functions. It is appropriate to ask people who know what cancer can be like, how much they are willing to pay to avoid a cancer. But asking them how much they care to reduce the emission of NO_x or dioxins is not, because they do not know how serious the impacts really are. After all, a rational policy must be based on real impacts, not on the fears of the moment.

1.6   External costs and their internalization

1.6.1   Definition of external costs

In recent years, the term external cost has been used widely to designate the damage costs of pollution. But we should point out the risk of confusion, because there are at least two definitions of external cost:

(1) costs imposed on nonparticipants, that are not **taken into account** by the participants in a transaction;

(2) costs imposed on nonparticipants, that are not **paid** by the participants in a transaction.

According to the first definition, a damage cost is internalized if the polluter reduces the emissions to a socially optimal level, for example, as a result of regulation that imposes an emission limit. The second definition requires, in addition, the polluter to pay for the damage by, for example, using a pollution tax to allow compensation of the victims. In either case, the level of emissions is brought to the social optimum. But the corresponding damage cost is external only according to the second, and not the first definition.

Many economics texts claim the first definition, according to which only the damage beyond the social optimum is an external cost. However, this is not practical, because to calculate an external cost according to that definition one would first have to determine the socially optimal emission level. Such a calculation is doubly difficult and the result doubly uncertain, because most abatement costs are difficult to ascertain. Furthermore, the question of what percentage of the damage cost is already internalized is not relevant in practice, because decisions should be based on CBA and minimization of total social cost (= private cost + damage cost). The percentage of damage cost that is already internalized is not needed for this minimization. In the wake of ExternE, the term external cost is now generally used in the sense of damage cost (definition 2), and we will follow that practice.

As for internalization, there is a major difference between these two definitions of external cost. With a limitation on the emission (or with tradable permits that are given free), the polluter pays only the cost of abatement, i.e. the externality is internalized according to the first definition. But with a tax equal to the marginal damage (or equivalent tradable permits that are auctioned by the government), the polluter pays the cost of abatement as well as the damage cost of the remaining emissions; that's internalization according to the second definition.

As for compensation of the victims, perfect compensation is not feasible, because identifying who suffers how much damage is too difficult and uncertain. However, a tax on air pollution can achieve compensation at least in an approximate average sense because we can all benefit from a reduction in air pollution.

There is a further complication with the definition of external costs. Some transactions impose costs on nonparticipants, but costs that are taken into account by the market. For example, an influx of new people to an area, buying homes, drives up house prices, making purchasing more difficult for the existing residents who have not yet bought a home. Another example is the shift to corn-based ethanol production, thus reducing the amount of corn for food and driving up the price of food for consumers who have nothing to do with the decision to produce ethanol. Such externalities are called pecuniary externalities, in contrast with the externalities due to pollution, which operate outside the market. To emphasize the distinction, the latter are sometimes called technological or real externalities. We will say a little more about pecuniary externalities in Section 6.3 of Chapter 6 on depletion of non-renewable resources.

To avoid ambiguities with the definition of external cost, we prefer to use the term, damage cost, for the externalities due to pollution. In any case one needs to know the damage costs, regardless of what fraction has been internalized according to which definition.

1.6.2   Policy instruments

It is now generally recognized, within democratic countries, that the costs of pollution should be internalized. Internalization involves making the polluters reduce their emissions as though they were also the victim of their pollution (this simple definition of internalization applies equally to each of the two definitions of external cost shown above). The Coase theorem implies that internalization can be achieved without government intervention but only if the transaction costs are negligible, a condition that is certainly not satisfied for pollutants (for a good discussion of the Coase theorem in the environmental context, see Graves (2013)). Therefore government intervention is required.

Table 1.1 *Policy instruments for reducing pollution. C = command and control, M = market-based instruments.*

Type	C or M	Examples
Limits on emission of pollutants for specific technologies	C	Max. mg SO_2 per m^3 of flue gas; max. g CO per km driven by cars.
National emission ceilings	C or M[a]	Limit on tonnes of SO_2, NO_x, VOCs, CO_2, etc. emitted by a country
Choice of technologies	C	Usually by demanding "best available technology" (BAT), e.g. flue gas desulfurization for coal or oil fired boilers.
Major technology choices	C	Decisions in certain countries not to use nuclear power or to phase out existing nuclear plants.
Subsidies for clean technologies	M	Tax credits for wind and solar Feed-in tariffs for electricity from wind and solar
Eco-labels	M	"Printed on recycled paper"; "no chlorine used"; "energy star" label for computers.
Pollution taxes	M	€/tonne of a pollutant.
Tradable permits	M	Government sets cap on number of permits (e.g. tonnes of SO_2), polluters can trade these permits.
Portfolio standards	M	Government sets minimum % for the market share of a clean technology (e.g. "zero emission" vehicles in California, or "green kWh" from solar energy) and industry adjusts the prices to achieve these goals.

[a] depending on how a country chooses to implement the legislation.

Table 1.1 lists the principal types of regulations that the government can use for this purpose. Some of these policy instruments act directly on the emissions, others such as eco-labels and portfolio standards can effect emissions indirectly by discouraging the use of polluting processes or energy sources; for example, a portfolio standard that requires a minimum percentage of "green electricity" raises the price of more polluting energy sources.

At one extreme are command–control regulations that impose rigid specific constraints, for instance, a limit on the concentration of SO_2 in the flue gas of power plants. At the other extreme, the government can impose certain general rules, for instance, a tax per tonne of SO_2, and let the market respond. The middle column of the table indicates whether the regulations are based on command–control (C) or on market mechanisms (M).

Command–control yields predictable results (e.g. the specified reduction of SO_2), but at greater cost than market mechanisms because all polluters must take the same action. Individual costs of pollution abatement are highly variable according to local circumstances, being for instance much higher for an industry that must install an expensive retro-fit, than for one that can include the pollution control equipment in the design of a new factory. Another drawback of command–control is the lack of incentives to do better than the regulation. Market-based mecha-nisms, in particular pollution taxes and tradable permits, avoid these drawbacks. For a more detailed comparison of command–control and market mechanisms, with examples, see Graves (2013).

A pollutant tax gives a clear signal to the polluter about the magnitude of the damage. Of course, in order to give a clear signal the tax has to be specific for each pollutant. The tax must not be a lumped tax, for example a tax per kWh on all of the pollutants emitted by power plants, otherwise it would be like a supermarket without price tags, where the consumer would only be given a final bill upon checkout. Of course, industrial polluters tend to pass the payment on to the consumers, who will only see a general price increase, but the key point is that each tax has to be specific for each pollutant.

Under a pollution tax each polluter can choose how much of the pollutant to remove by abatement equipment and pay tax on whatever remains. A pollution tax achieves reductions at the lowest possible cost per avoided kg of pollutant (highest economic efficiency), and it provides an incentive to reduce emissions as much as is economically justified at the specified level of the tax; however, the magnitude of the realized reduction is difficult to predict.

There is a policy instrument, tradable permits, that combines highest economic efficiency with predictable results – at least in theory. Under this system, the government issues permits for a specified quantity of a pollu-tant that may be emitted in a region, and industry is free to buy or sell these permits. There are several variants, the two main distinctions being whether the government sells the permits at an auction or gives them away (for instance to each polluter according to last year's emission). Obviously, industry prefers the latter. In the USA, tradable permits have already been in use for SO_2 since the beginning of the 1990s, with great success: the cost per avoided kg turned out to be much lower than under the previous regime of command–control. Tradable permits, given for free, are now also used in the EU for CO_2. If permits are provided free, their allocation involves difficult and problematic negotiations with the polluters concerned.

Wild price swings for CO_2 permits in the EU since their introduction, in particular with prices dropping below 5 €/t_{CO_2} in 2013, have brought to light a serious problem with the use of permits in practice. Most industries have little short-term flexibility in being able to change emissions, because the emissions are determined by the technologies that are in place. If demand for a CO_2-intensive product goes up (down) temporarily, the most cost-effective response is to buy (sell) additional permits, rather than implementing process changes that tend to require more time and investment. The recession, since 2008, has curtailed the demand for energy and thus for CO_2 permits. A relatively inelastic demand for permits, without adjustment to the supply of permits (politically difficult to negotiate) implies large price swings. A CO_2 tax would have avoided such price swings, which are in themselves problematic for the affected industries and for the trading system.

There has been some opposition to the idea of tradable permits, especially from people who misunderstand "permit" to mean unlimited license to pollute and who do not recognize that the most widely used regulation, namely emission limits, is in effect a permit that is given away free, but cannot be traded.

Some may wonder whether the moral limits to markets, so eloquently explained by Sandel (2012), are an argument against tradable permits or pollution taxes. As a striking example of carrying the idea of pollution taxes to an extreme, he considers a hypothetical fee for littering in the Grand Canyon. Such a fee would remove the moral stigma of littering, and people rich enough to pay and too lazy to carry their trash back out would be free to litter. When taxes or permits are used to limit the emission of pollutants, Sandel laments the loss of moral stigma. However, there are crucial differences between littering and the emission of pollutants. The ideal, zero litter, can easily be attained at negligible cost and the social control is sufficiently strong and direct to restrain potential violators. By contrast, the ideal of zero pollution would impose unacceptable costs on an industrial society, and the social control is far too weak. Furthermore, pollution is driven by the choice of the consumer, who in most cases does not understand the relation between his/her choice and the associated pollution. Even those rare consumers who feel a sufficiently strong moral imperative to accept personal sacrifice have trouble knowing which goods are really cleaner. So there is no meaningful connection with moral stigma.

The great advantage of tradable permits or pollution taxes is the transmission of correct information about the damage cost of pollution to the price of all goods. This presupposes, of course, that the damage costs are known and that the government sets the right level for the permits or pollution taxes. For greenhouse gases the opposition to increased prices, coupled with disputes about international equity, may well prevent imposition of the necessary regulations. To think about the consequences, we recommend reading, *Collapse: How Societies Choose to Fail or Succeed* by Jared Diamond (2005).

This brief discussion can give no more than a highly simplified overview, without regard to any of the many different ways that each of these approaches can be implemented, or to the advantages and disadvantages they may offer in particular situations. Low administrative costs and ease of verification must be taken into account in choosing the most appropriate regulations. For example, unlike large power plants, monitoring of lawn mower emissions is not practical. And last but not least, a regulation is worthless if compliance is not verified and enforced.

1.6.3   Who is responsible, who should be targeted?

Whereas the "polluter pays" principle seems straightforward, in practice, the question of who is the polluter and who should be targeted by an internalization instrument can be more complex. For example, in the case of pollutants that a consumer emits into the waste water (e.g. estrogen or testosterone from hormone replacement therapies, or pollutants from cleaning products), the consumer can hardly do anything about it (other than using less polluting products), but the sewage company and/or the product manufacturer can. If they are forced to do it, they will pass the cost on to the consumer, so all can be automatically and correctly internalized.

Similar considerations apply to consumer products such as batteries. Here, one of the policy options is to oblige the producer of batteries to use technologies without toxic components (Cd, Ni and Hg), even though that increases the cost; of course, the cost will be passed on to the consumer. Electronic equipment contains many potentially harmful materials that should be recycled or treated as toxic waste, and the most effective policy for keeping such materials out of municipal solid waste is to require manufacturers to take the equipment back when the consumer wishes to discard it. The key consideration in such cases is "who can repair or minimize the problem for the lowest cost?"

Note that in any case it is crucial to aim the policy correctly at the right target, namely the specific pollutant that causes the problem. For example, an aggregated tax per kWh does not tell a power plant which pollutant contributes how much of the tax, so the power plant does not know how to allocate its abatement expenditures in an optimal manner from a social perspective. By way of contrast, a tax per kg of each pollutant, emitted by each source, does give a clear and correct signal.

The ultimate purpose of environmental CBA is to help to formulate policies that provide the correct incentives for each actor in the economy to achieve the goal of optimal protection of our environment – in other words, make the incentives goal-compatible.

1.7   Scope of this book

Most of this book deals with air pollution, although much of the methodology is also applicable to other forms of pollution. There are several reasons for this focus on air pollution, in addition to the fact that the authors are most familiar with its analysis. Air pollutants cause serious problems for the majority of the world's population, especially in terms of global warming and health impacts. Even though the industrialized countries have succeeded in reducing the emission of several air pollutants, especially SO_2 and PM_{10}, limitation of the greenhouse gases has been far more difficult. In the developing world, a reduction in air pollution is all the more challenging as these countries try to attain ever higher standards of living while their populations are, in most cases, increasing.

Soil pollution, at least in industrialized countries, is mostly a legacy of the past, although continual vigilance is needed to prevent problems with new installations. Likewise, the emission of most water pollutants has been greatly reduced in industrialized countries, with the glaring exception of nitrates and phosphates from agriculture.

The analysis of damage costs is simpler for air pollution than for water and soil pollution, because air is a homogeneous medium where pollutants disperse over large areas, whereas water pollution is more localized and site-specific. Soil pollution is localized and very site-specific, requiring detailed local data for its analysis. Even though for the important sectors of energy production, waste management and transport, the dominant impacts arise from air pollution, according to current assessments, we do not want to imply that water and soil pollution can be forgotten.

A word about the outline. It might seem most logical to organize this book according to the steps of the impact pathway analysis (IPA), followed by chapters with results and applications. However, we found such an approach to be awkward, because of all the close links between the steps of an IPA. For example, calculation of exposure by dispersion models is much simpler if the ERF is linear without threshold, than if there is a threshold, and so we begin with a survey of the ERFs and the evidence for linearity. Also, we find it instructive to illustrate the "Uniform World Model," an especially simple model, with examples for which we need specific ERFs for health impacts. Therefore we present health impacts before dispersion models. As an alternative, one could organize the chapters according to impact categories, but then much material would need to be repeated to avoid the need for an excessive number of cross-references. Finally, we chose an intermediate approach that allows us to write the chapters as relatively self-contained units without too much overlap or cross-referencing.

References

Abt 2004. Power Plant Emissions: Particulate Matter-Related Health Damages and the Benefits of Alternative Emission Reduction Scenarios. Prepared for EPA by Abt Associates Inc., 4800 Montgomery Lane, Bethesda, MD 20814–5341.

Diamond, J. M. 2005. *Collapse: How Societies Choose to Fail or Succeed.* Viking Press.

Graves, P. 2013. *Environmental Economics: An Integrated Approach.* CRC Press/Taylor & Francis, Boca Raton, FL 33487.

IIASA 1998. Janusz Cofala and Sanna Syri. Sulfur emissions, abatement technologies and related costs for Europe in the RAINS model database. International Institute for Applied Systems Analysis, IIASA Interim Report, IR-98-35, Appendix 6: Sulfur dioxide emission abatement cost curves for FRANCE (page 6.14).

IIASA 2010. GAINS data sheet on greenhouse gas mitigation potentials. Version 2.1, April 2010. International Institute For Applied Systems Analysis, Laxenburg, Austria. www.gains.iiasa.ac.at/index.php/home-page

Keeney, R. L. 1994. Mortality risks induced by the costs of regulations. *Journal of Risk and Uncertainty* **8**, 95–110.

Lutter, R., Morrall, M. and Viscusi, W. K. 1999. The cost-per-life-saved cutoff for safety-enhancing regulations, *Economic Inquiry* **37**, 599–608.

Morrall, J. F. 2003. Saving lives: A review of the record. *Journal of Risk and Uncertainty* **27**(3):221–237.

NRC 2010. *Hidden Costs of Energy: Unpriced Consequences of Energy Production and Use.* National Research Council of the National Academies Press. National Academies Press, 500 Fifth Street, NW Washington, DC 20001.

ORNL/RFF 1994. External Costs and Benefits of Fuel Cycles. Prepared by Oak Ridge National Laboratory and Resources for the Future. Edited by Russell Lee, Oak Ridge National Laboratory, Oak Ridge, TN 37831.

Rabl, A. and van der Zwaan, B. 2009. Cost–benefit analysis of climate change dynamics: uncertainties and the value of information. *Climatic Change* **96**, No. 3 / October, 2009.

Rabl, A., Spadaro, J. V. and van der Zwaan, B. 2005. Uncertainty of pollution damage cost estimates: to what extent does it matter? *Environmental Science & Technology* **39**(2), 399–408.

Sandel, M. J. 2012. *What Money Can't Buy: The Moral Limits of Markets.* Farrar, Strauss & Giroux, New York.

Tengs, T. O. and Graham, J. D. 1996. The opportunity costs of haphazard social investment in life-saving, chapter 8, pp. 167–184, of *Risks, Costs and Lives Saved*, R. W. Hahn, ed. Oxford University Press, Oxford.

van der Zwaan, B. C. C. and Rabl, A. 2008. Uncertainties of external costs: how large are they and to what extent do they matter?, Section 1.16 in *The Social Cost of Electricity: Scenarios and Policy Implications.* Markandya, A., Bigano, A. and Roberto Porchia, R., eds. 2010. Fondazione Eni Enrico Mattei. Edward Elgar Publishing Ltd, Cheltenham, UK.

2 Tools for environmental impact and damage assessment

Summary

Countless tools, models and software packages have been developed for the analysis of environmental problems. This chapter focuses on tools that allow the assessment of environmental impacts and the comparison of technologies and policy choices. Impact Pathway Analysis (IPA) is presented in some detail because it is the correct approach for quantifying impacts and damage costs of pollution. Section 2.2 is an introduction to IPA; detailed discussions of the various elements follow in Chapters 3 to 9. We also discuss Life Cycle Assessment (LCA) and the relation between LCA and IPA. Difficulties and problems with the use of the various tools are addressed in Sections 2.4 and 2.5. Section 2.6 proposes an integrated framework for the analysis of environmental questions.

2.1 Overview of tools

2.1.1 Starting point: the DPSIR framework

There are a great number of tools, methods and models for the analysis of environmental problems. They differ in approach and objectives, but there is also much overlap and they are difficult to classify in a systematic scheme. We will not attempt a systematic survey but will focus instead on a few key features that are crucial for decision making, namely the ability to:
- define the appropriate scope for the analysis,
- model the dispersion of the pollutant(s) in the environment,
- calculate the exposure of the receptors,
- calculate the impacts,
- assign monetary values to the impacts,
- rank the options and identify the best choice(s).

21

Table 2.1 *Illustration of the DPSIR framework. A given demand can cause many different types of pressure which in turn can change many states with many different impacts; a few examples are shown here.*

	Example 1	Example 2
Objective	Improve water quality	Improve waste management
Demand	Nitrogen fertilizers in agriculture	Disposal of organic wastes
Pressure	Nitrate formation and leaching	Greenhouse gas emissions
State	Increased concentrations of nitrate in drinking water and aquatic ecosystems	Increased radiative forcing in the atmosphere, climate change
Impact	Damage to health and environment	Sea level rise, heat stress, spread of tropical diseases, energy demand, etc.
Response	Limit fertilizer use, remove nitrates at water treatment works	Reduce generation of organic wastes. Change management of wastes

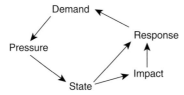

Fig. 2.1 The DPSIR framework.

As a starting point for the presentation of tools, it is instructive to consider the "DPSIR" (Demand – Pressure – State – Impact – Response) framework that has frequently been invoked. It is shown in Fig. 2.1 and illustrated with examples in Table 2.1.

Even though the DPSIR framework is not a tool for analysis, it is useful simply by presenting a link from demand through to response in a formalized way. This prompts users to think of:
- other forms of demand that generate the same pressure(s),
- other forms of demand that cause the same impacts,
- alternative response strategies,
- possible metrics for sustainability indicators to monitor progress.

Figure 2.2 illustrates the use of DPSIR for the problem of pollution by nitrogen compounds. It is good at highlighting the complexity of the problem, by showing the numerous links between human activities and impacts. But it does not identify what is needed for decision making,

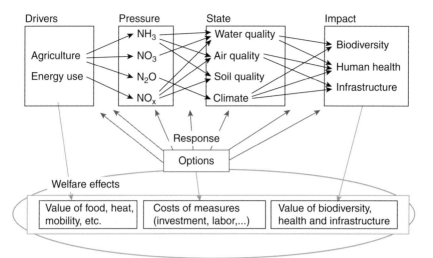

Fig. 2.2 DPSIR for the problem of pollution by nitrogen compounds. From Corjan Brink, with kind permission.

namely the impacts and costs associated with specific choices, such as the choice between nuclear and coal for energy production, nor does it indicate which tools could be used for the analysis.

2.1.2 Tools for analysis of impacts and for decision making

Among the many specific tools we will discuss the following:

- **Life Cycle Assessment (LCA)** LCA is a methodology for the stand-ardized analysis and comparison of systems (such as alternative methods for the production of wood pulp followed by the manufacture, use and disposal of paper) and their associated environmental burdens (pollutant emissions, resource use, waste generation, etc.). Frequently this approach is extended to include an assessment of impacts, called **Life Cycle Impact Assessment (LCIA)**. The relation between IPA and LCIA is discussed in Section 2.3.
- **Impact Pathway Analysis (IPA)** This approach, greatly refined through the ExternE Project Series of EC DG Research since 1991, follows the impact pathway:
  - from source of burden (e.g. noise, air pollution, accident risk),
  - to exposure receptors (e.g. population, plants, buildings), using pol-lution dispersion models,
  - to quantification of impacts, using exposure–response functions.
  - to monetary valuation, based on willingness to pay.

- It is the logical method to analyze the progression from burden to damage, and it generates a series of results about the magnitude of burdens, impacts and damage costs, each of which may be of policy interest.
- Since an IPA is an essential ingredient for any realistic assessment, we provide more information in Section 2.2 and the bulk of this book presents further details of the methodology as well as results.

- **Environmental Impact Study (EIS)** In most countries, environmental impact studies are a standard requirement before a new installation with potential impact on health or environment can be approved. However, in spite of the similarity in the names, the methods and goals of IPA and EIS are totally different. An EIS is concerned only with local impacts, typically within a few km of the installation. In contrast, an IPA seeks to quantify the totality of the impacts regardless of geographic range. Expected impacts are hardly ever calculated in an EIS; instead, in most cases, no-effect thresholds or legislated standards are used to argue that there is no significant risk for the surrounding population. The goal of an EIS is to give decision makers information on whether any section of society will be exposed to an unreasonable level of impact. It can also provide information to local residents, businesses, etc. of the anticipated extent of local impacts, which may in turn affect their support or opposition to a given scheme.

- **Cost–Benefit Analysis (CBA)** An environmental cost–benefit analysis compares the benefits of a proposed reduction in pollution with the abatement costs, i.e. the cost per unit of avoided pollutant, based on an IPA. In performing CBA one should seek to identify all possible consequences of a given action, including both co-benefits and tradeoffs. A full quantification of all effects may not be possible, but at least a CBA provides a fuller understanding of the advantages and disadvantages of a proposed action.

- **Multi-Criteria Analysis (MCA)** MCA involves consultation with concerned individuals (experts and stakeholders) to determine their preferences among a series of options that are clearly described. The individuals are asked to assign weighting factors that express the importance of all the impacts. The weighting factors are processed to determine a ranking of the options. If some of the impacts have already been valued in monetary terms, the weighting factors imply monetary values for the remaining impacts. This method is particularly useful for ranking options when monetary valuation has not been feasible.

2.2   Impact pathway analysis (IPA)

2.2.1   The steps of an IPA

To calculate the damage costs of polluting activities such as energy production, one needs to carry out an impact pathway analysis, tracing the passage of a pollutant from where it is emitted to the affected receptors (population, crops, forests, buildings, etc.). As illustrated in Fig. 2.3, the principal steps of an IPA can be grouped as follows:

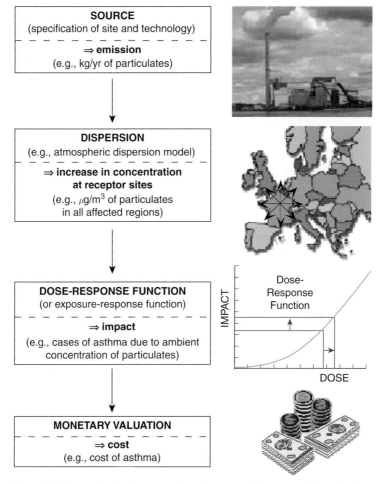

Fig. 2.3 The principal steps of an impact pathway analysis, for the example of air pollution.

- **Emission**: specification of the relevant technologies and pollutants, e.g. kg of NO_x per GWh_e emitted by a power plant at a specific site.
- **Dispersion**: calculation of increased pollutant exposure in all affected regions, e.g. incremental concentration of ozone, using models of atmospheric dispersion and chemistry for ozone formation due to NO_x (this step is also called *environmental fate analysis*, especially when it involves more complex pathways that pass through the food chain).
- **Impact**: calculation of impacts (damage in physical units) from this exposure, using exposure–response functions, e.g. cases of asthma due to this increase in ozone.
- **Cost**: a monetary valuation on these impacts, e.g. multiplication by the cost of a case of asthma.

The impacts and costs are summed over all receptors of concern. The work involves a multidisciplinary system analysis, with input from engineers, dispersion modelers, epidemiologists, ecologists and economists. The result of an IPA may be the damage cost per kg of emitted pollutant (as shown in Fig. 12.1), the benefits of a pollution control policy, an estimate of national damage from air pollution, etc. In the following subsections we present a brief description of the IPA methodology. More detail can be found in Chapters 3 to 9, and results are presented in Chapters 12 to 15. Whereas the principles of the methodology, i.e. the steps of the IPA, are timeless, the detailed models and parameters that are assumed evolve with the state of the science involved. Therefore, different implementations can yield fairly different results. And, with the exception of globally dispersing pollutants, the damage costs depend on location and type of source, and particularly its height above ground.

In many situations one needs to evaluate not only a single pollutant, but all pollutants emitted by a source (e.g. the damage cost per km driven by a car), or all pollutants emitted by an entire process (e.g. the production of a tonne of cement). If one wants to know the damage cost per km due to the pollution from driving a car in a city, one simply multiplies cost per kg by the respective emission per km for each pollutant and sums over all pollutants. For some issues, one needs to analyze an entire process, by means of an LCA, to take into account all pollutants emitted by all stages of the process chain (more precisely, one needs the inventory of an LCA, followed by an IPA for each of the pollutants; see Section 2.3). For the example of electricity from coal, the analysis can be broken down into the following stages:

- Mining of the coal.
- Transport of the coal.
- Construction of the power plant.

- Combustion of the coal at the power plant.
- Transmission of the electricity.
- Disposal of the wastes.

Often some of the stages can be omitted, if preliminary analysis shows that they have a negligible impact (power plant construction and electricity transmission in this example), whereas for others a breakdown into smaller stages may be appropriate.

Whether an IPA of a single source or an LCA of an entire chain is required depends on the policy decision in question. For evaluating the optimal limit for the emission of NO_x from a cement kiln, an IPA is sufficient, but the choice between different construction materials for house building involves an LCA.

2.2.2 *Dispersion of pollutants and exposure*

The principal greenhouse gases, CO_2, CH_4 and N_2O, stay in the atmosphere long enough to mix uniformly over the entire globe. No specific dispersion calculation is needed, but the calculation of impacts is complex, as we will discuss in Chapter 10.

For most other air pollutants, in particular PM_{10}, NO_x, SO_2 and VOCs, atmospheric dispersion is significant over hundreds to thousands of km, so both local and regional effects are important, and a combination of local and regional dispersion models is needed. At the regional scale one needs to take into account the chemical reactions that lead to the transformation of primary pollutants (i.e. the pollutants as they are emitted) to secondary pollutants, for example, the creation of sulfates from SO_2.

Whereas only the inhalation dose matters for classical air pollutants (PM_{10}, NO_x, SO_2 and O_3), toxic metals and persistent organic pollutants also affect us through food and drink. Here a much more complex IPA is required to calculate ingestion doses, taking into account the pathways in Fig. 2.4. An analysis of these pathways is described in Chapter 8. An interesting general finding is that when these pollutants are emitted into the air, the ingestion dose can be about two orders of magnitude larger than the dose by inhalation.

2.2.3 *Exposure–response functions*

Having calculated the exposure of the population or other receptors to a pollutant, one applies exposure–response functions (ERF) which indicate how much impact there is as a function of the exposure, for example, how many additional hospitalizations there are as a function of ambient PM_{10} concentration. Dose–response function (DRF) is another term that is

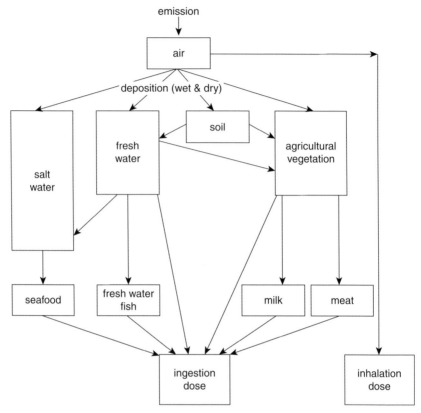

Fig. 2.4 Pathways taken into account for the health impacts of air pollutants. Direct emissions to soil or water are a special case where the analysis begins at the respective "soil" and "water" boxes.

frequently used, although sometimes with a slightly different definition, namely the impact as a function of the dose received or absorbed by an organism. For air pollutants the term concentration–response function (CRF) is also common. In this book, we prefer the term exposure–response function because it is more general, linking the impact directly to the exposure which has been calculated by dispersion models, without worrying about the actual dose. In actual practice, the terms ERF, DRF and CRF are frequently used interchangeably, and in some places in this book we may use "dose" instead of "exposure".

The ERF is a central ingredient in the impact pathway analysis and merits special attention. Damage can only be quantified if the corresponding ERF is known. Such functions are available for the impacts on human health, building materials and crops caused by a range of pollutants, such as primary and secondary (i.e. nitrate, sulfate) particles, ozone, CO, SO_2, NO_x, benzene, dioxins, As, Cd, Cr, Hg, Ni and Pb.

Unfortunately, for many pollutants and many impacts the ERFs are uncertain or not even known at all. In particular, for many substances and their health impacts the only available information covers thresholds, typically the NOAEL (no observed adverse effect level) or LOAEL (lowest observed adverse effect level). Knowledge of thresholds is not sufficient for quantifying impacts; it only provides an answer to the question of whether or not there is a risk. The principal exceptions are carcinogens and the classical air pollutants, for which explicit ERFs are available.

By definition, an ERF starts at the origin, and in most cases it increases monotonically with dose, as sketched schematically in Fig. 2.5. At very high doses the function may level off in S-shaped fashion, due to saturation (e.g. where most of a population is affected); but this case is generally not of

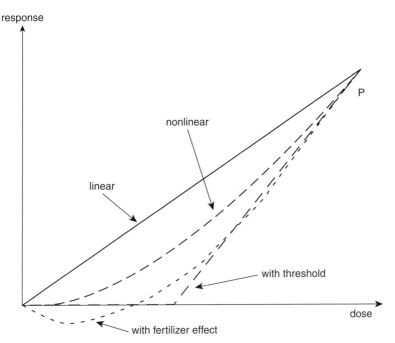

Fig. 2.5 Possible behavior of exposure–response functions at low doses. If P is the lowest dose where a nonzero impact has been observed, the extrapolation to lower doses is uncertain, but values higher than linear are unlikely.

interest although nonlinear ERFs may be relevant for regions with extreme pollution, e.g. cities in China.

A major difficulty for health impacts lies in the fact that one needs relatively high exposures in order to obtain observable nonzero responses, unless the sample is very large; such exposures are often far higher than the range over which one wants to apply the ERFs, namely, typical ambient exposures, or even lower exposures if one wants to know the benefit of additional abatement measures. Thus, there is a serious problem in how to extrapolate from the observed data towards low exposures. Figure 2.5 indicates, schematically, several possibilities for the case where point P corresponds with the lowest exposure at which a response has been measured. The simplest model is linear, i.e. a straight line from the origin through the observed data point(s). The available evidence suggests that an ERF is unlikely to go above this straight line in the low exposure limit. The straight line model does appear to be appropriate in many cases, in particular, for substances that initiate cancers and for the classical air pollutants.

Another possibility is the "hockey stick": a straight line down to some threshold, and zero effect below that threshold. Thresholds occur when an organism has a natural repair mechanism that can prevent or counteract damage up to a certain limit. The hockey stick is an idealization; in most cases a smooth function like the dashed line labeled "nonlinear" in Fig. 2.5 is more likely.

There is even the possibility of a benefit at low doses, as indicated by the line labeled "with fertilizer effect" in Fig. 2.5. This can be observed, for example, in the ERFs for the impact of NO_x and SO_2 on crops; a low dose of these pollutants can increase the crop yield, in other words the damage is negative. Generally, a fertilizer effect can occur with pollutants that provide trace elements needed by an organism, for instance the health effects of selenium. An intriguing example is alcohol, obviously harmful at high doses, but apparently beneficial at moderate consumption (mortality has a minimum at around one drink/day) (Thun et al., 1997).

In a wider sense, the entire chain of the impact pathway analysis can be considered to be an integrated (i.e. comprehensive) ERF that yields the damage cost directly as a function of the emission. In some cases, such integrated ERFs have been derived directly from empirical data. As an interesting example, we cite a study that links the incremental renovation costs of buildings along the main traffic arteries of the city of Neuchâtel, Switzerland, to the number of vehicle kilometers, relative to buildings on streets with less traffic (Stritt, 1993). Unfortunately not enough detail is provided in this study to permit transferability to other sites.

There is also the possibility of intermediate approaches. An example is our analysis of damage to buildings in France, in Section 4.5. Data on materials and surfaces of buildings were not available, but since cleaning and repair expenditures were declared in income tax returns in order to obtain tax reductions, we could correlate these expenditures with local ambient concentrations and derive an ERF linking exposure directly to damage costs (Rabl, 1999).

2.2.4 *Monetary valuation*

2.2.4.1 General principle
The goal of the monetary valuation of damages is to account for all costs, market and non-market. For example, the valuation of an asthma attack should include not only the cost of the medical treatment (a market cost), but also the willingness to pay (WTP) to avoid the residual suffering. It turns out that damage costs of air pollution are dominated by non-market goods, especially health and mortality. If the WTP for a non-market good has been determined correctly, it is like a price, consistent with prices paid for market goods. The tools that economists have developed for determining non-market costs, together with results, will be presented in Chapter 9.

There are two general approaches for non-market goods: revealed preference and stated preference. Revealed preference methods, also called hedonic methods, look at consumer decisions and try to infer an implicit valuation. Wage–risk studies for the determination of the value of a prevented fatality are a classic example: based on the idea that workers demand higher pay for more dangerous jobs, all else being equal. For a good reference on hedonic methods, see Graves (2013). The suitability of hedonic methods for pollution impacts is quite limited, and in many cases stated preference methods are required. Here the basic idea is to ask individuals how much they would be willing to pay for the good in question if they could buy it.

2.2.4.2 Global warming
The valuation of global warming damages is complicated by timescales, geographic scale, the range of impacts that can be attributed to gradual change, the likely increase in extreme events that are by their nature difficult to quantify, and the potential for adaptation. Not only is a quantification of the many possible physical impacts in all countries of the world extremely difficult, but their monetary valuation is problematic and controversial, because of ethical issues related to the choice of the discount rate for intergenerational costs, the valuation of mortality in developing countries and the question of equity weighting. These issues will be discussed in Chapter 10.

2.2.4.3 Valuation based on abatement costs

Some analysts have tried to calculate external costs of pollutants on the basis of abatement costs instead of damage costs, arguing that the observed abatement costs express the implicit valuation of damages by society (see e.g. Ottinger *et al.*, 1991). Marginal abatement costs would indeed be equal to marginal damage costs if the emission limits imposed by environmental regulations were already optimal. But policy makers do not know where the social optimum is, to say nothing about the twists and turns of the political processes that lead to the choice of regulations in practice. In reality, policy makers need information on damage costs in order to formulate the environmental regulations. Therefore, using abatement costs as proxy for damage costs merely begs the question.

Abatement costs have been used by ExternE for the cost of CO_2 in the years from about 2000 to 2005. Specifically, the cost was taken as the marginal abatement cost of measures needed to comply with the Kyoto Agreement, which the EU had signed; this cost was 19 $€/t_{CO_{2eq}}$ in the EU. Compliance being difficult, an extra tonne of CO_2 emitted in the EU implies a cost, because it necessitates abatement measures costing 19 $€/t_{CO_{2eq}}$. However, we emphasize the need for caution in the use of results from such an approach; their applicability is very limited and in particular they must not be used for cost–benefit analysis or for environmental regulations.

2.2.5   *Spatial and temporal boundaries of the analysis*

The boundaries of the analysis should be wide enough to include all or almost all of the impacts, say at least 95% of the total. For greenhouse gases, the entire globe has to be included, and the time horizon should be several decades for CH_4 and at least one or two centuries for CO_2 and N_2O. It should also be sufficiently long (compared with the time constants of the transfer processes) for toxic metals and persistent organic pollutants.

For the classical air pollutants, the impact is significant only in the short term (or years to decades for chronic health impacts), and the geographic range should extend to at least a thousand km from the source. This is illustrated in Fig. 7.14 in Chapter 7, which shows the relation between range of analysis and corresponding fraction of the total damage cost. The local impact becomes important if the source is close to a large population center and if the stack height is low, especially for vehicle emissions. But there are always significant impacts even far from the source, and one has to use both local and regional dispersion models.

Impacts beyond the present generation pose a problem, not only in the choice of an appropriate discount rate (see Section 9.1.3), but also in the assumptions for the evolution of the costs. Furthermore, impacts beyond national boundaries or in the far future may be awkward for policy makers. For example, to what extent should policy makers of the EU take into account global warming damages imposed on developing countries a century from now? There is no simple solution that would be accepted by all critics. Instead, it may be advisable to present a breakdown of impacts and costs according to separate spatial and temporal categories. An example will be shown in Section 13.5 and Table 13.8, for the nuclear fuel chain.

2.2.6   Software

For the calculation of damage costs, ExternE uses the *EcoSense* software package (Krewitt *et al.*, 1995), an integrated impact assessment model that combines the atmospheric models (WTM and ISC) with databases for receptors (population, land use, agricultural production, buildings and materials, etc.), exposure–response functions and monetary values. In the current (since 2008) version, the WTM for regional dispersion has been replaced by source–receptor matrices that have been calculated by EMEP.

In addition, there are two tools for simplified approximate assessments: EcoSenseLE and RiskPoll. EcoSenseLE (the LE stands for look up edition) is a free online tool that provides tables of typical damage costs for a variety of emission sites. RiskPoll is a package of several models with different input requirements and levels of accuracy. It is based on the interpolation of dispersion calculations by EcoSense, and with its simplest version yields results that are typically within a factor of two to three of detailed EcoSense calculations for stack heights above 50 m. A more complex model of RiskPoll includes the ISC Gaussian plume model for the analysis of local impacts and emissions at or near ground level. RiskPoll also contains a module for the multimedia pathways of Fig. 2.4. It can be downloaded without charge or restrictions from www.ari.rabl.org.

In the USA, the Air Pollution Emissions Experiments and Policy analysis model (APEEP) of Muller and Mendelsohn (2007) has been used; some results will be presented in Chapters 12 and 13. Another model is BenMAP,[1] a Geographic Information System (GIS)-based computer program to estimate the health and economic impacts when populations experience changes in air quality.

---

[1] www.epa.gov/air/benmap/

2.2.7   *Environmental impact study and local impacts*

Note that in spite of the similarity in names, an IPA is totally different from an environmental impact study (EIS), which is required before a proposed installation (factory, power plant, incinerator, highway, etc.) can be approved. The purpose of an EIS is to ensure that nobody is exposed to an "unacceptable" risk or burden. Since the highest exposures are imposed in the local zone, it is sufficient for an EIS to focus on local analysis, up to perhaps ten km, depending on the case. The criterion for unacceptable risk does not require quantification of impacts; examination of thresholds is sufficient (usually by showing that environmental quality standards or no-effect thresholds are not exceeded or that the lifetime cancer risk per person due to the installation is less than one in a million).

Damage costs are needed primarily by decision makers at the national or international level, and generally by anyone concerned with total impacts. They should be evaluated in any CBA with significant pollution impacts. Even for local projects that are decided at a local level, it would be desirable to evaluate any significant regional or global impacts, to make sure that the collective wellbeing is taken into account.

Policy instruments for the internalization of significant local impacts tend to be different from those for regional or global air pollutants. In particular, negotiation with local stakeholders is appropriate. Monetary values or external costs are not needed here, unless the negotiations involve financial compensation or cost–benefit analysis of abatement options. In the case of health impacts, the affected population usually rejects direct financial compensation as unacceptable because they consider their health non-negotiable.

Whereas burdens such as noise, odor and visual intrusion are limited to the local zone, for air pollutants the local impacts tend to make only a small contribution to the total damage costs (except for ground level emissions of primary pollutants in large cities, where much of the total can be imposed within the local zone). In any case, local impacts should be evaluated and presented even if they are small compared with the total damage costs, for the sake of protecting the rights of the local population.

2.3   Life cycle assessment

2.3.1   *Life cycle inventory*

Many environmental decisions involve complex systems, and if one looks only at one stage or process in such systems one can draw totally inappropriate conclusions. A classic example is the electric vehicle: whereas its utilization is perfectly clean, pollution is emitted by the power plants that provide the necessary electricity. It is a "pollution elsewhere vehicle." In recent years the importance of life cycle thinking has been brought to the fore by the controversy surrounding biofuels. Of course biofuels appear renewable and carbon neutral, if one neglects the use of resources (especially energy, water and fertilizer) and the associated pollution during the production of these fuels. But a correct analysis shows that for many biofuels production processes, the overall environmental effects are dubious or even downright harmful (see Section 15.1.5).

In order to provide correct assessments of complex systems, the LCA (life cycle assessment) approach has been developed. Life cycle assessment has become a well-established discipline, codified by the ISO 14000 series of standards. The analysis begins by establishing an inventory of all the energy and material flows, including pollutants, that are associated with a process or product. Typically, they are stated per unit of output (called functional unit), for instance per kWh of electricity or per vehicle km. For the inventory phase of LCA, several databases are available; one of the most complete is the ecoinvent (Frischknecht *et al.*, 2007) database.[2] In Table 2.2, we list some LCI (life cycle inventory) databases. For a more complete review of LCI databases worldwide, see Curran (2006).

One problem with most such databases is their retrospective nature: since they are based on existing production technologies, the data may be inappropriate for the analysis of future systems. To appreciate this point, note that industries have been greatly reducing their emission of most pollutants (apart from CO_2), even in some cases by factors of around 2 to 5 per decade, and this evolution is continuing, thanks to continuing political pressure. One should really use prospective databases to estimate the pollution that will be emitted by future processes and technologies. Only CO_2 has been very little affected by this evolution and will remain so until some mix of renewable energy technologies, nuclear, and carbon capture and sequestration is implemented on a wide scale. However, development of prospective databases seems to be difficult, as demonstrated by the attempt of Heck *et al.* (2009). Such work would be in addition to the tremendous effort required to assemble the data for conventional LCI tables.

It is interesting to see what sort of data one can find in databases that are free. As an example, we have downloaded files from the USLCI database, concerning production of fresh fruit and transport by three modes: diesel

---

[2] http://www.ecoinvent.ch

Table 2.2 *Some databases for life cycle inventories.*

Free databases

GEMIS
Global Emission Model for Integrated Systems
Öko-Institut www.oeko.de/service/gemis/en/
Energy, material, and transport

ELCD
European Reference Life Cycle Database
Joint Research Center http://lca.jrc.ec.europa.eu/lcainfohub/datasetArea.vm
End-of-life treatment, energy carriers and technologies, materials production, systems, transport services

USLCI
U.S. Life-Cycle Inventory (LCI) database
NREL (National Renewable Energy Laboratory) www.nrel.gov/lci/database/default.asp
(Note. This database no longer exists)
Transportation, chemical manufacturing, crop production, fabricated metal, product manufacturing, food manufacturing, forestry and logging, mining, nonmetallic mineral product manufacture, oil and gas extraction, petroleum and coal products manufacture, primary metal manufacturing, transportation equipment manufacturing, utilities, waste management and remediation service, wood product manufacturing

CPM LCA
Chalmers University of Technology http://cpmdatabase.cpm.chalmers.se/Start.asp
Industrial production processes, plants and supply chains

LCA Food Database www.lcafood.dk/
Basic food products produced and consumed in Denmark

EIOLCA
Economic Input-Output Life Cycle Assessment
Carnegie Mellon University, Green Design Institute www.eiolca.net/
LCA coupled to input–output analysis of economy

Commercial databases

SimaPro
PRé Consultants www.pre.nl/content/simapro-lca-software
Comprehensive software for LCA and LCIA
Includes the LCI databases ecoinvent v.2, US LCI, ELCD, US Input Output, EU and Danish Input Output, Dutch Input Output, LCA Food, Industry data v.2

Ecoinvent
Swiss Centre for Life Cycle Inventories www.ecoinvent.ch/
One of the most comprehensive international LCI databases

truck, ocean freighter and cargo plane. The functional unit for fruit production is 1 kg at the farm. For transportion, the functional unit in the database is tonne-km. In Table 2.3, we present these data, after expressing the transport data in terms of 1 kg of fruit transported over 1000 km by truck or over 10,000 km by ship or plane. We have chosen these units for the transport data to make it easier to see the role of

Table 2.3 *Data from the USLCI database, concerning production of 1 kg of fresh fruit and its transportation by three modes.*

Emissions to air (kg)	1 kg fresh fruit at farm[a]	Transport 1000 km, diesel truck[b]	Transport 10,000 km, ocean freighter[c]	Transport 10,000 km, aircraft, freight[d]
CO_2	8.56E-03	1.71E-01	1.60E-01	1.05E+01
CO	2.30E-06	2.46E-04	4.28E-04	4.41E-02
CH_4		4.13E-06	7.87E-06	
Hydrocarbons, unspecified				1.05E-02
VOC		8.42E-05	1.61E-04	
NMVOC	1.00E-06			
NO_x	5.50E-06	1.22E-03	4.34E-03	5.37E-02
N_2O		6.19E-06	3.89E-06	
PM unspecified	6.60E-05			
PM >2.5 μm and <10 μm		2.35E-05	1.08E-04	
SO_2	2.30E-06	3.77E-05	5.05E-04	

[a] USLCI data file SS_Harvesting, fresh fruit bunch, at farm.xls
[b] USLCI data file SS_Transport, single unit truck, diesel powered.xls
[c] USLCI data file SS_Transport, ocean freighter, residual fuel oil powered.xls
[d] USLCI data file SS_Transport, aircraft, freight.xls

transport emissions (often discussed under the banner of "food miles"), the respective distances being very roughly representative of these transport modes. Collecting the data for an LCI database requires an enormous amount of work, searching for reports and publications, reading them and interpreting the information. Usually the original reports and publications had not been intended for this purpose and the basic data are therefore not quite what is needed for LCI. For this example, the specifications for volatile organic compounds (VOC) and particulate matter (PM) are different in different files. In particular, it is awkward that the PM emissions of diesel trucks are shown as between 2.5 and 10 μm, even though in reality they are entirely $PM_{2.5}$; one wonders what these PM numbers mean. This distinction is important because only the finer particles penetrate to the deepest levels of the lung, where they can do the most harm.

The most reliable numbers in this table are probably the CO_2 emissions. Let us comment, in the context of two issues that have attracted some attention in the public media: "food miles" and "carbon footprint." A focus on food miles (the distance over which food is transported) is part of the movement towards buying local products, a reaction to the negative impacts of globalization.[3] Worrying about food miles and carbon footprint for each item that one buys in the supermarket is not a very productive activity: there are too many numbers to remember and too much cognitive effort required for too little benefit. If, instead, all external costs were correctly internalized, the price of each food item would automatically give the right signal about all impacts, including food miles and carbon footprint. To see what that would imply in this case, we anticipate the results of later chapters (Sections 10.4.1 or 12.2.1) and cite the damage cost of 21 €/t$_{CO_2}$ assumed by ExternE (2008). The CO_2 emissions are about the same for transport by truck over 1000 km as for transport by ocean freighter over 10,000 km, and would cost about 0.16 kg$_{CO_2}$/kg$_{fruit}$ * 21 €/t$_{CO_2}$ = 0.0034 €/kg$_{fruit}$, not a significant amount. Air transport over 10,000 km would add much more, 0.22 €/kg$_{fruit}$, but still not much compared with the basic cost of the kind of fruit that one would transport that way (e.g. cherries from Chile for the winter in Europe). To influence a purchase decision in this case, a carbon tax would have to be substantially higher than 21 €/t$_{CO_2}$. As we argue in Section 12.2.3, an appropriate carbon tax should be about three times higher, at around 65 €/t$_{CO_2}$, but even that may not significantly affect purchases of such food.

Obviously the food miles of 1 kg of fresh fruit do not amount to much. But how about our total food intake, roughly 800 kg/yr per person? If 800 kg are transported by truck over 1000 km, on average, the CO_2 emission is 0.14 t$_{CO_2}$/(pers.yr), a very small fraction of the total CO_2 emission per capita in the EU: approximately 10 t$_{CO_2}$/(pers.yr). At 21 €/t$_{CO_2}$, the total food miles would cost only 0.14 t$_{CO_2}$/(pers.yr) * 21 €/t$_{CO_2}$ = 2.94 €/(pers.yr).

2.3.2   Life cycle impact assessment

The results from an LCA inventory are usually too detailed to be suitable as a guide for decision making as they include burdens associated with a very large number of substances. Some kind of aggregation is needed to obtain a manageable number of impact indicators. While the inventory phase of LCA is straightforward and uncontroversial, there are several methods for LCIA (life cycle impact assessment), and LCIA methods are evolving. For more information on LCIA we refer to JRC (2010). Here we list the steps according to ISO 14044 (ISO 2006), which involves two mandatory and two optional steps:

---

[3] There are, as usual, winners and losers. On average the gains are far larger than the losses, a fact demonstrated by the enormous costs that a country has to bear when it is subjected to a trade embargo. But the gains of globalization tend to be diffuse, while the losses, especially in jobs due to outsourcing, are heavy for a minority (and unfortunately at the present time the compensation for such losses is totally inadequate in most cases).

(1) **Selection of impact categories and classification,** where the categories of environmental impacts, which are of relevance to the study, are defined by their impact pathway and impact indicator, and the elementary flows from the inventory are assigned to the impact categories according to the ability of the substances to contribute to different environmental problems. (Mandatory step according to ISO.)

(2) **Characterization,** where the impact from each emission is modeled quantitatively according to the underlying environmental mechanism. The impact is expressed as an impact score in a unit common to all contributions within the impact category (e.g. kg CO_2-equivalents defined in terms of global warming potential for each greenhouse gas contributing to the impact category climate change) by applying *characterization factors*. (Mandatory step according to ISO.)

(3) **Normalization,** where the different characterized impact scores are related to a common reference, e.g. the impacts caused by one person during one year, in order to facilitate comparisons across impact categories. (Optional step according to ISO.)

(4) **Weighting,** where a ranking and/or weighting is performed of the different environmental impact categories, reflecting the relative importance of the impacts considered in the study. Weighting is needed when tradeoff situations occur in LCAs used for comparisons. (Optional step according to ISO.)

The terms used by LCA and LCIA sound very different from those of IPA (and we find them a little strange and less self-explanatory than those of IPA). Above all, most LCA and LCIA methods refuse monetary valuation, using instead more or less subjective weighting factors for the different impact categories.

2.3.3 Relation between IPA and LCA

In principle, the damages and costs for each source of pollution (or other environmental burden) in the life cycle should be evaluated by a site-specific IPA, as indicated in Fig. 2.6. But in practice, almost all LCAs have taken the shortcut of first summing the emissions over all stages and then multiplying the result by site-independent impact indices. Also, most practitioners of LCA reject the concept of monetary valuation, preferring instead to use about ten non-monetary indicators of "potential impact," one for each of the major impact categories. "Potential impact" takes little account of measures in place to limit harm, for example the controls on operation of landfill sites in the EU. These "potential impact" indices are a far cry from the realistic IPA of ExternE. In most LCA methods there is little or no modeling of the environmental dispersion or use of ERFs.

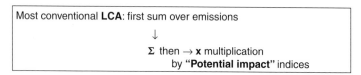

| IPA: → **real impact** for each stage (site-specific)   **Goal**: evaluate the entire matrix |

Steps of IPA → **Stage of process chain ↓**	Emission	Dispersion	Exposure-response function	Monetary valuation
Production of the materials				
Assembly of the materials				
Fuel feedstock				
Fuel supply				
Utilization of the car				
Disposal of the car				

Most conventional **LCA**: first sum over emissions

↓

Σ then → **x** multiplication
by **"Potential impact"** indices

Fig. 2.6 Relation between impact pathway analysis and current practice of most LCAs, illustrated for the example of electricity production.

The impact indices of most LCAs are what are called midpoint evaluations for an entire impact category (e.g. human toxicity), in contrast with the detailed evaluation of endpoints (e.g. mortality, chronic bronchitis, hospitalizations, cancers, asthma attacks, etc.) carried out by an IPA. An interesting intermediary between the IPA of ExternE and conventional LCA is the Impact 2002+ method of Jolliet *et al.* (2003) for LCIA (life cycle impact assessment) which does use explicit ERFs and fairly detailed pathway modeling. However, it refrains from monetary valuation, instead aggregating the human health impacts in terms of DALYs. Most of the ERFs of Impact 2002+ are derived from the NOAEL or LOAEL thresholds of toxicology, rather than epidemiology, using a method proposed by Pennington *et al.* (2002). The basic idea is to assume linearity without threshold, with slope inversely related to NOAEL or LOAEL (this use of a threshold may seem somewhat paradoxical, but it expresses the idea that the lower the NOAEL or LOAEL, the higher the toxicity). With such a method one can evaluate the entire, very large set of substances for which threshold data are available, by contrast with the rather limited set of substances for which epidemiologists have been able to determine ERFs. Unfortunately, the uncertainties of the method of Pennington *et al.* are very large and systematic validation is lacking. In fact, linearity without threshold is wrong for some substances, as demonstrated for example by the "mega-mouse experiment" of Frith *et al.* (1981).

Table 2.4 shows the impact categories that have been addressed by LCIA and by ExternE. LCIA covers a vast list of pollutants, against which

Table 2.4 *Pollutants and impact categories that have been addressed by LCIA and by ExternE. Whereas most LCIA provides only midpoint indicators for its impact categories, Impact 2002+ provides detailed evaluation of the corresponding endpoints.*

	LCIA *Impact 2002+*	IPA *ExternE*
Monetary valuation	No	yes
Pollutants considered	All for which emissions data are available	CO_2, CH_4, N_2O, PM, SO_2, NO_x, VOC, As, Cd, Cr, Hg, Ni, Pb, dioxins, benzene, radionuclides
Impact categories		
Human toxicity	X	X
Global warming	X	X
Ionizing radiation	X	X
Photochemical oxidation	X	X
Terrestrial acidification	X	X
Land use	X	X
Ozone layer depletion	X	
Aquatic ecotoxicity	X	
Terrestrial ecotoxicity	X	X
Aquatic eutrophication	X	
Consumption of non-renewable energy	X	
Mineral extraction	X	
Agricultural losses		X
Buildings and materials		X
Accidents		X
Energy supply security		X
Reduction of visibility		

the list addressed by ExternE may appear very limited. However, ExternE includes the most important pollutants. We believe that a realistic assessment of the most important pollutants is more appropriate than an assessment that aims for completeness, while using dubious methods.

The inclusion of economics in ExternE is reflected in several impact categories that have not been addressed at all by LCIA: agricultural losses, damage to buildings and materials, accidents and energy supply security. Consumption of non-renewable energy and mineral resources is an important category for LCIA, whereas ExternE has argued that this is already internalized by the market. The truth is probably somewhere between these two extremes: there is some internalization by the market, but without a sufficiently long-term perspective (see Section 6.4). Ozone layer depletion has not been considered by ExternE because it is essentially a problem of the past, the emission of ozone depleters having been reduced to negligible proportions thanks to the Montreal Protocol. Reduction of visibility has been found to be a significant damage cost in the USA, but ExternE has not yet tried to estimate this impact, which has received very little attention in Europe (see Section 6.1).

Vad (2008) compared the health impacts between two LCIA methods (Impact 2002+ and CML-IA) and ExternE, in terms of DALY per kg of pollutant. Her results are shown in Fig. 2.7. For most pollutants, the results differ by several orders of magnitude between each of the three

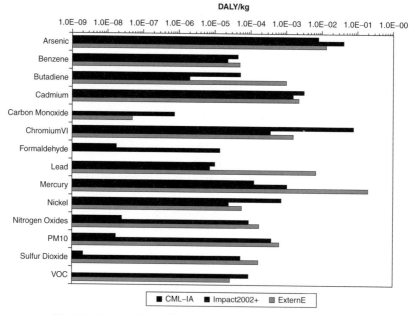

Fig. 2.7 Comparison of health impacts, in terms of DALY/kg, between Impact 2002+, CML-IA and ExternE, based on Vad (2008).

methods. We hasten to add that such a comparison is problematic for CML-IA and necessitates assumptions that are somewhat arbitrary, since CML-IA reports only midpoint indicators. Nevertheless, at the logarithmic scale of Fig. 2.7, any arbitrariness in such a comparison is insignificant and pales before the vast differences between the results of these methods. Comparison between Impact 2002+ and ExternE is more direct and less problematic than for CML-IA, because Impact 2002+ calculates endpoint indicators, using ERFs that are in many cases close to those of ExternE. The latter are those of ExternE (2008) (converted at 40,000 €/DALY) except for CO, benzene and butadiene, which are from ExternE (2000).

2.4   Difficulties for the analysis

2.4.1   Lack of data

As has been already pointed out above, a damage can be quantified only if the corresponding exposure–response function (ERF) is known. Unfortunately, for many pollutants and many impacts, the ERFs are either very uncertain or not even known at all. Many pollutants may have significant ecosystem impacts, but we lack ERFs to quantify them beyond some information on threshold levels. An especially troubling example of substances without known ERF are pesticides. Some studies of the damage cost of pesticides can be found in the literature, but almost all are simply based on willingness to pay (WTP) to avoid pesticides, determined by contingent valuation. Since people do not know the real damage, their WTP expresses a vague fear rather than an informed judgment about the real damage, which is unknown because of the lack of ERFs. The only exception is a recent assessment by Fantke et al. (2012) who use ERFs that have been estimated by extrapolation from toxicity tests on animals.

For several potentially important impacts one does not have sufficient information for monetary evaluation. Particularly important examples are ecosystem impacts and, for many developing countries, the valuation of premature mortality.

If there is no adequate basis for estimating an ERF and/or monetary value for a potentially significant impact, the analyst should indicate this. Since a list containing all potential impacts would be unmanageably long, the analyst should merely point out the ones that might be significant. It would be good to remind the reader that any quantification of pollution damage is necessarily incomplete if not all impacts are known, and thus the real costs could be larger (however, they could also be overestimated due to uncertainties). For pollutants that have been studied extensively at low dose, in particular the classical air pollutants, we believe that all significant impacts have been taken into account. But for some toxic metals surprises may yet turn up.

2.4.2   *Boundaries of the analysis*

Another difficulty lies in the need to evaluate the consequences of all relevant alternatives. In many cases a choice may decrease a risk that is evident, but induces increased risks elsewhere; so-called countervailing risks, that may be important even though less visible.

An example of countervailing risk is the use of the diesel engine for passenger cars, which has been encouraged in several European countries, especially France, because of its higher energy efficiency. Unfortunately, the diesel engine emits high levels of toxic particulates. To reduce these emissions the particulate filter was developed and has now become standard equipment. While very effective at reducing PM emissions, the particulate filters tend to increase the NO_2/NO ratio of the NO_x emissions. This lowers the amount of ozone scavenging near the source and thus tends to increase ozone levels in cities. On balance, the benefit is probably far greater than the countervailing risk of increased ozone, but this example highlights the ubiquitous possibilities of countervailing risks, even in seemingly the most beneficial developments.

Induced effects can also be beneficial. For example, many options for the abatement of greenhouse gases also have co-benefits for the classical air pollutants (e.g. shifting electricity production from coal to natural gas combined cycle). Failure to take co-benefits into account can cause serious bias, both in cost-effectiveness analysis (see Section 1.3) and in CBA.

2.4.3   *Nonlinearity of impacts*

Some damage costs are strongly nonlinear functions of the burden, for example O_3 damage due to NO_x emissions. Nonlinearities are also important for ecosystem impacts, e.g. eutrophication. The correct procedure for treating nonlinearities is somewhat technical and complicated. As far as we know, it has not yet been implemented correctly in damage cost studies.

The goal is to estimate marginal damage costs, because the socially optimal level of pollution control corresponds with the point where the sum of marginal damage and marginal abatement costs equals zero. However, if this seemingly simple statement is interpreted carelessly, it could lead to absurd policy recommendations for impacts that are a nonlinear function of the emission. To illustrate this problem, consider Fig. 2.8, which shows a pollutant whose damage increases with emission at low emission levels, but decreases again if the emission is high. Such a

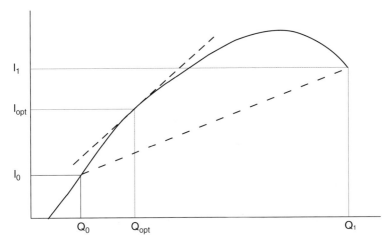

Fig. 2.8 Pollutant whose impact I increases with emission Q at low levels, but decreases again if the emission is high. The slope of the dashed tangent at Q_{opt} is the appropriate marginal damage, i.e. at optimal emission Q_{opt} (unknown). The slope of the chord from pre-industrial (Q_0, I_0) to current (Q_1, I_1) would be a much better estimate of the appropriate marginal damage than the marginal damage at current emission Q_1.

situation actually occurs with O_3 impacts as a function of one of the precursor emissions, NO (note that most NO_x is emitted as NO). The case of O_3 damage due to NO is the most extreme (complicated even more by the strong dependence of the curve on the other precursor VOC), but the problem also occurs in a milder form with aerosols created by NO_x and SO_2 emissions.

With careless interpretation one would find a negative marginal damage (tangent at the current emission level Q_1), implying that the policy response should be to encourage even greater emission of this pollutant. That would miss the real optimum at Q_{opt}. To provide the correct information to policy makers, one needs to examine carefully what the marginal damage costs will be used for and how they should be calculated.

The key observation is that the optimization condition (marginal damage cost + marginal abatement cost = 0) requires knowledge of these marginal costs in the vicinity of the optimal emission level. Both the damage cost and the abatement cost can vary with emission site, and so does the optimal emission level. Ideally, a policy maker should know the entire set of cost curves for marginal damage and abatement at each site. In the case of NO_x, SO_2 and VOC, the damage costs are complicated site-dependent functions of not only the pollutant under consideration, but also the simultaneous emission of several other pollutants, with due consideration of all of their respective emission sites. Optimization requires solution of the coupled optimization equations.

Of course this poses a problem in practice, since policy makers want simple numbers rather than complicated functions, to say nothing of the computational difficulties of determining the complete functions. If one wants a single number, it should be reasonably close to the value at the optimum. This begs a question since the optimum is not known.

The best one can do is to proceed iteratively. With an initial guess of the optimal emission levels one can derive a first estimate of the appropriate marginal damage costs. Comparing these with the abatement costs one can then improve the estimation of optimal emission levels. In view of the uncertainties in the abatement costs (if they are even known in the required range), estimates of the optimal emission levels are likely to remain very rough, with the ensuing additional uncertainties of the appropriate marginal damage costs. Fortunately there seems to be a fair amount of tolerance to errors in the determination of optimal emissions, as shown in Section 11.7, so even an initial estimation of the optimum may suffice for the purpose of calculating the damage costs for nonlinear impacts.

As a starting point for applications in the EU, one could take the estimates of optimal emission levels by Rabl et al. (2005), who find that the emissions of NO_x and of SO_2 should be reduced to levels of between 0.2 and 0.8 times (depending on pollutant and country) their level in 1998. But the estimation of optimal levels would have to be extended to VOC and NH_3, paying due attention to the coupling between VOC and NO_x through their contributions to O_3 formation.

Optimal NO_x emissions are much more uncertain than those for SO_2, for several reasons. Not only is the damage cost due to nitrate aerosols uncertain because of a lack of information on their toxicity, but the optimum also depends on the damage costs due to O_3, because the optimization for NO_x involves setting the marginal abatement cost equal to the total marginal damage cost, not the individual cost components due to nitrates and ozone. The O_3 damage due to NO_x depends in turn on the background emissions of VOC. So far, the optimal emission levels for VOC have not been estimated, and in any case iterations would be needed because of the coupled nature of the equations.

In view of such complications, one may need radical simplifications to estimate the marginal damage cost of O_3 formation due to NO_x. Referring to Fig. 2.8, we recommend taking the slope of the chord from pre-industrial (Q_0, I_0) to current (Q_1, I_1) conditions, where Q refers to NO_x emissions.

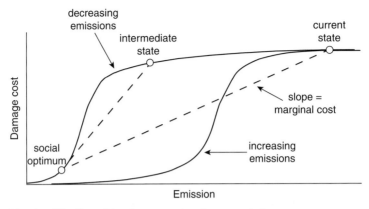

Fig. 2.9 Nonlinearities that can arise, for example in ecosystems:
- there is a threshold zone above which the impacts increase sharply with the burden (e.g. emission of nitrates to water), reaching a plateau at high levels of the burden;
- to bring the system back to the original low impact level, the burden must be reduced to the threshold for decreases, which is lower than the threshold for increases – a phenomenon known as hysteresis.

The appropriate marginal damage cost for policy applications is the slope of the dashed line from the current state to the social optimum.

Figure 2.9 illustrates another example of nonlinearities; this can arise in ecosystems, for example with eutrophication of water bodies. Two features are very different from most other impact types, in particular from health impacts of air pollutants:
- there is a threshold zone above which the impacts increase sharply with the burden (e.g. emission of nitrates to water), reaching a plateau at high levels of the burden;
- to bring the system back to the original low impact level, the burden must be reduced to the threshold for decreases which is lower than the threshold for increases – a phenomenon known as hysteresis.

Again, the calculation of external costs requires knowledge of the social optimum: the marginal external cost is the slope of the line from the optimum to the current state. As the burden is reduced, the marginal external cost increases. Since the social optimum is not known in advance, this process necessitates iterations: one begins by guessing where the optimum might be and calculates the corresponding marginal external cost. Comparison with the marginal abatement cost curve then yields a new estimate of the optimum, to be used for the next iteration.

2.4.4   *Alternative tools for problematic impacts*

2.4.4.1 **Problematic impacts**

Whereas for many impacts fairly reliable damage cost estimates are available and for others progress is being made, some impacts defy quantification. Especially troubling are the risks of nuclear power: the risk of a large accident, the storage of high level waste, and the risks linked to nuclear proliferation and terrorism are especially controversial. They are the reason why debates about nuclear power tend to remain inconclusive, despite recognition by most experts that the impacts from normal operation of the nuclear fuel chain are negligible, especially when compared with fossil fuels. People's attitudes are a reflection of optimism or pessimism about mankind's ability to manage this technology.

Risks of damage from storage of toxic waste are uncertain because they depend on the future management of the storage site. In principle, they can be avoided completely if the site can be permanently maintained in a safe condition without any leaks, and the waste is stored in a retrievable manner so that it can be reprocessed and rendered less harmful once future technologies allow. Thus the analysis requires an assumption of scenarios, and the results are of the "what if" type rather than simple definite numbers.

Some impacts have attributes that can pose problems for quantification in monetary terms. In particular, an irreversible impact is more serious than one that is reversible. Consider, for example, the destruction of a landscape under two scenarios: (a) the damage lasts only for a finite duration, (b) the damage is permanent. According to standard economic practice, the damage cost for the latter can be calculated from the former by the rules of discounting; for typical discount rates and durations this is not very much larger than the former. For instance, at a 3% discount rate a loss during 30 years is 19.6 times the annual loss, whereas a permanent loss is 33.3 times the annual loss, only 1.7 times larger. Yet most people would rightly feel that the permanent damage is much more serious.

In cases where there is a probability distribution of damages with a long tail of high values, the maximum potential damage could be very large and the expectation value does not capture the full severity. Global warming provides a good illustration of these considerations: it cannot be reversed for many generations and there is a risk of catastrophic damages (see e.g. Weitzman, 2009).

Unlike irreversibility and potentially catastrophic damage, the mere fear of risk,[4] without objective justification, is not an appropriate criterion for

---

[4] For a good reference on risk analysis and risk perception we recommend the book by Wilson and Crouch (2001).

increasing a damage cost estimate. For example, the health impacts of dioxins received so much media attention during the eighties and nineties that they were dreaded far beyond their real magnitude. A policy based on risk aversion might forbid waste incinerators because of their dioxin emissions, i.e. it would in effect set the damage cost at a prohibitively high level. But would that be rational if, as is the case now, these emissions have been reduced to the point where the damage is no longer significant?

Impacts that are potentially important but have not been quantified must not be forgotten in the final report. They should be listed with a sufficiently detailed description to allow decision makers to take them into account in a multicriteria analysis (MCA).

2.4.4.2 Multicriteria analysis

Multicriteria analysis involves the consultation with concerned individuals (experts and stakeholders) to determine their preferences among a series of environmental outcomes that are clearly described. The individuals are asked to assign weighting factors that express the importance of all impacts. The weighting factors are processed to determine which outcomes provide a best balance with those preferences. If some of these impacts have already been valued in monetary terms, the corresponding weighting factors imply monetary values for the remaining impacts. This method is particularly useful for ranking options when complete monetary valuation has not been possible or is too controversial. For examples of the application of MCA, see e.g. Hokkanen and Salminen (1997), Haastrup et al. (2002) and Vaillancourt and Waaub (2002).

Carrying out an MCA can be difficult and labor intensive. Much effort is needed to explain the issues to the stakeholders and to help them to choose their weighting factors. A major difficulty lies in getting a sample of stakeholders that would be a fair representation of the interests of society. The very definition of what is representative poses problems: to what extent should individuals who do not care much about an issue be represented with equal votes to those who feel very concerned? How can one even decide whether a particular group of stakeholders is "representative"?

2.4.4.3 Risk comparisons

A helpful tool for assessing problematic impacts is a comparison of risks. Such a comparison can be qualitative or quantitative. At the qualitative level, one prepares a listing of all the important risks and benefits for the choices under consideration. Typically, that would be shown as a table with columns for each choice, and similar items would be shown in the same row. For risks that are sufficiently similar, for instance mortality, one can proceed to a quantitative risk comparison. Risk ladders are commonly

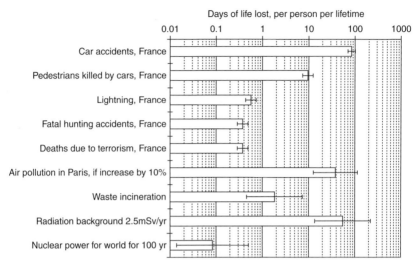

Fig. 2.10 An example of risk comparisons: the loss of life expectancy due to various causes.

used to compare the risk of dying from different causes, usually presented as a bar chart with a logarithmic scale. As an example, we show in Fig. 2.10 a comparison of the loss of life expectancy due to various causes. In later chapters we explain how these numbers were calculated:

For car accidents, Section 15.1.6;
For air pollution, Section 3.4.3 (Example 3.1);
For waste incineration, Section 14.6 (Example 14.1);
For nuclear power, Section 13.5.1 (Example 13.2).

2.4.4.4 The precautionary principle

Sometimes the precautionary principle has been invoked in attempts to solve environmental problems. It is not a tool, but merely a guiding principle. Basically it is an admonishment to "think before you act!"

Whereas the precautionary principle is valuable as a general guideline, the difficulty lies in its application to specific situations, because there are no implementation rules that are sufficiently explicit. In fact, it would be problematic, if not downright impossible, to formulate specific imple-mentation rules because each situation requires an analysis of the specific circumstances. In choosing between two actions, A and B, one person may invoke the principle to justify A, while the other may feel just as strongly that B would be indicated by the principle; the difference arising from different perceptions of the respective risks. For example, in the choice between nuclear and fossil fuels, there is no generally accepted ranking of the risks associated with nuclear power on the one hand and global warming on the other, and either technology could be preferable in the name of precaution. Subjective judgment is unavoidable.

Some people invoke the precautionary principle to argue for overestimating the impacts of a pollutant. This is totally inappropriate because it neglects the countervailing risks of excessive abatement costs, as already explained in Section 1.1. Rather, one needs to find the social optimum, and that requires the best estimate of the damage costs.

Sometimes the statement, "absence of proof of a harmful effect is no proof of absence," is cited to justify the prohibition of a substance in the name of precaution. Although this statement is true it is very misleading: absence of proof is in fact the only evidence for absence of an effect that we can ever find. Sometimes the principle is used as a powerful weapon by people who are afraid of change or who are opposed to a particular project: who would want to be labeled "enemy of precaution"? For reasons such as these there is a serious danger of misuse of the precautionary principle.

2.4.5 Estimation of abatement costs for CBA

There is a general tendency to overestimate abatement costs in CBA, for several reasons. Since most CBA of pollution abatement concerns new technologies that have not yet been implemented on a large scale, the costs are difficult to foresee, even for honest and well-informed experts. The task is all the more challenging since what matters are the costs after routine implementation, and those costs decrease due to learning and mass-production, decreases that are difficult to estimate in advance.

The experts most qualified for this task are those working for the concerned industries and they have an incentive to err on the high side. If they underestimate the costs, they will be reprimanded by their employers for having brought about a costly regulation, whereas the risk of penalties for overestimation is low: how could the government prove that they exaggerated, if the estimate is not blatantly wrong? The government could get honest estimates by competitive bids, but that is suitable only for government funded pilot installations, not for routine implementation.

Quite generally, industry has an incentive to fight any regulation that puts them at a disadvantage vis-à-vis competitors from countries without regulation. Unfortunately, without the necessary international agreements there is no level playing field. Even without the threat of foreign competition, industry tends to fight regulations that increase the cost of a product, for fear that the increase would lower demand and profits. However, not all industrial lobbying is against environmental improvements. Some forward looking companies correctly anticipate future public demand for a cleaner environment, and by lobbying for stricter regulations they can gain a competitive advantage by being the first to implement a new abatement technology.

2.5   Problems with the use of these tools

2.5.1   Common problems

To err is human – and analysts are human. From review of a large number of studies, we have identified several problems in the use of tools and errors that have been made by various analysts. Table 2.5 identifies a number of areas where things can go wrong during the analysis.

2.5.2   Checklist for avoiding problems

Before analysis:
(1) What is the question that decision makers need answering?
(2) Does the analyst understand the question or need more information?
(3) What information do decision makers need for evaluating the question?
(4) Is the proposed comparison fair and reasonable?

During method development:
(5) What are the correct system boundaries with respect to:
  (a) life cycle stages?
  (b) burdens?
  (c) timeframe?
  (d) location?
(6) Are the defined boundaries complete?
(7) Do data sources reflect the timeframe of interest, accounting for evolution in technology and legislation?
(8) Which method(s) is(are) best suited to answering the question?
(9) Have relevant data sources been identified?
(10) Are proposed methods already accepted or should stakeholders be invited to comment?

During analysis:
(11) How large should the results be? (Test with simple scoping analysis, back-of-the-envelope calculation or reference to existing sources.)
(12) Have input data been checked?
(13) Have units been checked for consistency?
(14) Have calculations been checked?

Table 2.5 *Summary of common errors.*

Area	Problem	Tool	Consequence
Analyzing the wrong problem.	Needless expansion of analysis (e.g. to life cycle stages that are not affected by the policy options under consideration).	LCA, IPA	Wrong frame of reference for policy makers, leading to a lack of focus, allocation of effort to areas where it is not needed and possible development of "blunt" instruments.
	Failure to consider all significant consequences of an action or measure under investigation and to assess which of these consequences are relevant to the question asked.	All	Again, wrong frame of reference for policy makers, with the effect of making good options appear less worthwhile (or bad options more worthwhile) than they really are.
	Inappropriate comparison (e.g. old coal versus new nuclear).	All	Undue bias against one specific option. Very common, and very misleading.
	Failure to account for evolution in legislation and technologies.	All, but particularly LCA	Development of policy that is outdated before it becomes law.
Human error in calculation.	Misunderstanding of units (e.g. converting the 'm' (milli, 10^{-3}) prefix to the 'M' (Mega, 10^{6}) prefix).[5]	All numeric tools	Potential for errors covering several orders of magnitude, so the conclusions can be very wrong.
	Failure to check calculations, allowing simple errors to pass through the analysis. May arise through: • Mis-typing data • Combination of inconsistent units • etc.	All numeric tools	Error in results. Where such errors are detected by stakeholders there will be a loss of confidence in the analysis, possibly undermining attempts to develop new legislation.
Choice of appropriate data, models, etc.	Assumption that one method is best for all questions.	All	Results biased to those impacts that can be dealt with well by the method chosen.

[5] An error we have frequently encountered in this field is a factor of 1000. Sometimes it results from font conversion μ → m during copy and paste.

Table 2.5 (*cont.*)

Area	Problem	Tool	Consequence
	Use of out of date information.	All	Development of policy outdated before it becomes law.
	Confusion between discounting and accounting for price changes.	IPA, CBA	Data errors.
	Confusion of damage costs and abatement costs.	CEA, IPA, CBA	Results cannot be used for CBA.
	Artificial truncation of boundaries, leading to the exclusion of relevant burdens and impacts.	IPA, MCA	Introduces unknown bias to the analysis.
	Use of inappropriate discount rates and other data for long-term assessments, including the assumption that one discount rate is suitable for all circumstances.	IPA, CBA	Too high a rate will make future impacts appear negligible, especially over long time periods.
	Giving inadequate information to stakeholders concerning the type and severity of effects.	MCA	Stakeholders express preferences in relation to a situation different from the one under investigation.
Accounting for uncertainty.	Failure to account for uncertainty (e.g. using sensitivity analysis).	All	Can make results appear more robust than they really are.
	Focusing uncertainty analysis on issues that make little difference to the analysis.	All numeric tools	Can make results appear more robust than they really are.
	Lack of understanding of techniques for reporting uncertainty.	All numeric tools	Lack of information on the robustness of results, misleading uncertainty assessment.
Misrepresentation of results.	Use of misleading metrics to report effects.	IPA, CBA	Misunderstanding of impacts.
	Use of insufficient metrics to report results, e.g. cost–benefit difference but not cost–benefit ratio also.	All, but particularly CBA	Misunderstanding of the sensitivity of results to uncertainty.

Table 2.5 (*cont.*)

Area	Problem	Tool	Consequence
	Inappropriate use of expectation values, upper bounds, etc.	All numeric tools	Misunderstanding of the scale of impacts.
	Failure to declare debatable assumptions underlying the analysis.	All numeric tools	Appearance that results are more robust than they really are.
	Assuming that results for one system (technology, location, etc.) are representative of others.	All	Can lead to large errors.
	Assuming that estimated costs are far more robust than estimated benefits.	CBA	Conservative view on strategies for environmental improvement.
Convincing others of the quality of analysis	Failure to consult a representative group of stakeholders on methods and data used.	All	Potential bias in methods and data, favoring the position of those that have been consulted.
	Failure to consult a representative group of stakeholders on MCA weighting.	MCA	Bias to the position of the stakeholders who have been consulted.
	Failure to say how results have been checked or verified.	All	Lack of acceptance of results, even if analysis is correct.
Applying the precautionary principle.	Blocking a proposed development by naïve invocation of the precautionary principle.	All	Risk of unintended consequences in other sectors that are indirectly affected, e.g. net increase in environmental damage by blocking the development of new installations that would replace more polluting ones.
	Failure to alert the reader to the possibility of risks that have not been quantified.	All	Acceptance of unknown risks.

Reporting results and uncertainties:
(15) How should results be reported?
(16) How should uncertainties be quantified and reported?
(17) What are likely to be the key uncertainties for sensitivity analysis?
(18) Is the presentation of results, including uncertainties, accurate and reasonable?
(19) Are key assumptions provided alongside results?
(20) How do results compare with expected outputs based on the scoping analysis in point 11?

Interpretation of the analysis:
(21) Is guidance given on the limits of application of results?
(22) Do decision makers understand the outputs?
(23) Have decision makers checked that their interpretation of study outputs is valid?
(24) Would it be useful to apply the precautionary principle, particularly in relation to unquantified risks?

2.6   An integrated framework for the analysis of environmental questions

2.6.1   The framework

The proposed integrated framework, in its full generality, involves the following steps (although in simple cases not all of these steps may be required, or they could be abbreviated):
(1) Define the policy options and identify the stakeholders who will be affected.
(2) Define the boundaries of the analysis appropriate for the policy issue under consideration.
(3) Use LCA, with the boundaries thus defined, to assemble a systematic database of the burdens (e.g. emission of pollutants) imposed by the activities concerned by this policy issue.
(4) Define impacts in ways that can be understood by expert and non-expert alike – this is clearly fundamental to any MCA type process where stakeholders are asked to compare across impacts in order to develop weighting schemes.
(5) Carry out an IPA to quantify, as much as possible, the physical impacts (e.g. years of life lost) and external costs (= damage cost) for each of these burdens.
(6) Examine options that reduce the burdens, and obtain data for their costs (abatement costs as well as induced costs, such as impact on consumer spending).

(7) Estimate uncertainties, and identify impacts and costs that defy quantification.

(8) Compare the quantified costs and benefits of the policy options (distinguishing short-term and long-term costs, identifying winners and losers and examining distributional effects).

(9) Develop indicators for impacts and costs that have not been quantified (e.g. sustainability indicators).

(10) Involve the concerned stakeholders to estimate weighting factors for the various indicators to be used in the MCA.

(11) Carry out a MCA to derive aggregate scores for the considered policy options by combining the developed indicators with the weighting factors provided by the stakeholders.

(12) Select the preferred policy option resulting either from the average score or by applying a majority rule, or through a constructive negotiation procedure between stakeholders.

2.6.2   Presenting the results

Serious thought needs to be given to the presentation and discussion of results. Questions need to be asked at each stage of the analysis. Going through the above series of stages, one at a time, the following questions may arise:

(1) In relation to the study objectives, are there other policy questions that could be addressed using the data and broad methodological framework presented here?

(2) In relation to policy options, are there alternatives that are not being discussed, and if so, why are they omitted? For example, policies to reduce air pollutant emissions may focus on technical solutions, but why not include structural measures also, such as regulation of energy markets? Who bears the cost and benefits? Are there actions that one might undertake to correct distributional problems that may arise?

(3) Are the chosen system boundaries complete?

(4) What do we learn from the outputs of the LCA? Does it appear to give clear messages about preferred options?

(5) What are the impacts related to the systems under investigation? Are they well understood, or do they need additional explanation? What are the dominant causes of each type of impact? Does it appear justified to concentrate resource into the current policy area if it does not address a leading cause of the effects of most concern?

(6) What do we learn from the impact pathway analysis? How big are impacts and how big are costs? What policy conclusion would we

reach based on the IPA results alone? What added value does IPA give over LCA?

(7) How reliable are estimates of the costs of reducing pollution and other environmental burdens? Are they likely to lead to overestimation of abatement costs? If so, why, and by how much? Are there additional effects of these measures that should be taken into account?

(8) Which uncertainties cause the biggest problems, which ones can be ignored to provide a clearer focus?

(9) How do costs and benefits compare? To what extent do uncertainties in the analysis complicate the cost–benefit comparison?

(10) How important are unquantified effects for the comparison of costs and benefits? Do any of them seem likely to make a difference to the results quantified to this point?

(11) Is the range of stakeholders brought into the MCA process representative of those who may be affected by any decision taken? How should results be reported – averaged across all stakeholders, for each group of stakeholders individually, etc.? Do all stakeholders have a good understanding of what is meant by the different impacts under discussion?

(12) What are the conclusions of the MCA? How do they compare with the conclusions drawn at other parts of the analysis? What added value has been obtained by going beyond LCA, IPA and CBA?

2.6.3 Summarizing results

The final results often need to be presented in a very concise manner. This can be quite a challenge: a typical request from policy makers is "no more than one page, and don't forget to explain all the assumptions." The appropriate format will clearly vary from case to case. In cases where several different methods have been used in an integrated fashion, it may be worth reporting results for each component separately as well as in a united fashion. So, for example, a study that has evaluated several options could use a table such as Table 2.6, to show the ranking according to the outputs of the different methods used. Criteria for deciding which option fared best in the LCA would of course need to be specified (e.g. summed ranking across impact categories).

Such a table would be able to show the consistency of findings according to the method used. The more consistent the outcome, the more confident the decision maker can be in accepting the results of the analysis. An outcome where the preferred option varies according to the different methods highlights the need to take an integrated approach to

Table 2.6 *Possible format for presenting the ranks of different options.*

	Rank		
	LCA	**CBA**	**MCA**
Option 1	1	1	2
Option 2	4	2	1
...	...	...	...
...	...	...	...
Option n	5	4	4

the analysis, and to probe the findings more deeply in order to understand why different outcomes have been obtained. This format of reporting can be more widely convincing than simple reporting of the final outcome of the analysis, as different stakeholders will be familiar with different ways of doing things.

2.7 Conclusions

The main conclusions of this review of tools are:

(1) There is a wide variety of tools that are being used for the analysis of environmental problems.
(2) Each of these methods has its strengths and weaknesses. None is so widely applicable, or so generally strong, that it negates the need for the other tools.
(3) For externalities analysis we recommend an integrated framework, using LCA, impact pathway analysis, cost–benefit analysis and, in problematic cases, multicriteria analysis, in combination to provide better guidance than would be possible from the use of each tool in isolation. The need to apply the full framework will vary in practice, according to the particular situation under investigation, though an appreciation of each of the tools reviewed here will be beneficial for policy makers seeking to ensure that supporting analysis is appropriately robust.
(4) To get the most from this framework it is essential that careful thought be given to the way in which the results of the analysis are presented. Clear and complete documentation is crucial.
(5) The tools highlighted here are those that seem most applicable to investigation of policy choices. Other tools, for example ecological footprint analysis, have their uses, but not as part of this.

The ground rule of the framework we have presented is quantify as much as possible, and if any remaining impacts are too uncertain or defy quantification entirely, then involve stakeholders in multicriteria analysis (MCA).[6]

Finally, we emphasize that in many if not most environmental decisions, quantified costs and benefits are not the only criteria. For example, decisions about nuclear power need to address the risks of proliferation. For another example, if internalization of the damage costs of energy leads to large price increases for consumers, the poor may no longer be able to heat their residences sufficiently; in such a case considerations of equity may necessitate modification of proposed regulation. Such criteria lie outside the analysis of external costs, but they should be taken into account at the decision stage by means of an MCA with the involvement of stakeholders.

References

Curran, M. A. 2006. Report on Activity of Task Force 1 in the Life Cycle Inventory Programme: Data Registry – Global Life Cycle Inventory Data Resources. *International Journal of Life Cycle Assessment*, 11(4), 284–289.

ExternE 2000. External Costs of Energy Conversion – Improvement of the ExternE Methodology and Assessment of Energy-related Transport Externalities. Final Report for Contract JOS3-CT97-0015, published as *Environmental External Costs of Transport*. R. Friedrich and P. Bickel, eds. Springer Verlag, Heidelberg, 2001.

ExternE 2008. With this reference we cite the methodology and results of the NEEDS (2004–2008) and CASES (2006–2008) phases of ExternE. For the damage costs per kg of pollutant and per kWh of electricity we cite the numbers of the data CD that is included in the book edited by Markandya, A., Bigano, A. and Porchia, R. in 2010: *The Social Cost of Electricity: Scenarios and Policy Implications*. Edward Elgar Publishing Ltd, Cheltenham, UK. They can also be downloaded from http://www.feem-project.net/cases/ (although in the latter some numbers have changed since the data CD in the book).

Fantke, P., Friedrich, R. and Jolliet, O. 2012. Health impact and damage cost assessment of pesticides in Europe. *Environment International* 49, 9–17.

Frischknecht, R., Jungbluth, N., Althaus, H.-J. et al. 2007. Overview and Methodology. Final report ecoinvent data v2.0, No. 1. Swiss Centre for Life Cycle Inventories, Dübendorf, CH.[7]

Frith, C. H., Littlefield, N. A. and Umholtz, R. 1981. Incidence of pulmonary metastases for various neoplasms in BALB/cStCrlfC3H/Nctr female mice fed N-2-fluorenylacetamide. *Journal of the National Cancer Institute* 66, 703–712.

---

[6] In the SusTools project we tested this integrated framework with case studies: one on options for the treatment of waste and one on options to reduce the impacts of nitrogen fertilizer. The detailed documentation can be found in the Full Reports, at the SusTools page of www.arirabl.org.

[7] www.ecoinvent.ch

Graves, P. 2013. *Environmental Economics: An Integrated Approach.* CRC Press/ Taylor & Francis, Boca Raton, FL 33487.

Haastrup, P., Maniezzo, V., Mattarelli, M. *et al.* (2002). A decision support system for urban waste management. *European Journal of Operational Research* **109**, 330–341.

Heck, T., Bauer, C. and Dones, R. 2009. Development of parameterisation methods to derive transferable life cycle inventories – Technical guideline on parameterisation of life cycle inventory data. Report RS1a D4.1, NEEDS (New Energy Externalities Developments for Sustainability). European Commission. (www.needs-project.org/2009).

Hokkanen, J. and Salminen, P. 1997. Choosing a solid waste management system using multicriteria decision analysis. *European Journal of Operational Research* **98**, 19–36.

ISO 2006. Environmental management – Life cycle assessment – Requirements and guidelines. International Organization for Standardization. Available at www.iso.org/iso/catalogue_detail?csnumber=38498

Jolliet, O., Margni, M., Charles, R., *et al.* (2003). IMPACT 2002+: A new life cycle impact assessment methodology. *International Journal of LCA*, **8**(6), 324–330.

JRC 2010. *ILCD Handbook: Analysing of existing Environmental Impact Assessment methodologies for use in Life Cycle Assessment.* First edition. European Commission, Joint Research Centre, Institute for Environment and Sustainability. http://lct.jrc.ec.europa.eu/pdf-directory/ILCD-Handbook-LCIA-Background-analysis-online-12March2010.pdf

Krewitt, W., Trukenmueller, A., Mayerhofer, P. and Friedrich, R. 1995. EcoSense – an Integrated Tool for Environmental Impact Analysis. in: Kremers, H. and Pillmann, W. (Ed.): *Space and Time in Environmental Information Systems.* Umwelt-Informatik aktuell, Band 7. Metropolis-Verlag, Marburg.

Muller, N. Z. and Mendelsohn, R. 2007. Measuring the damages of air pollution in the United States. *Journal of Environmental Economics and Management* **54** (2007) 1–14.

Ottinger, R. L. *et al.* 1991. *Environmental Costs of Electricity.* Oceana Publications, New York.

Pennington, D., Crettaz, P., Tauxe, A. *et al.* 2002. Assessing human health response in life cycle assessment using ED10s and DALYs: part 2 – noncancer effects. *Risk Analysis* **22** (5), 947–963.

Rabl, A. 1999. Air Pollution and Buildings: an Estimation of Damage Costs in France. *Environmental Impact Assessment Review* **19**(4), pp. 361–385.

Rabl, A., Spadaro, J. V. and van der Zwaan, B. 2005. Uncertainty of Pollution Damage Cost Estimates: to What Extent does it Matter? *Environmental Science & Technology* **39**(2), 399–408.

Stritt, M. A. 1993. Coût des salissures causées aux bâtiments par le trafic routier: aspects méthodologiques et résultats empiriques pour la ville de Neuchâtel (Suisse) (Cost of soiling of buildings caused by road traffic: methodological aspects and empirical results for the town of Neuchâtel (Switzerland)). *Science of the Total Environment* **134**, issue 1–3 (June 25, 1993), p. 31–38.

Thun, M. J., Peto, R., Lopez, A. D. *et al.* 1997. Alcohol consumption and mortality among middle-aged and elderly U.S. adults. *New England J Medicine* **337** (24), 1705–1714.

Vad, K. A. 2008. Life Cycle Impact Assessment and Impact Pathway Analysis: an analysis of their differences and insights to be gained from their integration. Graduation Project September 2008. Master of Science Industrial Ecology, Institute of Environmental Sciences (CML), Leiden University, Leiden, The Netherlands.

Vaillancourt, K. and Waaub, J-P. 2002. Environmental site evaluation of waste management facilities embedded into EUG_EENE model: A multicriteria approach. *European Journal of Operational Research* **139**, 436–448.

Weitzman, M. L. 2009. On Modeling and Interpreting the Economics of Catastrophic Climate Change. *Review of Economics and Statistics*, **91** (1): 1–19.

Wilson, R. and Crouch, E. A. C. 2001. *Risk-Benefit Analysis*. Harvard University Press, Cambridge, MA.

3 Exposure–response functions for health impacts

Summary

This chapter is fairly long and detailed because health impacts weigh heavily in the estimation of damage costs. It begins with an overview of the health impacts of air pollution. It then describes the methods used for measuring the health impacts of pollution. The key ingredient in the calculation of damage costs is the exposure–response function (ERF), and we discuss its general features in Section 3.3. The rest of the chapter presents ERFs for specific pollutants and end points. Section 3.4 discusses mortality and life expectancy, and Section 3.5 presents morbidity impacts of the classical air pollutants. Finally, Section 3.6 addresses other pollutants, especially the toxic metals. A summary of the ERFs used by ExternE (2008) will be provided in Table 12.3 in Chapter 12.

A word of caution should be given in relation to the contents of this chapter. There is a great deal of research going on into the health effects of air pollution at the current time. The core position defined here reflects relatively recent consensus, but this will inevitably be revised as more evidence becomes available. The two areas where this is most likely to make a difference concern quantification of the long-term (chronic) effects of exposure to ozone, and the effects of exposure to NO_2. For the latter, there are significant questions of causality being considered – are the effects linked to NO_2 a true effect of the pollutant, or is the pollutant simply an indicator of other stresses? Readers should refer to the final reports of the REVIHAAP and HRAPIE studies led by WHO-Europe on behalf of the European Commission, once they become available, for an updated perspective. Whilst we accept that new findings will influence the choice of response functions, the principles described in this chapter are likely to remain robust.

Major changes can also be expected for some of the toxic metals for which recent epidemiological studies have found more severe health effects.

3.1   Overview of health impacts of pollution

That air pollution damages health has been known for a long time, especially since the infamous smog episodes of Donora, PA, in 1948 and London in 1952. For some of the first assessments of impacts and costs, see Lave and Seskin (1977) and Graves and Krumm (1981). Health impacts are especially important because they contribute by far the largest part of the total damage cost of air pollutants that has been quantified (other than greenhouse gases). A consensus has been emerging among public health experts that air pollution, even at current ambient levels, aggravates morbidity (especially respiratory and cardiovascular diseases) and leads to premature mortality (e.g. Wilson and Spengler, 1996, WHO, 2003 and 2005, Holland *et al.*, 2005, Zmirou *et al.*, 2007, Chen *et al.*, 2008). For an overview, see Table 3.1. There is less certainty about specific causes, but most studies have identified particulate matter (PM) as a prime culprit.[1] Ozone has also been implicated directly. The largest contribution to the damage cost comes from mortality due to PM. Another important contribution arises from chronic bronchitis due to PM (Abbey *et al.*, 1995). In addition, there may be significant direct health impacts of SO_2 and perhaps also of NO_2; for the latter, the evidence has been considered until now to be less convincing, although this may be changing.

The reason for the question marks in the lines for SO_2, sulfates, NO_2 and nitrates is the lack of specific evidence for their toxicity. Sulfates and nitrates constitute a large percentage of ambient PM, but most of the available epidemiological studies are based simply on the mass of PM, without any distinction between components or characteristics (acidity, solubility, surface area, chemical composition, . . .). In particular, there is a lack of epidemiological studies of nitrate aerosols, because this pollutant has not normally been monitored by air pollution monitoring stations.

Quite generally, it is difficult for epidemiologists to attribute a particular health impact to a particular pollutant, because populations are exposed to a mix of different pollutants that tend to be highly correlated with each other. The conclusion that air pollution damages health is much more certain than the attribution of damage to a particular pollutant. For that reason, some epidemiologists emphasize that any individual pollutant is merely an indicator of pollution and that attribution of an impact to any specific pollutant is very uncertain. Many epidemiologists have tended to attribute the damage mostly to PM, although in recent years they have also

---

[1] Usually PM has a subscript, indicating the largest diameter (in μm) of the particles that are included.

Table 3.1 *Pollutants and their effects on health.*

Primary pollutants[a]	Secondary pollutants[b]	Impacts
Particulate matter PM[c] (PM$_{10}$, PM$_{2.5}$, black smoke)		cardio-pulmonary morbidity (cardiovascular and respiratory hospital admissions, heart failure, chronic bronchitis, upper and lower respiratory symptoms, aggravation of asthma) mortality
SO$_2$		respiratory morbidity mortality
	sulfates (due to SO$_2$)	like PM?
NO$_2$		morbidity? mortality?
	Nitrates (due to NO$_x$)	like PM?
	O$_3$ (due to NO$_x$+VOC)	respiratory morbidity mortality
CO		cardiovascular morbidity mortality
diesel soot, PAH, formaldehyde, benzene, 1,3,-butadiene		cancers
dioxins		cancers, endocrine disruption
As		cancers, loss of lung function, neurotoxic morbidity, mortality
Cd		cancers, osteoporosis, kidney disease, mortality
Cr-VI		cancers
Ni		cancers
Hg		neurotoxic and cardiovascular morbidity
Pb		neurotoxic and cardiovascular morbidity, anaemia, mortality
radionuclides		cancers

[a] Emitted by pollution source.
[b] Created by chemical reactions in the atmosphere.
[c] The subscript of PM indicates the largest diameter (in μm) of the particles that are included.

recognized the possibility of a larger role for the gaseous pollutants. Separation of impact is possible from studies in which the concentrations of a number of pollutants have been assessed, and for which it is possible to define "single pollutant models" of impact and "models of impact adjusted for other pollutants." At the time of writing (June 2013), this is particularly relevant to the assessment of chronic effects of exposure to $PM_{2.5}$, ozone and NO_2 on mortality. A problem arises because very few epidemiological studies consider as many pollutants as one would like.

Unfortunately, in order to calculate damage costs one needs specific ERFs for each pollutant; saying that such and such a pollutant is merely an indicator of pollution is too vague for this purpose. One could define a weighted average of the concentrations of different pollutants, but how should one choose the weights? Such an approach begs questions, because one would need specific ERFs to determine the weights. Even worse, the weights could vary from site to site because the composition, and hence toxicity, of PM can vary. Therefore we present specific ERFs for specific air pollutants, while noting the uncertainties about the causal relation between specific pollutants and specific end points.

We also emphasize the possibility that new studies may find impacts that are more important than what has been assessed until now. That is especially likely for pollutants for which biomarker data have been measured in the NHANES (National Health and Nutrition Examination Survey) program of the Centers for Disease Control and Prevention (www.cdc.gov/nchs/nhanes.htm). With biomarker data for each individual, the exposure (or an indicator proportional to exposure) is known with far greater accuracy than has been possible in most traditional studies. One of the authors (AR) has just started a new assessment of ERFs for toxic metals and preliminary results suggest that the damage costs may be far larger than what are currently presented in this book; unfortunately the results are not yet sufficiently certain to be included here.

3.2   Methods for measuring impacts of pollutants

3.2.1   Epidemiology, toxicology and experiments

There are three approaches for measuring the impact of a pollutant: epidemiology, laboratory experiments and toxicology. They all have strengths and limitations, and they can complement each other.

(1) **Epidemiology** involves comparing human populations with different exposures ("exposure contrast"), all else being the same. Using careful statistical analysis, one looks for a correlation (called "association") between exposure and impact (also called "health end point," or simply "end point"). Such associations may indicate a causal effect.

However, in most cases the uncertainties are very large. There is always the nagging question of whether the observed impact is really due to the pollutant and not to other factors that have not been taken into account, called "confounders." Smoking is a notorious confounder: if the population with high exposure has a high rate of smoking, the impact may be due to smoking rather than the pollutant.[2] In many cases a practical problem lies in the difficulty of determining the real exposures of the individuals in the study populations. But in contrast with toxicology the great advantage of epidemiology is the ability to measure impacts on real human populations.

(2) **Laboratory experiments with humans** avoid the uncertainties in the exposure, by exposing individuals in test chambers to a controlled concentration of air pollutants, but this approach is of course very limited because of ethical constraints.

(3) **Toxicology** involves exposing animals, or in some cases tissue cultures, to a pollutant. As test animals one usually takes rats or mice, and the sample sizes are usually very small compared with epidemiological studies (several tens of individuals per exposure level). By contrast with human populations, the animals are selected to be as homogenous as possible in order to obtain reproducible results. Obviously, extrapolation to real human populations introduces large uncertainties, to say nothing about the extrapolation from tissue cultures. One of the advantages of toxicology is the ability to identify mechanisms of action of a pollutant, because one can dissect and analyze the affected animals. And whereas epidemiology is limited to substances to which people have been exposed, with toxicology one can test anything. For many substances, tests with animals are the only way to identify carcinogenic or other harmful effects. Toxicology can also suggest new questions to be investigated by epidemiology. Thus epidemiology and toxicology are complementary.

3.2.2 Types of epidemiological studies for air pollution

Exposure of a population to air pollution is easy to measure and it tends to vary rapidly from day to day, unlike exposure via ingestion, which is difficult to measure. This fact is the basis of time series (TS) analysis,

---

[2] There are countless examples of correlations that are not causal relations; for instance, the positive correlation between declining birth rates and declining stork populations can hardly be taken as evidence that storks bring children.

the most commonly used methodology for air pollution epidemiology. One looks for correlations between the daily frequency of an end point and the daily concentration to which the population is exposed, as measured by the monitoring stations of the local air quality network. The exposure contrast, crucial for epidemiology, comes from the differences between daily concentrations, while one can assume that potential confounders do not change from day to day; for example, that smoking patterns remain the same. Time series studies are relatively simple and inexpensive because the required data are readily available without any need for the collection of individual data to eliminate confounders. Time series studies are possible only in sufficiently large cities, with populations of at least several hundred thousand, otherwise the number of impacts per day would be too small to measure significant correlations with pollution.

A limitation of conventional TS studies is that they can only identify acute impacts, i.e. short-term impacts that occur within a few days of exposure. This is the origin of the rather strange term, "acute mortality": it is the mortality impact observed by TS immediately after a change in exposure. "Chronic mortality" by contrast, is the mortality impact due to chronic exposure. Measuring chronic impacts requires long-term observations that are much more difficult and costly than TS because of the need to eliminate confounders. An important method for chronic impacts is the cohort study, where one follows age cohorts, i.e. groups of individuals of the same age, e.g. between 60 and 61 at the start of the study. Within each cohort, one needs individuals with different exposures. An important example of a cohort study is the analysis of the American Cancer Society data set by Pope et al., (2002) which has been the basis of the mortality ERF used by many studies, including ExternE. This set contains fairly detailed personal data (address, educational level, smoking habit, etc.) for about half a million individuals who have been followed over a period of 16 years; all of the major metropolitan areas in the USA are included.

3.2.3   Difficulties with epidemiological studies

The uncertainties in most studies of pollution are very large, for several reasons. First of all, the health impacts are small at typical exposures – fortunately for us: we are not all dropping dead from pollution. By the same token, it is difficult for epidemiologists to measure the impacts.

Secondly, for most pollutants the exposures are difficult to determine. For air pollution studies one only has the data measured by a few monitoring stations, not the actual exposures for each individual. This is not too serious a problem for PM_{10} and $PM_{2.5}$ because their concentrations inside buildings tend to be closely correlated with those measured at stations in the same city (apart from buildings with smokers).[3] But NO_x and O_3 can be quite different from the monitoring station data. For pollutants such as toxic metals, most older studies are based on workers who were exposed to high concentrations, often long before the toxicity was sufficiently recognized. In such cases, the exposures have to be estimated from whatever scant data can be found, such as length of employment and type of work in a factory.

Thirdly, populations are exposed to a mix of different pollutants that tend to be highly correlated with each other. Therefore it is difficult to establish definite links between an end point and a particular pollutant. Many studies have looked at several air pollutants, comparing the results of different regressions, but very few are able to provide statistically significant regression models with several pollutants at once, and rarely more than two.

Fourthly, by contrast with the extreme complexity of the underlying biological processes, epidemiological studies can only take into account certain simple gross features, for example, variation in respiratory hospital admissions as a function of the SO_2 concentration to which the study population is exposed. There may be large individual differences, for instance in sensitivity to a particular pollutant. But even the most detailed studies cannot take into account more than a few limited characteristics of any situation.

Finally, there is the problem of developing a thorough understanding of the range of impacts that a pollutant has on health. For ozone, for example, there are many studies that deal with the effects of acute exposures and mortality and respiratory hospital admission. Data for these effects are collected routinely, so are available for comparison with air quality data. However, data on other effects (chronic bronchitis, restricted activity days, etc.) may require a concerted information gathering effort and thus tend to be much less readily available. However, without such data one may only have a very limited perspective of the true burden of pollutants on morbidity.

3.2.4   *Differences between results from different studies*

In the following paragraphs we illustrate these difficulties by looking at differences between the results of different epidemiological studies of air

---

[3] For time series studies this is no problem because smoking habits do not vary from day to day, but for cohort studies (see Section 3.4.2) one needs individual smoking data.

pollution. We describe the situation as it was more or less a decade ago, though also with reference to more recent findings. At the time of writing (June 2013), a major reassessment of response functions is underway in Europe through the REVIHAAP and HRAPIE studies,[4] results of which will be made available through WHO-Europe. Another perspective will be provided by WHO's Global Burden of Disease Project. Even though in the meantime some of these differences have been resolved, we find these examples instructive as a warning not to put too much faith in any single study unless it is confirmed by additional evidence.

Exposure–response functions obtained from different populations might reflect different sensitivities, but they could also be due to differences in the local pollution mix or in the way the exposures have been measured,[5] to say nothing of differences in methodology (for example, the choice of the time lag between concentration and end point in a TS analysis).

Thus it is not surprising that different studies find different results; some, for instance, find strong effects from SO_2 while others do not. In some cases a reanalysis of the same data with improved methodology has obtained results that are appreciably different. For example, when the results of the APHEA Project were published (Katsouyanni et al., 1997), it seemed that the acute mortality ERF for PM_{10} in Europe had only about half the slope of the average found by numerous studies in North America. The latter had mostly been done for single locations, and not always following the same protocol. But then an improved and more comprehensive analysis of pooled data in the USA (HEI 2001) obtained essentially the same slope as APHEA, which had used a standardized protocol for 15 cities. To cite from HEI, "... HEI's US-wide National Morbidity, Mortality and Air Pollution Study (NMMAPS) (Samet et al., 2000) found a 0.5% increase in total nonaccidental mortality associated with a 10 $\mu g/m^3$ increase in PM_{10} in the 90 largest US cities where daily average PM_{10} ranged from 15 to 53 $\mu g/m^3$. This result agrees closely with that of the European APHEA study (0.6% per 10 $\mu g/m^3$) (Katsouyanni et al., 1997) and with a recent meta-analysis of 29 studies in 23 locations in Europe and North and South America (0.7% per 10 $\mu g/m^3$) (Levy et al., 2000)."

Another illustration of the danger of relying on a single study can be seen in the APHEA results for Eastern Europe. Initially, they seemed to

---

[4] REVIHAAP: Review of Evidence of Health Aspects of Air Pollutants. HRAPIE: Health Risks of Air Pollution in Europe.

[5] There can be significant differences between the concentrations measured by monitoring stations and the concentrations to which the individuals in a city are exposed, especially for NO_x and O_3.

Table 3.2 *Variation of acute mortality due to PM$_{10}$ in Europe. Data from Katsouyanni et al. (2001). (in parentheses 95% CI).*

	% increase per 10 µg/m^3
Average, Europe	0.60% (0.40–0.80%)
City with low average NO$_2$	0.19% (0.00–0.41%)
City with high average NO$_2$	0.80% (0.67–0.93%)
Cold climate	0.29% (0.16–0.42%)
Warm climate	0.82% (0.69–0.96%)
City with low standardized mortality rate	0.80% (0.65–0.95%)
City with high standardized mortality rate	0.43% (0.24–0.62%)

imply a much smaller ERF slope than for Western Europe, but a re-analysis by Samoli *et al.* (2001) found "... *The ratio of western to central-eastern cities for estimates was reduced to 1.3 for BS (previously 4.8) and 2.6 for SO$_2$ (previously 4.4). We conclude that part of the heterogeneity in the estimates of air pollution effects between western and central-eastern cities reported in previous publications was caused by the statistical approach used and the inclusion of days with pollutant levels above 150 µg/m^3....*"

Interesting results have been published that shed some light on the variability of ERFs between different regions. For example, the result for acute mortality varies between different regions of the USA, being more than twice as large in the north east as in the south west (with the exception of southern California where it is almost as high as in the north east) (HEI, 2001). Table 3.2 summarizes results on regional variability in Europe, from the APHEA2 study (Katsouyanni *et al.*, 2001). The range of variation is large and roughly comparable to HEI (2001). But these results are not sufficient to allow generalization to other locations.

A reanalysis of the NMMAPS data with improved models (GAM and GLM) resulted in lower ERFs: about 0.21% to 0.27% per 10 µg/m^3 PM$_{10}$ (HEI, 2003). However, the WHO (2003) meta-analysis of European studies recommended an ERF very close to the original estimates, 0.6% per 10 µg/m^3 for all-cause mortality, all ages, due to PM$_{10}$.

Some authors of health impact assessments favor country or region specific ERFs, sometimes even to the exclusion of the international literature. What is the justification? There could indeed be different ERFs due to differences in the local mix of air pollutants (with different synergistic effects, or different composition of PM). And different populations could have different sensitivities to pollution, due to lifestyle or genetic makeup. However, the uncertainties of air pollution epidemiology are so large that we have come to the conclusion that a synthesis of the international literature is more reliable than a limitation to country or region specific studies. A meta-analysis of all available studies is the preferred approach for such a synthesis. As an example, we cite the meta-analysis of chronic mortality studies by Chen *et al.* (2008).

Epidemiological studies always indicate their uncertainties, usually by showing 95% confidence intervals (CI). An association is considered statistically significant only if the CI does not include zero. However, these are the CIs of the statistical analysis, assuming random errors; the true uncertainties could be much larger. For example, suppose that regressions had consistently found a statistically significant health impact of SO_2 emissions, but that in reality this were due to trace metals that are emitted together with SO_2 in all the cities where the studies have been carried out. Then the true impact of SO_2 could be zero, well outside the CI.

3.2.5   *Hill's criteria for causality*

As one can gather from the above paragraphs, the uncertainties of epidemiology are large.[6] Of course, epidemiologists have given much thought to the question of when an association between exposure and impact demonstrates a causal relation. In particular, the great epidemiologist Hill formulated a set of criteria, summarized here in Table 3.3. Column one lists the criteria, column two gives a brief assessment of the case of air pollution.

3.3   Exposure–response function (ERF): general remarks

3.3.1   *ERFs and calculation of impacts*

A dose–response function (DRF) relates the quantity of a pollutant that affects a receptor (e.g. population) to the physical impact on this receptor (e.g. incremental number of hospitalizations). In the narrow sense of the term, it should be based on the dose actually absorbed by a receptor. However, the term dose–response function is often employed in a wider sense, where it is based on the concentration of a pollutant in the ambient air, accounting implicitly for the absorption of the pollutant from the air into the body; the term concentration–response function (CRF) is often used in this sense. In this book, we prefer exposure–response function

---

[6] It's been said that one can define an environmental catastrophe as something so big that even epidemiologists can see it!

Table 3.3 *Hill's criteria for causality of associations found in epidemiological studies.*

Hill's criterion	Situation for air pollution
Strength of the association (is effect weak or strong?)	Effects at typical exposures are weak, only observable in large populations (large uncertainty in individual studies)
Consistency of the association (is effect same at different times and places?)	Mostly yes, consistent results in North America, Europe, Asia, South America
Specificity of the association (is exposure to specific pollutant associated with specific effects, with no other likely explanation?)	Respiratory and cardiovascular illness Problem: composition of PM is not well defined; association with specific pollutants is not clear
Temporality (does effect occur after exposure?)	Yes
Dose–response (does effect increase with exposure?)	Yes
Biological plausibility (is effect plausible in terms of biological mechanisms?)	In recent years more and more studies have identified mechanisms
Coherence (of whole body of data, including animal studies etc)	More or less
Experimentation (does removal of exposure remove effect?)	Limited data, yes (shutdown of steel mill Utah Valley 1986–87 (Pope, 1989); fuel change in Dublin (Clancy *et al.*, 2002) and in Hong Kong 1990 (Hedley *et al.*, 2002))
Analogy (is effect plausible on the basis of analogous situations?)	Smoking: among the most harmful constituents of tobacco smoke is PM due to combustion. The health effects of smoking have been documented beyond any doubt by countless studies, in particular the impressive 50 yr follow-up of 34,000 doctors in the UK (Doll *et al.*, 2004)

(ERF) because "exposure" is a more general term that can refer to dose as well as concentration.

The ERF is a central ingredient in impact pathway analysis (IPA) and merits special attention. An impact can be quantified only if the corresponding ERF is known. Such functions are available for the impacts on human health, building materials and crops, caused by a range of pollutants such as primary and secondary (i.e. nitrates, sulfates) particles, O_3,

CO, SO_2, NO_2, benzene, dioxins, As, Cd, Cr, Hg, Ni and Pb. The most comprehensive reference for health impacts is the IRIS database of EPA,[7] although it is not always up to date. For application in an IPA, the information often has to be expressed in a somewhat different form, accounting for additional factors such as the incidence rate. One has to be careful about the units because different databases may state an ERF in different ways, for instance, cases per lifetime exposure at a specified dose rate or cases per ingested dose.

For many pollutants and many impacts the ERFs are very uncertain or not known at all. For most substances and non-cancer impacts, the only available information indicates thresholds, typically the NOAEL (no observed adverse effect level) or LOAEL (lowest observed adverse effect level).[8] Knowing thresholds is not sufficient for quantifying impacts; it only indicates whether or not there can be an impact. However, it turns out that for the most important pollutants, in particular the classical air pollutants and many carcinogens, explicit ERFs are known and impacts can indeed be calculated.

Some analysts worry that the ERFs of available studies do not take into account synergistic effects of exposure to multiple pollutants. The possibility of interactions between pollutants is indeed a classic problem in epidemiology. Sometimes the combined effects are calculated by means of multiplicative rather than additive models. In the case of air pollution, all epidemiological studies involve a mix of pollutants, and thus synergistic effects are implicitly taken into account in any ERF for a single pollutant. However, what happens when several pollutants change at the same time is less clear. In any case, the uncertainties are so large and ERFs for simultaneous changes of several pollutants are so rare) (for an example, see Katsouyanni et al., 2001) that it seems better to assume independence of the different pollutants and to simply add the effects of different pollutants without worrying about interactions.

3.3.2   Form of the ERF and extrapolation to low exposures

Frequently one needs to apply an ERF at exposures below the range where it has been determined. This is especially true for ERFs that have been

---

[7] www.epa.gov/iriswebp/iris/index.html

[8] Pennington et al. (2002) have proposed a promising method for using LOAEL or NOAEL data for estimating ERFs. The basic idea is that there is a correlation between toxicity and threshold: the lower the threshold, the more toxic the substance. However, their results are not yet sufficiently complete and validated to be used for damage cost calculations.

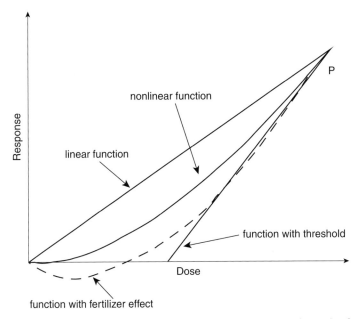

Fig. 3.1 Possible behavior of ERFs at low exposure or dose: the four functions shown have the same value at P (the lowest exposure where a nonzero impact has been observed).

determined from industrial exposures, but the problem also arises for the classical air pollutants if one wants to estimate the benefit of reducing the exposures below current levels. How can one extrapolate from the observed data towards low exposures? In many cases, the range of the available data is not sufficient for the desired extrapolation and the uncertainties are large. Figure 3.1 indicates schematically several possibilities for the case where the point P corresponds to the lowest exposures at which a response has been measured.

The simplest is the linear model, i.e. a straight line from the origin through the observed data point(s). The available evidence suggests that an ERF is unlikely to go above this straight line in the low dose limit. The simple straight line model does appear to be appropriate in many cases, in particular for the classical air pollutants at concentrations typical of, for example, western Europe and North America, and for many cancers. Linearity without threshold provides a wonderful simplification for the calculation of impacts: all is proportional to exposure or dose, and the results are readily transferred to other exposures. If there is a threshold, the calculations become complicated and would have to be redone for each change of exposure. The analysis of mercury impacts in Section 8.3 illustrates some of the complications. It should, however, be noted that Pope *et al.* (2009) suggest a nonlinear function for application over greater ranges of air pollution.

Another possibility is the "hockey stick": a straight line down to some threshold, and zero effect below that threshold. Thresholds occur when an organism has a natural repair mechanism that can prevent or counteract damage up to a certain limit. A hockey stick with a sharp transition at the threshold is probably not realistic, and a smooth nonlinear function is far more likely.

There is even the possibility of a "fertilizer effect" at low doses, as indicated by the dashed line in Fig. 3.1. This can be observed, for example, in the dose–response functions for the impact of NO_x and SO_2 on crops: a low dose of these pollutants can increase the crop yield, in other words the damage is negative. Generally, a fertilizer effect can occur with pollutants that provide trace elements needed by an organism. Chromium is an interesting example to illustrate the complexity of possible effects: as an essential element for the human body it is beneficial at low doses, but at high doses it is toxic; in addition it is carcinogenic if in oxidation state VI. Another example is alcohol: it seems to reduce mortality at low dose (optimum around one drink/day) but is harmful in excess (see e.g. Thun *et al.*, 1997).

As for thresholds, there is a difference between the homogeneous populations of nearly identical animals studied by toxicologists and the populations studied by epidemiologists. The absence of a no-effect threshold for population-level ERFs is plausible for real populations, because they always contain individuals with widely differing sensitivities; for example, at any moment about 1% of a population with a 75-year life expectancy is within the last nine months of life and for the most part very frail. Thus the question of thresholds depends on what impact and what type of population it is based upon. Exposure–response functions for individuals or for a group of fairly similar individuals are more likely to have a no-effect threshold. For example, in a group of young and healthy individuals a moderate air pollution peak will not induce any premature deaths, by contrast with a population that contains very frail individuals. The ERFs used by health impact assessments and in this book are based on entire populations.

3.3.3   Linear ERFs

Fortunately the ERFs for the most important health impacts seem to be approximately linear without threshold over the range of application in

Europe and North America; in particular, the ones for PM and for cancers or neurotoxic impacts of the toxic metals.[9] Such linearity is also the most plausible model for the ERFs for radionuclides. Let us look at some of the evidence.

The ERFs for PM that have been or are used by most impact assessments, including ExternE, are assumed to be linear without threshold. In support of linearity one can cite the following:

- The meta-analysis of 107 studies by Zmirou et al. (1997), which states in its English abstract, "The dose–response functions seem linear in the range of observed concentrations" for PM, O_3, SO_2 and NO_2.
- The study of air pollution mortality in nine cities in France by Zeghnoun et al. (2001) which states in its English abstract, "These associations were linear without threshold."
- The review by Daniels et al. (2000) who conclude that "... linear models without a threshold are appropriate for assessing the effect of particulate air pollution on daily mortality even at current levels."
- The review by Daniels et al. (2004) who reconfirm "... that linear models without a threshold are appropriate for assessing the effect of air pollution on daily mortality."
- A cohort study by Schwartz et al. (2008) of chronic mortality due to $PM_{2.5}$ which tested a large number of possible functional forms and found that the straight line without threshold gave the best fit.

For O_3, the case for linearity without threshold is less clear because of the large uncertainties of O_3 studies. Some experts prefer to assume a no-effect threshold for O_3. WHO (2003) concluded that there was no evidence for a threshold in the ERFs for ozone, although they admitted that there was little evidence for a firm judgment. Consequently, WHO in its recommendations for the Clean Air for Europe (CAFE) Program of the EC (http://ec.europa.eu/environment/archives/cafe/general/keydocs.htm) decided that, in the main analyses, the effects on mortality should be quantified only at maximum eight-hour mean O_3 concentrations above 35 ppb (70 µg/m^3). The cumulative exposure above 35 ppb during the entire year was called SOMO35. WHO emphasized that the use of a cut-off should not be interpreted as acceptance of a threshold, and recommended that, as sensitivity analyses, effects should be estimated also without threshold. The NEEDS phase of ExternE (2008) assumes linearity with a threshold of 35 ppb for all the ERFs for O_3.

---

[9] The Global Burden of Disease Project (Lim et al., 2012) applied nonlinear functions following the approach of Pope et al. (2011). However, the GBD study had to estimate effects globally, including areas where pollution levels are far greater than those typical of Europe and North America.

For the classical air pollutants, the background in most industrialized countries is above the level where effects are known to occur. When evaluating large reductions of exposure, to levels below those that have been observed in epidemiological studies, the possibility of a no-effect threshold cannot be ruled out. However, in the case of $PM_{2.5}$, a new analysis of the low concentration regime by Krewski *et al.* (2009) renders the possibility of a no-effect threshold very unlikely, because they find evidence that the ERF slope may even increase as exposures are lowered. This has been strengthened by the results of Crouse *et al.* (2012) in a large Canadian cohort study where the mean $PM_{2.5}$ level was only 8.7 µg/m³.

For cancers, most authors assume linearity without threshold. This is the most plausible model on theoretical grounds for substances that initiate a cancer (and also for radionuclides), as explained below. For substances that only promote the growth of a cancer the ERF could have a no-effect threshold, but there are few specific data. In view of the importance of the threshold question for impact analysis, a classic study was carried out by Frith *et al.* (1981) in an attempt to find out whether or not there are thresholds for carcinogens. This study seems worth citing, even though it used a substance not found among the pollutants of power plants. Some 24,000 mice were exposed to the carcinogen, 2-acetyl-amino-fluorene over several different dose levels. The response for liver tumor was linear, whereas the one for bladder tumor showed a threshold. This demonstrates conclusively that there is no general rule: some ERFs have a threshold, others do not.

Note that for the calculation of incremental damage costs there is no difference between the linear and the hockey stick function with the same slope, if the background concentration is everywhere above this threshold; only the slope matters. Since a straight line through the origin is uniquely characterized by its slope, we state all ERFs in terms of their slope, S_{ERF}.

Quite generally, if a pollutant acts additively with any already ongoing process (background), then the abscissa of the ERF is the sum of the background and the contribution of the pollutant. If the ERF is a smooth function, one can expand it in a Taylor series around the background exposure, and for small anthropogenic additions to the background one can approximate it by a linear term. That is how Crawford and Wilson (1996) argued that low dose linearity is the most plausible model when a pollutant acts additively to a natural background process, for instance when the mechanism of action is the same as a background mechanism in the body. Such is the case for carcinogens that act by damaging DNA. Since DNA damage is an ongoing process due to various causes, the damage due to a particular carcinogen is an addition to a background and the ERF for that carcinogen is linear without threshold at low dose.[10]

By the same token, linearity without threshold is plausible for PM and O_3 because important modes of their action are additive to natural background processes in the body, namely inflammation and/or oxidative stress. If the addition due to pollution is small compared with the natural background, the Taylor series argument implies that the increment is approximately proportional to exposure.

Some people cite hormesis in the context of the threshold question. Hormesis is the hypothesis that a very small dose of a toxic substance can have beneficial effects. It is controversial because it is very difficult to prove or disprove by experiment (see review by Calabrese et al. (1999)). Hormesis is plausible if a brief exposure to a toxic substance stimulates the defensive mechanisms of an organism. But we do not find hormesis plausible at all for chronic exposures, and therefore irrelevant for damage costs of pollution (except possibly for accidental exposures). And even if there were hormesis for chronic exposures, it would already be implicit in the observable ERFs.

Natural selection is sometimes invoked to argue that the human body has developed a certain resistance to toxic metals such as Hg because they occur naturally in the environment. This argument rests on a misunderstanding of natural selection. The health damage due to such exposures is so small as to be entirely negligible compared with other factors that affect survival, especially if it manifests itself only after the production of offspring. Evolution has not lasted long enough for each and every little factor to exert its influence.

3.3.4   Accounting for a threshold in an IPA

How to account for a threshold depends on whether the ERF is for inhalation or for ingestion. The concentration of a pollutant in the air varies rapidly from hour to hour, whereas the concentration in drinking water or specific food items varies very slowly. Individual intake by ingestion varies from one meal to another, but is fairly constant on average.

For ERFs based on inhalation, one counts only those times when the concentration is above the threshold. This is straightforward if the impact pathway analysis (IPA) uses dispersion software with hourly output. But even if one has only the average inhalation dose, as for example in the UWM, one can take into account a threshold by noting that concentrations in the air tend to have a lognormal distribution. The examples presented

---

[10] In the case of 2-acetyl-amino-fluorene in the above mentioned study by Frith, Littlefield and Umholtz (1981), the modes of action seem to be different for the two cancer types.

by Ott (1995) suggest that the geometric standard deviation for the concentrations of air pollutants is around $\sigma_g = 1.5$. Knowing the distribution one can calculate the fraction of time that the concentration is above a specified threshold.

For ERFs based on inhalation one can, in the calculation of total impacts due to current total exposures, simply multiply the no-threshold result by the fraction f_{th} of the population that is above the threshold. For the calculation of marginal impacts, however, simple multiplication by f_{th} is not exact, although it may be an acceptable approximation for air pollutants. The reason lies in the difference between the distributions of total and of marginal dose among the population, and the calculation is more complicated, as we illustrate for the example of mercury impacts in Section 8.3.

3.3.5   Relative risk and ERF

Most epidemiological studies report their results in terms of relative risk RR, defined as the ratio of the incidence observed at two different exposure levels. For air pollution, the risk increase ΔRR is often expressed as % change in endpoint per 10 $\mu g/m^3$ of the pollutant. Some studies use the case-control design, observing subgroups that may or may not be representative of the entire population, and they report an odds ratio OR. To understand the relation between OR and RR, note that the general design of epidemiological studies involves the comparison of populations with and without exposure (to a pollutant, risk factor, medical treatment, etc.). With two levels of exposure, one observes the number of individuals $N_{exp,eff}$ for each exposure level that show or do not show the effect under study, as indicated in this table:

	Effect: yes	Effect: no
Exposure 1	$N_{1,yes}$	$N_{1,no}$
Exposure 0	$N_{0,yes}$	$N_{0,no}$

The odds ratio is

$$OR = \frac{N_{1,yes}/N_{0,yes}}{N_{1,no}/N_{0,no}} = \frac{N_{1,yes}N_{0,no}}{N_{0,yes}N_{1,no}} \tag{3.1}$$

and the relative risk is

$$RR = \frac{\dfrac{N_{1,yes}}{N_{1,no}+N_{1,yes}}}{\dfrac{N_{0,yes}}{N_{0,no}+N_{0,yes}}}. \tag{3.2}$$

One can calculate one from the other if each of the four numbers has been reported. However, if the effect, e.g. the incidence of asthma, is rare, one has $N_{0,yes} \ll N_{0,no}$ and $N_{1,yes} \ll N_{1,no}$ and OR is close to RR. This is an acceptable approximation for most air pollution studies.

To quantify damages one needs to translate RR in terms of an ERF for the incremental cases per exposure increment. The number of cases is the product of ΔRR and the baseline or reference level of incidence, I_{ref}. Most epidemiological studies do not provide data for I_{ref}, and other sources must be consulted. For many health endpoints, data on background rates of morbidity in the country or region of interest are not readily available. Sometimes one can find such information in other general epidemiological studies of that health endpoint, for example, the International Study of Asthma and Allergies in Children (ISAAC) and, for adults, the European Community Respiratory Health Study (ECRHS).

In this book, we find it convenient to define all ERFs in terms of cases per year per average person per $\mu g/m^3$, because then they can be applied directly to the entire population without worrying about affected sub-groups. Thus they include a factor for the fraction of the population that is affected ("risk group fraction").

Transferring ERFs to other countries requires data for the respective I_{ref}, together with an assumption about the RR: is the RR the same for populations in different countries? Often the data are insufficient for a firm answer.

3.3.6   Which pollutant causes how much health damage?

Several health impact assessments, for example by the World Health Organization (WHO, 2003), have estimated the health impacts of exposure to ambient levels of air pollution. This is sufficient for informing policy makers about the benefit of reducing the concentration values recommended as guidelines for ambient air quality. For such an assessment, the results of epidemiological studies can be used without any hypotheses about the toxicity of different components of ambient PM: both the studies and the assessments are based directly on typical compositions of ambient PM.

By contrast, IPA is a bottom-up methodology and starts from the source of the pollutants, calculating the damage attributable to each emitted pollutant. The need for this kind of information becomes obvious when one recognizes that, in order to actually attain lower ambient concentrations, specific regulations must be put in place to force the polluters to reduce their emissions. For the optimal formulation of such regulations one needs to compare the benefits of reducing the emission of a pollutant and the cost of such a reduction for all abatement technologies under consideration. In some cases tradeoffs must be made between the reductions of different pollutants; for example, certain automotive technologies reduce the emission of PM while increasing the emission of NO_2. Thus the optimal formulation of environmental policies requires more detailed information on the health effects of specific pollutants: one needs to know the incremental impact of an incremental kg of each pollutant that is emitted by a particular source such as a power plant or a car.

Separating the roles of SO_2, NO_2 and PM_{10} is particularly problematic, given that they tend to be correlated. It is not entirely clear to what extent the apparent effects of PM are in reality a reflection of the effects of NO_2 or SO_2, or vice versa, or whether the presence of other pollutants affects the toxicity of PM. Thus there are uncertainties in applying ERFs in a situation where the ambient pollutant mixture is different from the one where the original epidemiological study was carried out.

The current position of most assessments, including ExternE, is to use only ERFs for PM and O_3, but none for SO_2 or NO_x, a choice also made in other health impact assessments (WHO, 2003, Abt, 2004, NRC, 2010). The health impacts of SO_2 or NO_x are assumed to be due to their transformation to sulfate and nitrate aerosols. However, the situation is not clear and opinions could change as further evidence comes to light. In particular, the Hong Kong intervention study showed a sustained benefit of reduced mortality following a reduction of SO_2 emissions (Hedley et al., 2002, Rabl et al., 2011). There could indeed be significant direct effects of SO_2, also found by Elliott et al. (2007). But, conceivably, they could be due to transition metals, in particular Ni and V, that are emitted by the dominant SO_2 sources, namely combustion of oil or coal. Such metals have been identified in some studies as possible agents that increase the toxicity of ambient PM (e.g. Lippmann et al., 2006).

The focus on PM is understandable because the most consistent results, worldwide, have been found for PM; one of the reasons is that the indoor concentration is close to that measured outdoors at monitoring stations (apart from contributions due to indoor sources, without an effect on the results because they are uncorrelated with outdoor concentrations), by contrast with most of the gaseous pollutants, for which the indoor/outdoor ratio is highly variable. In particular, multipollutant analyses have usually found PM to be the most significant. This despite the fact that PM is an ill defined mixture of pollutants (anything, solid or liquid, that accumulates in a particle detector, including sulfuric acid and ammonium nitrate) whose composition can be quite different between different sites. Studies on the relative toxicity of different components of PM have, so far, not been sufficiently conclusive to draw firm conclusions, although sulfates do appear in quite a few significant associations, in particular in Pope et al. (2002). For example, some studies find associations with acidity, others do not (Lippmann et al., 2000). Several studies have found that crustal particles, a major constituent (typically 10 to 50%) of ambient PM_{10}, appear to be harmless (Laden et al., 2000, Pope et al., 1999, Schwartz et al., 1999); most of the damage seems to be caused by combustion particles. Reiss et al. (2007) emphasize the lack of sufficient evidence for the toxicity of sulfate and nitrate aerosols.

As for O_3, its health impacts are very difficult to measure because of the poor correlation between concentrations at measuring stations and personal exposures. Roughly speaking, tropospheric O_3 is created when there is a combination of light, NO_x and VOC. But it involves complex processes with hundreds of chemical reactions. Most NO_x is emitted as NO, and rapidly oxidized to NO_2 by consuming O_3, thus causing depletion of O_3 near sources of NO_x. This can lead to strong local variations in O_3 concentrations. Since O_3 is rapidly depleted indoors, personal exposure depends on lifestyle, especially the fraction of time spent outdoors. For O_3, the evidence from epidemiology is far less convincing than for PM, although by now a sufficient fraction of studies have demonstrated credible associations. Furthermore, toxicological observations have identified oxidative damage and inflammation of the lungs due to O_3 (Donaldson, 2006) and lend further support to the assumption of a causal link.

The main question remaining for O_3 concerns the possibility of chronic impacts. This is crucial for the monetary valuation, because for PM the contribution of most acute impacts is small compared with that of chronic ones, especially chronic bronchitis and chronic mortality. Thus, the study of Gauderman et al. (2004) is important because it followed the lung function of a cohort of children over an eight-year period. These authors find significant associations of lung function reduction with chronic exposure to $PM_{2.5}$, elemental carbon, acid vapor and NO_2, but not with O_3. There is also evidence of chronic effects of ozone exposure on mortality, for example by Jerrett et al. (2009).

At the time of writing (June 2013), there was much interest through the REVIHAAP and HRAPIE studies, led by WHO-Europe for DG Environment of the European Commission, in extending the analysis for use in Europe and North America to include some impacts related to NO_2. These effects included both acute and chronic mortality impacts (e.g. Samoli et al., 2006; Hoek et al., 2013) as well as some morbidity impacts and so could add substantially to the damage estimates of ExternE and CAFE. However, the application of (particularly) the chronic function created some potential for double counting. Readers should consult the final reports of REVIHAAP and HRAPIE when they become available.

3.4   ERFs for mortality

3.4.1   *Loss of life expectancy and number of deaths*

Many studies have attempted to quantify the impacts of mortality due to air pollution (ORNL/RFF, 1994, Rowe *et al.*, 1995, ExternE, 2000, ExternE, 2008, Levy *et al.*, 1999, Abt, 2004, Kuenzli *et al.*, 2000, and others). Whereas all studies before 1996 calculated a number of premature deaths, and applied a "Value of Prevented Fatality" (VPF) to obtain a monetary value of these deaths, there has been a growing recognition that it is more meaningful to look at loss of life expectancy (LE) (see e.g. Wilson and Crouch (2001)). In particular, the ExternE project series has, since 1998, based the monetary valuation of air pollution mortality on LE change and the value of a life year (VOLY).

Since this is not yet universally accepted and some analysts, especially in the USA, continue to use number of deaths as the impact indicator, we list the main reasons why the VPF approach is wrong for air pollution:

- one cannot simply add the number of deaths due to different contributing causes (such as air pollution, smoking or lack of exercise), because one would end up with numbers far in excess of total mortality;
- the number of deaths fails to take into account a crucial aspect for the monetary valuation, namely the magnitude of the loss of life per death; very different between typical air pollution deaths and typical accidents;
- by contrast with primary causes of death (such as accidents), the total number of premature deaths attributable to air pollution is not observable;
- the method that has been used for calculating the number of deaths for cohort studies is wrong.

Each of these reasons is sufficient by itself. The first two are obvious, the third and fourth are explained in Rabl (2003), who examined what exactly has been measured in epidemiological studies of mortality. He shows that studies of chronic mortality cannot distinguish whether the observed result is due to everybody losing a little LE or only some individuals losing much LE. Analogy with studies of smokers suggest that everybody's life is shortened to some extent by air pollution; in this case the VPF approach would have to say that every death is an air pollution death – not a very meaningful conclusion.

There are several reasons for the popularity of calculations of a number of premature deaths due to air pollution. One is the emotional impact, an important consideration for journalists: "deaths due to pollution" sounds more dramatic than "years of life lost." Another reason is that such calculations seem natural and plausible, in view of the fact that epidemiological studies report an increase in the number of deaths per time interval. The reason for the error in such calculations for chronic mortality is not entirely obvious, as explained by Rabl (2003); it lies in the neglect of dynamic effects (if a pollution peak causes some individuals to die now, ΔLE sooner than otherwise, the number of deaths during the ensuing period of ΔLE will be lower because people cannot die twice). As far as damage costs are concerned, the results of the VPF and of the VOLY approach are not radically different, as found in a comparison made by the CAFE project (Hurley *et al.*, 2005): a VPF calculation with VPF = 2M€ yields damage costs that are only about 50% higher than a VOLY calculation with VOLY = 52,000 €.

Loss of LE avoids the above problems; it is a meaningful and appropriate impact indicator for all risk factors, even those that are not observable as the cause of an individual death. In particular, it can be added across different risk factors (at least in the limit of small risks).

Number of deaths seems appropriate only for acute mortality and for infant mortality; however, for monetary valuation these impacts are problematic because the epidemiological studies provide no information on the LE loss per death for these end points. In particular, acute mortality of adults is determined by time series methods, and the only way to get information on the associated LE loss is to include a very large number of lags in the regression, going back at least several years, as shown by Rabl *et al.* (2011).

There has been much confusion about the interpretation of air pollution mortality studies. Such studies report changes in the observed death rates in terms of relative risk RR, and when it comes to different studies, people tend to think that mortality is mortality and RR is RR. But in reality, death rates depend on the timescale used. Consider the increase in the number of deaths per day that has often been observed during heatwaves. If most individuals who succumb to a heatwave are extremely frail and would have died a few days or weeks later even during mild weather, the initial increase is followed by a decrease in the daily death rate during the weeks after the heatwave. Not enough is known about the real LE loss due to heatwaves, but for the sake of illustration, let us suppose that the loss of LE is at most a few weeks; in this case, the total number of deaths during the year (i.e. the annual death rate) is exactly the same with and without the heatwave.

Next, consider a permanent decrease in air pollution, starting at time t_0. It is followed by a decrease in the daily death rate. But not for ever. The decrease in the daily death rate is a reflection of the fact that some individuals live longer. Of course they do not live for ever, and when they do die, the death rate increases again. A long time after t_0, the death rate returns to the original value, because everybody dies exactly once. Under steady-state conditions the death rate of the population is equal to the birth rate, whatever the various factors that affect life expectancy. In these last two paragraphs we have taken the death rate as the number of deaths per day per total population. But if the death rate is normalized, for instance, as deaths per day per million, the rate decreases in proportion to 1/LE, because people live longer and for a given birth rate the population size increases in proportion to LE.

For another look at this, consider the age-specific mortality rate $\mu(x)$, defined such that someone who has reached age x has a probability $\mu(x)\,\Delta x$ of dying between x and $x + \Delta x$ (data are usually stated in terms of $\Delta x = 1$ year). The data in Table 21 of Krewski et al. (2000) suggest that air pollution increases the age-specific mortality for all ages by approximately the same amount. Now consider how that permanent decrease in air pollution affects the number of deaths observed in a population. In a lifetable calculation (see Section 3.4.3), one sees that the number of deaths in a given population decreases after t_0 for all but the oldest age groups. Beyond a certain age, the number of deaths increases despite the decrease of $\mu(x)$, because more individuals survive to these higher ages.

Note that a relative change in $\mu(x)$ implies a far smaller relative change in LE. This is a consequence of the rapid increase of $\mu(x)$ with age x. Beyond about age 30 years, $\mu(x)$ increases exponentially with x, as indicated by Eq. (3.7). With a simple lifetable calculation for typical populations of industrialized countries, one can show that LE decreases only by a factor of about $\Delta RR/7$ when $\mu(x)$ increases by a small amount $\Delta RR\,\mu(x)$. This is easy to understand by considering a hypothetical extreme, namely a population where everybody dies between 65 and 66. In such a population, $\mu(x) = 0$ for all x < 65, and if $\mu(x)$ is increased by a factor, regardless of how large, the LE cannot drop below 65.

Thus, one has to be careful about what has been assumed in a study of air pollution mortality. The resulting RRs depend on the timescale for the death rates and on the exposures that have been taken into account. Acute mortality is based on short-term exposures and daily death rates, chronic mortality on long-term exposure and age-specific mortality $\mu(x)$.

3.4.2   Studies of chronic mortality

It is very difficult and costly to measure the total impacts (short- plus long-term) of air pollution, and there are relatively few long-term studies available. These studies are called cohort studies, because they analyze the survival of a cohort of individuals over a long period, at least several years, and take into account individual characteristics. Several important cohort studies have succeeded in measuring the long-term impacts of air pollution on mortality. Two of these (Dockery *et al.*, 1993, Pope *et al.*, 1995) found positive associations between exposure to particles and total mortality, while the third (Abbey *et al.*, 1999) found a positive association with mortality for men, but not for women. Confirmation of long-term mortality impacts has been provided by studies in other countries, for example in the Netherlands (Hoek *et al.*, 2002). For a general review and meta-analysis of chronic mortality studies we refer to Chen *et al.* (2008).

Because chronic mortality studies are of crucial importance for the formulation of environmental policy, EPA requested a reanalysis by an independent team. This reanalysis was carried out by Krewski *et al.* (2000); it confirmed the validity of the data and the analysis of the original studies by Dockery *et al.* and Pope *et al.* In addition, Krewski *et al.* performed a large number of sensitivity studies. In the meantime, the cohort study of Pope *et al.* (1995) has continued and new results have been published by Pope *et al.* (2002), based on an observation period of about 16 years. Of the long-term mortality studies, that of Pope *et al.* has by far the largest sample, about half a million individuals. The observation period for this cohort has been further extended to 18 years by Krewski *et al.* (2009).

Without going into the details of the statistical analysis of cohort studies, one can say that such studies measure a change in $\mu(x)$. Using the Cox proportional-hazards regression model, the survival data are fitted by assuming that exposure to a concentration increment Δc increases the mortality relative to the baseline $\mu_0(x)$ by

$$\mu(x) = \mu_0(x) \exp(k \, \Delta c), \qquad (3.3)$$

with the parameter k to be determined. The factor $\exp(k \, \Delta c)$ is reported as mortality risk ratio or relative risk RR,

$$RR = \exp(k \, \Delta c). \qquad (3.4)$$

When Δc is small, one can approximate the risk increase $\Delta RR = RR - 1$ by

$$\Delta RR = \exp(k \, \Delta c) - 1 \approx k \, \Delta c. \qquad (3.5)$$

Pope *et al.* (2002) report two different numbers, 1.04 and 1.06, for the relative risk due to a 10 μg/m^3 increment of PM$_{2.5}$, depending on assumptions about the relevant exposure period. ExternE (2008) uses 1.06,

$$RR = 1.06 \text{ for an increment } \Delta c = 10 \, \mu g/m^3 \text{of PM}_{2.5}. \qquad (3.6)$$

The risk increase is approximately independent of age, as shown by the data in Table 21 of the re-analysis by Krewski *et al.* (2000). Note that Pope *et al.* (2002) also report significant associations of all-cause mortality with the concentrations of sulfates and of SO$_2$.

3.4.3   Loss of life expectancy for chronic mortality of adults

Since this relative risk refers to an increase in age-specific mortality, a more elaborate calculation, involving lifetable data, is required to find the corresponding YOLL (years of life lost). For the 1998 version of ExternE, a relatively simple steady-state calculation was done (Rabl, 1998). The most comprehensive dynamic analysis was published by Leksell and Rabl (2001), who evaluated the sensitivity of the result to the underlying assumptions.

Here we will explain the steady-state calculation. A key element for the analysis is the RR for the age-specific mortality μ(x), reported by the epidemiological studies. As an example of μ(x) data, we show in Fig. 3.2 the data for France in 2005, with separate lines for males, females and the total population. The general pattern is similar in other countries. Mortality rates are higher for males than for females. The mortality starts relatively high at birth but drops rapidly to a minimum at around 10 years of age. Then it increases towards a plateau at around 25, followed by an exponential increase with age after age 30 (i.e. a straight line at the logarithmic scale of Fig. 3.2). This exponential increase had already been noted by Gompertz, who in 1825 proposed the formula,

$$\mu(x) = \alpha \, \exp{(\beta x)} \qquad \text{for } x > 30 \text{ yr}, \qquad (3.7)$$

with parameters α and β that depend on the specific population.

Next, one defines the survival function S(x,x′) as the fraction of a cohort of age x that survives at least to age x′. Since the fraction that dies between x′ and x′ + Δx′ is ΔS(x,x′) = − S(x,x′) μ(x) Δx, one gets the differential equation,

$$dS(x, x') = - S(x, x') \, \mu(x') \, dx'; \qquad (3.8)$$

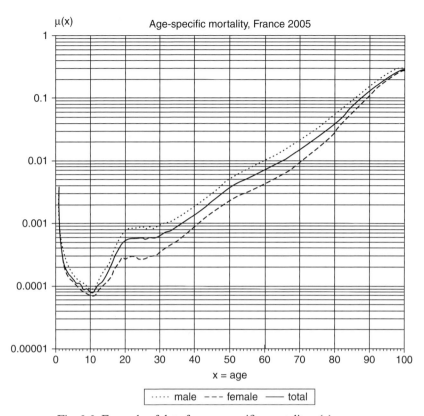

Fig. 3.2 Example of data for age-specific mortality $\mu(x)$.

the boundary condition is $S(x,x) = 1$. One readily finds the solution,

$$S(x, x') = \exp\left[-\int_x^{x'} \mu(x'') \, dx''\right]. \tag{3.9}$$

Figure 3.3 shows the survival function for France, for the total population with $\mu(x)$ of Fig. 3.2.

For example, a fraction, 0.6 of the population survives to 81 yr and a fraction, 0.5 to 84 yr. Thus a fraction, $0.6 - 0.5 = 0.1$ of the population dies between these ages, at about $x' = 82.5$ on average; this corresponds to a horizontal slice under the survival curve between $S(0,x') = 0.5$ and 0.6. Dividing the entire area under the survival curve into a large number of horizontal slices, one sees that each slice has an area equal to the fraction $\Delta S(0,x')$ times x', and that the average of these areas is the life expectancy of the population. Since the fractions sum to unity, the sum over these

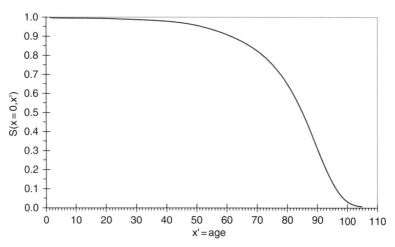

Fig. 3.3 Survival function $S(x = 0, x')$ for France, for the total population with $\mu(x)$ of Fig. 3.2.

areas is equal to their average. Therefore the life expectancy is equal to the area under the survival curve and can be calculated as the integral of $S(0, x')\, dx'$.

More generally, the remaining life expectancy $L(x)$ of a cohort of age x is

$$L(x) = \int_x^\infty S(x, x')\, dx'. \tag{3.10}$$

If $\mu(x)$ changes, for example due to air pollution, $S(x, x')$ and $L(x)$ change accordingly. The resulting change $\Delta L(x)$ for a cohort of age x is the difference between $L(x)$ calculated without and with this change,

$$\Delta L(x) = \int_x^\infty [S_{\mu 0}(x, x') - S_\mu(x, x')]\, dx', \tag{3.11}$$

where $S_{\mu 0}(x, x')$ is the survival curve for the baseline mortality $\mu_0(x)$ and $S_\mu(x, x')$ the survival curve for the changed mortality. The impact on the entire population is obtained by summing $\Delta L(x)$ over all affected cohorts, weighted by the age distribution. Only ages above 30 have been included in the calculations, because the underlying cohort studies did not include younger people. Since the absolute mortality is very low between the end of infancy and age 30, assuming the same RR also for this age group would increase the total population LE loss by only about 10%. Infant mortality is a different matter, and it is treated separately, requiring different methods for the epidemiological studies and for the analysis.

The above calculation is directly applicable to steady-state exposures, and Eq. (3.11) yields the LE loss due to constant exposure during an entire lifetime. The results of Pope *et al.* also correspond with steady-state exposure, because time variation of the concentration data was not considered. But it is also of interest to consider time varying exposures, especially for applications to environmental policy. Furthermore, the real exposure of the cohorts studied by Pope *et al.* declined during the study period, and it is not obvious how this may have affected the results.

For these reasons, Leksell and Rabl (2001) extended the above framework by accounting for time varying concentrations and impacts, in order to derive a correction factor for the exposure changes of Pope *et al.* (1995). They do this by introducing the time constant(s) for the decrease of risk after exposure, based on estimates from studies of smoking (such a decrease of risk occurs if the body repairs some of the damage). Even though the uncertainty of time constant(s) inferred from smoking studies is large, it turns out to have almost no effect on the resulting LE loss, as Leksell and Rabl confirm by a sensitivity analysis.

Leksell and Rabl show that the LE loss is almost exactly proportional to exposure, defined as the time integral of the concentration weighted by an exponential decay factor; the time distribution of the concentration itself does not matter much. For comparison purposes it is therefore convenient to state the LE loss for exposure during one year to a concentration increment of 1 $\mu g/m^3$ of PM_{10}; the result can readily be scaled to other exposures. For adults, a good approximation can be obtained using the steady-state analysis, if one divides the loss due to a constant lifetime exposure by the life expectancy at birth.

Leksell and Rabl also evaluated how the result could change for different population data. As shown by the examples in Table 3.4, variations due to differences in lifetable data have only a relatively small effect. This table also shows the mortality data for the respective populations, as expressed by the parameters α and β, fitted by Leksell and Rabl to the lifetable data for the age-specific mortality $\mu(x)$ at age x, according to the Gompertz model, Eq. (3.7). In spite of a more than twofold difference between male and female mortality in the USA, the LE loss is much the same, 2.56E-04 YOLL for females versus 2.73E-04 YOLL for males, per person per $\mu g/m^3$ of PM_{10} per yr. The largest differences in Table 3.4 are between China (2.04E-04 YOLL) and

Table 3.4 *Coefficients α and β of the Gompertz model, and ERF for LE loss from chronic mortality for several populations. LE loss in YOLL per person for an exposure to 1 μg/m³ of PM₁₀ during one year, as calculated with the real age distribution of each population, assuming that the relative risk of Pope et al. (1995) applies equally to all populations. Adapted from Leksell and Rabl (2001) by converting from PM₂.₅ to PM₁₀.*

Population	α [per yr²]	β [per yr]	ERF YOLL/(person·yr·μg/m³), PM₁₀
USA, natural causes, male+female[a]	5.38E-05	8.78E-02	2.69E-04[f]
USA, natural causes, male[a]	7.76E-05	8.59E-02	2.73E-04[f]
USA, natural causes, female[a]	3.19E-05	9.20E-02	2.56E-04[f]
EU15, all causes, male + female[b]	3.70E-05	9.24E-02	2.56E-04
Sweden, natural causes, male+female[c]	9.67E-06	1.10E-01	2.25E-04[f]
France, all causes, male + female[b]	6.66E-05	8.50E-02	2.77E-04
Russia, all causes, male + female[d]	3.96E-04	6.78E-02	3.59E-04
China, all causes, male + female[e]	5.89E-05	9.15E-02	2.04E-04

[a] data for 1995.
[b] data for 1995.
[c] data for 1993–1997.
[d] data for 1996.
[e] data for 1998.
[f] includes correction factor 1/0.975 for nonviolent deaths.

Russia (3.59E-04 YOLL). Note that differences in LE loss are caused not only by differences in mortality but also by differences in the age distribution.[11]

Based on the relative risk, RR = 1.06 of Pope *et al.* (2002), Eq. (3.6), ExternE (2008) finds a loss of 6.51E-4 YOLL/(pers·yr·μg/m³) for PM₂.₅, so the ERF slope for chronic mortality is

$$S_{ERF} = 6.51E\text{-}4 \text{ YOLL}/(yr \times \mu g_{PM2.5}/m^3).$$ (3.12)

Example 3.1 In Paris in 2010, the PM₂.₅ concentration has been about 20 μg$_{PM2.5}$/m³. What lifetime LE gain can one expect for a 10% reduction?

---

[11] The lifetable data for China and Russia for this calculation originated under communist regimes, notorious for faking their statistics; but the message of Table 3.4 is not affected, namely that LE loss differences due to differences in lifetable data are small compared with the overall uncertainties of chronic mortality.

Solution: The LE gain is

$$6.51\text{E-}4 \, \text{YOLL}/(\text{yr.}\mu\text{gPM}_{2.5}/\text{m}^3) \times 20 \, \mu\text{gPM}_{2.5}/\text{m}^3 \times 0.1 \times 80 \, \text{yr}$$
$$= 38 \, \text{days.}$$

Example 3.2 PM_{10} concentrations in Chinese cities averaged about 300 $\mu\text{gPM}_{10}/\text{m}^3$ during the 1970s, and by 2005 they had decreased to 100 $\mu\text{gPM}_{10}/\text{m}^3$ (Matus *et al.*, 2011). What is the current loss of LE per year of exposure?

Solution: Not having data for $PM_{2.5}$, we assume a typical conversion factor of about 0.6 for the ratio of $PM_{2.5}$ and PM_{10} concentrations. The ERF slope should be calculated with lifetable data for China (which we do not have for the current population, and for which we do not trust the data used for Table 3.4). Not having these data, we assume Eq. (3.12). Thus the loss is $0.6 \times 100 \, \mu\text{gPM}_{10}/\text{m}^3 \times 6.51\text{E-}4 \, \text{YOLL}/(\text{yr·}\mu\text{gPM}_{2.5}/\text{m}^3) = 0.039 \, \text{YOLL} = 0.47$ months per year of exposure. For a constant exposure at this level during an entire 73 yr lifespan (the LE in China at around 2010), the loss would be $0.039 \, \text{YOLL/yr} \times 73 \, \text{yr} = 2.9$ years of life lost.

Example 3.3 Because of traffic, the concentrations of $PM_{2.5}$ in cities are higher than in the surrounding areas. For example in Paris in 2010, the annual mean has been about 20 $\mu\text{gPM}_{2.5}/\text{m}^3$ in most areas, except near busy streets, where values around 25 $\mu\text{gPM}_{2.5}/\text{m}^3$ are common. In the more distant suburbs (15 km from center, where one of the authors, AR, lives) the annual means are around 15 $\mu\text{gPM}_{2.5}/\text{m}^3$. How much longer can AR expect to live by having chosen the suburb instead of Paris?

Solution: The correct calculation would require additional data, namely the age when AR moved to the suburbs, his present age, and the fraction of time spent in Paris (his place of work, the Ecole des Mines is in the center of Paris). With such data, a lifetable calculation should be done, using the RR of Eq. (3.6). Instead of such a life table calculation we simplify and use Eq. (3.12) for the population average as a rough approximation. For an exposure difference of 5 $\mu\text{gPM}_{2.5}/\text{m}^3$ the LE gain due to life in the suburbs is $5 \, \mu\text{gPM}_{2.5}/\text{m}^3 \times 6.51\text{E-}4 \, \text{YOLL}/(\text{yr·}\mu\text{gPM}_{2.5}/\text{m}^3) = 3.3\text{E-}3$ life years for each year of exposure.

For 35 years lived in the suburb, the gain is 0.11 life years or 40 days. Anticipating a result from Chapter 9, such a gain would be worth $40,000 \, € \times 0.11 = 4,400 \, €$, if valued at the average value of a life year in the EU.

In passing, we will mention that the LE change in Fig. 2.10 is for a 10% decrease, i.e. a decrease of 2 $\mu g_{PM2.5}/m^3$. The result in Fig. 2.10 follows directly by multiplying 40 days by 2/5 for the concentrations, and 80/35 for the exposure duration.

Example 3.4 Many people enjoy the ambiance of candlelight and incense. How about the contribution to indoor air pollution? Data for candles have been measured by Guo *et al.* (2000) and for incense by Jetter *et al.* (2002). The $PM_{2.5}$ emissions are extremely variable from one sample to another. Sometimes a candle burns cleanly, sometimes it produces a lot of soot. Guo *et al.* found emission rates from about 0.04 to about 5.3 mg/hr per wick; typical values seem to be in the range 0.1 to 0.5 mg/hr per wick. Five candles emitting 0.3 mg/hr each in a 30 m³ room with one air change per hour increase the $PM_{2.5}$ concentration by 50 μg/m³. During the brief moment when the flame is blown out, the emission rates shoot up, being equivalent to about an hour of normal burning. For incense sticks, Jetter *et al.* found emission rates in the range 7 to 108 mg/hr. What is the health impact of spending an hour in a 51.5 m³ room with one air change/hr where five candles and an incense stick are burning?

Solution: 5 candles with 0.3 mg/hr each emit 1.5 mg/hr, little compared to a typical 50 mg/hr from incense. The steady-state concentration c is the ratio of emission rate and air flow rate,

$$c = (51.5 \text{ mg/hr})/(51.5 \text{ m}^3/\text{hr}) = 1000 \text{ μg/m}^3.$$

By comparison, the limit value for outdoor $PM_{2.5}$ concentrations in the EU will be 25 μg/m³, beginning in 2015.

Applying the ERF of Eq. (3.12) and noting that 1 yr = 8760 hr, we find an LE loss of (1000 μg/m³) × 1 hr × 6.51E-4 YOLL/(yr·$\mu g_{PM2.5}/m^3$) = 6.51E-1 YOLL/8760 = 0.74E-4 YOLL = 0.651 hr of life. The LE loss is about two-thirds of the exposure duration. This is reminiscent of the rule of thumb that each cigarette shortens life by about the duration of the smoke.

3.4.4 *Relevant exposure – cessation lag*

An important question about air pollution concerns the persistence of the effects of past exposures: to what extent and over what time span can the body repair the damage of past exposures? This question is often phrased in terms of relevant exposure or cessation lag. In other words, which exposures of the past are still relevant today, or what is the lag between a past exposure and the cessation of its effect? Or, what is the time needed for the body to repair the damage?

A related issue is the latency period for cancers, i.e. the time between exposure to a carcinogen and the appearance of a cancer. Most cancers develop very slowly, with latency periods in the range 10 to 20 years. Since some of the mortality due to $PM_{2.5}$ comes from lung cancers (Pope *et al.*, 2002), complete cessation of impacts may take decades, even if the cessation lag for cardio-vascular mortality may be much shorter.

The question of cessation lags for air pollution is very difficult to answer. Whereas most air pollution studies merely regress an effect at time t against some overall indicator of exposures before t, for this question one also has to identify how much of the effect is due to each of the exposures at different times in the past. Even for mortality, the end point with the best available data, there is no conclusive answer. Probably the most solid result is for smoking rather than air pollution. In particular, the impressive study by Doll *et al.* (2004) which followed a cohort of 34,000 doctors in the UK over 50 years, with very detailed questionnaires about lifestyle and health status. To cite from their paper: *"Men born in 1900–1930 who smoked only cigarettes and continued smoking died on average about 10 years younger than lifelong non-smokers. Cessation at age 60, 50, 40, or 30 years gained, respectively, about 3, 6, 9, or 10 years of life expectancy."* Thus the cessation lag for smoking is on the order of decades.

Among air pollution studies there is a wide range of answers. At one extreme is Schwartz *et al.* (2008), who carried out an extended follow-up of the Harvard Six Cities cohort study (Dockery *et al.*, 1993). One of the features of that work is an analysis by exposure window ("lag"), with exposure intervals of one to five years preceding death. Schwartz *et al.* found that the excess risk for mortality due to $PM_{2.5}$ decreases rapidly from one year to the next, and is negligible beyond the second year. That result is very different from another cohort study, by Puett *et al.* (2009), which finds that the coefficients for all-cause mortality due to $PM_{2.5}$ increase for exposures in previous years up to three years and begin to decline only slightly for the fourth year (the longest exposure considered in that study). Krewski *et al.* (2009) tried to find an answer by analyzing exposure windows of 0–5 yr, 6–10 yr and 11–15 yr for the ACS cohort, but were unable to draw firm conclusions about the relevant exposure windows. Rabl *et al.* (2011) took a totally different approach, namely a time series with lags up to 1095 days, and found that past exposure is significant for at least three years. The results of Elliott *et al.* (2007) imply that past exposures are significant for at least 16 years (the coefficients for the effect of exposures 12 to 16 years in the past are still significantly different from zero).

In view of this situation and the many commonalities (combustion particles, cardiopulmonary disease and lung cancer) between air pollution and smoking, we conclude that cessation lags for air pollution are at least several years and probably a decade or two.

3.4.5   Results for acute mortality of adults

It is interesting to compare the LE loss from chronic mortality with that from acute mortality. Even though acute mortality (short-term) studies do not provide any information about the YOLL per death, one can get a rough idea by assuming, for the sake of illustration, six months per death, as an average.

For acute mortality due to PM, the most comprehensive source is HEI (2001), because it lists the ΔRR results of the three most significant studies at the time:

- 0.05% per $\mu g/m^3$, by Samet et al. (2000), based on the 90 largest US cities;
- 0.06% per $\mu g/m^3$, by Katsouyanni et al. (1997), based on APHEA results for 12 European cities; and
- 0.07% per $\mu g/m^3$, by Levy et al. (2000), based on a meta-analysis of 29 studies in 23 locations in Europe and North and South America;

all three referring to PM_{10}. Here we take the central value of $\Delta RR = 0.06\%$ per $\mu g/m^3$ of PM_{10}.

This is essentially the same as the WHO meta-analysis of European studies (Anderson et al., 2004). Taking 10,000 deaths/yr per million as a typical value of the reference mortality rate, this implies an average loss of

$$10,000 \text{ deaths/yr per million} \times 0.0006 \text{ per } \mu g/m^3 \times 0.5 \text{ YOLL/death}$$
$$= 3E\text{-}6 \text{ YOLL per person per } \mu g/m^3 \text{of } PM_{10} \text{ per yr,}$$
$$\text{if loss is 6 months per death.} \qquad (3.13)$$

This would be only about 1% of the total found by long-term studies. With any reasonable assumption for the YOLL per acute death, one finds that the mortality observed by short-term studies is at most a small contribution to the total impact – and in any case it is already included in the results of the long-term studies by their very design and should therefore not be added. The smallness of acute mortality is entirely plausible when one considers what time series studies would be able to observe about mortality from smoking if applied in a hypo-thetical country where cigarette sales and smoking are forbidden on Sundays.

Table 3.5 *Calculation of LLE for postneonatal mortality (between 1 and 12 months of age). The LLE is the upper bound because infants who die because of pollution may have had a less than normal life expectancy (here assumed as 76 yr).*

	Woodruff et al. (1997)				Bobak and Leon (1999)	
	SIDS, NBW	Resp, NBW	Resp, LBW	**All causes**	Resp	**All causes**
Rel. risk RR	1.12[a]	1.20[a]	1.05[a]	1.04[a]	1.95[b]	1.19[b]
Δc (μ/m^3 PM_{10})	10	10	10	10	38.25[c]	38.25[c]
k (per (μ/m^3))	0.0113	0.0182	0.0049	0.0039	0.0175	0.0045
deaths per 1000 births				3.89[d]		3.89[d]
LLE (YOLL/(yr·µg/m³), per birth)				≤0.00115		≤0.00133
LLE (YOLL/(yr·µg/m³) per million)				**≤16.6**[e]		**≤19.3**[e]

NBW = normal birth weight; LBW = low birth weight; SIDS = sudden infant death syndrome; Resp = respiratory causes.
[a] Table 3 of Woodruff *et al.* (1997).
[b] Table 4 of Bobak and Leon (1999).
[c] Assuming a PM_{10}/TSP ratio of 0.77.
[d] Table 30 of NCHS (1999), subtracting perinatal from total infant mortality.
[e] For birth rate 14.5 per yr per 1000 population (NCHS 1999).

3.4.6 Infant mortality

Woodruff *et al.* (1997) analyzed cohorts consisting of about 4 million infants born in the USA between 1989 and 1991. They found statistically significant associations between PM_{10} concentrations and several causes of postneonatal mortality, namely SIDS (sudden infant death syndrome) and respiratory mortality. The corresponding relative risks are shown in Table 3.5, together with the corresponding Δc. Woodruff *et al.* looked only at the postneonatal period (between the ages of 27 days and 1 year), because postneonatal deaths are thought to be more influenced by the infant's external environment than deaths during the neonatal period.

Bobak and Leon (1999) carried out a case-control study of infant mortality in the Czech Republic from 1989 to 1991. Among the associations shown in their paper, the most relevant for the present purpose appear to be those for respiratory and for all-cause postneonatal mortality due to TSP (total suspended particles), shown here in Table 3.5 as RR together with the corresponding Δc. The latter has been converted to PM_{10}, assuming a PM_{10}/TSP ratio of 0.77, the center of the range 0.57 to 0.96, observed in the Czech Republic (M. Bobak, personal communication 14 Feb. 2002). Contrary to Woodruff *et al.*, Bobak and Leon did not find a significant association for SIDS.

The authors note that ". . . the associations with mortality were approximately linear." Applying a dose–response relation of Bobak and Leon to other countries involves uncertainties, because of different levels of air pollution and infant mortality. However, the concentrations during their study were only about a factor of two higher than those found in urban environments of the EU and USA, not a large difference in view of the approximate linearity observed by Bobak and Leon (the mean exposure in the study was 55 $\mu g/m^3$ of PM_{10}, with a range from 9 to 139, if one converts with a PM_{10}/TSP ratio of 0.77). The infant mortality rates are not very different: 9.6 per 1000 for the data set of Bobak and Leon and 7.2 per 1000 in the USA (NCHS, 1999).

Assuming the usual exponential relation of Eq. (3.4) between relative risk and concentration increment Δc, one finds the values of the constant k shown in Table 3.5. Since k × 1 $\mu g/m^3$ is very small compared with unity, k is essentially the fraction by which the respiratory infant mortality increases per $\mu g/m^3$. By definition, only the cohort having less than 1 yr of age is affected. If one assumes that in the absence of air pollution a baby would have lived a normal life span of 76 years, one finds an LE loss of

$$76 \, yr_{life} \times 3.89 \, deaths/1000 \, births \times 0.0039 \, per \, \mu g_{PM10}/m^3$$
$$= 0.00115 \, YOLL/(\mu g/m^3) \, per \, birth, \qquad (3.14)$$

as indicated on the second to last line of Table 3.5 for all-cause mortality, according to Woodruff *et al.* This is an upper bound, but it is perhaps not too unrealistic, since many babies end up living a normal healthy life despite a fragile infancy. The result for LLE, per million persons, is shown in the last line of Table 3.5. The numbers, for Woodruff *et al.*, 16.6 YOLL/(yr·$\mu g_{PM10}/m^3$) per million and for Bobak and Leon, 19.3 YOLL/(yr·$\mu g_{PM10}/m^3$) per million of the general population, are remarkably close; their average is approximately 18. Thus the ERF slope is,

$$S_{ERF} = 1.8E\text{-}5 \, YOLL/(yr \cdot \mu g_{PM10}/m^3) \, for \, infant \, mortality. \qquad (3.15)$$

Compared with adult mortality, the LLE due to infant mortality is an order of magnitude smaller. Even though there are large uncertainties in the epidemiological studies, further compounded for comparison by uncertainties in the PM_{10}/TSP and $PM_{2.5}/PM_{10}$ ratios, it seems unlikely that infant mortality would make a large contribution to the total air pollution mortality of the population. To explain the smallness of infant mortality, despite the maximal assumption of 76 years of life lost per infant death, note that the age-specific mortality around 50 is comparable to the rate for infants, and increases exponentially with age; for infant mortality only a one-year cohort is affected by a 1-yr pulse, whereas the mortality impact of adults includes all cohorts above 30.

3.4.7   Mortality impacts of O_3

For O_3, only acute mortality has been measured so far with sufficiently firm results. The WHO meta-analysis (Anderson et al., 2004) provided the following ERF of an increase in all-cause mortality:

0.3% (95% CI 0.1–0.43%) per 10 $\mu g/m^3$ increase in the daily maximum 8-hour mean O_3.

As always with time series results of acute mortality, the question arises of whether the loss of life expectancy (LE) is significant. Zanobetti and Schwartz (2008) found that the LE loss is certainly more than a few days. But so far the evidence is not yet sufficient for a chronic mortality ERF, comparable to that of PM.

Taking 10,000 deaths/yr per million as a typical value of the reference mortality rate, we obtain the ERF slope,

$$S_{ERF} = 3.0E\text{-}6 \text{ deaths}/(yr \cdot \mu g_{O3}/m^3) \text{ deaths due to acute mortality.}$$

$$(3.16)$$

For the NEEDS phase of ExternE (2008), the ERFs for O_3 have been used by assuming linearity above a no-effect threshold of 35 ppb, i.e. SOMO35. Hurley et al. (2005) adjudged that, on average, one year of life would be lost per death to enable results to be valued using the VOLY. They considered the likely distribution would be strongly skewed left, with most people losing only a few days of life. However, for a few people, a substantial amount of life shortening could result, for example, in the case of someone with a treatable heart condition.

Given the finding that the chronic effects of PM greatly outweigh the acute effects, there is increased interest in quantification of long-term effects of exposure to ozone. Jerrett et al. (2009) provide evidence of such an effect. Difficulty arises when considering how the resulting response function can be combined with results from the use of a function against $PM_{2.5}$. Holland (2013) provides illustrative results, indicating that inclusion of the chronic effect would increase ozone related mortality damage by a factor of 9. However, this part of the analysis was performed only as a sensitivity analysis.

3.5   ERFs for morbidity

3.5.1   *Morbidity due to PM*

3.5.1.1 New cases of chronic bronchitis

Chronic bronchitis (CB) was one of the end points examined in the US Seventh Day Adventist Study (AHSMOG: Adventist Health Smog). The Seventh Day Adventists are of special interest for air pollution epidemiology, because their religion does not allow them to smoke. A group of several thousand Seventh Day Adventists in California was studied on two occasions about ten years apart, in 1977, and again in 1987/88. Chronic bronchitis was defined as the occurrence of a chronic cough or sputum, on most days, for at least three months of the year, for at least two years. New cases of CB were defined as those that had these symptoms in 1987/88, but not in 1977. Using a RR from Table 6 of Abbey *et al.* (1995) together with a background incidence rate (adjusted for remission of chronic bronchitis symptoms) of 0.378%, estimated from Abbey *et al.* (1995), Hurley *et al.* (2005) derived an ERF of:

New cases of chronic bronchitis per year per $100,000$ adults aged $27+$
$$= 26.5\,(95\%\mathrm{CI} - 1.9, 54.1)\ \mathrm{per}\ 10\,\mu\mathrm{g}/\mathrm{m}^3 \mathrm{PM}_{10}.$$

Multiplying by the risk group factor, 0.7, we obtain the ERF slope,

$$S_{ERF} = 1.86\text{E-}5\ (\mathrm{cases/yr})/(\mu g_{PM10}/\mathrm{m}^3)\ \text{of chronic bronchitis.} \quad (3.17)$$

To apply this ERF for damage cost calculations, one should note that there is a wide range of levels of severity of CB, some being quite light, others totally debilitating, with a corresponding range of unit costs. The distribution of severity levels has been taken into account in the monetary valuation of CB by ExternE.

There has been subsequent analysis by others, for example, under the recent SAPALDIA study in Switzerland (Schindler *et al.*, 2009). This generates a higher risk factor than that described above, leading to the view that Eq. (3.17) may be an underestimate.

The term "chronic bronchitis" may seem out of date to some health professionals, given the current emphasis on COPD (chronic obstructive pulmonary disease). However, the studies on which the response function is based used "chronic bronchitis" and so it would be misleading to refer to it otherwise.

3.5.1.2 Hospital admissions

Data for respiratory hospital admissions (RHA) can be derived from the APHEIS-3 study by Medina *et al.* (2002), based on eight European cities. Using the results for all ages, both for RR and for background rates, Hurley *et al.* (2005) derived an ERF,

7.03 RHAs (95% CI 3.83, 10.30) per 10 μg/m^3 PM$_{10}$ per 100,000 people (all ages) per year.

The ERF slope based on the general population is

$$S_{ERF} = 7.03E\text{-}6 \, (\text{cases/yr})/(\mu g_{PM10}/m^3)$$

of respiratory hospital admissions.   (3.18)

For cardiac hospital admissions (CHA), Hurley *et al.* (2005) used results of APHEA-2 for eight cities in Western and Northern Europe (Le Tertre *et al.*, 2002) and an annual rate of cardiac hospital admissions estimated as the arithmetic mean of rates from eight European cities in the Appendices of the APHEIS-3 report of Medina *et al.* Together they imply an ERF of

4.34 CHAs (95% CI 2.17, 6.51) per 10 μg/m^3 PM$_{10}$ per 100,000 people (all ages) per year.

The ERF slope based on the general population is,

$$S_{ERF} = 4.34E\text{-}6 \, (\text{cases/yr})/(\mu g_{PM10}/m^3) \text{ of cardiac hospital admissions.}$$

(3.19)

3.5.1.3 Restricted activity days (RADs) and associated
health end points

Ostro (1987) and Ostro and Rothschild (1989) analyzed data from the US Health Interview Study (HIS), a multi-stage probability sample of 50,000 households from metropolitan areas of all sizes and regions throughout the USA, with adults aged 18–64, during six consecutive years (1976–81). Days of restricted activity were classified according to three levels of severity as (i) bed disability days; (ii) work or school loss days and (iii) minor restricted activity days (MRADs). The latter involve some noticeable limitation of normal activity, but without work loss or bed disability. With this data set, Ostro (1987) studied both RADs and work loss days (WLDs).

Restricted activity days (RADs)

Linking the resulting weighted mean coefficient for RADs to an estimated average background rate of 19 RADs per person per year (ORNL/RFF, 1994), one obtains an ERF of,

$$902 \text{ RADs } (95\% \text{ CI } 792, 1013) \text{ per } 10 \text{ } \mu g/m^3 PM_{2.5} \text{ per year}$$
$$\text{per } 1,000 \text{ adults at age } 15-64.$$

With a risk group fraction of 0.672 one obtains the ERF slope based on the general population as,

$$S_{ERF} = 6.06\text{E-}2 \text{ (cases/yr)}/(\mu g_{PM2.5}/m^3) \text{ of restricted activity days.}$$
$$(3.20)$$

ExternE (2008) applied this ERF to people aged 15–64. It is probably indicative of a similar effect above 65, but the monetary valuation would be lower.

Minor restricted activity days (MRADs) and work loss days (WLDs)

Based on these studies, Hurley *et al.* (2005) also derived the following ERFs for work loss days (WLDs) and minor RADs:

$$207 \text{ WLDs } (95\% \text{CI } 176 - 238) \text{ per } 10\mu g/m^3 PM_{2.5} \text{ per year}$$
$$\text{per } 1000 \text{ adults aged } 15-64$$

and

$$577 \text{ MRADs } (95\% \text{ CI } 468 - 686) \text{ per } 10\mu g/m^3 PM_{2.5} \text{ per year}$$
$$\text{per } 1000 \text{ adults aged } 18-64$$

The ERF slopes based on the general population are,

$$S_{ERF} = 1.39\text{E-}2 \text{ (cases/yr)}/(\mu g_{PM2.5}/m^3) \text{ of work loss days} \quad (3.21)$$

and

$$S_{ERF} = 3.69\text{E-}2 \text{ (cases/yr)}/(\mu g_{PM2.5}/m^3) \text{ of minor restricted activity days.}$$
$$(3.22)$$

3.5.1.4 Medication (bronchodilator) usage by asthmatics

Many epidemiologists assume a causal relationship between air pollution and aggravation of asthma in children. Based on the WHO meta-analysis by Anderson *et al.* (2004) ExternE uses the following ERFs:

Children aged 5–14 years

Combining the RRs from Anderson *et al.* (2004) with estimates of the mean daily prevalence of bronchodilator usage among panels of school-children who meet the PEACE study criteria, ExternE finds an ERF of:

$$\text{days of bronchodilator use per year}$$
$$= 180 \; (95\% \; \text{CI} - 690, \; 1060) \; \text{per} \; 10 \; \mu g/m^3 PM_{10}$$

per 1000 children aged 5–14 years meeting the PEACE study criteria.

Data from the International Study of Asthma and Allergies in Childhood (ISAAC Steering Committee, 1998) were used to estimate that approximately 15% of children in Northern and Eastern Europe and 25% in Western Europe met the PEACE study criteria. Taking a risk group fraction of 0.022, one obtains an ERF slope of,

$$S_{ERF} = 3.96\text{E-}4 \; (\text{cases/yr})/(\mu g_{PM10}/m^3) \; \text{of days of bronchodilator use.} \tag{3.23}$$

Adults above 20 years

Likewise, the ERF for adults was obtained as:

$$\text{days of bronchodilator use per year}$$
$$= 912 \; (95\% \; \text{CI} - 912, \; 2774) \; \text{per} \; 10 \; \mu g/m^3 PM_{10}$$

per 1000 adults aged 20 + with well-established asthma.

4.5% of the adult population is estimated to suffer from asthma, so the risk group fraction is 0.036, implying an ERF slope of,

$$S_{ERF} = 3.28\text{E-}3 \; (\text{cases/yr})/(\mu g_{PM10}/m^3) \; \text{of days of bronchodilator use.} \tag{3.24}$$

3.5.1.5 Lower respiratory symptoms (LRS) in adults with chronic respiratory disease

Using data from ECRHS (1996), together with estimates of the mean daily prevalence of LRS, including cough, ExternE obtained this ERF:

1.30 (95% CI 0.15, 2.43) symptom days per year
(LRS, including cough) per 10 μg/m^{3}PM$_{10}$
per adult with chronic respiratory symptoms.

ExternE estimates that approximately 30% of the adult population has chronic respiratory symptoms, implying an ERF slope of

$$S_{ERF} = 3.25E\text{-}2 \; (\text{cases/yr})/(\mu g_{PM10}/m^3) \text{ of lower}$$
$$\text{respiratory symptoms, adults.} \qquad (3.25)$$

3.5.1.6 Lower respiratory symptoms (LRS) in children

ExternE combined the RRs of the review by Ward and Ayres (2004) for effects of PM$_{10}$ on respiratory symptoms in children with an estimate of the mean daily prevalence of LRS, including cough, from two studies of Dutch children (van der Zee et al., 2000; Hoek and Brunekreef, 1995), to obtain this ERF:

1.86 (95% CI 0.92, 2.77) symptoms days per year
per child aged 5–14, per 10 μg/m^{3}PM$_{10}$.

With a risk group fraction of 0.11 the ERF slope is,

$$S_{ERF} = 2.05E\text{-}2 \; (\text{cases/yr})/(\mu g_{PM10}/m^3) \text{ of days of bronchodilator use.}$$
$$(3.26)$$

3.5.2   Morbidity due to O$_3$

3.5.2.1 Hospital admissions

Based on results from five cities in western Europe, Anderson et al. (2004) estimate the change in all respiratory hospital admissions (RHAs) in various age groups as a function of 8-hr daily average O$_3$. They find a statistically significant effect only for elderly people: a RR of 0.5% (95% CI -0.2%, 1.2%) per 10 μg/m^3 O$_3$ (daily maximum 8-hr moving average) in people above age 65. Combining this with background rates from the APHEIS second year report, one obtains a CRF of

Annual RHAs per 100,000 people at age 65+
= 12.5 (95% CI-5.0, 30.0) per 10 μg/m^3 of 8-hr daily maximum O$_3$.

The risk group fraction is 0.16 and the ERF slope,

$$S_{ERF} = 2.00E\text{-}6 \ (\text{cases/yr})/(\mu g_{O3}/m^3) \ \text{respiratory hospital admissions.}$$
$$(3.27)$$

As for cardiovascular hospital admissions, or other cardiovascular morbidity endpoints for that matter, there is no strong or quantifiable evidence that they are increased by ozone. The same goes for emergency room visits.

3.5.2.2 Minor restricted activity days (MRADs)

A minor restricted activity day (MRAD) can reflect a variety of unspecified illnesses. Ostro and Rothschild (1989) studied associations between MRADs and ozone for current urban workers. They reported associations between MRADs and two-week averages of the daily 1 hr max O_3 concentration, for workers aged 18–64. Linking their results with a mean background rate of 7.8 MRADs per year among employed people gives the CRF,

115 (95% CI 44, 186) MRADs per year
per 10 μg/m^3 8-hr daily maximum O_3 per 1000 adults aged 18–64.

The risk group fraction is 0.64 and the ERF slope,

$$S_{ERF} = 7.36E\text{-}3 \ (\text{cases/yr})/(\mu g_{O3}/m^3) \ \text{of minor restricted activity days.}$$
$$(3.28)$$

3.5.2.3 Medication (bronchodilator) usage for asthma

The medication for asthma involves bronchodilators. Hiltermann *et al.* (1998) gave results for an association between daily maximum 8-hr moving average O_3 and daily prevalence of bronchodilator usage; it was not statistically significant at a lag of one day, although when 7-day cumulative ozone was considered, the effect was stronger and statistically significant. Linking this to background rates, from Hiltermann *et al.* and from the European Community Respiratory Health Survey (ECRHS, 1996), Hurley *et al.* obtained this CRF:

Annual days of bronchodilator use of
730 (95% CI -255, 1570) per 10 μg/m³O₃ (8-hr daily max)
per 1000 adults aged 20 + with well-established asthma
(approximately 4.5% of the adult population).

The risk group fraction is 0.036 and the ERF slope,

$$S_{ERF} = 2.63E\text{-}3 \ (cases/yr)/(\mu g_{O3}/m^3) \text{ of days of}$$
$$\text{bronchodilator use for asthma.} \qquad (3.29)$$

3.5.2.4 Lower respiratory symptoms in children

According to work by the Committee on the Medical Effects of Air Pollutants (COMEAP) in the UK, O_3 increases lower respiratory symptoms (LRS), including cough, and this is not restricted to just people with chronic respiratory symptoms such as asthma (Heather Walton, 2004, personal communication). For the CAFE project, Hurley *et al.* (2005) considered the study by Declerq and Macquet (2000), who looked at a small general population of 91 children in Armentieres, Northern France. Using background rates from Hoek and Brunekreef (1995), Hurley *et al.* found the following CRFs:

0.93 (95% CI−0.19, 2.22) cough days per year
per child aged 5–14 years (general population), per 10 μg/m³O₃
(8-hr daily max),

and

0.16 (95% CI−0.43, 0.81) days of LRS (excluding cough) per year
per child aged 5–14 years (general population), per 10 μg/m³O₃
(8-hr daily max).

The risk group fraction is 0.11 and the ERF slopes are,

$$S_{ERF} = 1.02E\text{-}2 \ (cases/yr)/(\mu g_{O3}/m^3) \text{ of cough}$$
$$\text{days per year, children} \qquad (3.30)$$

and

$$S_{ERF} = 1.76E\text{-}3 \ (cases/yr)/(\mu g_{O3}/m^3) \text{ of}$$
$$\text{LRS (excluding cough), children.} \qquad (3.31)$$

3.6   ERFs for toxic metals, dioxins and other pollutants

3.6.1   General remarks

Considering both toxicity and emitted quantities, the most harmful metals are As, Cd, Cr (in oxidation state 6, designated as CrVI), Hg, Ni and Pb. They have a variety of adverse health impacts, but at the present time the only end points that have been quantified are cancers due to As, Cd, CrVI and Ni, and neurotoxic impacts due to Pb and Hg. Among the impacts of dioxins are endocrine disruption and cancers, but only the latter can be quantified at the present time. We also consider cancers due to inhalation of benzene, formaldehyde, butadiene and benzo(a)pyrene.

For many of these pollutants, in particular dioxins and toxic metals, the dose from ingestion of food is about two orders of magnitude larger than the inhalation dose. However, the health impact per dose can be different depending on the intake mode: for example, according to current knowledge Cd, CrVI and Ni are carcinogenic only via inhalation. The impact can also depend on the chemical form: methyl mercury, for instance, is the most toxic form of Hg, far more toxic than the metallic vapor.

For ERFs determined by epidemiological studies, the question arises of whether the effect of the ingestion dose should be added to that of inhalation. This depends on what exactly was measured in the epidemiological study. Often the study population was exposed simultaneously via inhalation and ingestion. Even if the result of a study is stated as ERF in terms of ambient air concentration, it may in fact reflect the total dose. But if the ratio of inhalation and ingestion for the general population is different from that of the study population, one does not know how to apply the ERF, unless one can make reasonable assumptions about the separate inhalation and ingestion doses of the study population and the relative effectiveness of these two dose routes.

3.6.2   Cancers

With the assumption of linearity, the ERFs for the ingestion of carcinogens are often stated as slope factor (SF), in particular by the EPA. Slope factors are usually estimated from human epidemiological or animal studies using mathematical models, most commonly the linearized multistage model. The SFs of EPA are expressed in units of mg intake per kg body mass per day. They represent the 95% upper confidence limit of the probability of a carcinogenic response per daily unit intake of a substance over a lifetime of 70 years. Thus the slope factor is an upper-bound estimate. This is awkward for external costs, because one really needs the expectation value rather than the 95% upper confidence limit. Unfortunately, information for the correct SF is often hard to find because one has to go back to the original epidemiological studies. In this book we use the SFs of the EPA for simplicity, noting that the resulting damage costs due to cancers are therefore upper bounds.

This approach of the EPA stems from the old idea that there are thresholds and that the goal of environmental regulations should be to ensure that there are no harmful impacts. Such a goal would indeed be achievable if there are thresholds: just make sure that the emission of a pollutant is so low that nobody is exposed above a threshold. Of course, the case of cancers is different because linearity is assumed, even by the EPA. However, often a threshold is in effect introduced after the calculation of a cancer risk, by the requirement that nobody should be exposed to a cancer risk that exceeds a specified threshold, typically one in a million during a lifetime exposure.

For cancers due to inhalation, the ERF is stated by the EPA as unit risk factor (URF), defined as the increase in the probability of getting a cancer from lifetime exposure to a concentration of 1 $\mu g/m^3$. Data on SFs and URFs can be found in the IRIS database of the EPA (www.epa.gov/iris/index.html). Cancer risk due to drinking water is also stated as URF, as the probability of getting a cancer from lifetime exposure to drinking water with a concentration of 1 mg/L, implicitly assuming a consumption of 2 L/day.

The following paragraphs show how to convert URFs and SFs to the format we use for ERF slopes (if one treats these factors as central values rather than upper limits of the confidence intervals), using the example of As. Since we state our ERFs for inhalation as impact per year due to exposure at a constant concentration of 1 $\mu g/m^3$, we have to divide the URF by 70 and obtain,

$$S_{ERF} = URF/70 (cases/yr)/(\mu g/m^3) \text{ cancers due to inhalation}$$
$$\text{of pollutant.} \tag{3.32}$$

For ingestion, we find it convenient to state the ERFs as impact per year due to ingestion at a dose rate of 1 mg/yr, so we have to convert the dose rate from mg/day per kg body mass, in addition to dividing by 70. For the average body mass, we assume 70 kg_{body} to obtain,

$$S_{ERF} = SF/(70 \text{ kg}_{body} \times 365 \text{ days/yr} \times 70) \text{ (cancers/yr)/(mg/yr)}$$
$$\text{due to ingestion of pollutant.} \tag{3.33}$$

In the following, we discuss the most important environmental carcinogens. The percentage of fatal outcomes is approximately 90% for lung cancers; very high because most lung cancers are detected too late for effective treatment. For other cancers, we have assumed the average survival rate of about 50%. Arsenic causes both lung cancers and other cancers.

In addition to mortality, one should also value the years lived with disease (YLD). The YLD can be estimated if one has data for both incidence and prevalence of cancers. Incidence data are new cases per year, whereas prevalence is the total number of cases in a given population at any time. YLD is the ratio,

$$YLD = prevalence/incidence. \qquad (3.34)$$

Patients exit from the prevalence pool, either by being cured or by dying.

Example 3.5   The prevalence of colon cancer in the Netherlands is 169 cases per 100,000, the incidence is 34 cases/yr per 100,000. What is the average YLD?

Solution: YLD = prevalence/incidence = 169/(34/yr) = 5.0 yr.

For the monetary valuation, one has to take into account the latency period, i.e. the time between exposure to a carcinogen and the appearance of a cancer. Most cancers develop very slowly, with a latency period in the range 10 to 20 years.

3.6.3   Arsenic

The effects of arsenic in humans are best known from studies of populations that have been exposed to arsenic in drinking water; some information has also been derived from studies of exposed workers. The most clearly identified impacts are lung cancers, due to inhalation and ingestion. Other effects include skin lesions, other cancers (liver, kidney, bladder and skin), cardiovascular mortality, hypertension, diabetes and neurological effects; however, for these end points, the ERFs at low exposures were less reliable and not considered sufficient for inclusion in an external cost assessment by ExternE.

For cancers due to inhalation, the EPA IRIS lists a URF of 4.3×10^{-3} cancers per $\mu g_{As}/m^3$. As per Eq. (3.32) we obtain,

$$
\begin{aligned}
S_{ERF} &= 4.3 \times 10^{-3} \text{cancers per } \mu g_{As}/m^3/(70\,yr) \\
&= 6.14\text{E-}05 \left(\text{cancers}/yr\right)/\left(\mu g_{As}/m^3\right) \\
&\qquad \text{cancers due to inhalation of As.} \qquad (3.35)
\end{aligned}
$$

Ingestion is considered carcinogenic, with slope factor 1.5 per mg_{As}/(kg_{body}.day), but only for inorganic As. For organic As compounds, the EPA and the International Agency for Research on Cancer do not indicate any evidence for carcinogenicity. However, most of the ingestion dose is organic, with the exception of drinking water which is inorganic,

$$S_{ERF} = 6.14E\text{-}05 (cases/yr)/(mg_{As}/yr) \text{ cancers due to ingestion of}$$
$$\text{inorganic As.} \tag{3.36}$$

Example 3.6   In many areas of the world, the ground water contains traces of As from natural sources. The WHO recommends an upper limit of 10 μg_{As}/L for the concentration of As in drinking water, and this standard has been implemented in the USA and in the EU. How large is the risk of cancer at 10 μg_{As}/L?

Solution: At a typical water intake of 2 L/day the yearly dose is 365×2 L/yr $\times$ 10 μg_{As}/L = 7.3 mg_{As}/yr. The As in water is inorganic and according to Eq. (3.36), the cancer risk is 7.3 mg_{As}/yr $\times$ 6.14E-05 (cases/yr)/(mg_{As}/yr) = 4.48 E-04 cases/yr. Since children consume less and a cancer can take years or decades to develop, let us assume an exposure period of 50 yr. Then the probability for a person to develop a cancer is 50 yr $\times$ 4.48 E-04 cases/yr = 2.2%, not at all negligible.

3.6.4   Cadmium

The effects of Cd in humans have been established from studies of populations in Japan who consume Cd in rice grown on contaminated soil, of populations who live near major industrial sources in northern Europe and who are exposed in air and in food, and of populations who are exposed to Cd in the work place. Most studies have assessed effects in relation to urinary or blood Cd levels. Other studies have assessed effects in relation to concentrations of Cd in food or estimated lifetime intake of Cd in food.

Among the numerous end points are lung cancer, osteoporosis, renal dysfunction, diabetes and mortality. Of these, only lung cancer has met the criteria of reliability of the ERF and availability of monetary values of ExternE. The EPA IRIS shows a URF for lung cancer due to inhalation of Cd, $1.8 \times 10^{-3}/\mu g/m^3$ in ambient air, and it seems unlikely that dietary Cd contributes much to lung cancer risk. So the ERF slope is,

$$S_{ERF} = 2.57E\text{-}05 \left(cases/yr\right)/\left(\mu g_{Cd}/m^3\right) \text{ lung cancers due to}$$
$$\text{inhalation of Cd.} \tag{3.37}$$

3.6.5   Chromium

The effects of Cr in humans are best known from studies of workers exposed to airborne Cr. There is a relatively well-established relationship between exposure to hexavalent Cr (Cr^{VI}) and adverse respiratory effects including cancer. There is no information about effects in populations exposed to Cr in water or food. Limited information about the toxicity of ingested Cr is available from animal studies. Most of the Cr present in ambient air, water and food is in the trivalent rather than the hexavalent state.

The ERF information available for Cr is predominantly for Cr^{VI}, with the exception of some limited information about "mild ill health" that is linked to total Cr exposure. There is no evidence that trivalent Cr is associated with cancer, but trivalent Cr could have adverse effects on kidney function. The relative potency of trivalent Cr in comparison with other heavy metals has not been established. The information for effects on the respiratory system is based on exposure to airborne Cr^{VI}, and it seems unlikely that ingested Cr would have a substantial effect on respiratory health. The EPA IRIS shows a URF for lung cancer due to inhalation of Cr^{VI} as $1.2 \times 10^{-3}/\mu g/m^3$ in ambient air. The ERF information for other health end points is based on ingested Cr and includes nasal damage, mild ill health, effects on fetal development and impaired kidney function. But it has not been sufficiently reliable for inclusion in ExternE. The ERF slope for lung cancers is,

$$S_{ERF} = 1.71E\text{-}04(cases/yr)/\left(\mu g_{Cr^{VI}}/m^3\right) \text{ lung cancers due to}$$
$$\text{inhalation of } Cr^{VI}. \tag{3.38}$$

3.6.6   Lead

An extensive literature links exposure to lead to adverse effects in humans, including studies of workers exposed to lead in workplace air, and children and others exposed to lead in the general environment, often through ingestion of contaminated drinking water. In addition, ingestion of non-food items such as paint is a potentially important source of exposure for young children. Most studies have used concentrations of Pb in blood as an index of exposure. The WHO provides some information linking Pb in blood to exposure, although the relationships are variable and do not apply well at extreme (low or high) levels of exposure. Children appear to absorb more Pb than adults for a given level of exposure and also seem to experience adverse effects at lower blood Pb levels than adults.

There are a variety of impacts on the nervous system, such as cognitive impairment in adults, hearing impairment in children, cognitive impairment in children, reduced IQ of children, effects on nerve conduction, amyotrophic lateral sclerosis (a degenerative motor neuron disease), and brain damage. There is considerable overlap between these end points and for most of them there are no monetary values. However, loss of IQ is included in most of these and it can be measured, even for young children; furthermore the monetary valuation for this end point is relatively firm. For these reasons we take reduced IQ of children as proxy for all neurotoxicity of Pb.

Other end points include anemia, hypertension, renal dysfunction, spontaneous abortion, and possible effects on male fertility, but the ERFs were not considered by ExternE to be sufficiently well established at low doses.

The ERF for reduced IQ of children is quite well determined, thanks to numerous studies, including a meta-analysis by Schwartz (1994), who found a decrement of 0.026 IQ points for a 1 µg/L increase of Pb in blood; a relation that appears to be linear without threshold. A more recent study, designed to identify effects at the lowest doses, found an even larger effect, 0.055 IQ points per 1 µg/L, without any threshold (Lanphear et al., 2000). Here, we continue to use 0.026 IQ points per 1 µg/L, being based on a meta-analysis rather than a single study.

To relate blood level to exposure we use a relation between blood level Pb and ingestion dose, published by WHO (1995). Surprisingly, the blood level per ingested quantity is higher at low doses, perhaps because of increased excretion at higher dose or storage in bones. Here we use a lower value in the range of blood levels per dose, 1.6 µg/L per ingested µg/day. Together with the 0.026 IQ points per 1 µg/L of blood Pb, this implies a loss of 0.026 IQ points/(µg/L) × 1.6 µg/L/(µg/day) × 1 yr/365 days = 1.14E-04 IQpoints/(µg/yr) for a child who ingests a dose of 1.0 µg for one year.

One also has to consider the time window during which an exposure causes damage. The sensitivity of the brain to Pb is greatest during pregnancy and the first years of life, although the precise time distribution of the damage is not known. However, as we explain in this paragraph, this does not matter since the result of Schwartz expresses the total impact in a population due to a constant exposure. Furthermore, the half life of Pb in blood and other soft tissues is relatively short, about 28–36 days (although much longer in bones) (WHO, 1998–2001). Thus, for the purposes of damage calculation, one can equally well assume that the damage is incurred during a one-year exposure by infants between the ages of 0 and 1 only, or during a three-year exposure, between the ages of 0 and 3. To see that the effect is the same, note that the fraction of the population between 0 and 3 is essentially three times the fraction between 0 and 1 (the latter being 1.1% in the EU15, but the precise value does not matter for this argument). If the sensitive period is only one year, the loss due to a one-year exposure is 1.14E-04 IQ points/(µg/yr) × 1.1% of the population of the EU. If the sensitive period is three years, the affected cohort is essentially three times as large, but the damage rate three times smaller, so the loss due to a one-year exposure is (1.14E-04 IQ points/(µg/yr))/3 × (3 × 1.1% of the population of the EU), essentially the same.

Since the annual intake is likely to increase with age, we write the ERF in terms of an average over the age group 0 to 3 years. Thus we write the slope S_{ERF} of the ERF as,

$$S_{ERF} = 1/3 \times f_{0-3yr} \times 1.14\text{E-}04 \text{ IQpoints}/(\mu g_{Pb}/yr)$$
$$\text{IQ loss due to ingestion of Pb}, \tag{3.39}$$

where f_{0-3yr} is the fraction of the population between 0 and 3 years of age; this S_{ERF} is to be multiplied by the average annual dose of this age group. For an age pyramid like the one in the EU, $f_{0-3yr} = 3.3\%$.

Example 3.7   Estimate the average IQ loss due to intake of Pb from food in Europe.

Solution: As usual it is difficult to find all the required information, and so we try an estimate with what we have been able to find. According to EFSA (2010), the average intake of Pb from food by average adult consumers in 19 European countries ranged from 0.36 to 1.24 $\mu g_{Pb}/kg_{body}/$day (lower bound for country with lowest average exposure – upper bound for country with highest average exposure). Not having explicit data for intake by children, we assume the average of these numbers, multiplied by an average body weight of 13 kg for the age group 0 to 3 years. The resulting Pb intake is 0.80 $\mu g_{Pb}/kg_{body}/$day × 13 kg_{body} × 365 day/yr = 3.8 mg_{Pb}/yr.

Multiplying by S_{ERF} from Eq. (3.39), with $f_{0-3yr} = 3.3\%$, we obtain an IQ loss of 0.011 × 1.14E-04 IQ points/(μg_{Pb}/yr) × 3.8 mg_{Pb}/yr = 0.011 × 1.14 × 3.8 = 0.0048 IQ points.

This is the average loss per person due to such an exposure during childhood.

Of course the dose is extremely variable from one individual to another. As one of many sources of such variability, consider the leaching of Pb from so-called crystal glass, for people who like to drink their port wine from crystal decanters. Loss of IQ should not be of concern, since one should abstain from alcohol during pregnancy and not serve port wine to infants. Without reliable ERFs for other end points we cannot evaluate the health impacts for adults, but we can compare the dose with background rates to get an idea whether this could be a problem.

Example 3.8   How large is the incremental dose due to taking 100 drinks/yr of 0.1 L each of port wine stored an average of four months in a crystal decanter?

Solution: Data for such leaching have been measured by Graziano and Blum (1991), who report that the Pb content of wine rose to about 3.5 mg_{Pb}/L after four months. So the dose to such drinking habit is 100 × 0.1 L × 3.5 mg_{Pb}/L = 35 mg_{Pb}/yr, quite a bit larger than the average adult dose in Europe, 0.80 $\mu g_{Pb}/kg_{body}$/day × 70 kg_{body} × 365 day/yr = 20 mg_{Pb}/yr.

3.6.7   Mercury

Among the various possible health impacts of Hg at low doses, the neurotoxic impacts on fetuses and infants have been investigated most thoroughly and they appear to be the most worrisome and the least uncertain. The adult brain is far less sensitive to Hg and no significant association with neurotoxic impacts on adults has been found at low doses (Weil et al. 2005). Evidence is also accumulating for another impact at low doses, coronary heart disease for adults, see e.g. the review by Virtanen et al. (2007), although the case is less clear. Measuring harmful cardiovascular effects of Hg is challenging because the principal intake of Hg comes from seafood, which at the same time provides omega-3 oils that protect the heart. Disentangling the beneficial from the harmful effects of seafood is very difficult. In this book, we only consider neurotoxic effects of Hg, noting however that Rice and Hammitt (2005) find that cardiovascular impacts can make a very large contribution to the damage cost.

The most important studies on neurotoxic impacts have followed cohorts of children among three populations (in New Zealand, the Seychelles, and the Faroe Islands) whose diet contains an especially large portion of seafood; here significant associations between exposure and neurotoxic impacts have been observed.

Based on the findings in New Zealand, the Seychelles, and the Faroe Islands, Trasande et al. (2005) have estimated the social cost of the IQ decrement that can be attributed to Hg ingestion in the USA. In view of the large uncertainties they consider several possible forms of the ERF, both linear and logarithmic, with a no-effect threshold of 8.2 µg/L$_{cord}$ in the cord blood of the newborn infant. In an update (Trasande *et al.*, 2006) they revise their ERFs, the new values being much lower than in their 2005 paper.

In the present paper, we take the ERF of Axelrad *et al.* (2007), because it is derived by an integrative analysis of the New Zealand, the Seychelles, and the Faroe Islands studies, with a method that uses the maximum of information from all three studies. Thus it holds the promise to be much more reliable than any single study. They assume a linear ERF and their central estimate of the slope is 0.18 IQ points per ppm increase of maternal hair mercury (95% confidence interval 0.009 to 0.378). Sensitivity analyses produce estimates in the range from 0.13 to 0.25 IQ points per ppm.

To apply this ERF, we need to relate the Hg concentration in maternal hair to the MeHg dose rate $\dot{D}$ by ingestion. Axelrad *et al.* indicate a concentration ratio of hair/cord blood of 200. The concentration in cord blood is higher than in maternal blood, but there is considerable uncertainty about the ratio of cord blood concentration/maternal blood concentration. Here we assume a ratio of 1.65, the mean of the meta-analysis by Stern and Smith (2003). These authors find that the distribution of values for this ratio is lognormal with median (= geometric mean) of 1.45, which implies a geometric standard deviation $\sigma_g = 1.66$, since one has (see Eq. (11.29) in Chapter 11),

$$\sigma_g = \mathrm{Exp}\left[\left(2\,\mathrm{Ln}(\mathrm{mean}/\mathrm{median})\right)^{0.5}\right] \text{ for the lognormal distribution.}$$

$$(3.40)$$

To relate blood concentration to dietary intake, we note that according to UNEP (2002), the ratio of the steady-state blood concentration c (in µg/L$_{mat}$) and the average dietary MeHg intake $\dot{D}$ (in µg/day) is in the range c/$\dot{D}$ = 0.3 to 0.8. Here we assume the model of Stern (2005) for the relation between intake dose $\dot{D}$ of MeHg and concentration c namely,

$$c = 0.61 \times \dot{D}.$$

$$(3.41)$$

The coefficient 0.61 accounts for blood volume, absorption and elimination rate.

Multiplying these factors we obtain an ERF slope of,

$$S_{ERF} = 0.18 \text{ IQpoints/ppm}_{hair} \times 0.2 \text{ ppm}_{hair}/(\mu g/L_{cord})$$
$$\times 1.65 \mu g/L_{cord}/(\mu g/L_{mat}) \times 0.61 \mu g/L_{mat}/(\mu g/day)$$
$$= 0.036 \text{ IQpoints}/(\mu g/day). \tag{3.42}$$

As for a possible threshold, EPA (2001) noted "*no evidence of a threshold arose for methylmercury-related neurotoxicity within the range of exposures...*". Axelrad *et al.* (2007) also argue for linear ERF without threshold. We find the possibility of a straight line without threshold not only plausible but probable. Hg is a neurotoxicant that damages the developing brain and reduces the IQ, just like Pb. Also, like Pb it is a substance that has only harmful effects, by contrast with other metals (for instance Cr and Se) that are toxic at high doses, but of which the organism needs a certain minimum to survive. Furthermore, whereas in the past the ERF for IQ decrement due to Pb was believed to have a threshold, more recent studies have found that at the lowest doses the ERF for Pb is at least as high as the extrapolation of the high dose points, and quite possibly even higher (Lanphear *et al.*, 2005). Nevertheless, in Section 8.3 we also evaluate the impacts if there is a no-effect threshold dose rate $\dot{D}_{th}$, assuming the same slope S_{ERF}. As threshold we take the oral reference dose RfD of EPA (2001), noting, however, that it is not a no-effect level, but intended as a guideline for protecting the population with a sufficient margin of safety.

3.6.8   *Nickel*

The effects of nickel in humans are reasonably well established for workers exposed to airborne nickel, although reported effects are variable and appear to be different for different nickel compounds. There is also a well-established association between nickel and contact dermatitis, but there is no monetary valuation for this end point. There is very little information about the effects of community exposure to ingested or inhaled nickel, apart from some information linking dietary exposure to dermal sensitization to nickel. The health end points associated with both ingestion and inhalation are presumed to be similar, with the exception of lung cancer, which is expected to be primarily associated with inhalation. There is relatively little information about the relative toxicity of Ni due to inhalation and ingestion. WHO (1988–2001) cites limited information that suggests an absorption rate of about 75% for inhaled Ni, about 25% for Ni ingested in water and about 1% for Ni in food. The EPA IRIS indicates a URF of 2.4×10^{-4} per $\mu g/m^3$ for lung cancer due to inhalation, implying an ERF slope of:

$$S_{ERF} = 3.43\text{E-}06\,(\text{cases/yr})/(\mu g_{Ni}/m^3)\ \text{cancers due to inhalation of Ni.}$$

$$(3.43)$$

Besides lung cancer, Ni may cause neurotoxic effects such as lethargy and ataxia, neonatal mortality, kidney dysfunction, and worsening of eczema in sensitive subjects, but the ERFs are not sufficiently well established.

3.6.9   Dioxins and PCBs

Dioxins (more precisely polychlorinated dibenzo-*p*-dioxins) are a class of 75 individual compounds, some of which are extremely toxic, acting as carcinogens and as endocrine disrupters. Closely related are the furans (more precisely polychlorinated dibenzofurans), a class of 135 different compounds, and the PCBs (polychlorinated biphenyls), a class of 209 compounds. They are "dioxin-like," i.e. they have similar chemical structure, similar physical–chemical properties and invoke a common battery of toxic responses. Implicated in the Seveso accident 1976 and blamed for health effects from the defoliation with Agent Orange in Vietnam, dioxins have acquired a frightful reputation.

Dioxins are an undesirable byproduct of certain processes where chlorine and organic matter can combine, for example the incineration of chlorinated plastics. Other sources of dioxins include the steel industry and the production of pesticides. Thanks to stringent environmental regulations, the emission of dioxins in Europe and the USA has been drastically reduced. However, because of their hydrophobic nature and resistance towards metabolism, these chemicals persist in the environment for many years. In particular they bioaccumulate in fatty tissues of animals and humans. Most of the human dose comes from ingestion.

PCBs had been used in the past as transformer oil, but that was outlawed many years ago and there are no new sources. However, a remainder of PCB pollution is still lingering in the environment, especially in the sea.

Dioxin is one of the most thoroughly studied of all of the pollutants. Several human epidemiological studies and numerous studies in experimental animals have been carried out. There can be acute as well as chronic effects. Dioxins cause changes in laboratory animals that may be associated with developmental and hormonal effects; however, the mechanism of carcinogenicity is unclear. To what extent such biochemical changes may result in adverse health effects in people, and at what concentrations, is not very well known. Among the acute effects at high dose is chloracne, a skin disease. However, the most troubling impacts are cancers and developmental changes, the latter because dioxins have certain similarities with hormones.

It is convenient to characterize the relative toxicity of different compounds in terms of the toxic equivalency factor (TEF), defined as the toxicity of a compound divided by that of the most toxic one, 2,3,7,8-tetrachlorodibenzo-*p*-dioxin (TCDD). Using the TEF, the toxicity of a mixture of dioxins or dioxin-like substances can be expressed as the toxic equivalent quantity (TEQ) of the total dioxin emission. Usually the emissions, doses etc. of dioxin mixtures are stated as TEQ. For example, for municipal solid waste incineration, the TEQ is about 1/60 times the total mass of emitted dioxins (EPA 1994a, vol.II, p.3.104).[12]

In laboratory experiments with animals, TCDD has been found to be one of the most potent known toxins, with LD50 ranging from 0.6 to 3000 μg per kg of body mass for different mammals (LD50 is the dose that kills half of a test group) (Tschirley, 1986). This wide range of values suggests that extrapolation from one animal species to another is very uncertain.

However, Tschirley (1986) cites other bits of evidence that are directly relevant to humans. In particular, there is an experiment on prisoners, performed in days past when such experiments were not yet considered immoral. In one such experiment 60 prisoners were exposed, twice within two weeks, to a TCDD dose of 3 to 114 ng per kg_{body}. No symptoms were observed. An interesting additional data point, in the same reference, comes from another experiment with 10 volunteer prisoners who were exposed to a much larger dose of 107 μg per kg_{body}. Eight of them developed chloracne, but no other symptoms were noted. These numbers indicate that man is not among the most sensitive species as far as acute dioxin toxicity is concerned.

The hormonal effects of dioxins interfere with the endocrine system, but we have found no ERFs for that. The only quantifiable impact is cancer. Dioxins (2,3,7,8-TCDD and HxCDD) were said by the EPA to be "the most potent carcinogen(s) evaluated by the EPA's Carcinogen Assessment Group." The slope factor is 1.0×10^6 cancers/(mg/(kg_{body}·day)) TEQ (EPA, 2000) and the ERF slope,

$$S_{ERF} = 5.45E\text{-}01 \, (cases/yr)/(mg/yr) \text{ cancers due to ingestion}$$
$$\text{of dioxin (TEF).} \tag{3.44}$$

---

[12] The EPA Scientific Reassessment of Dioxin, initiated in 1991 was published in preliminary draft form in August of 1994, in three volumes. The preliminary draft is labeled "review draft (do not cite or quote)," but as yet, a final draft or final health assessment document has not been produced. An update of the dose–response functions was issued in EPA (2000), which also carries the "do not cite or quote" label.

For PCBs, www.epa.gov/iris/index.html states, "Joint consideration of cancer studies and environmental processes leads to a conclusion that environmental PCB mixtures are highly likely to pose a risk of cancer to humans" and lists the following ERFs for cancer:

URF for exposure to air $= 1 \times 10^{-4}$ per $\mu g/m^3$,
URF for drinking water $= 1 \times 10^{-5}$ per $\mu g/m^3$ per mg/L, and
Slope factor for ingestion $= 4 \times 10^{-2}$ per $mg/kg_{body}/day$.

They are an average over different congeners. The cancers found in epidemiological studies are hematologic cancer, cancer of the liver, gall bladder, gastrointestinal tract and biliary tract, and skin cancer. The mortality rate for such cancers may be around 50%.

3.6.10   Benzene, butadiene, benzo(a)pyrene and formaldehyde

3.6.10.1 Benzene

Benzene is recognized as a human carcinogen, for instance IARC (1995) classifies it as Category 1 (human carcinogen). There are many occupational studies investigating exposure to benzene and development of cancer, especially leukemia. However, a quantification of risks is complicated by uncertainties due to lack of quantitative data, short follow up at low exposure concentrations, and co-exposures to other potential carcinogens. Furthermore, there may be large individual variation in susceptibility or metabolism, the latter being relevant because the body breaks down benzene to metabolites which seem to be more toxic than the parent substance. There is no convincing evidence of chronic non-cancer effects at ambient concentrations.

As so often, different risk estimates have been derived, using different assumptions about the pattern of exposures, the shape of the ERF, and the extrapolation to low concentrations. In 1990, the EPA gave a URF of 8×10^{-6} cancers/($\mu g/m^3$), not too different from the estimates of Crump (1994), who gives a range of 4.4 to 7.5×10^{-6} cancers/($\mu g/m^3$) for the URF of leukemia. Taking the average of that range we obtain the ERF slope,

$$S_{ERF} = 8.6\text{E-}08\,(cases/yr)/(\mu g \text{ benzene}/m^3) \text{ cancers due to inhalation}$$
$$\text{of benzene.} \tag{3.45}$$

3.6.10.2 1,3 Butadiene

1,3 butadiene is potentially carcinogenic to both the white and red cell systems. Animal studies have shown that it is carcinogenic in both rats and in mice. There is, however, a wide discrepancy in metabolism between different species, complicating extrapolation to humans. Although the available animal evidence for 1,3-butadiene, and comparison with substances of similar chemical structure would support the classification of butadiene as a human carcinogen, the available human data is limited; and 1,3-butadiene is classified by IARC as Category 2A (= probable human carcinogen).

1,3-Butadiene is a major ingredient of synthetic rubber, and being volatile, the route of absorption is primarily inhalation. The epidemiological evidence consists mostly of mortality studies, which use qualitative estimates or exposure categories rather than estimates of actual lifetime exposures, and with limited consideration of other workplace exposures. The URF of 3×10^{-5} cancers/($\mu g/m^3$) of the EPA is based on multi-stage modeling of animal (mice) experimental data. The corresponding ERF slope is,

$$S_{ERF} = 4.3\text{E-}07\,(\text{cases/yr})/(\mu g \text{ butadiene}/m^3) \text{ cancers due to inhalation}$$
$$\text{of 1,3 butadiene.} \tag{3.46}$$

3.6.10.3 Polycyclic aromatic hydrocarbons

Polycyclic Aromatic Hydrocarbons (PAH) result from the incomplete combustion of organic material. They cover a wide range of substances including benzo[a]pyrene (BaP). The relationship between BaP and other PAH emissions can vary with type of source, but the concentrations are relatively similar in the ambient air of several towns and cities.

There is strong evidence, including from epidemiological studies (e.g. Hurley et al., 1983; Armstrong and Tremblay, 1994) to suggest that certain PAHs, and especially benzo[a]pyrene, are carcinogenic in humans; and that nitroaromatics as a group pose a hazard to health. In 1986, IARC and the US National Cancer Institute concluded that PAHs are a risk factor for lung cancer in humans. Benzo[a]pyrene specifically, rather than PAHs as a group, is labeled as a probable human carcinogen.

Benzo[a]pyrene is the only PAH for which a suitable database is available, allowing quantitative risk assessment. The EPA unit risk factor of lung cancer for BaP is 1×10^{-7} per $\mu g/m^3$. Limitations in the use of benzo[a]pyrene as an indicator of PAH toxicity in air pollution are that some PAH is bound to particulates, and that some of the gaseous components are not included. WHO (1987) estimated a URF of 8.7×10^{-8} per $\mu g/m^3$, almost the same as that used by US EPA. For the URF of EPA the ERF slope is,

$$S_{ERF} = 1.4E\text{-}09\,(\text{cases/yr})/(\mu g_{BaP}/m^3) \text{ cancers due to inhalation}$$
$$\text{of benzo[a]pyrene.} \qquad (3.47)$$

3.6.10.4 Formaldehyde

Formaldehyde in buildings comes mainly from outgasing of materials used in furniture. Formaldehyde is highly water soluble and most of the inhaled formaldehyde is deposited in the lining of the nose. It is a potent irritant, and no clear threshold has been found for such effects. At typical ambient exposures, formaldehyde could produce irritant symptoms to the eyes and respiratory tract in a sensitive subgroup of the general population. Thus, an occasional mild effect, e.g. a symptom day, among sensitive people cannot be ruled out at relatively high exposures. However, such effects are likely to be small and are difficult to quantify; and no ERFs are proposed here.

It is possible that formaldehyde could exacerbate symptoms of asthma or contribute to its development, especially where there is co-exposure to environmental tobacco smoke (Krzyzanowski et al., 1990), but the evidence is not sufficiently well established for quantification.

IARC (1995) classifies formaldehyde as Category 2A (probable human carcinogen). There is, however, no convincing evidence of an effect at low ambient exposures, and possible mechanisms suggest that in the absence of damage to the respiratory tract tissue, cancer risks at low ambient concentrations are negligible. In 1990, the EPA gave a URF of 1×10^{-5} per $\mu g/m^3$ for formaldehyde, not very different from the current estimate of 1.3×10^{-5} per $\mu g/m^3$ (EPA, 2006). The corresponding ERF slope is,

$$S_{ERF} = 1.86E\text{-}07\,(\text{cases/yr})/(\mu g \text{ formaldehyde}/m^3) \text{ cancers due to}$$
$$\text{inhalation of formaldehyde.} \qquad (3.48)$$

3.6.11 Radionuclides

For the 1995 version of ExternE, a detailed analysis of the nuclear fuel cycle was carried out by Dreicer et al. (1995), the results of which are summarized in Section 13.5. The aggregated population exposure (in person/Sv) due to the emission of radionuclides was calculated by modeling their dispersion in the environment. Then the health effects were obtained by multiplying this exposure by the risk coefficients for cancer and hereditary effects, taken from the International Commission on Radiological Protection (ICRP, 1991). A linear no-threshold ERF was assumed, the most plausible assumption and in line with ICRP. To cite the United Nations Scientific

Committee on the Effects of Atomic Radiation in their conclusion of a major review of the biological effects of low-dose ionizing radiation: "an increase in the risk of tumor induction and of hereditary disease proportionate to the radiation dose is consistent with developing knowledge and it remains, accordingly, the most scientifically defensible approximation of low-dose response" (UNSCEAR, 2000, annex G).

We find the linear no-threshold hypothesis for radiation most plausible, for the following reasons:

(i) radiation damages DNA,
(ii) there is a persistent background rate of DNA damage due to radiation and other harmful influences,
(iii) the body repairs most, but not all of the DNA damage,
(iv) it is most unlikely that the relation between damage and repair should change drastically right around the background radiation level (the argument of Crawford and Wilson (1996)). It is the incremental dose above background that matters for public policy.

With further research the risk coefficients can change. A new set was published by ICRP (2007). The old and the new values are listed in Table 3.6. Those for cancers decreased slightly, those for hereditary effects decreased by a factor of 6 for the whole population and by a factor of 8 for the subset of adults.

Overall, ICRP (1991) had estimated a lifetime risk of cancer death of 5% per Sv. All of these numbers are very uncertain, because they are extrapolated from effects observed at much higher doses to dose rates relevant for the general population. In view of the uncertainties, ICRP continues to recommend 5% per Sv as an overall estimate of lifetime risk of fatal cancer, and so we take the ERF for nuclear radiation for the general population as,

$$S_{ERF} = 0.05 \text{ cancer deaths per person/Sv.} \tag{3.49}$$

Table 3.6 *Risk factors for cancer and hereditary effects of ICRP (2007), as probability of occurrence per person in units of 10^{-2}/Sv. The older values of ICRP (1991) are shown in parentheses.*

	Cancers	Hereditary effects
Whole population		
ICRP (2007)	5.5	0.2
(ICRP (1991))	(6.0)	(1.3)
Adults		
ICRP (2007)	4.1	0.1
(ICRP (1991))	(4.8)	(0.8)

References

Abbey, D. E., Hwang, B. L., Burchette, R. J., Vancuren, T. and Mills, P. K. 1995. Estimated long-term ambient concentrations of PM10 and development of respiratory symptoms in a nonsmoking population. *Arch Env Health* **50**: 139–152.

Abbey, D. E., Nishino, N., McDonnell, W. F. *et al.* 1999. Long-term inhalable particles and other air pollutants related to mortality in nonsmokers. *Am. J. Respir. Crit. Care Med* **159**, 373–382.

Abt 2004. *Power Plant Emissions: Particulate Matter-Related Health Damages and the Benefits of Alternative Emission Reduction Scenarios.* Prepared for EPA by Abt Associates Inc. 4800 Montgomery Lane. Bethesda, MD 20814-5341.

Anderson, H. R., Atkinson, R. W., Peacock, J. L., Marston, L. and Konstantinou, K. 2004. Meta-analysis of time-series studies and panel studies of particulate matter (PM) and ozone (O3). Report of a WHO task group. World Health Organization. (www.euro.who.int/document/e82792.pdf; accessed November 2004).

Armstrong, B. and Tremblay, C. 1994. Lung cancer mortality and polyaromatic hydrocarbons. *American J Epidemiology* **139** (3), 250–262.

Axelrad, D. A., Bellinger, D. C., Ryan, L. M. and Woodruff, T. J. 2007. Dose-response relationship of prenatal mercury exposure and IQ: an integrative analysis of epidemiologic data. *Environ Health Perspect* **115**(4): 609–615.

Bobak, M. and Leon, D. A. 1999. The effect of air pollution on infant mortality appears specific for respiratory causes in the postneonatal period. *Epidemiology* **10**(6), 666–670.

Calabrese, E. J., Baldwin, L. A. and Holland, C. D. 1999. Hormesis: A highly generalizable and reproducible phenomenon with important implications for risk assessment. *Risk Analysis* **19** (2), pp. 261–281.

Chen, H., Goldberg, M. S. and Villeneuve, P. J. 2008. A systematic review of the relation between long-term exposure to ambient air pollution and chronic diseases. *Reviews On Environmental Health* **23** (4), 243–297.

Clancy, L., Goodman, P., Sinclair, H. and Dockery, D. W. 2002. Effect of air-pollution control on death rates in Dublin, Ireland: an intervention study. *Lancet* **360**, October 19.

Crawford, M. and Wilson, R. 1996. Low-dose linearity: the rule or the exception?, *Human and Ecological Risk Assessment* **2**, 305–330.

Crouse, D. L., Peters, P. A., van Donkelaar, A. *et al.* 2012. Risk of non accidental and cardiovascular mortality in relation to long-term exposure to low concentrations of fine particulate matter: A Canadian national-level cohort study. *Environ Health Perspect* **120**: 708–714.

Crump, K. 1994. Risk of benzene induced leukemia: a sensitivity analysis of the pliofilm cohort with additional follow-up and new exposure estimates. *J. Toxicology and Environmental Health* **42**, 219–242.

Daniels, M. J., Dominici, F., Samet, J. M. and Zeger, S. L. 2000. Estimating particulate matter-mortality dose-response curves and threshold levels: an analysis of daily time-series for the 20 largest US cities. *Am J Epidemiol* **152** (5): 397–406. See also Comment in: *Am J Epidemiol* **152**(5): 407–412.

Daniels, M. J., Dominici, F., Zeger, S. L. and Samet, J. M. 2004. The National Morbidity, Mortality, and Air Pollution Study, Part III: PM10 Concentration–Response Curves and Thresholds for the 20 Largest US Cities. Health Effects Institute Research Report Number 94, Part III, May 2004.

Declerq, C. and Macquet, V. 2000. Short-term effects of ozone on respiratory health of children in Armentieres, North of France. *Rev Epidemiol Sante Publique* **48** Suppl 2: 2S37–43.

Dockery, D. W., Pope, C. A. III, Xiping, Xu *et al.* 1993. An association between air pollution and mortality in six US cities. *New England J of Medicine* **329**, 1753–1759 (Dec. 1993).

Doll, R., Peto, R., Boreham, J. and Sutherland, I. 2004. Mortality in relation to smoking: 50 years' observations on male British doctors. *British Medical Journal* **328**(7455): 1519–1527.

Donaldson, K. 2006. Lecture at COST633 conference on health effects of PM constituents, Vienna, 3–5 April 2006.

Dreicer, M., Tort, V. and Margerie, H. 1995. Nuclear fuel cycle: implementation in France. Final report for ExternE Program, contract EC DG12 JOU2-CT92-0236. CEPN, F-92263 Fontenay-aux-Roses. This report is included in Rabl *et al.* (1996).

ECRHS 1996. European Community Respiratory Health Survey: Variations in the prevalence of respiratory symptoms, self-reported asthma attacks, and use of asthma medication in the European Community Respiratory Health Survey (ECRHS). *Eur Respir J* **9**: 687–695.

EFSA 2010. Scientific Opinion on Lead in Food. EFSA Panel on Contaminants in the Food Chain (CONTAM), European Food Safety Authority (EFSA), Parma, Italy. *EFSA Journal* 2010; **8**(4):1570.

Elliott, P., Shaddick, G., Wakefield, J. C., de Hoogh, C. and Briggs, D. J. 2007. Long-term associations of outdoor air pollution with mortality in Great Britain. *Thorax* **2007** (0), 1–8.

EPA 1994. Estimating exposure to dioxin-like compounds, and Health Assessment Document for 2,3,7,8-Tetrachlorodibenzo-p-Dioxin (TCDD) and Related Compounds. Report EPA/600/BP-92/001 a,b, and c. USEPA, Washington, DC.

EPA 2000. Exposure and Human Health Reassessment of 2,3,7,8-Tetrachlorodibenzo-p-Dioxin (TCDD) and Related Compounds: Part III: Integrated Summary and Risk Characterization for 2,3,7,8-Tetrachlorodibenzo-p-Dioxin (TCDD) and Related Compounds. Report EPA/600/P-00/001Bg, September 2000. United States Environmental Protection Agency. Washington, DC 20460. Available at www.epa.gov/ncea.

EPA 2001. Oral Reference Dose for Methylmercury. United States Environmental Protection Agency. Integrated Risk Information System (IRIS). Office of Research and Development, National Center for Environmental Assessment, Washington, DC.

EPA 2006. IRIS Database for Risk Assessment. www.epa.gov/iris/index.html

ExternE 1998. ExternE: Externalities of Energy. Vol.7: Methodology 1998 Update (EUR 19083); Vol.8: Global Warming (EUR 18836); Vol.9: Fuel Cycles for Emerging and End-Use Technologies, Transport and Waste (EUR 18887); Vol.10: National Implementation (EUR 18528). Published by European Commission, Directorate-General XII, Science Research and Development.

Office for Official Publications of the European Communities, L-2920 Luxembourg. Results are also available at http://ExternE.jrc.es/publica.html

ExternE 2000. External Costs of Energy Conversion – Improvement of the ExternE Methodology and Assessment of Energy-Related Transport Externalities. Final Report for Contract JOS3-CT97-0015, published as *Environmental External Costs of Transport*. R. Friedrich & P. Bickel, eds. Springer Verlag Heidelberg 2001.

ExternE 2008. Results of the NEEDS phase of ExternE. Available at www. externe.info

Frith, C. H., Littlefield, N. A. and Umholtz, R. 1981. Incidence of pulmonary metastases for various neoplasms in BALB/cStCrlfC3H/Nctr female mice fed N-2-fluorenylacetamide. *Journal of the National Cancer Institute* **66**, p. 703–712.

Gauderman, J. M., Avol, E., Gilliland, F. *et al.* The effect of air pollution on lung development from 10 to 18 years of age. *N Engl J Med* **351**: 1057–1067.

Graves, P. E. and Krumm, R. J. 1981. Health and air quality: evaluating the effects of policy. Volume 322 of AEI Studies in economic policy. Enterprise Institute for Public Policy Research, 1981.

Graziano, J. H. and Blum, C. 1991. Lead exposure from lead crystal. *Lancet* **337** (8734), 141–142.

Guo, Z., Mosley, R., McBrian, J. and Fortmann, R. 2000. Fine particulate matter emissions from candles, Engineering Solutions to Indoor Air Quality Problems Symposium, Air & Waste Management Association, VIP-98, pp. 211–225.

Hedley, A. J., Wong, C-M, Thach, T. Q. *et al.* 2002. Cardiorespiratory and all-cause mortality after restrictions on sulphur content of fuel in Hong Kong: an intervention study, *Lancet* **360**, November 23.

HEI 2001. Airborne particles and health: HEI epidemiologic evidence. HEI Perspectives, June 2001. Health Effects Institute, Charlestown Navy Yard, 120 Second Avenue, Boston, MA 02129-4533. Available at www.healtheffects.org/

HEI 2003. Revised Analyses of Time-Series Studies of Air Pollution and Health. Special Report. Health Effects Institute, Boston MA.

Hiltermann, T. J. N., Stolk, J., Zee, S. C. *et al.* 1998. Asthma severity and susceptibility to air pollution. *European Respiratory Journal* **11**: 686–693.

Hoek, G., Brunekreef, Bert, Goldbohm, Sandra, Fischer, Paul and van den Brandt, Piet A. 2002. Association between mortality and indicators of traffic-related air pollution in the Netherlands: a cohort study. *Lancet* **360** (Issue 9341, 19 October), 1203–1209. See also "Mortality and indicators of traffic-related air pollution", *Lancet* **361**, Issue 9355, 1 February 2003, Page 430.

Hoek, G. and Brunekreef, B. 1995. Effect of photochemical air pollution on acute respiratory symptoms in children. *Am J Respir Crit Care Med* **151**: 27–32.

Hoek, G., Krishnan, R. M., Beelen, R. *et al.* 2013. Long-term air pollution exposure and cardio-respiratory mortality: a review. *Environ Health* **28**;12(1): 43.

Holland, M. 2013. Cost-benefit Analysis of Policy Scenarios for the Revision of the Thematic Strategy on Air Pollution: Version 1 Corresponding to IIASA TSAP Report #10, Version 1. March 2013. Contract report to European Commission DG Environment.

Holland, M., Hunt, A., Hurley, F., Navrud, S. and Watkiss, P. 2005. *Methodology for the Cost-Benefit Analysis for CAFE: Volume 1: Overview of Methodology.* Didcot. UK: AEA Technology Environment. Available: http://europa.eu.int/comm/environment/air/cafe/pdf/cba_methodology_vol1.pdf

Hurley, F., Archibald, R. McL., Colling, P. L. *et al.* 1983. The mortality of coke workers in Britain. *Ann. J. Ind. Med* **4**, 691–704.

Hurley, F., Hunt, A., Cowie, H. *et al.* 2005. *Methodology for the Cost-Benefit Analysis for CAFE: Volume 2: Health Impact Assessment.* Didcot, UK: AEA Technology Environment. Available: http://europa.eu.int/comm/environment/air/cafe/pdf/cba_methodology_vol2.pdf

IARC 1995. IARC Monographs on the Evaluation of Carcinogenic Risks of Chemicals to Humans. International Agency for Research on Cancer.

ICRP 1991. *1990 Recommendations of the International Commission on Radiological Protection.* Publication ICRP 60.

ICRP 2007. *The 2007 Recommendations of the International Commission on Radiological Protection.* ICPR Publication 103. Elsevier.

ISAAC 1998. The International Study of Asthma and Allergies in Childhood Steering Committee. Worldwide variations in the prevalence of asthma symptoms: the international study of asthma and allergies in childhood (ISAAC). *Eur Respir J* **12**: 315–335.

Jerrett, M., Burnett, R. T., Pope, C. A. III *et al.* 2009. Long-term ozone exposure and mortality. *N Engl J Med* **360**(11): 1085–1095.

Jetter, J. J., Guoa, Z., McBrian, J. A. and Flynn, M. R. 2002. Characterization of emissions from burning incense. *Science of the Total Environment* **295**: 51–67.

Katsouyanni, K., Touloumi, G., Samoli, E. *et al.* 2001. Confounding and effect modification in the short-term effects of ambient particles on total mortality: results from 29 European cities within the APHEA2 project. *Epidemiology* **12**(5): 521–531.

Katsouyanni, K., Touloumi, G., Spix, C. *et al.* 1997. Short-term effects of ambient sulphur dioxide and particulate matter on mortality in 12 European cities: Results from time series data from the APHEA project. *British Med. J* **314**: 1658–1663.

Krewski, D., Burnett, R. T., Goldberg, M. S. *et al.* 2000. *Particle Epidemiology Reanalysis Project.* Health Effects Institute, Cambridge MA. Available at www.healtheffects.org/pubs-recent.htm.

Krewski, D., Jerrett, M., Burnett, R. T. *et al.* 2009. Extended Follow-Up and Spatial Analysis of the American Cancer Society Study Linking Particulate Air Pollution and Mortality. Report 140. Health Effects Institute, Charlestown Navy Yard, 120 Second Avenue, Boston, MA 02129-4533.

Krzyzanowski, M., Quackenboss, J. J, and Lebowitz, M, D. 1990. Chronic respiratory effects of indoor formaldehyde exposure. *Environmental Research* **52**, 117–125.

Kuenzli, N., Kaiser, R. Medina, S. *et al.* 2000. Public health impact of outdoor and traffic-related air pollution: a European assessment. *Lancet* **356** (Sept.), 795–801.

Laden, F., Neas, L. M. Dockery, D. W. and Schwartz, J. 2000. Association of Fine Particulate Matter from Different Sources with Daily Mortality in Six

U.S. Cities. *Environmental Health Perspectives – New Series* **108** – issue 10, 941–948.

Lanphear, B. P., Dietrich, K., Auinger, P. and Cox, C. 2000. Cognitive deficits associated with blood lead concentrations <10 µg/dL in US children and adolescents. *Public Health Reports* **115**(6), 521–529.

Lanphear, B. P., Hornung, R., Khoury, J. *et al.* 2005. Low-level environmental lead exposure and children's intellectual function: an international pooled analysis. *Environ Health Perspect* 2005 Jul; **113**(7): 894–899.

Lave, L. B. and Seskin, E. P. 1977. *Air Pollution and Human Health.* Republished in 2010 by Taylor & Francis.

Le Tetre, A., Medina, S., Samoli, E. *et al.* (2002). Short-term effects of particulate air pollution on cardiovascular diseases in eight European cities. *J Epidemiology and Community Health* **56**: 773–779.

Leksell, L. and Rabl, A. 2001. Air pollution and mortality: Quantification and valuation of years of life lost. *Risk Analysis* **21** (5): 843–857.

Levy, J. I., Hammitt, J. K. and Spengler, J. D. 2000. Estimating the mortality impacts of particulate matter: What can be learned from between-study variability? *Environ Health Perspect* **108**(2): 109–117.

Levy, J. I., Hammitt, J. K., Yanagisawa, Y. and Spengler, J. D. 1999. Development of a new damage function model for power plants: methodology and applications. *Environmental Science & Technology* **33**(24): 4364–4372.

Lim, S. S., Vos, T., Flaxman, A. D. *et al.* 2012. A comparative risk assessment of burden of disease and injury attributable to 67 risk factors and risk factor clusters in 21 regions, 1990–2010: a systematic analysis for the Global Burden of Disease Study 2010. *Lancet*, **380**(9859): 2224–2260.

Lippmann, M., Ito, K., Hwang, J-S., Maciejczyk, P. and Chen, L-C. 2006. Cardiovascular effects of nickel in ambient air. *Environmental Health Perspectives* **114**(11): 1662–1669.

Lippmann, M., Ito, K., Nádas, A. and Burnett, R. T. 2000. *Association of Particulate Matter Components with Daily Mortality and Morbidity in Urban Populations.* Research Report 95, Health Effects Institute. Cambridge, MA. Available at www.healtheffects.org/

Matus, K., Nam, K-M., Selin, N. E. *et al.* 2011. *Health Damages from Air Pollution in China.* Report No. 196, MIT Joint Program on the Science and Policy of Global Change. Massachusetts Institute of Technology, Cambridge, MA.

Medina, S., Plasència, A., Artazcoz, L. *et al.* 2002. *APHEIS Health Impact Assessment of Air Pollution in 26 European Cities. Second year report,* 2000–2001. Institut de Veille Sanitaire, Saint-Maurice, France.

NCHS 1999. National Center for Health Statistics, Hyattsville, Maryland. *National Vital Statistics Reports*, Vol. 47, No. 19, June 30, 1999.

NRC 2010. *Hidden Costs of Energy: Unpriced Consequences of Energy Production and Use.* National Research Council of the National Academies Press. National Academies Press, 500 Fifth Street, NW Washington, DC 20001.

ORNL/RFF 1994. External Costs and Benefits of Fuel Cycles. Prepared by Oak Ridge National Laboratory and Resources for the Future. Edited by Russell Lee, Oak Ridge National Laboratory, Oak Ridge, TN 37831.

Ostro, B. D. 1987. Air Pollution and Morbidity Revisited: a Specification Test, *J. Environ. Econ. Manage.* **14**: 87–98.

Ostro, B. D. and Rothschild, S. 1989. Air pollution and acute respiratory morbidity; an observational study of multiple pollutants. *Environmental Research* **48**: 238–247.

Ott, W. 1995. *Environmental Statistics and Data Analysis.* Lewis Publishers. CRC Press *LLC*, 2000 N.W. Corporate Blvd., Boca Raton, FL 33431.

Pennington, D., Crettaz, P., Tauxe, A. *et al.* 2002. Assessing human health response in life cycle assessment using ED10s and DALYs: part 2 – noncancer effects. *Risk Analysis* **22** (5): 947–963.

Pope, C. A. 1989. Respiratory disease associated with community air pollution and a steel mill, Utah Valley. *Am J Public Health.* May; 79(5): 623–628.

Pope, C. A., Burnett, R. T., Thun, M. J. *et al.* 2002. Lung cancer, cardiopulmonary mortality, and long term exposure to fine particulate air pollution. *J. Amer. Med. Assoc.* **287**(9): 1132–1141.

Pope, C. A., Burnett, R. T., Krewski, D. *et al.* 2009. Cardiovascular mortality and exposure to airborne fine particulate matter and cigarette smoke: Shape of the exposure-response relationship. *Circulation* **120**: 941–948.

Pope, C. A. 3rd, Burnett, R. T., Turner, M. C. *et al.* 2011. Lung cancer and cardiovascular disease mortality associated with ambient air pollution and cigarette smoke: shape of the exposure-response relationships. *Environmental Health Perspectives,* **119**(11): 1616–1621.

Pope, C. A., Hill, R. W. and Villegas, G. M. 1999. Particulate air pollution and daily mortality on Utah's Wasatch Front. *Environmental Health Perspectives* **107**(7): 567–573.

Pope, C. A., Thun, M. J., Namboodri, M. M. *et al.* 1995. Particulate air pollution as a predictor of mortality in a prospective study of US adults. *Amer. J. of Resp. Critical Care Med* **151**: 669–674.

Puett, R. C., Hart, J. E., Yanosky, J. D. *et al.* 2009. Chronic fine and coarse particulate exposure, mortality, and coronary heart disease in the nurses' health study. *Environmental Health Perspectives* **117** (11).

Rabl, A. 1998. Mortality risks of air pollution: the role of exposure-response functions. *Journal of Hazardous Materials* **61**: 91–98.

Rabl, A. 2003. Interpretation of air pollution mortality: number of deaths or years of life lost? *J Air and Waste Management* **53**(1): 41–50.

Rabl, A., Curtiss, P. S., Spadaro, J. V. *et al.* 1996. *Environmental Impacts and Costs: the Nuclear and the Fossil Fuel Cycles.* Report to EC, DG XII, Version 3.0 June 1996. ARMINES (Ecole des Mines), 60 boul. St.-Michel, 75272 Paris CEDEX 06.

Rabl, A., Thach, T. Q., Chau, P. Y. K. and Wong, C. M. 2011. How to determine life expectancy change of air pollution mortality: a time series study. *Environmental Health* **10**: 25.

Reiss, R., Anderson, E. L., Cross, C. E. *et al.* 2007. Evidence of health impacts of sulfate- and nitrate-containing particles in ambient air. *Inhalation Toxicology* **19**: 419–449.

Rice, G. and Hammitt, J. K. 2005. Economic Valuation of Human Health Benefits of Controlling Mercury Emissions from US Coal-Fired Power Plants. Northeast States for Coordinated Air Use Management (NESCAUM). Boston, MA. February 2005.

Rowe, R. D., Lang, C. M., Chestnut, L. G., Latimer, D., Rae, D., Bernow, S. M. and White, D. 1995. *The New York Electricity Externality Study*. Oceana Publications, Dobbs Ferry, New York.

Samet, J. M., Dominici, F., Zeger, S. L., Schwartz, J. and Dockery, D. W. 2000. The National Morbidity, Mortality and Air Pollution Study, Part I: Methods and Methodologic Issues. Research Report 94, Part I. Health Effects Institute, Cambridge MA. Available at www.healtheffects.org/

Samoli, E., Aga, E., Touloumi, G. *et al.* 2006. Short-term effects of nitrogen dioxide on mortality: an analysis within the APHEA project. *Eur Respir J* 27: 1129–1137

Samoli, E., Schwartz, J., Wojtyniak, B. *et al.* 2001. Investigating regional differences in short-term effects of air pollution on daily mortality in the APHEA project: a sensitivity analysis for controlling long-term trends and seasonality. *Environ Health Perspect*, **109**(4): 349–53.

Schindler, C., Keidel, D., Gerbase, M. W. *et al.* 2009. Improvements in PM10-exposure and reduced rates of respiratory symptoms in a cohort of Swiss adults (SAPALDIA-study). *Am J Respir Crit Care Med* **179**: 579–587.

Schwartz, J., Coull, B., Laden, F. and Ryan, L. 2008. The effect of dose and timing of dose on the association between airborne particles and survival. *Environmental Health Perspective* **116**(1): 64–69.

Schwartz, J., Norris, G., Larson, T. *et al.* 1999. Episodes of high coarse particle concentrations are not associated with increased mortality. *Environmental Health Perspectives* **107**(5): 339–342.

Schwartz, J. 1994. Low-level lead exposure and children's IQ: a meta-analysis and search for a threshold. *Environmental Research* **65**: 42–55.

Stern, A. 2005. A revised estimate of the maternal methyl mercury intake dose corresponding to a measured cord blood mercury concentration. *Environmental Health Perspectives* **113**(2): 155–163.

Stern, A. H. and Smith, A. E. 2003. An assessment of the cord blood: maternal blood methylmercury ratio: Implications for risk assessment. *Environmental Health Perspectives* **111**(12): 1465–1470.

Thun, M. J., Peto, R., Lopez, A. D. *et al.* 1997. Alcohol consumption and mortality among middle-aged and elderly U.S. adults. *New England Journal of Medicine* **337**(24): 1705–1714.

Trasande, L., Landrigan, T. J. and Schechtes, C. 2005. Public health and economic consequences of methyl mercury toxicity to the developing brain. *Environmental Health Perspectives* **113**: 590–596.

Trasande, L., Schechter, C., Haynes, K. A. and Landrigan, P. J. 2006. Applying cost analyses to drive policy that protects children. *Ann NY Acad Sci* **1076**: 911–923.

Tschirley, F. H. 1986. Dioxin. *Scientific American* **254**, 29. Feb.1986.

UNEP 2002. United Nations Environment Programme. Global Mercury Assessment. UNEP Chemicals, Geneva, Switzerland.

UNSCEAR (2000) Report Vol. II Sources and effects of ionizing radiation United Nations Scientific Committee on the Effects of Atomic Radiation UNSCEAR 2000 Report to the General Assembly, with scientific annexes Volume II: Effects, Annex G Biological effects at low radiation doses.

van der Zee, S., Hoek, G., Boezen, H. M. *et al.* 2000. Acute effects of urban air pollution on respiratory health of 50–70 yr old adults. *Eur Respir J* **15**: 700–709

Virtanen, J. K., Rissanen, T. H., Voutilainen, S. and Tuomainen, T-P. 2007. Mercury as a risk factor for cardiovascular diseases. *Journal of Nutritional Biochemistry* **18**: 75–85.

Ward, D. J. and Ayres, J. G. 2004. Particulate air pollution and panel studies in children: a systematic review. *Occup Environ Med.* **61**(4): e13

Weil, M., Bressler, J., Parsons, P. *et al.* 2005. Blood mercury levels and neurobehavioral function. *JAMA* **293**(15): 1875–1882.

WHO 1987. Air Quality Guidelines for Europe, European Series No.23. World Health Organization, Regional Publications, Copenhagen.

WHO 1988–2001. WHO 1988, Chromium. Environmental Health Criteria 61. WHO 1990, Methyl Mercury. Environmental Health Criteria 101. WHO 1991, Inorganic Mercury. Environmental Health Criteria 118. WHO 1991, Nickel. Environmental Health Criteria 108. WHO 1992, Cadmium. Environmental Health Criteria 134. WHO 1995, Inorganic Lead. Environmental Health Criteria 165. WHO 2001, Arsenic and arsenic compounds. Environmental Health Criteria 224. World Health Organization, Geneva, Switzerland.

WHO. 2003. Health aspects of air pollution with particulate matter, ozone and nitrogen dioxide, Report on a WHO Working Group, Bonn, Germany, 13–15 January 2003. World Health Organization: Available at www.euro.who.int/ document/e79097.pdf; accessed November 2004.

WHO 2005. WHO Air quality guidelines for particulate matter, ozone, nitrogen dioxide and sulfur dioxide – Global update 2005 – Summary of risk assessment. World Health Organization report WHO/SDE/PHE/OEH/06.02. http:// whqlibdoc.who.int/hq/2006/WHO_SDE_PHE_OEH_06.02_eng.pdf

Wilson, R. and Crouch, E. A. C. 2001. *Risk-Benefit Analysis.* Harvard University Press, Cambridge, MA.

Wilson, R. and Spengler, J. D. eds 1996. *Particles in Our Air: Concentrations and Health Effects.* Harvard University Press, Cambridge, MA.

Woodruff, T. J., Grillo, J., and Schoendorf, K. C. 1997. The relationship between selected causes of postneonatal infant mortality and particulate air pollution in the United States. *Environ Health Perspect* **105**(6): 608–612.

Zanobetti, A. and Schwartz, J. 2008. Mortality displacement in the association of ozone with mortality: an analysis of 48 cities in the United States. *Am J Respir Crit Care Med* **177**(2): 184–9.

Zeghnoun, A., Eilstein, D., Saviuc, P. *et al.* 2001. Surveillance of short-term effects of urban air pollution on mortality. Results of a feasibility study in 9 French cities. *Rev Epidemiol Sante Publique* **49**(1): 3–12.

Zmirou, D., Balducci, F., Dechenaux, J. *et al.* 1997. Méta-analyse et fonctions dose-réponse des effets respiratoires de la pollution atmosphérique (Meta-analysis and dose–response functions of air pollution respiratory effects). *Rev Epidemiol Sante Publique* **45**(4): 293–304.

4   Impacts of air pollution on building materials

Summary

This chapter describes two methods for quantifying air pollution damage of buildings in physical and economic terms; one is bottom-up, the other top-down. We begin by showing how the amenity cost can be obtained from the repair cost, without the need for a contingent valuation. Then we describe the effect of air pollution on the main building materials and we show the corresponding exposure–response functions. Sections 4.4 and 4.5 describe the bottom-up and the top-down methods. The results suggest that typical damage costs in the EU are in a range 0.1 to 0.4 €/kg of SO_2; this is a very small percentage (about 1 to 4%) of the costs of health damages due to SO_2. By contrast with the rather detailed calculations for SO_2, only very preliminary estimates have been made for the damage costs from soiling caused by particulate emissions. These suggest values in the order of 0.07 to 0.3 €/kg of particulates emitted by combustion; like for SO_2, this is only a very small percentage of the corresponding costs of health damages. For damage to historical buildings and monuments, despite the fact that this was one of the early motivations towards dealing with acid rain, we have regrettably no good numbers, merely a very rough estimate for France.

4.1   Introduction

Air pollution can damage materials, especially those used in buildings because of their long life. Damage to other objects tends to be less important: most cars, for instance, are replaced long before air pollution damage has become significant. The phenomenon of the degradation of buildings is complex because of the numerous factors that intervene. There are factors of natural origin such as sun, rain and the frost/thaw cycle, in addition to man-made atmospheric pollution. Poor design and building techniques may also play a role, a famous example being the

Houses of Parliament in London (HMSO, 1926) for which many fine carvings suffered failure and rapid erosion within a few years of construction. It is often difficult to distinguish the individual share of each factor. However, it is generally recognized that man-made pollutants have greatly increased the degradation rate of buildings. Of particular importance are soiling due to particles (especially soot) and corrosion or erosion due to SO_2. In Europe and North America, the role of SO_2 is now far less important than it once was, with concentrations in urban areas having fallen by more than 90% in recent decades, though this is not the case in many other parts of the world. Damage in some poorer countries may be exacerbated by the use of lower quality building materials that are more susceptible to degradation by air pollutants.

We begin with a comment on the economic valuation. For buildings there are three cost components:

- expenditures to repair the damaged object, e.g. by cleaning or repainting,
- preventive measures, e.g. the cost of anti-ozonant additives to improve the ozone resistance of tires,
- loss of amenity, e.g. the displeasure of seeing a dirty building.

The total damage cost is the sum.

The first of these is the most straightforward, because cleaning and repair of buildings involve market transactions. By contrast, the cost of preventive measures is very difficult to determine. There is ongoing development of materials, technologies and management, with such a variety of costs and benefits that it would be difficult if not impossible to evaluate. For instance, over the years the paint of cars has been improved so much that damage by air pollution has become negligible. What has been the extra cost, if any, and what are the benefits in addition to protection from corrosion? Often the cost of protective measures, e.g. the extra cost of better paint, is included in the cost of repair; trying to separate the protective component seems neither easy nor necessary.

The third component, loss of amenity, involves subjective perception and might therefore appear to necessitate contingent valuation. However, this subjective valuation is reflected in market decisions about cleaning and repair, and, as we show in Section 4.2, the amenity loss can be inferred by a simple rule from the cleaning and repair costs. This rule says that the cost of amenity loss is approximately equal to the cost of cleaning and repair, and it is probably as accurate as contingent valuation, in view of the known uncertainties of the latter. However, this rule has not been applied consistently in all of the estimates of ExternE.

For buildings of aesthetic or cultural merit, such as ancient cathedrals, estimating the value of damage is problematic. There are some contingent valuation studies, but just as for the preservation of species, they suffer from severe imbedding bias: is the declared willingness to pay for a particular monument, all monuments in a city, all monuments in a country, or all monuments in the world? Consideration needs to be given to the amenity and existence values for such buildings, as replacement costs do not adequately reflect the full cost of, for example, the loss of fine carvings, several hundred years old. Damage costs for monuments are extremely site-specific, both in terms of the merit of the item in question, and in the way in which it can be treated. A further difficulty lies in the lack of adequate inventories to describe the stock at risk. So far, damage to historical monuments and buildings has not yet been quantified by ExternE, except for an extremely rough estimate in France (Rabl, 1999).

There are two main approaches to estimating the damage costs of buildings and materials: bottom-up and top-down. The bottom-up approach, implemented by ExternE, uses ERFs to estimate repair frequencies, and combines these frequencies with inventories of buildings and materials to calculate the total repair cost according to the formula

$$\text{Total repair cost} = \sum_i \text{Surface area}_i \times \text{Repair frequency}_i \times \text{Repair cost}_i$$

$$(4.1)$$

the sum running over all surface types in the inventory. The repair frequency is calculated at each building site by using an atmospheric dispersion model for the air pollutant(s) and applying the appropriate ERF. Unfortunately, for many countries, no inventory data are available.

As an alternative, some studies use a top-down approach, based directly on observed data from renovation frequencies and costs, as has been done by Newby et al. (1991) for the UK and by Rabl (1999) for France. As it happens, in France, there are fairly detailed data on renovation costs because they can be deducted from the income tax, and certain aggregated data on tax returns are accessible to the public. For the city of Paris, data are available for cleaning frequencies and costs. These data make it possible to determine an integrated ERF, directly yielding the cost as a function of the concentration of a pollutant. A summary of the top-down approach for France is presented in Section 4.5.

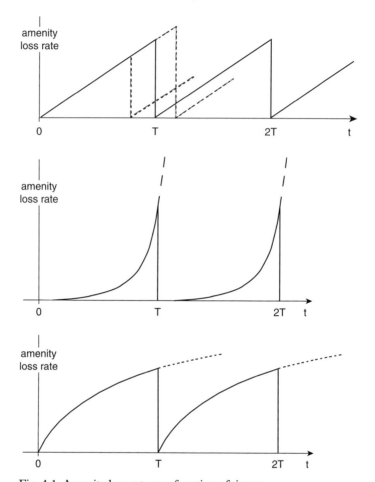

Fig. 4.1 Amenity loss rate as a function of time t.
(a) If a linear increase with time.
    Solid line = renovation period T, dashed lines = renovation period
    shorter or longer than T.
(b) Amenity loss due to corrosion.
(c) Amenity loss due to soiling.

4.2   Amenity cost

4.2.1   *Repair costs and amenity loss*

During the time between renovation actions, there is a loss of amenity as
the appearance of the building degrades, even if renovation will ultimately
restore the original condition of the building (if there were no loss of

amenity, there would be no desire to clean the buildings). As a first approximation, consider the case where the instantaneous amenity loss increases linearly with the time, t, since the last renovation; let us also neglect discounting for now. If T is the time between renovation actions, the total amenity loss during one period T is,

$$L = \int_0^T \alpha\, t\, dt = \frac{\alpha}{2} T^2, \tag{4.2}$$

where α is a constant of proportionality. This is illustrated in Fig. 4.1a.

The total cost during the period is the sum of amenity loss L and renovation cost R. Of the parameters of this model, the only one over which a building owner has control, is the length of the period, T. A rational owner will choose T so as to minimize the total cost. This is equivalent to minimizing the total average cost C_{av} during one period T,

$$C_{av} = \frac{L}{T} + \frac{R}{T}. \tag{4.3}$$

Inserting Eq. (4.2) and taking the derivative of C_{av} with respect to T, we obtain the condition for the optimal period T_o,

$$0 = \frac{\alpha}{2} - \frac{R}{T_o^2}, \tag{4.4}$$

or,

$$T_o = \sqrt{\frac{2R}{\alpha}}. \tag{4.5}$$

Inserted into Eq. (4.2), this implies a very simple result,

$$L = R; \tag{4.6}$$

the amenity loss L is equal to the renovation cost R.

As a simple generalization of the linear increase in Fig. 4.1a, let us consider a power law of the form,

$$L = \int_0^T \alpha\, t^n\, dt = \frac{\alpha}{n+1} T^{n+1}. \tag{4.7}$$

With $n > 1$, as shown in Fig. 4.1b, this could be a reasonable approximation for corrosion damage, since in many cases the need for repair does not become visible until the damage is quite substantial, and the need for repair becomes evident. For example, a tin roof that is corroding away may entail negligible amenity loss until it begins to leak. For soiling, on the other hand, n < 1 seems more relevant, since the loss of reflectance is rapid at first, but slows as dirt builds up; see Fig. 4.1c: the building cannot get much blacker than black. Now the above optimization leads to a generalization of Eq. (4.6),

$$L = \frac{R}{n} \text{ for zero discount rate.} \qquad (4.8)$$

The amenity loss L is smaller than the expenditure R for n > 1 (corrosion), and larger than R for n < 1 (soiling).

4.2.2   Discounting

Discounting reduces the weight of future costs relative to present costs, and likewise for benefits. The decision when to renovate is based on discounted costs and benefits. A rational building owner chooses the period T between renovations so as to minimize the present value P of the total life cycle cost. Using continuous discounting with rate r, one finds that P is given by,

$$P = R + R\,e^{-r\,T} + R\,e^{-2r\,T} + R\,e^{-3\,r\,T} + \ldots$$
$$+ \int_0^T \alpha\,t\,e^{-r\,t}\,dt + \int_T^{2T} \alpha\,(t - T)\,e^{-r\,t}\,dt + \int_{2T}^{3T} \alpha\,(t - 2T)\,e^{-r\,t}\,dt + \ldots \qquad (4.9)$$

for the case where the amenity loss rate increases linearly with time t. Summing the geometric series and designating the amenity loss during the first period by L,

$$L = \int_0^T \alpha\,t\,e^{-r\,t}\,dt = \frac{\alpha}{r^2}\left[1 - (1 + r\,T)e^{-r\,T}\right]. \qquad (4.10)$$

One can write P in the form,

$$P = \frac{R + L}{1 - e^{-r\,T}}. \qquad (4.11)$$

Analogous to the derivation of the optimal period T_o in Eq. (4.5), we now take the derivative of P with respect to T and set this derivative equal to zero. This yields T_o as the solution of the equation,

$$\frac{r\,R}{\alpha} = T_o + \frac{1}{r}(e^{-rT_o} - 1).\qquad(4.12)$$

Here we do not need to solve this transcendental equation. All we want is the relation between L and R, and that is readily found by using Eq. (4.10) (evaluated at $T = T_o$) to eliminate α. The result is,

$$L = R\frac{1 - (1 + r\,T_o)e^{-r\,T_o}}{r\,T_o + e^{-r\,T_o} - 1}.\qquad(4.13)$$

Plausible values of rT_o range from 0 (at r = 0%) to 3 (at r = 10% and T_o = 30 yr). The corresponding ratio L/R varies from 1 to 0.4, and a typical value might be around 0.7 for rT_o = 1 (at r = 5% and T_o = 20 yr). Thus, discounting has the effect of reducing the amenity loss L relative to the zero discount rate result of Eq. (4.8).

4.2.3   A simple rule for amenity cost

To sum up, we have shown that amenity loss due to building damage can be inferred from renovation expenditures. The key assumptions are:
(1) the amenity loss is restored by renovation;
(2) people minimize total cost;
(3) the decision to clean or repair is made by the people who suffer the amenity loss.
For (2), we note that most building owners are acutely aware of costs, and budget constraints impose strong pressures to minimize cost. There may be errors of judgment or inability to obtain the necessary information, but being random, such errors tend to cancel and on average the period T is chosen more or less correctly. The third assumption may seem questionable in cities, where the majority of people who get to see a building are not the owner. There is, however, a certain collective pressure that forces the owners to internalize the amenity loss of the public. Such collective pressure is particularly strong in cities like Paris, where the appearance of the buildings has a direct effect on tourism (a crucial industry for France), and where legislation has been passed to enforce a minimal renovation frequency (see Section 4.5.2).

On the other hand, Eq. (4.6) is likely to give an underestimate for the amenity loss if, for example, building owners cannot afford to maintain or repair their property at the optimal time (e.g. during a period of economic crisis), or operate to a fixed maintenance schedule.

It is of course difficult to guess the magnitude of this underestimate, even with contingent valuation. As a practical compromise, it is perhaps best to adopt the simple rule that amenity loss is approximately equal to renovation cost and that

$$\text{total damage cost} = 2 \times \text{renovation cost.} \qquad (4.14)$$

As suggested by Eqs. (4.7) and (4.8), with zero discount rate the factor may be larger than 2 for soiling and smaller than 2 for corrosion. Discounting tends to reduce the factor somewhat. On balance, we find a factor in the range 1.7 to 2 reasonable, rounded to 2 for simplicity and in view of the uncertainties.

The above arguments also apply to historical monuments and buildings, in the sense of collective decision making: public expenditures for the restoration of historical buildings are a reflection of our collective willingness to pay. We argue that it is meaningful to look at restoration costs, despite the often heard opinion that historical monuments are "priceless" (similar to the widespread refusal of a monetary value for mortality risks). The purpose of quantifying such costs is pragmatic: to establish consistent reference values for the efficient allocation of limited financial resources; it is not the purpose to pass judgment on the intrinsic value of the damaged objects. Thus we take the expenditures for the renovation of historical buildings as de facto expression of society's valuation, and we use Eq. (4.14) to account for loss of amenity.

To close this section, we note that only part of the expenditures for renovation can be attributed to pollution, the rest being due to various natural factors. Of course, only the part due to pollution is to be counted in Eq.(4.14).

4.3   The effects of air pollution on materials

4.3.1   General remarks

In general the following types of impact are possible:
- discoloration;
- structural failure;
- degradation of material;
- soiling.

For the costs of discoloration, there are no valuation studies or material inventories, but such costs are probably very small. Structural failure due to pollution seems unlikely, except in cases of gross neglect of maintenance. Therefore most studies have focused on soiling or on degradation of material, especially corrosion or erosion due to acidic deposition. The latter is due both to direct effects of sulfur dioxide and to effects of acid deposition resulting from both SO_2 and NO_x emissions. Degradation of materials occurs even in the absence of pollution, because of a background of natural processes, including rain, attack by micro-organisms, particularly the fungi that cause wet and dry rot, freeze–thaw cycles and sea salt (in coastal regions). However, measured deterioration rates due to pollution can be a factor of 10 to 100 higher.

For several materials, the dry deposition of SO_2 exerts the strongest corrosive effect of atmospheric pollutants. Wet deposition of pollutants, as acid rain, also has a corrosive but generally weaker effect on certain materials. The role of NO_2 is not entirely clear. Although laboratory studies have found a strong synergistic effect with SO_2, this has not yet been confirmed in the field. O_3 damages some polymeric materials, for instance paints, plastics and rubbers (Lee *et al.*, 1996). Damage to rubber seems especially important (Holland *et al.*, 1998; 2007). O_3 can also act synergistically with SO_2 (Kucera *et al.*, 1993a; Kucera, 1994).

Damage has been evaluated for the most sensitive materials commonly used by the construction industry, including calcareous stone, mortar, paint, concrete, aluminium and galvanized steel. Non-galvanized steel is assumed to be painted and is considered as part of the paint inventory. A possible loss of transparency of glass has not been considered, because modern glass is very resistant to attack. Table 4.1 shows a summary of the materials and their susceptibility to air pollution.

4.3.2   The stock at risk

To quantify the impacts of corrosion on building materials, one needs data for the stock at risk, meteorological conditions and ambient pollution levels. The stock at risk is derived from building surveys. Such surveys have been carried out for individual cities, and they are extrapolated to the national level. For countries where such data are not available, in particular France and most countries of southern Europe, ExternE has extrapolated values from other countries; this introduces significant uncertainty for the materials damage assessment. However, for quantification of the overall damage linked to air pollution, the added uncertainty is small given the limited contribution of materials damage to total pollutant damage. Sources of data are as follows:

Table 4.1 *Sensitivity of materials to air pollution and stock at risk in the EU.*

Material	Sensitivity to air pollution	Stock at risk in the EU
Glass	negligible degradation, but soiling can be quite visible	very large
Brick	very low	very large
Mortar	moderate to high	very large
Concrete	low	very large
Natural stone (sandstone, limestone, marble)	high (severely affected by SO_2)	large (especially culturally valuable objects)
Unalloyed steel	high (severely affected by SO_2)	very small
Stainless steel	very low	medium
Nickel and nickel-plated steel	high (especially in SO_2-polluted environment)	very low
Zinc and galvanized steel	high (especially in SO_2-polluted environment)	medium
Aluminium	very low	medium
Copper	low	low
Lead	very low	low
Rubber	high for unprotected natural rubber, very low for some synthetics	large (especially vehicles)
Paint	uncertain, given reformulation of paints in recent years	very large
Organic coatings	uncertain	very large

for Prague; Stockholm, Sweden; and Sarpsborg, Norway: Kucera *et al.* (1993b) and Tolstoy *et al.* (1990)
for UK: ECOTEC (1986);
for Greece: NTUA (1997);
for Dortmund and Köln: Hoos *et al.* (1987).

These data on stock at risk are now quite dated, as most attention has now turned towards health impacts.

4.3.3   *Specific materials*

4.3.3.1 **Natural stone**

Among the types of stone commonly used for buildings and monuments are granite, sandstone, limestone, marble and slate. Their durability depends on their composition and porosity. Granites are composed mainly of silica as quartz; they have low porosities and are generally very durable. Limestones consist mostly of calcium carbonate (calcite); they are much more porous and susceptible to attack. Sandstones consist of quartz grains bonded together by siliceous or calcareous "cement"; the latter is similar in durability to limestone. Marble has a dense crystalline structure and consists of almost pure calcite. Reviews (Harter, 1986; Lipfert, 1987; Lipfert, 1989; UKBERG, 1990; NAPAP, 1990) indicate that acid deposition damage to siliceous stones is negligible, and so the present discussion is limited to calcareous stones, i.e. limestone, marble and calcareous sandstones, materials that have been widely used for buildings in Europe, particularly cultural heritage.

Weathering of stone occurs naturally, primarily because of CO_2 in the atmosphere. Carbon dioxide dissolves in rainwater, making it slightly acidic. Acidic rainwater slowly reacts with the calcium carbonate in calcareous stone to form calcium bicarbonate, which is soluble and readily washed away. Bicarbonate solution that remains on the surface or in pores and subsequently evaporates will re-precipitate calcium carbonate on the surface (Cooke and Gibbs, 1994). The amount of calcite removed depends on CO_2 concentration, temperature and the physical characteristics of the stone, such as its porosity. The overall long-term effect is a thinning of the stone. Water running off the surfaces is usually not uniform, giving rise to dirt streaks and erratic etching. Besides this natural chemical attack, there are other natural damage mechanisms, in particular stresses from the freeze–thaw cycle of water and from salt crystallization cycles in the stone, which can lead to blistering and exfoliation. Rainfall and windblown particles can cause surface abrasion.

If there is SO_2, a much faster chemical attack can occur via dry deposition of pollutants or via additional dissolution by acid rain, accelerating the reactions and causing the formation of calcium sulfate, which is more soluble than calcium carbonate. For low porosity stones whose surface is frequently washed by rain, the deterioration products do not accumulate and are continuously washed away. The surface looks unaltered and clean, but a very thin layer of stone is removed and the surface consists of re-crystallized calcite. In more porous stones, inward diffusion of acidic solution can cause re-crystallization of calcium sulfate within the pores. This process not only damages directly, but also indirectly by increasing potential for damage from freeze/thaw cycling.

Dry deposition of SO_2 also attacks calcium carbonate. In an urban environment, dry deposition on the stone surface may be more than ten times greater than wet deposition of ions. A small amount of water can later activate these dry deposited ions, forming an aggressive solution which dissolves and progressively transforms the carbonate surface into the crystalline gypsum (hydrated calcium sulfate). The mechanism is complex and depends very much on the atmospheric conditions; particulate matter and O_3 appear to catalyze the process. NO_x may also enhance these reactions.

Where stone is exposed to rainfall or run-off, the calcium sulfate will be washed away during wet periods. In sheltered areas, the calcium sulfate accumulates in surface pores of the stone. When the water evaporates, a gypsum crust is formed. Carbonaceous particles are also incorporated into the crust making it black. Depletion of calcite under the crust leaves a weakened layer. During alternate wet/dry cycles the crust expands/contracts, and the resulting stresses can cause the crust to separate from its substrate (exfoliation).

The deterioration of stone can be broken down into three stages:

- **Stage I** (short term): simple dissolution of calcium carbonate, including (i) normal dissolution of calcite in rain from CO_2, (ii) acceleration due to acid rain, (iii) dry deposition of gaseous pollutants especially SO_2. These processes are relatively easy to characterize in terms of ERFs.
- **Stage II** (medium term): the dissolution of calcium carbonate plus the fall-out of less soluble granular particles within the matrix e.g. for calcareous sandstones; the removal of small amounts of the calcium carbonate matrix may loosen sand grains, leading to more severe surface erosion.
- **Stage III** (long term): where calcium sulfate is not washed away by rain or run-off, there is a build-up of salts, causing the formation of a crust which may be followed by exfoliation. In contrast with formation of crusts which is slow, exfoliation is very damaging. No ERFs are available for these late processes of stage III.

4.3.3.2 Brickwork, mortar and rendering

Whereas brick itself, which is a calcium–aluminium silicate ceramic, is believed to be immune to sulfur dioxide attack, the mortar component of brickwork is not. Mortar consists of sand, calcium hydroxide and other carbonate phases. Mortar erosion is primarily due to acid attack on the calcareous cement binder (UKBERG, 1990; Lipfert, 1987). As there are no specific ERFs for mortar, preliminary estimates have been made by using the ERFs for calcareous stone. However, this will probably underestimate the damage, since mortars tend to be more porous than calcareous stone.

4.3.3.3 Concrete

Portland cement, the major binding agent in most concrete, is an alkaline material which is susceptible to acid attack, leading to soiling/discoloration, surface erosion, spalling and enhanced corrosion of embedded steel. However, these damages (except for surface erosion) are more likely to occur as a result of natural carbonation and ingress of chloride ions, rather than interaction with pollutants such as SO_2.

The durability of concrete is affected mainly by the corrosion of embedded steel reinforcement. In new concrete, the steel is protected from acid corrosion by the alkaline characteristics of the cementitious component of the concrete. If the concrete layer covering the steel is thick enough and the concrete of good quality, the acid pollution only etches the surface without impact on the structural integrity. However, in badly prepared concrete, carbonation can lead to steel corrosion. Cracks can then develop and make the material more accessible to attack by SO_2, since the corrosion products of steel occupy a greater volume than the original steel (Webster and Kukacka, 1986).

4.3.3.4 Paint and polymeric materials

Paints contain polymers that can be damaged by acid deposition and by photochemical oxidants. The effects include loss of gloss, erosion of polymer surfaces, soiling, loss of paint adhesion to substrates, and interaction with sensitive pigments and fillers. Contamination of substrates before painting can cause premature failure and mechanical deterioration such as embrittlement and cracking, in particular of elastomeric materials. The direct action of acid pollutants on the pigments and fillers in paint can accelerate erosion. Especially serious is the effect of SO_2 on paints with calcium carbonate fillers. NO_x has been found to have only a minor effect on paints (Spence et al., 1975; Haynie and Spence, 1984). NAPAP (1990) provides ERFs for the erosion of paint work. Haynie (1986) provides functions for carbonate and silicate based paints, finding that there is a 10-fold difference in acid resistance between the two paint types.

There has been a dearth of published work on the susceptibility of paint to air pollution in recent years. Paint manufacturers are constantly carrying out research and reformulating their products. The most obvious change in the last two decades has been from the use of oil based paints to water based paints. The overall effect is that the earlier functions are likely to be unreliable. Holland et al. (1998; 2007) found little or no sensitivity to ozone of paints typical of the mid to late 1990s. Whilst further research would be useful, we no longer recommend the use of the functions provided by Haynie or NAPAP.

4.3.3.5 Metals

Atmospheric corrosion of metals is an electrochemical process that occurs only when the surface is wet. The corrosion rate depends on several variables, especially humidity, precipitation, temperature and concentration of pollutants. Sulfur dioxide causes most damage, but in coastal regions chlorides are also important. The role of NO_x and O_3 in the corrosion of metals is uncertain; Kucera (1994) indicates that O_3 may accelerate some reactions.

There are ERFs for many metals, but good inventory data are generally available for only two that are widely used in construction, i.e. steel and aluminum. Among common building materials, aluminum is the most corrosion resistant because when exposed to air it gets covered by a thin, dense oxide coating, which is highly protective down to a pH of 2.5. In clean environments aluminum will retain its appearance for years. Steel is almost always used either galvanized (i.e. zinc coated) or coated with paint; in the latter case the ERFs for paint are appropriate.

Zinc is not used as a construction material itself, but rather as a protective coating for steel (galvanized steel), because it has a lower corrosion rate. In an unpolluted atmosphere, the initial reaction of zinc results in the formation of zinc oxide and zinc hydroxide, which in turn are converted to the relatively insoluble zinc carbonate. Even though such films are slowly washed away, they do significantly inhibit the corrosion process.

4.3.4   Exposure–response functions

Exposure–response functions (ERFs) for building materials are determined by field studies of real buildings or by studies of idealized test materials, either in the field or the laboratory. Relying too heavily on test materials is problematic because they may not correctly represent the variety of real building materials. Also, the mechanisms and the rates of damage can vary with ambient conditions. On the other hand, studies of real building materials do not allow sufficient flexibility in the control of all relevant factors. For the most reliable ERFs, one should combine controlled studies and field measurements.

To derive ERFs one has to make assumptions about the appropriate functional form. For example, one often assumes linearity for the dependence on time and on concentrations. In some cases the justification for such assumptions may be weak, because of insufficient empirical data, although they may be justified on theoretical grounds by models of damage mechanisms.

A major program to develop ERFs has been carried out for the UNECE (United Nations Economic Commission for Europe) by Kucera (1993a, b; 1994). It is based on results from 39 test sites across several European countries, as well as three in North America. Covering a very wide range of pollution levels and climates, the results should therefore be quite generally

Table 4.2 *ERFs for materials. Annual rates for functions giving mass loss (ML) or mass increase (MI) are derived by dividing by 4 (the number of years for which samples were exposed).*

Natural stone
ICP – unsheltered limestone (4 years), stage I:

$ML = 8.6 + 1.49 \text{ TOW SO}_2 + 0.097 \text{ H}^+$

ICP – unsheltered calcareous sandstone or mortar (4 years), stage II:

$ML = 7.3 + 1.56 \text{ TOW SO}_2 + 0.12 \text{ H}^+$

ICP – sheltered limestone (4 years), stage III (without later damages such as exfoliation):

$MI = 0.59 + 0.20 \text{ TOW SO}_2$

ICP – sheltered sandstone (4 years), stage III (without later damages such as exfoliation):

$MI = 0.71 + 0.22 \text{ TOW SO}_2$

Paint
Haynie – carbonate paint:

$\Delta ER/t_c = 0.01 \text{ P } 8.7 \ (10^{-pH} - 10^{-5.2}) + 0.006 \text{ SO}_2 \text{ f}_1$

Haynie – silicate paint:

$\Delta ER/t_c = 0.01 \text{ P } 1.35 \ (10^{-pH} - 10^{-5.2}) + 0.00097 \text{ SO}_2 \text{ f}_1$

Zinc and galvanized steel
ICP – unsheltered zinc (4 years):

$ML = 14.5 + 0.043 \text{ TOW SO}_2 \text{ O}_3 + 0.08 \text{ H}^+$

ICP – sheltered zinc (4 years):

$ML = 5.5 + 0.013 \text{ TOW SO}_2 \text{ O}_3$

Key:
ER = erosion rate (μm/year)
P = precipitation rate (m/year)
SO_2 = sulfur dioxide concentration (μg/m^3)
O_3 = ozone concentration (μg/m^3)
H^+ = acidity (meq/m^2/year)
ICP = International Cooperative Programme
R_H = average relative humidity, %
$f_1 = 1 - \exp[-0.121 \ R_H/(100 - R_H)]$
TOW = fraction of time relative humidity exceeds 80% and temperature >0 °C
ML = mass loss (g/m^2) after 4 years
MI = mass increase (g/m^2) after 4 years

applicable. ERFs have also been determined by others, in particular Lipfert (1987 and 1989) and Butlin *et al.* (1992).

Table 4.2 lists ERFs that have been used by ExternE, in standardized format after some modifications of the format in the original publications.

In particular, the original H+ concentration term (in mg/l) of the ICP functions (International Cooperative Programme (ICP) on Integrated Monitoring of Air Pollution Effects on Ecosystems) has been replaced by an acidity term using the conversion $P H^+$ (mg/l) = 0.001 H^+ (acidity in meq/m^2/year), and the erosion rate for stone and zinc has been converted to mass loss by assuming respective densities of 2.0 and 7.14 tonnes/m^3.

4.4   Estimation of damage costs: bottom-up

4.4.1   Calculation of repair frequency

Most ERFs yield a loss of weight or thickness as a function of time. To calculate impacts, one has to convert these losses to frequencies of repair or replacement. For this one needs to know how large the loss has to be before the damage is repaired. Unfortunately there is a dearth of reliable information, and often one invokes engineering judgment or estimates based on common experience, with the attendant large uncertainties. Table 4.3 shows a summary of the assumptions of ExternE for critical thickness loss for maintenance and repair.

4.4.2   Estimation of repair costs

Repair cost estimates of ExternE have been taken from different sources. For the UK they are based on data from ECOTEC (1996) and Lipfert (1987). For Germany, they were obtained through inquiries with German manufacturers. Costs data in a study for Stockholm, Prague and Sarpsborg (Kucera *et al.*, 1993b) were also considered. Table 4.4 summarizes the damage costs assumed by ExternE.

Table 4.3 *Assumptions of ExternE for critical thickness losses for maintenance or repair.*

Material	Critical thickness loss
Natural stone	4 mm
Rendering	4 mm
Mortar	4 mm
Zinc	50 μm
Galvanized steel	50 μm
Paint	50 μm

Table 4.4 *Repair and maintenance costs [€/m²].*

Material	€/m²
Zinc	25
Galvanized steel	30
Natural stone	280
Rendering, mortar	30
Paint	13

Table 4.5 *ExternE results for cost of damage to materials due to SO₂ for various sites in Europe.*

Site	m€/kg$_{SO_2}$	% of total SO₂ cost
Albi (Fr)	98	1.3
Barcelona (Es)	148	1.5
Bordeaux (Fr)	299	2.5
London (GB)	524	4.1
Nantes (Fr)	88	1.0
Paris (Fr)	335	2.2
Piacenza (It)	310	2.5
Stuttgart (De)	386	3.0
Vienna (Au)	345	3.6
Average	**281**	**2.4**

In each grid cell of the dispersion calculation, the frequency of replacement and the total area of each material were multiplied by these unit costs, and the results were summed over all grid cells in Europe. Table 4.5 shows some results.

4.4.3   Estimation of soiling costs

Exposure–response functions for soiling due to PM are often stated in terms of reduced surface reflectance. Beloin and Haynie (1975) provide such functions for a variety of materials. Hamilton and Mansfield (1992) proposed such functions for exposed painted wood and sheltered painted wood. However, the ERFs for soiling were not used by ExternE, because a simple first estimate indicated costs that were so low that a detailed bottom-up analysis did not seem necessary.

Instead we cite two top-down studies. The first, by Newby *et al.* (1991), assumed that the total impact of building soiling in the UK can be attributed to emissions in the UK. The total UK building cleaning market has been estimated to be £80 million annually. UK emissions of black smoke in 1990 were 453,000 tonnes (DOE, 1991). This implies an average cost for building cleaning of around 300 €/tonne of black smoke. The second study, by Rabl (1999), described in Section 4.5.1, found that the average cost of soiling in France was approximately 210 €/tonne of PM_{10}.

4.5   Estimation of damage costs: top-down

4.5.1   Data from tax deductions

Here we describe the top-down study of Rabl (1999) in France, because it is one of the few that obtain results for soiling (the monetary values have been converted from the then current French francs at the exchange rate of 6.56 FF/€ without any adjustment for inflation). From the income tax service of France, we were able to obtain data on deductions for building renovation expenditures in 1994, per taxpayer and aggregated over administrative regions such as cities or Départements (mainland France consists of 95 Départements). We have also obtained data for ambient pollutant concentrations in some of the corresponding urban areas (ADEME, 1992, Stroebel et al., 1995). Unfortunately, the data for cities are rather limited, and only for 15 cities have we been able to get complete data for renovation expenditures, particle concentrations and SO_2 concentrations; for an additional two cities we have renovation expenditures and particle concentrations. One reason for the data limitation stems from the fact that two different instruments have been used in France for measuring particle concentrations; one measures black smoke, the other PM_{13}. Most urban areas have chosen only one or the other. Since there is no site-independent conversion factor, we have chosen the PM_{13} data because they are closer to PM_{10}, the standard international measure for ambient concentrations.

After trying many different regressions, as reported in Rabl (1999), we found that the most reasonable model for the renovation cost R appears to be the one with the β coefficients, of Table 4.6,

$$R = \beta_O + \beta_{Inc} \text{ Income} + \beta_{Inc*PM} \text{ Income} \times c_{PM}, \qquad (4.15)$$

where Income is in 1000 €/(person·yr), c_{PM} is the concentration of PM_{13} in $\mu g/m^3$ and R has units of €/(person·yr). This equation is in effect an ERF and the coefficient of PM_{13} is the slope, apart from a correction factor, to be discussed in Section 4.5.5. At the mean income of 6630 €/(person·yr) the coefficient is

Table 4.6 *Linear regression of renovation expenditures versus Income and Income × concentration of PM_{13}. $R^2 = 0.65$.*

Variables	T1>Parameters	Units	Student t
Constant	$\beta_0 = -10.2$	€/person/yr	-2.0
Income	$\beta_{Inc} = 3.02$	€/1000 €	4.9
Income×PM_{13}	$\beta_{Inc×PM13} = 0.0158$	€/(1000 €.µg/m^3)	1.0

$$\frac{\partial R}{\partial c_{PM}} = \beta_{Inc \cdot PM} \times \text{Income} = 0.105 \text{ €/(person} \cdot \text{yr} \cdot \text{µg/m}^3). \quad (4.16)$$

For Paris, the city with the highest per capita income in France, the coefficient is 0.15 €/(person·yr·µg/m^3). A major source of uncertainty lies in the lack of sufficiently good and detailed environmental data and the fact that the concentrations have been changing markedly over the years.

The lack of a positive correlation with SO_2 is disturbing, since numerous studies have established that SO_2 damages building materials. However, it is plausible that renovation expenditures in France may be driven more by soiling than by corrosion. Also, in our data, the SO_2 damage may be covered up by the correlations between PM_{13}, SO_2 and income.

That particles emitted by combustion of fossil fuels drive up cleaning costs is as obvious as the blackness of soot. Therefore, we find the positive correlation of renovation costs with PM_{13} entirely plausible, despite the low t statistic. The only question concerns the magnitude of the effect.

The average renovation expenditure for our data is 12.76 €/(person·yr). If all of that were due to particles (with average PM_{13} concentration 32.4 µg/m^3), the slope of the ERF would be

$$\frac{12.76 \text{ €/}person \cdot yr}{32.4 \, \mu g/m^3} = 0.40 €/(\text{person} \cdot \text{yr} \cdot \text{µg/m}^3). \quad (4.17)$$

The coefficient of PM_{13}, Eq. (4.16), is much smaller, 0.105 €/(person·yr·µg/m3). This suggests that only about a quarter of the renovation expenditures,

$$\text{fraction attributable to pollution} = 0.105/0.40 = 26\% \quad (4.18)$$

is attributable to particles; of course this fraction is very uncertain in view of the low t statistic in Table 4.6. The value of 0.40 €/(person·yr·µg/m^3) is certainly an upper bound.

Incidentally, the income tax service has also given us data for the number of renovation actions per year; they imply a renovation frequency per taxpayer. With reasonable assumptions about the number of buildings per taxpayer, this implies an average frequency of about once per 50 years for France and once per 30 years for Paris.

4.5.2   Cleaning data for Paris

For verification, it is interesting to look at independent data that we have been able to obtain for Paris.

4.5.2.1 Legislation

In 1852, Napoléon III passed a law that required all building owners in Paris to clean the façades exposed to the street at least once every ten years. This measure was extended, in June 1904, to all façades (Brandela, 1992). During the twentieth century, application of the law was somewhat irregular. The legal requirement of a ten-year cleaning frequency remains in effect, although it is enforced only if a building is considered too dirty. The city can take legal action against recalcitrant owners, with fines from 150 to 3000 €. A survey of cities in France realized by Virolleaud and Laurent (1990) indicates that 30 of a sample of 100 impose a renovation obligation similar to Paris.

4.5.2.2 Building renovation statistics for Paris

Table 4.7 shows the key data we have been able to obtain. More detailed information on individual renovation actions is not generally available, e.g. exposed surface, type of construction, type of work done and cost. If we assume that the cleaning frequency is ten years, as implied by the law, more than 11,000 buildings must be cleaned per year. The actual frequency, averaged over 1990–1994, is only about a quarter of that.

Table 4.7 *Data for the cleaning of buildings in Paris.*

Number of buildings in Paris	110 588
Number of renovation permits/yr	2,714
Cleaning frequency, once per	41 yr

Sources: Council of the City of Paris, and Nicolas Roy, "Sous-Direction du Permis de Construire", Mairie de Paris.

Table 4.8 *Price list for cleaning of façades (without taxes).*

Operation	Price ($€/m^2$)
Cleaning	
Dry scrubbing	5^a
Gumming	$14-18$
Sand jet	12^a
Jet and brush cleaning	11^a
Steam cleaning	15^a
High pressure cleaning with detergents	9^a
Chemical removal of paint	14^a
Surface treatment	
Painting	23
Waterproof coating	$9-12^{b,c}$
Plaster	16^a
Lime	20^a
Cement	25^a
Other	
Scaffolding (rental, including set-up and removal)	$11-25^{a,b,c}$
Protective covering (rental, including set-up and removal)	$3-6^{a,d}$

[a] Moniteur (1994).
[b] C.P.P, St. Ouen; Mr. Katalinic, personal communication.
[c] Sarpie, St. Michel-sur-Orge; Mr. Antoine, personal communication.
[d] Technie Puts, St. Ouen; Mr. Garnalec, personal communication.

4.5.2.3 Cleaning cost

Data for the costs of façade cleaning or renovation are presented in Table 4.8. They are based on the "Publications du Moniteur," which publishes each year a statement of prices for the construction industry. We obtained some additional cost data by telephone inquiry of several firms, as indicated by the references in Table 4.8. The numbers suggest a total cost of about 38 € per m^2 of façade.

4.5.2.4 Annual cost of façade renovation in Paris

Usually only the side facing the street is cleaned, the major cause of pollution in Paris being exhaust from diesel vehicles (during the nineties between a third and half of all passenger cars were diesel, and buses and taxis, the vehicles with the greatest utilization, are always diesel). We might point out a clear demonstration of this effect in the very building where we have worked. One side of the Ecole des Mines faces a large park, the other a busy street; the street side is much more dirty even though it has been cleaned more recently. This is one of the reasons why we believe that soiling due to particles from

152 Impacts of air pollution on building materials

vehicles is the major cause of air pollution induced renovation expenses in the cities of Europe where the percentage of diesel vehicles is high.

If we assume, for the sake of illustration, an average building height of 20 m and an average width of 20 m as typical for buildings in Paris, we find an area of 400 m^2. For the cost per area, Table 4.8 suggests a value of around 38 €/m^2. Together with the number of buildings cleaned per year (Table 4.7), we find,

$$\text{Cost/yr} = \frac{\text{Cost}}{\text{Area}} \times \frac{\text{Area of facade}}{\text{building}} \times \text{Number of buildings cleaned/yr}$$
$$= 400 \, \text{m}^2/\text{building} \times 38 \, €/\text{m}^2 \times 2714 \, \text{buildings/yr}$$
$$= 41.3 \, \text{M€/yr}. \tag{4.19}$$

Since the population of the City of Paris is 2.15 million, this implies 19.21 €/(person·yr). This is close to the 20.27 €/(person·yr) for Paris in our data set.

4.5.3   Comparison with other studies

It is interesting to compare these results with those of other studies. In Table 4.9 we list the ones that appear most comparable. Two of these

Table 4.9 *Results of several studies for the cost of building renovation due to pollution.*

	Exposed surface per inhabitant m^2/person	Renovation cost per inhabitant €/(person·yr)	Renovation cost per exposed surface €/(m^2·yr)	Pollution μg/m^3 SO$_2$	Pollution μg/m^3 PM$_{13}$
Germany[a]	55.6	56	1.01	> 30	
Prague[b]	83	115	1.39	70	
Sarpsborg[b]	165	55	0.33	20 – 60	
Stockholm[b]	132	21	0.16	< 20	
UK[c]		3			
Paris[d]	21[e]	5	(0.94[e])	19	26

[a] Isecke *et al.* (1991): corrosion and soiling; survey of the real-estate owners and managers.
[b] Kucera *et al.* (1993b): corrosion; survey of real-estate owners and consultation of construction guides, with calculated repair frequencies, and assuming Swedish prices for all three cities.
[c] Newby *et al.* (1991): soiling; survey of the stone cleaning market.
[d] Rabl (1999): real renovation costs from tax records, estimated surface area; assuming 26% of the total renovation cost of 20 €/(person·yr) attributable to pollution.
[e] In Paris, most renovation activities concern only the façade, so we count only the façade area.

studies, in Germany and in three European cities (Stockholm, Prague and Sarpsborg), have determined the cost of building renovation and the share of the atmospheric pollution.

The German study (Isecke *et al.*, 1991) analyzes the case of the city of Dortmund, then extrapolates the results to evaluate the cost for the former West Germany. Frequencies of renovation are determined, from an inquiry among property managers and construction firms. Two cases are distinguished: polluted areas (annual average concentration of $SO_2 > 30$ μg/m^3) and non-polluted areas (annual average concentration of $SO_2 < 30$ μg/m^3). The cost is calculated only in areas where $SO_2 > 30$ μg/m^3, as the author considered the SO_2 impact to be negligible below this concentration. Results are presented in Table 4.9, converted with the exchange rate of 1991 and corrected for inflation in France since 1991.

The Scandinavian study (Kucera *et al.*, 1993b) looked at three cities: Prague (Czech Republic), Stockholm (Sweden) and Sarpsborg (small industrial city of Norway), to determine the cost of the corrosion on buildings. Four pollution intervals were distinguished:
- concentration of $SO_2 < 20$ μg/m^3,
- 20 μg/m$^3 <$ concentration of $SO_2 < 60$ μg/m^3,
- 60 μg/m$^3 <$ concentration of $SO_2 < 90$ μg/m^3
- concentration of $SO_2 > 90$ μg/m^3.

Renovation frequencies were determined mainly by field inspection and/ or different sources of published or well-documented guidelines for renovation of buildings (Kucera *et al.*, 1993b). As the model for the additional cost of corrosion in polluted areas they used

$$K_a = K\,S\left(\frac{1}{L_P} - \frac{1}{L_c}\right), \qquad (4.20)$$

with

K_a = additional cost for renovation
K = cost of renovation per area
S = exposed material surface
L_p = interval between renovations in polluted zones
L_c = interval between renovations in unpolluted zones.

This permits us to determine the gain that could be made if the pollution were decreased to the lowest SO_2 pollution level (< 20 μg/m^3). Results are presented in Table 4.9.

Unfortunately, Isecke *et al.* (1991) and Kucera *et al.* (1993b) give no information on the ambient concentration of particles. Instead they use the concentration of SO_2, a parameter for corrosion but less good as an indicator for soiling, although areas with high SO_2 levels often have a high concentration of particles as well.

In the Scandinavian study, costs are determined by using a price list for Swedish builders in 1991 for all three cities. Thus, by comparing the values of the annual cost with square meters of exposed material, it is possible to judge the importance of pollution, because at the assumed constant prices it is the main parameter responsible for the variation in cost.

However, the high cost for the city of Prague is not only due to the high pollution level, but also to the fact that 24% of the materials exposed are plaster and 40% metals; and the cost of repairing plaster is three times the cost of that of any other construction material in this study; and the cost of metal renovation is also high. Therefore, the cost of renovation is not only influenced by the level of pollution but also by the type of material. Thus, much of the factor three between the annual renovation cost per surface in West Germany and in Sarpsborg, despite an equivalent pollution level, is due to the different materials. The surfaces in Sarpsborg are mostly painted wood, whereas the former West Germany has many buildings with façades of stone, mortar or plaster.

The German study is closer to real costs, because it is based on the renovation intervals practised by the building owners, while the Scandinavian study estimates these intervals from construction guides data whose rules are not necessarily applied.

The costs in Isecke et al. (1991) and Kucera et al. (1993b) are an order of magnitude higher than the 3 €/(person·yr) obtained by Newby et al. (1991) in their economic evaluation of the soiling of English buildings, based on the cleaning market in the UK. Our result of 5 €/(person·yr) is much closer to the latter than the former, especially when one recalls that expenditures in Paris are about 50% higher than in the rest of France.

4.5.4   Costs of renovation of historical monuments

It is difficult to get a generic price for the cleaning of a square meter of a historical monument façade, since each monument will require a specific treatment due to its characteristics such as shape, age, material, location and exposure. Also the cleaning of a historical monument is only one operation among several typically carried out during a restoration project, and we do not know what fraction of the total cost can be attributed to pollution.

The budgets allocated by the French Ministry of Culture for restoration and renovation of the national heritage were 183 M€/yr from 1988 to 1992 and 239 M€/yr from 1994 to 1998 (MdC 1994). The contribution of the

Table 4.10 *Expenditures for restoration of historical monuments in France. From Ministère de Culture (MdC 1994).*

Number of historical monuments	12,500
Total expenditures for restoration	approx. 460 M€/yr (private + public)
Population of France	58 million
Total expenditure/(person·yr)	approx. 8 €/(person·yr)
Contribution of air pollution, if 26% of total	very approx. 2 €/(person·yr)

Government covers only 40% to 50% of the expenses. Therefore the total annual expenditure is approximately 460 M€/yr, which corresponds to 8 €/(person·yr).

This includes historical monuments, but also historical objects, furniture and gardens. Only a fraction of the expenditure can be attributed to air pollution. If we estimate the share of air pollution to be 26%, by analogy to what we found for utilitarian buildings in Section 4.5.1, we find that the damage to historical monuments is around 2 €/(person·yr). These numbers are summarized in Table 4.10. Comparing the total expenditures for renovation, 8 €/(person·yr) in Table 4.10 with the above value of 12.76 €/(person·yr) for utilitarian buildings, Eq. (4.17), suggests that air pollution damage to historical monuments in France is significantly smaller, even if one multiplies by a factor of two for loss of amenity.

4.5.5 *Damage per kg of particulate emission*

In Eq. (4.15) we have a linear ERF for damage cost as a function of PM_{13} in ambient air. For the calculation of damage costs due to particle emissions one needs a correction factor, because there is a difference in composition between the primary particles emitted by combustion equipment and the particles in the ambient air. Only a portion of the latter, perhaps 20 to 50%, is due to primary particles from combustion, the rest is composed of soil particles, and of nitrate and sulfate aerosols. Most soil particles are less black than soot, and the contribution of nitrate and sulfate aerosols is white. For that reason, the coefficient of PM_{13}, Eqs. (4.15) and (4.16), should probably be multiplied by a factor of 2 to 4 when it is used for damage calculations per kg of emitted combustion particles. Here we will take 3 as a typical value. Furthermore we multiply by a factor of 2 to account for amenity loss as per Eq. (4.14). Thus the ERF for the total damage cost to utilitarian buildings in France has the slope,

$$S_{ERF} = \frac{\partial R}{\partial c_{PM}} \times 3 \times 2$$
$$= 0.105 \, \text{€}/(\text{person} \cdot \text{yr} \cdot \mu g/m^3) \times 6 = 0.63 \, \text{€}/(\text{person} \cdot \text{yr} \cdot \mu g/m^3).$$
$$(4.21)$$

This is the value for the average income, and for simplicity we use it without trying to account for different incomes in different regions of France. Applying it to typical industrial emission sources, using the UWM (Uniform World Model) of Section 7.4, we find a damage cost per kg of particulate emissions of

$$D_{uni} = 0.21 \, \text{€}/\text{kg} \, PM_{10}. \qquad (4.22)$$

For transport emissions, especially in large cities, this has to be multiplied by the correction factors of the UWM.

4.6   Conclusions

The results suggest that typical damage costs are in a range 0.1 to 0.4 €/kg of SO_2; this is a very small percentage (about 1 to 4%) of the costs of health damages due to SO_2. Interestingly, these estimates are rather similar (once adjusted for inflation and the sulfur content of coal) to a very old estimate of £0.17 to £0.42 per ton of coal, reported by Clinch (1955), which are understood to be based on emissions in London at the time. It is likely that these results were based on a top-down assessment, performed in much the same way as some of the analysis presented in this chapter for France.

By contrast with the rather detailed calculations for SO_2, only very preliminary estimates have been made for the damage costs from soiling caused by particulate emissions: they suggest values in the order of 0.07 to 0.3 €/kg of particulates emitted by combustion; as for SO_2, this is only a very small percentage of the corresponding costs of health damages.

For historical buildings and monuments, a category of great importance in Europe, we have regrettably no good numbers. At least for France we have a very rough estimate, based on data for total national expenditures for renovation. These imply that the cost of pollution damage is somewhat smaller than for utilitarian buildings, if one assumes that the fraction of renovation expenditures attributable to pollution is the same.

References

ADEME 1992. Pollution Atmosphérique en France. Report December 1992. ADEME, Service des Réseaux de Mesure. 27 rue Louis Vicat, F-75737 Paris CEDEX 15.

Beloin, N. J. and Haynie, F. H. 1975. Soiling of building materials. *Journal of the Air Pollution Control Association* 25: 393–403.

Brandela, J. P. 1992. Présentation du cadre administratif et réglementaire du ravalement à Paris, Journée Ravalement, Direction de l'Architecture et Direction de la Construction et du Logement, Ville de Paris.

Butlin, R. N. *et al.* 1992. Preliminary Results from the Analysis of Stone Tablets from the National Materials Exposure Programme (NMEP). Atmospheric Environment, 26B 189, and Preliminary Results from the Analysis of Metal Samples from the National Materials Exposure Programme (NMEP). *Atmospheric Environment* 26B, 199.

Clinch, H. G. 1955. *Atmospheric Pollution in London and the Home Counties: A Report on Known Facts.* London and Home Counties Smoke Abatement Advisory Council.

Cooke, R. U. and Gibbs, G. B. 1994. *Crumbling Heritage? Studies of Stone Weathering in Polluted Atmospheres.* National Power. Swindon.

DOE 1991. *Digest of UK Environmental Protection and Water Statistics.* Department of the Environment, HMSO, London.

ECOTEC 1986. Identification and assessment of materials damage to buildings and historic monuments by air pollution. Report to the UK Department of the Environment.

ECOTEC 1996. *An evaluation of the benefits of reduced sulphur dioxide emissions from reduced building damage.* ECOTEC Research and Consulting Ltd., Birmingham, UK.

Hamilton, R. S. and Mansfield, T. A. 1992. The soiling of materials in the ambient atmosphere, *Atmospheric Environment* 26A, 18: 3291–3296.

Harter, P. 1986. Acidic Deposition – Materials and Health Effects. IEA Coal Research TR36.

Haynie, F. H. and Spence, J. W. 1984. Air pollution damage to exterior household paints. *Journal of Air Pollution Control Association* 34: 941–944.

Haynie, F. H. 1986. Atmospheric Acid Deposition Damage due to Paints. US Environmental Protection Agency Report EPA/600/M-85/019.

HMSO. 1926. *Memorandum on the Defective Condition of the Stonework at the Houses of Parliament and Proposals for its Restoration.* His Majesty's Stationery Office, London.

Holland, M. R., Haydock, H., Lee, D. S. *et al.* 1998. The effects of ozone on materials. Contract report for the UK Department of the Environment, Transport and the Regions, London.

Holland, M. R., Haydock, H., Cape, J. N. *et al.* 2007. Ozone damage to paint and rubber goods in the UK. *Pollution Atmospherique*, Special Edition October 2007, 73–86.

Hoos, D., Jansen, R., Kehl, J., Noeke, J. and Popal, K. 1987. Gebäudeschäden durch Luftverunreinigungen – Entwurf eines Erhebungsmodells und

Zusammenfassung von Projektergebnissen. Institut für Umweltschutz, Universität Dortmund, 1987.

Isecke, B. *et al.* 1991. *Volkswirtschaftliche Verluste durch umweltverschmutzungsbedingte Materialschäden in der Bundesrepublik Deutschland*, Umweltbundesamt, Texte 36/91.

Kucera, V. 1994. The UN ECE International Cooperative Programme on Effects on Materials, Including Historic and Cultural Monuments. Report to the working group on effects within the UN ECE in Geneva, Swedish Corrosion Institute, Stockholm, 1994.

Kucera, V., Henriksen, J., Leygraf, C. *et al.* 1993a. Materials Damage Caused by Acidifying Air Pollutants – 4 Year Results from an International Exposure Programme within UN ECE. International Corrosion Congress, Houston, September 1993.

Kucera, V., Henriksen, J., Knotkova, D. and Sjöström, Ch. 1993b. Model for Calculations of Corrosion Cost Caused by Air Pollution and Its Application in Three Cities. Report No. 084, Swedish Corrosion Institute, Roslagsvägen, 1993.

Lee, D. S., Holland, M. R. and Falla, N. 1996. The potential impact of ozone on materials. *Atmospheric Environment* **30**: 1053–1065.

Lipfert, F. W. 1987. *Effects of acidic deposition on the atmospheric deterioration of materials*, National Association of Corrosion Engineers, Materials Performance, July 1987, pp. 12–19.

Lipfert, F. W. 1989. Atmospheric damage to calcareous stones: Comparison and reconciliation of recent findings. *Atmospheric Environment* **23**: 415.

MdC 1994. René Dinkel, personal communication. Direction du patrimoine, Ministère de Culture, Paris.

Moniteur 1994. Borderau de prix du bâtiment tout corps d'état. Editions du Moniteur. 17 rue Uzès, 75002 Paris.

NAPAP 1990. National Acid Precipitation Assessment Programme. 1990 Integrated Assessment Report. NAPA, Washington D.C.

Newby, P. T., Mansfield, T. A. and Hamilton, R. S. 1991. Sources and economic implications of building soiling in urban areas, *The Science of the Total Environment* **100**: 347–365.

NTUA 1997. National Technical University of Athens, Danae Diakoulaki, personal communication.

Rabl, A. 1999. Air pollution and buildings: An estimation of damage costs in France. *Environmental Impact Assessment Review* **19**(4): 361–385.

Spence, J. W. *et al.* 1975. Effects of gaseous pollutants on paints: A chamber study. *Journal of Paint Technology* **47**: 57–63.

Stroebel, R., Berthelot, V. and Charré, B. 1995. La qualité de l'air en France 1993–94 (Air quality in France 1993–94). ADEME, 27 rue Louis Vicat, F-75015 Paris; and Ministère de l'Environnement, 20 ave. de Ségur, F-75007 Paris.

Tolstoy, N., Andersson, G., Sjöström, Ch. and Kucera, V. 1990. External Building Materials – Quantities and Degradation. Research Report TN:19. The National Swedish Institute for Building Research, Gävle, Sweden, 1990.

UKBERG. 1990. *The Effects of Acid Deposition on Buildings and Building Materials.* UK Building Effects Review Group. HMSO, London.

Virolleaud, F. and Laurent. 1990. Le ravalement, guide technique, réglementaire et juridique, Ed. du Moniteur.

Webster, R. P. and Kukacka, L. E. 1986. Effects of acid deposition on Portland concrete. In: *Materials Degradation Caused by Acid Rain.* American Chemical Society, 1986, pp. 239–249.

5   Agriculture, forests and ecosystems

Summary

In this chapter, we discuss the impacts of the classical air pollutants on agriculture and ecosystems; we also consider some impacts of agriculture, in particular those due to the use of nitrogen fertilizer. Exposure–response functions for agricultural impacts are established and monetary valuation of the losses is straightforward, at least for marginal changes, as described in Section 5.2. The current practice of agriculture also has significant impacts on the environment, and in Section 5.3 we estimate the damage costs due to the use of nitrogen fertilizer, followed by damage costs of pesticides in Section 5.4. By contrast with the effect of pollution on agricultural crops, most ecosystem impacts are far more difficult to quantify, and more so, to express in monetary terms. In Section 5.5.1 we explain why monetary valuation, in particular via contingent valuation, is so problematic for ecosystem impacts. Some examples of impacts are described in Section 5.5.2. In Section 5.5.3 we present an interesting cost–benefit analysis of pollution abatement, to reduce the eutrophication of the Baltic Sea. This is followed in Section 5.5.4 by the estimation of ecosystem impacts and costs carried out in the NEEDS phase of ExternE. Finally, we mention an interesting assessment of the total value of ecosystem services; even though it does not enable the determination of marginal damage costs, it is a compelling reminder of the dangers of destroying our ecosystems.

5.1   Introduction

Pollution can have serious impacts on ecosystems and on agriculture. Along with the effects of greenhouse gases and global warming, the effects of nitrogen are a serious threat to ecosystems (Sutton *et al.*, 2013). Several of the classical air pollutants, especially O_3 (Mills and Harmens, 2011), are harmful to plants and have significant impacts on agriculture. Because of the importance of agriculture, much research has been carried out to determine exposure–response functions (ERFs), and results are available for the effects of NO_x, SO_2 and O_3 on many economically important species. The monetary valuation can be straightforward for scenarios where the effects are sufficiently small to be approximated by first-order changes. Thus it suffices to multiply the change in output of each crop by the corresponding price. ERFs and costs for agriculture are discussed in Section 5.2.

Estimations of ecosystem damage costs, by contrast, have remained extremely uncertain, if they have been attempted at all. Here forest decline is one of the simplest categories, because data are available for the monetary valuation, both for the market goods (production of wood) and for the non-market goods (recreational and aesthetic value). However, the ERFs are not particularly well established, especially when considering long-term exposures which are relevant to forest systems over their life cycle. Forest decline turns out to be a complex process, caused by a combination of stressors including drought, parasites and pollution. Conditions are too variable from one site to another to measure the contribution of specific pollutants, unlike agricultural plants that can be grown in test chambers with carefully controlled growing conditions and pollution exposures.

As a precautionary note about the estimation of ecosystem impacts, recall that during the 1980s, acid rain was thought by many to be the chief cause of forest decline. Dying forests (especially in the north east of the USA, Canada, Scandinavia, Germany and the 'Black Triangle')[1] became a focal point of the environmental movement. Earlier observations had demonstrated a clear causal link between industrial emissions and the loss of forest in North America as well as Europe (e.g. NRCC, 1939; Knabe, 1970; Farrar et al., 1977). Cost estimates at the time suggested that this impact was the main contribution to the total damage cost of air pollution; today this might appear to be forgotten. One of the major reasons for this is that since 1980 the emission of SO_2 has been greatly reduced (Reis et al., 2012). Between 1990 and 2008, ambient SO_2 concentrations in the USA decreased by about 60%, and in the EU27, SO_2 emissions have been reduced by about 75%. Analysis now suggests that, thanks to these abatement efforts, thresholds for acidification (critical loads for deposited acidity and critical levels for air concentrations of SO_2) are exceeded in rather limited areas of Europe and North America. There was also a tendency for any decline to be blamed on pollution, when in reality many factors were at play (ExternE, 1995, chapter 9), as noted above. One very positive response of the concern

---

[1] The area where Poland, Czechoslovakia and East Germany met, and which was subject to very high rural as well as urban SO_2 levels.

over pollution was the initiation of systematic surveys of forests across Europe, leading to a better understanding of forest health.

In general, the estimation of external costs for ecosystems is problematic, and Section 5.5 on this topic is relatively brief. However, research in this field is gathering pace, particularly through the development of the ecosystem services approach. A positive outcome of this approach, even where it does not extend to valuation, is that scientists can use it to better convey reasons in favor of ecological protection, providing policy makers with a better understanding of the consequences of their decisions.

5.2   Impacts on agriculture

5.2.1   Exposure–response functions (ERFs) for crop losses

At typical ambient concentrations, NO_x does not seem to have any harmful effects on crops: the leaves are sufficiently insensitive, and the plant as a whole benefits from the nitrogen deposited on the ground. Thus a farmer could (at least in theory) reduce the purchase of commercial nitrogen fertilizer. ExternE calculated the reduction in fertilizer requirement as,

$$\Delta F = 14.0 \, \text{g/mol} \cdot A \cdot \Delta DN, \tag{5.1}$$

where

ΔF = reduction in fertilizer requirement in kg_N/yr
A = agricultural area in km^2, and
ΔDN = annual nitrogen deposition in $meq/m^2/year$ (meq = molar equivalent).

Example 5.1 Compare the contribution of nitrogen deposition in Europe with typical fertilizer use, 100 to 200 $kg_N/ha/yr$.

Solution: Data on nitrogen deposition in Europe for the time period 1989–94 have been provided by Holland et al. (2005). They indicate a range of values from 2.6 to 8.1 $kg_N/ha/yr$; in most of central Europe the annual wet deposition flux is about 5 $kg_N/ha/yr$. Since this is already in units of kg_N, there is no need to use Eq. (5.1) and convert to molar equivalent m_{eq}. Thus, the 5 $kg_N/ha/yr$ (which is due to air pollution) can be compared directly with the fertilizer use of 100 to 200 $kg_N/ha/yr$: it is a relatively small benefit, but not negligible. Since the NO_x emissions in the EU27 have decreased by about a factor of 0.6 between 1992 and 2009, the current deposition data are roughly a factor of 0.6 lower. A complication

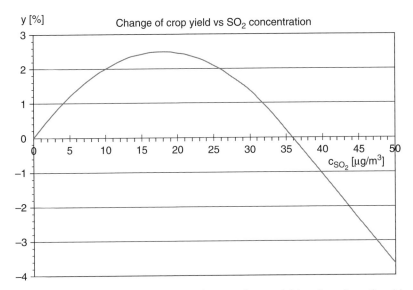

Fig. 5.1 The ERF for y = change of crop yield as function of ambient SO_2 concentration.

arises in that the response of crops to N fertilizer is not linear (it has an inverted U shape, like the ERF for SO_2 in Figure 5.1). Yield peaks at some level dependent on the crop and growing conditions, and at some point further on it starts to decline (DEFRA, 2010). It is therefore possible that yield could decline with added N deposition. However, for practical purposes, the assumption that deposited N can be valued as an avoided fertilizer requirement is reasonable. At the very least, the inclusion of this effect as a benefit takes away from the criticism that analysis is only concerned with negatives.

A similar fertilizer effect also occurs for SO_2, but the point of equilibrium between harm to the leaves and fertilizer benefit lies in the range of typical ambient concentrations, because the mass of sulfur needed by plants is far smaller than the mass of nitrogen. Thus the ERF is U-shaped. Based on Baker $et\ al.$ (1986), ExternE uses the following ERF for the relative change in crop yield for wheat, barley, potato, sugar beet and oats:

$$y = 0.74\, c_{SO_2} - 0.055\, c_{SO_2}{}^2 \qquad \text{for } 0 < c_{SO_2} < 13.6\,\text{ppb}$$

and   (5.2a)

$$y = 9.3 - 0.69\, c_{SO_2} \qquad\qquad \text{for } c_{SO_2} > 13.6\,\text{ppb},$$

or, after converting according to $1\ ppb_{SO_2} = 2.66\ \mu g_{SO_2}/m^3$

$$y = 0.278\ c_{SO_2} - 0.0078\ c_{SO_2}{}^2 \qquad \text{for } 0 < c_{SO_2} < 36\mu g/m^3$$

and   (5.2b)

$$y = 9.3 - 0.259\ c_{SO_2} \qquad\qquad \text{for } c_{SO_2} > 36\mu g/m^3,$$

where

$\quad\quad$ y = yield change in %, and

$\quad\quad c_{SO2}$ = ambient SO_2-concentration.

This function is plotted in Fig. 5.1. Up to about 20 $\mu g/m^3$, the marginal impact, i.e. the slope of the ERF, is positive, meaning that the production increases with SO_2 concentration. In the EU, the ambient levels of SO_2 have been dropping, thanks to stringent regulations for emissions, and currently the annual average concentrations are everywhere below 10 $\mu g/m^3$ and almost everywhere below 5 $\mu g/m^3$, as can be seen from a map of pollution data from the European Environment Agency (www. eea.europa.eu/themes/air/airbase/interpolated, accessed 19 Aug 2011). Strictly speaking, one needs the SO_2 concentrations during the growing season, not the annual averages, but since the concentrations are higher in winter because of the use of coal, oil and wood for heating, the growing season values are even lower.

Example 5.2   The wheat production of France in recent years has been about 38 million tonnes/yr. The SO_2 concentrations are everywhere below 10 $\mu g/m^3$ and almost everywhere below 5 $\mu g/m^3$. How much does the value of the wheat production of France change if the SO_2 concentrations are reduced by 1 $\mu g/m^3$?

Solution: in the relevant range of concentrations, the ERF is not very different from a straight line and so we estimate the yield change by evaluating the ERF at 3 and 4 $\mu g/m^3$ (as for the growing season, even for winter wheat, it extends from Spring until July and excludes the high SO_2 concentrations of winter).

At 3 $\mu g/m^3$, Eq. (5.2b) gives, y = 0.76% and at 4 $\mu g/m^3$ y = 0.99%. Thus the wheat production declines by 0.99 − 0.76 % = 0.23% or 87,400 tonnes/yr. Multiplied by the wheat price of 137 €/tonne in Table 5.2 below, we find that such a reduction in SO_2 pollution would cause a loss of almost 12 million €/yr.

It will be noted that the literature referred to here is almost 30 years old, and specific only to wheat. However, there is little more recent data available from the literature, and little prospect of new studies, certainly in North America and Europe, given the low rural SO_2 levels observed.

Table 5.1 *Sensitivity factors (Eq. (5.3))*
for different crop species.

Crop species	Sensitivity factor α
Rice	0.4
Tobacco	0.5
Sugar beet, potato	0.6
Sunflower	1.2
Wheat	1.7

For the assessment of ozone impacts, ExternE uses a linear relation between yield loss and AOT40 value (Accumulated Ozone concentration above a Threshold of 40 ppb) calculated for the growth period of crops (May to June) (Fuhrer, 1996, Mills *et al.*, 2003). The relative yield is calculated using the following equation together with the sensitivity factors in Table 5.1:

$$y = 99.7 - \alpha \times \text{AOT40}_{\text{crops}}, \tag{5.3}$$

where

y = relative yield in %

α = sensitivity factor (%/ppm.hr), and

$\text{AOT40}_{\text{crops}}$ = Accumulated Ozone concentration above a Threshold of 40 ppb (80 $\mu g/m^3$), for crops, i.e. during growing season May–June (ppm.hr).

Even though y should be 100% at $\text{AOT40}_{\text{crops}} = 0$, Eq. (5.3) shows a different value because it is a curve fit to be used for typical concentrations.

Crops for which no ERFs are indicated have been assumed by ExternE not to be sufficiently sensitive to pollution, or else their contribution to the economy is not sufficiently important.

A more sophisticated approach is now being recommended, using estimates of ozone flux into leaves rather than simple concentration. A detailed description of methods is provided by Mills and Harmens (2011). This approach is preferred as water stressed plants can be exposed to very high concentrations (in the equations above, AOT40 levels) without experiencing any damage. Under such conditions, the stomata on leaves are shut to avoid further water loss and there is no way for ozone to enter the leaves (i.e. flux into the leaves is zero) and cause damage to sensitive receptors. Unfortunately, few flux-based response functions are yet available. In Europe, we have agreed functions only for wheat and tomato:

$$\text{Wheat : relative yield} = 1.00 - 0.038 \times \text{POD6}$$
$$\text{Tomato : relative yield} = 1.00 - 0.024 \times \text{POD6} \quad , \tag{5.4}$$

where POD6 = phytotoxic ozone dose over a flux of 6 mmol.m^{-2}_projected_leaf_area.s^{-1}.

This raises the question of how to develop an estimate that is reasonably comprehensive and reflects the wider literature on ozone damage. In practice, this means that sensitivity data derived from a broader literature, including information using the AOT40 metric are adopted.

A problem for policy analysis with both the AOT40 and PODx measures is that results depend critically on the accurate quantification of exceedance of the implied threshold in both metrics. Indeed, there is currently a debate amongst European analysts as to what level to set the implied threshold for flux based assessment to, with POD1, POD3 and POD6 under discussion, seeking to balance the quality of the response function with that of the estimated threshold exceedances. However, from the perspective of this book, other effects, such as health impacts, are likely to be more important, and the consequences for policy of error in the crop damage quantification would be small. On the other hand, there is much interest in the quantification of effects that feed directly into GDP, so these impacts may carry additional weight.

There are frequent episodes of ozone levels being sufficiently high in some areas to cause serious damage to crops, ranging from spring onions to lettuces to melons (again, see Mills and Harmens, 2011). However, it appears that such episodes affect rather small areas. Losses to individual farmers may thus be very high, perhaps involving the total loss of a crop (D. Velisariou, personal communication), but associated externalities will be small given the limited area impacted. We are not aware of any attempts to model such impacts. There are three major problems linked to a lack of data on:

(1) Observations of damage – there is no systematic mechanism in place to collate information, so most incidents go unreported.
(2) High resolution ozone mapping.
(3) Growing conditions at the time that damage takes place.

The experiments that have been used to derive response functions generally factor out the effects of pests and pathogens, that will also affect crop yield. This is understandable, as both will interfere with identification of the response of prime interest (the direct impact of pollutants on plants). However, there is a significant, albeit rather dated, literature on interactions between pollution and pests and crop diseases. Bolsinger and Flukiger (1989) went so far as to propose a mechanism for effects, with air pollution stimulating the production of certain amino acids by plants that were favorable to aphid growth and abundance. Riemer and Whittaker (1989) provided a more extensive review of studies demonstrating similar effects. Whilst quantification of this interaction is not possible, the literature is certainly suggestive that omission of this and similar interactions will lead to an underestimation of crop damage.

A further effect that is not currently factored into quantification of yield loss concerns changes in the quality of crops (e.g. with respect to protein content of wheat, sugar content of potatoes or oil quality in oilseed rape (Mills and Harmens, 2011)). Again, the literature is limited, but certainly suggestive of an effect that would again add to the estimates of damage derived from application of simple concentration or dose–effect functions.

Crop production accounts for only around 50% of agricultural production, with production of meat, milk, eggs, wool, etc. accounting for the remainder. The mechanism here is one of ozone damaging the productivity of grasslands, or added nitrogen increasing it. The farming response can go in two ways:

(1) No response, in which case production of animal products could decline for ozone or increase for nitrogen. We say "could" decline/ increase, as there are further issues to consider, such as the extent to which farmers have matched grass production to stocking density.

(2) Change the amount of feed supplements provided to livestock to compensate for the change in grass production.

Mills and Simpson (2013) provide a partial illustration of how the impact pathway for this effect could be implemented, taking the example of UK lamb production from unimproved (upland) pastures. They demonstrate how ozone can affect the quality of grassland and link this to the amount of additional feed needed to maintain productivity. Valuation can then be performed against the cost of feed. Whilst this analysis was not complete at the time of writing (June 2013), it provides a promising avenue for future assessment.

Acidification of agricultural soils is another harmful impact, but it can be mitigated by liming. Ideally, the analysis of liming should be restricted to non-calcareous soils, but this refinement has not been considered by ExternE, where only a simple upper-bound estimate of additional liming needs is calculated according to

$$\Delta L = 50 \, kg/meq \cdot A \cdot \Delta D_A, \tag{5.5}$$

where

ΔL = additional lime requirement in kg/year
A = agricultural area in ha, and
ΔD_A = annual acid deposition in meq/m^2/year.

The result turns out to be very small compared with other externalities. It is acceptable to value only in terms of additional lime application, given that lime is already routinely applied to agricultural fields to counteract acidification resulting from cropping (given the resulting loss of base cations such as potassium, calcium and magnesium from the system).

5.2.2   Valuation of crop losses

ExternE assumes that impacts of crop loss on crop prices are negligible. Thus the change of production due to O_3 or SO_2 can be multiplied by the market price of the crop to obtain the respective damage cost contribution. As an example, we list in Table 5.2 the prices assumed by ExternE (2005). The assumption of negligible price change due to crop loss may not be justified. It depends on the overall scale of yield changes. Hence, it may be quite acceptable for the assessment of impacts of a single power station, but not for comparison of scenarios involving a major change in pollution levels. However, this may be of little consequence, given that crop damage is unlikely to generate a large part of overall impacts for such a scenario, given likely effects also on human health (Holland, 2013).

5.2.3   Impacts on forest growth

Forests provide many functions for society, including timber production, regulation of water and carbon cycles, recreational amenity and cultural value. This section deals only with the effect of pollution on timber production and carbon storage, the others are linked to the discussion of ecosystem services below.

The understanding of the relationships between the condition of forest ecosystems and the impacts of air pollution is still incomplete. The picture of air pollution effects on forests is rather complex and subject to spatial variation. In many instances, air pollution at the regional level is just one factor impacting on forest health. In the UK, for example, a more serious concern is the introduction of new diseases, such as Dutch Elm Disease in the 1970s and ash dieback in the 2010s. Both have a widespread and likely permanent impact on what were two of the country's most abundant trees.

Effects of SO_2 on forests have been noted above (e.g. Knabe, 1970; Farrar et al., 1977) and reference to the problems of the "Black Triangle." However, in Europe and North America, these are effects of a largely bygone age, with current levels of SO_2 insufficient to cause damage on this scale (this does not of course mean that similar effects are not occurring elsewhere in the world where SO_2 levels are high and increasing).

Table 5.2 *Prices of major crops, as used by ExternE (2005).*

Crop	Price, €/tonne
Sunflower	273
Wheat	137
Potato	113
Rice	200
Rye	99
Oats	132
Tobacco	2895
Barley	93
Sugar beet	64

Similarly, the effects of acidification on forests in North America and Europe are of much less importance than previously, given the great reduction in emissions of (particularly) SO_2 and hence a substantial reduction in the area of forest subject to exceedance of the critical load (compare the results for forest exceedance linked to acidification in IIASA (1998, Figure 2.2)).

Wittig and Ainsworth (2007) estimate that ozone may have led to an 11% reduction in photosynthesis since the Industrial Revolution, and this in turn may have reduced tree productivity by about 7% in Europe (Wittig and Ainsworth, 2009). Karlsson *et al.* (2005) provided a first attempt to assess the economic impacts of ozone on forest production in Europe. This was done for the Estate Östads Säteri in southwestern Sweden. Harvests were estimated to be reduced by 1.8% because of ozone damage. The economic return, defined as the difference between revenues and costs from harvests, was reduced by 2.6%. The greatest uncertainty in the estimates of ozone impacts on forest production was the up-scaling of ozone effects on growth, measured on young trees under experimental conditions, to mature trees under field conditions. The results arrived at for the Estate Östads Säteri are extrapolated to the whole of Sweden and all EU Member States. The authors state that this does not lead to consistent outcomes, since the effects of ozone may differ due to different exposure levels. However, doing so provides an idea of the potential economic effects of ground-level ozone on forest production on a wider scale. The total loss to Sweden would be in the order of €56 million/yr. For Europe, this would result in a total loss of about €316 million/yr. A more extensive analysis by Karlsson was reported in Harmens and Mills (2012) and by Mills *et al.* (2013), which focused on the effect of ozone on forest carbon sequestration. This took a more detailed account of forest stocks and productivity in Northern Europe. Ozone exposure–response relationships were derived from the earlier study by Karlsson *et al.* (2005). The forest stem increment growth was calculated as

$$y = h / \left(100 + (i * j) / 100\right), \qquad (5.6)$$

where y = annual increment growth ($m^3 \, y^{-1}$),

h = annual increment growth under current ozone exposure levels ($m^3 \, y^{-1}$),

i = AOT40 (ppm h),

j = the slope for the correlation between AOT40 and growth reduction (% (ppm h)$^{-1}$, negative values imply growth reductions). This is equal to −0.26 / −0.13 for conifers up to/over 60 years and −0.49 and −0.25 for broadleaved species up to/over 60 years.

Although this work relies on functions expressed against the concentration-based ozone metric of AOT40, there are moves underway to adopt flux-based functions. Readers are advised to look to the webpage of the ICP Vegetation (the International Collaborative Program on Vegetation under the UN/ECE Convention on Long Range Transboundary Air Pollution) for future updates.

5.3   Damage costs due to nitrogen fertilizer

5.3.1   Impact categories

Many people have a bucolic image of agriculture as something natural and environmentally benign. However, in reality, most contemporary agriculture is an industry with severe environmental impacts, including soil erosion, destruction of landscapes (replacement of regions with once rich biodiversity by monocultures), odors (especially from pig farms), and depletion of scarce water resources (for example the Colorado river, once called Grand, has been reduced to a ghost of its former self). But as with other industries, the means exist to clean up the act. Appropriate solutions, identified by cost–benefit analyses, should be implemented. A comprehensive analysis of the impacts of agriculture is beyond the scope of this book, and we limit ourselves to one item, the use of synthetic nitrogen fertilizer (von Blottnitz *et al.*, 2006).

Considering the life cycle of synthetic nitrogen fertilizer, the following impact categories may be significant:

- global warming due to the production of fertilizer;
- damages due to air pollutants emitted by the production of fertilizer;
- global warming due to the application of fertilizer;
- eutrophication due to leaching of applied fertilizer;
- pollution of drinking water due to leaching of applied fertilizer; and
- damages due to release of volatile substances (especially NH_3) from applied fertilizer.

But acidification of soils should not arise if good farming practices are followed.

5.3.2 Impacts due to fertilizer production

The upstream impacts are mainly due to the emission of greenhouse gases and other air pollutants during production of fertilizer and its raw materials, ammonia and nitric acid. Apart from greenhouse gases, only NO_x and NH_4NO_3 emissions make a significant contribution. The emission rates in Table 5.3 are typical of current technologies (Gruncharov et al. 2000). For the damage cost per kg_{NOx}, we take the numbers of Fig. 12.1 in Chapter 12, calculated for LCA applications in EU27 by ExternE (2008).

For greenhouse gas emissions, the value of 7.0 kg CO_{2eq}/kg_N is taken from Wood and Cowie (2004, their Table 5) as the average value for European nitrogen fertilizer production, based on a review of eight life cycle assessments. It is made up mainly of CO_2 emissions from ammonia production, and N_2O emissions from nitric acid production, but also includes other emissions of these gases and of CH_4 in the production of fertilizers. The greenhouse gases are expressed in terms of CO_2 equivalent, using their global warming potential (GWP). The final stage of fertilizer manufacture requires electricity input, about 0.2 kWh/kg_N, and mostly from base load plants burning coal or lignite, with the emission of PM_{10}, NO_x and SO_2. The damage cost of these pollutants is typically in the range 0.01 to 0.04 €/kWh, implying a contribution of about 0.002 to

Table 5.3 *Emissions and damage costs due to fertilizer production, per kg of nitrogen.*

Pollutant	kg_{poll}/kg_N	€/kg_{poll}	€/kg_N
NO_x	2.4 E-3	7.10 €/kg_{NOx}	0.017
NH_4NO_3	3.7E-3	12.7 €/kg_{NH4NO3}	0.047
CO_{2eq}	7.0	0.021 €/kg_{CO2}	0.147

0.008 €/kg$_N$. Since this is rather small compared with the uncertainties of the estimates in Table 5.3, we do not take it into consideration. For the damage cost of greenhouse gases, we assume 21 €/t$_{CO2eq}$, see Section 12.2.1.

5.3.3   Global warming due to N fertilizer use

Soil bacteria decompose nitrates and emit N_2O, a powerful greenhouse gas. Its global warming potential (GWP) is about 300, meaning that 1 kg of N_2O causes as much warming as 300 kg of CO_2. Thus, despite the relatively small quantities that are emitted, it contributes currently about 5% of the total global warming. The emission of N_2O has increased dramatically above the natural background, ever since the 1960s when the massive worldwide use of synthetic nitrogen fertilizer began.

The emission of N_2O from cultivated land is highly variable in space and time. Variations due to soil management, cropping systems and rainfall are larger than those due to the type of mineral fertilizer. According to IPCC (1995) and Mosier et al. (1998), a typical average value is

N_2O flux [in kg$_N$/ ha] $= 1.0 + 0.0125 \times$ N input [in kg$_N$/ha]

(uncertainty range for the factor multiplying N input: 0.0025 to 0.0225).

$$(5.7)$$

Typical fertilizer input rates are in the range 100 to 200 kg$_N$/ha, implying that the anthropogenic contribution is larger than the background flux (the 1.0 in Eq. (5.7)).

All flows of nitrogen-containing compounds are reported in mass units of N (molar weight 14). To convert the N_2O flux from kg$_N$ to kg N_2O, recognizing that one mole of N makes half a mole of N_2O and multiplying by the molar masses, one finds that the incremental flux due to anthropogenic input is

$(2 \times 14 + 16)/2\,(g_{N2O}/mol)/(14 g_N/mol) = 22/14 g_{N2O}/g_N = 1.57 g_{N2O}/g_N.$

Multiplying by the GWP of N_2O, one obtains the equivalent CO_2 emission per kg of fertilizer,

$$300\,g_{CO2eq}/g_{N2O} \times 1.57\,g_{N2O}/g_N \times 0.0125$$
$$= 5.89\,kg_{CO2eq}/kg_N \text{ of fertilizer.}$$

At a damage cost of 0.021 €/kg_{CO2eq}, the corresponding damage is

$$5.89\,kg_{CO2eq}/kg_N \times 0.021\,€/kg_{CO2eq} = 0.124\,€/kg_N\,\text{of fertilizer.}$$

5.3.4   Health impacts

In the past the main health impact of concern has been methemoglobinemia, a serious and often fatal illness in infants due to conversion of nitrate to nitrite by the body, which can reduce the oxygen-carrying capacity of blood. Adults are not affected. To eliminate this risk, the current legislation for water quality in the EU and in the USA imposes a limit of 50 $mg_{nitrate}$/L in drinking water (EC, 1998). With this limit, the concentration is below the no-observed-adverse-effect level (NOAEL) for methemoglobinemia reported in epidemiological studies, hence no health risk is expected from treated water (www.epa.gov/iris/index.html).

The epidemiological studies that established the limit value are somewhat dated and were based in rural USA in around 1950. More recent work has established that the risk of methemoglobinemia is greatly enhanced by concurrent microbial infections and the observed cases were probably associated with poor hygiene and inadequate control of microbes in drinking water. The risk of methemoglobinemia is hence likely to be close to zero in countries with good water quality and concentrations below 50$mg_{nitrate}$/L. However, the risk could be significant in developing countries and in some regions of Eastern Europe. For developing countries, a study in India highlights the risk (Gupta *et al.*, 1999).

More recently, colon cancer has been identified as a possible impact of nitrogen compounds in drinking water, with potentially very significant damage costs (see e.g. van Grinsven *et al.*, 2010). However, the epidemiological evidence is not yet sufficient to derive reliable estimates.

5.3.5   Eutrophication

Eutrophication arises when the concentrations of phosphorus, nitrogen and other plant nutrients in an aquatic ecosystem become too high. This leads to excessive growth of certain species, especially blooms of algae (e.g. phytoplankton), and creates conditions that interfere with the recreational use of lakes and estuaries and the health and diversity of indigenous fish, plant and animal populations. Algal blooms hurt the system by clouding the water and blocking sunlight, and by increasing oxygen demand when the algae die and decompose. Much of the life at lower depths dies out for lack of light and oxygen.

Although aquatic eutrophication is a natural process in the aging of lakes and some estuaries, human activities can greatly accelerate the problem by increasing the rate at which nutrients and organic substances enter aquatic ecosystems from their surrounding watersheds, due to agricultural run-off, urban run-off, leaking septic systems, sewage discharges, eroded stream banks and similar sources. Atmospheric emissions of NO_x can also contribute to eutrophication. Some terrestrial systems can also be perturbed by eutrophication.

Eutrophication depends on the relative proportions of nitrates and phosphates. If the supply of one nutrient is limited, adding more of another nutrient has little effect on the growth because organisms need the right balance. Thus the impact of an incremental emission of NO_x or NH_3 depends on the existing ratio of nitrates and phosphates. Where the water is limited in phosphates, the most effective abatement strategy is phosphate abatement. By contrast, where the limiting nutrient is nitrogen, it is more effective to reduce nitrogen input.

Run-off of nutrients from agriculture is the dominant contribution to eutrophication in most areas, and it has been recognized as an important environmental problem in Europe. But no estimates are available for the eutrophication damage costs per emitted pollutant. Thus, even though several studies provide monetary information for eutrophication, the link to the emitted pollutants is missing. For example, Pretty et al. (2003) estimate that the total cost of freshwater eutrophication in the UK is between $105 and $160 million per year. However, this is a total, without any link to the individual causes, such as nitrate deposition following atmospheric emissions of NO_x, discharges from water works, inadequate management of animal manure, nitrate effluents from crop land and phosphate effluent from crop land.

One really needs an impact pathway analysis for the cost of eutrophication due to specific sources of nitrogen or phosphates, but that is extremely complex and difficult. Even where the physical impacts can be analyzed, the monetary valuation is often missing. As a very crude shortcut, let us take the results of Pretty et al. and assume that each of the nitrogen sources individually accounts for at least 10% and not more than 50% of the total damage estimate. Then the eutrophication cost of the 1228 kt_N of N-fertilizer used in 2001 in the UK lies about in the range 0.01−0.065 €/kg_N, with a central value of 0.03 €/kg_N.

Table 5.4 *Damage costs of nitrogen fertilizer, in €/kg$_N$.*

Impact category	€/kg$_N$
Greenhouse gases from fertilizer production	0.147
NO$_x$ from fertilizer production	0.017
NH$_4$NO$_3$ from fertilizer production	0.047
N$_2$O from fertilizer in soil[a]	0.124
NH$_3$ emissions from fertilizer in fields	Not quantified
Eutrophication[b]	0.03?
Health impacts due to nitrates in drinking water	?
Total of quantified costs	**0.36?**

[a] site-specific, numbers shown are global average.
[b] very site-specific, number shown is rough estimate of average freshwater eutrophication in UK.

5.3.6   Summary of the damage costs

The damage costs ("external costs") that we have quantified are summarized in Table 5.4, expressed per kg of N in the fertilizer. We do not consider NH$_3$ releases from applied fertilizer.

Even with this incomplete assessment, the total damage cost is not at all negligible compared with the market price of fertilizer, about 0.5 €/kg$_N$. If it were internalized by a pollution tax, it would cause a major increase in the price. If, as we argue in Section 12.2.3, the damage cost of greenhouse gases should be 0.065 €/kg$_{CO2eq}$, the damage cost of fertilizer would increase to 0.93 €/kg$_N$, and a corresponding tax would almost triple the market price.

5.4   Damage costs of pesticides

Health impacts of pesticides are a serious concern. For many people the fear of pesticides is the motivation for eating organic food. Unfortunately, an assessment of the real impacts is problematic because of the lack of reliable ERFs. Results of toxicological studies are reported only in terms of thresholds, insufficient for quantification of impacts. Epidemiological studies are practically impossible, because the exposures or doses cannot be ascertained with sufficient accuracy. For workers, the incidence of effects is high enough to be observable, but the exposures are far too uncertain. For the general public, both exposures and effects are almost always too small. At best, epidemiological studies might detect effects, but the results would not be usable for quantification. Furthermore, pesticides for which a high risk became apparent, have been outlawed and replaced by new formulations of lower toxicity, even more difficult to study by epidemiology.

An interesting example is Kepone, also known as chlordecone. Its use proved to be so disastrous that it is now generally prohibited (in the USA, since 1977). In the French Antilles it was used until 1993, and because of its long lifetime in the environment, on the order of decades, the remaining pollution of soil and ground water is still causing significant health impacts. It is one of the few pesticides for which epidemiological studies are becoming available, because the French government is looking for the most suitable solutions for dealing with this pollution.

As an alternative approach, Huijbregts et al. (2005) developed ERFs from general ideas about modes of action. Fantke and his colleagues used these ERFs in a comprehensive modeling approach for pesticide exposures of the general public and derived estimates of the damage costs for all the major pesticides in current agricultural use (Fantke et al., 2011a, Fantke et al., 2011b and Fantke et al., 2012b). In particular, Fantke et al. (2012a) published global estimates of LE loss in the EU. Their results are reproduced in Table 5.5.

Table 5.5 *Best estimate of human health impact scores for the five most extensively used pesticides on each crop class, as well as sum over crops across EU25 countries in 2003, with upper and lower 95% confidence interval limits for total impacts and damage costs in parentheses.*
From Fantke et al. *(2012a), with permission of Elsevier.*

Crop class	DALY/yr	DALY/person/yr	M€/yr	€/person/yr
Cereals	6.78	1.5E−08	0.27	5.9E−04
Maize	3.77	8.3E−09	0.15	3.3E−04
Oil seeds	8.82	1.9E−08	0.35	7.7E−04
Potato	1.35	3.0E−09	0.05	1.2E−04
Sugar beet	0.34	7.4E−10	0.01	3.0E−05
Grapes/vines	724.02	1.6E−06	28.96	0.06
Fruit trees	113.36	2.5E−07	4.53	9.9E−03
Vegetables	1100.58	2.4E−06	44.02	0.10
Total	1959.01	4.3E−06	78.36	0.17
	(4.75 to 838,505)	(1.0E−08 to 1.8E−03)	(0.18 to 33,540)	(4.0E−04 to 73.45)

Example 5.3 Compare the total annual LE loss per person, 4.3E-06 DALY, with the LE loss due to $PM_{2.5}$ exposure.

Solution: For a typical ambient concentration of 20 μg/m^3, the LE loss is, according to Eq. (3.12) (and taking DALY = YOLL), 20 μg/m^3 × 6.51E-4 YOLL/(yr·μg$_{PM2.5}$/m^3) = 1.3E-2 YOLL per year of exposure.

The LE loss from conventional air pollution is more than three orders of magnitude higher than the central value due to pesticides; even the upper limit 1.8E-03 is lower.

5.5   Ecosystems

5.5.1   Difficulties with the valuation

To calculate the contribution of ecosystem impacts to the damage cost of a pollutant, one needs, as always, to determine the physical impact per kg of pollutant and the monetary value. For most ecosystem impacts, the required information is not available. Because of the wide variety of different impact types it is difficult to define appropriate indicators of the damage. At least for forests one can use the number or proportion of healthy trees as a simple proxy, although even that may be too crude, because it does not account for the contribution of individual species, to say nothing of possible impacts on the fauna.

Land use can also have serious environmental impacts. Open-pit mines are an obvious example, in many cases with horrendous local impacts. Even if the mine operator is required by law to restore the land after closure of the mine, such restoration is likely to be far from perfect and full recovery may take many decades or even longer, depending on climate and geology. Another type of land use impact occurs when a road cuts an existing ecosystem into two separate parts. Since the number of species in an ecosystem increases with its size, the total number of species in the two separate systems will end up lower than the original.

Even if one could determine how many members of each species type are lost due to the emission of a pollutant or due to land use change, how should one value it in monetary terms? Initially, some economists carried out contingent valuation (CV) studies of species preservation by asking questions such as "how much would you be willing to pay to avoid the loss of salmon in the Columbia River?" or "how much ... for the loss of all fish in the Columbia River?" or "how much ... for the loss of all fish in the United States?" But if one asks one question at a time, rather than all three questions in the same questionnaire, it turns out that the willingness to pay (WTP) is almost the same, regardless of the scope of the impact. It looks as though people have in their mind a certain amount that they are willing to donate for worthy causes, such as the protection of the environment, and that's what they respond with, without any attempt to allocate such a lump sum to specific items. In a classic paper, Kahneman and Knetsch (1992) examined the role of CV for the valuation of public goods and concluded that CV responses reflect the willingness to pay for the moral satisfaction of contributing to public goods, not the economic value of these goods. Some of the early literature on species valuation was reviewed for ExternE (1995). A clear problem identified there was how to aggregate results. Given a typical valuation of $5/person/species/year, one immediately hits problems when multiplying by the number of species that this could hypothetically be applied to. There is also the question of applying responses to marginal changes when they simply deal with the presence or absence of a species.

Preservation of species and biodiversity are extremely difficult to value directly because they lack meaning. Sure, one can do a CV. But what do the results mean when the interviewees do not understand what the loss of biodiversity implies? Preservation of mosquitoes would get a much lower vote than mammals. But many mammals would suffer if mosquitoes disappeared, because the latter are part of the food chain of the former. If even the physical effects cannot be quantified adequately, how can economists provide meaningful monetary values? In an interesting review of the valuation literature, Nunes and van den Bergh (2001) conclude that the available estimates made to that time provide only a very incomplete perspective on biodiversity changes.

A precondition for obtaining reliable valuations directly from the economics literature is that the individuals affected by a change have a clear understanding of its consequences and can express their preferences. For the valuation of nature preserves, a classic approach is the travel cost method. The basic idea is that the total amount that people spend (transportation, food, lodging, entrance fees, value of the time needed for traveling, . . .) to visit the site in question is a measure of the utility they derive from the existence of the site. The required analysis can be quite complex, for example, it has to take into account the existence of alternative sites; but if done right, the results are appropriate for existing sites. Of course, in practice one is usually more interested in proposed new sites and the extrapolation from existing sites can be problematic.

For certain species of interest to people who like to hunt or fish, one can do a CV by, for example, asking hunters how much they are willing to pay for the possibility to hunt Canada geese (one could also ask how much they would have to be paid to give up such a possibility – and for the interpretation there are the usual questions about the appropriateness of

WTP versus willingness to accept values). Such CV studies can give a meaningful lower bound for the value of preserving the species in question; it is a lower bound because it includes only the value for hunters, not the value of the species for the general population. CV studies are also meaningful for certain environmental interventions that are under consideration, for example, the creation of new nature preserves or changes in the management of reservoirs. Here the respondents can understand the consequences, if well explained, and express their preferences. Again the results may be only a lower bound, if there are benefits beyond the ones the respondents take into account. For instance, they might appreciate a new nature reserve just for its hiking trails, without recognizing the biodiversity benefits. Likewise, CV studies of the affected population can be appropriate for the valuation of damages due to oil spills. However, whilst the results of such a survey tell us something about values, it is hard to see how they could be applied in situations beyond somewhat localized decision-making (e.g. in respect of an application to develop a hydroelectric scheme), declaration of an area for nature preservation, etc. Results are not well suited to addressing the problems of pollution occurring at larger scales.

We present in Section 5.5.3 a CBA for the abatement of eutrophication in the Baltic Sea, to illustrate the type of environmental modeling that is required and the use of CV results. In this case the CV respondents can correctly evaluate the most visible benefit, namely improved clarity of the water. The fact that here too, the ecosystem benefits are probably underestimated does not affect the conclusion of the CBA because the abatement costs are low enough to justify the proposed changes in any case.

Some studies mention restoration costs as an alternative valuation method. This can create confusion, because abatement costs are not damage costs: they appear on different sides of the cost–benefit equation. Nevertheless, in some situations the restoration cost of ecosystems can be used for valuation, to ensure consistency of policy choices. Specifically, if governments have implemented a variety of ecosystem restoration projects, they reflect implicit valuations: the restoration cost is paid because it is lower than the perceived benefit. Thus the cost of an implemented restoration project is a lower bound of the collective valuation of the benefit of this project. An assumption implicit in the use of restoration costs is, of course, that there is a willingness to pay for restoration. In Section 5.5.4 we describe an attempt at using such costs.

The concept of valuation against "ecosystem services" has gained much attention in recent years. In a sense, this is simply an extension of some of the methods described above that deal with the production values of agriculture and forestry. However, it goes substantially further than that, addressing the question of why we value non-productive and indirectly productive functions of ecosystems as well, adding in the broad categories of cultural, regulatory and supporting services. Cultural services reflect a broad range of functions from recreational value (e.g. for hiking or fishing) through to non-use values (i.e. the value that one may associate with the largely unspoiled wilderness of the Antarctic). Regulatory services deal with the benefits we derive from ecosystems in terms of regulation of (for example) the water and carbon cycles. Supporting services concern functions such as soil formation and pollination. The ecosystem services approach is discussed in more depth at the end of the chapter.

5.5.2   Ecosystem impacts of pollution

Local impacts may be appreciable for air or water pollution from coal mines, from old power plants without flue gas treatment, and from poorly managed waste sites. But exposure to air pollution from the normal operation of modern power plants does not seem to have significant direct impacts on ecosystems (as opposed to impacts via acidification and eutrophication by sulfates and nitrates). At first glance, this claim may appear strange since human health impacts are significant and one might indeed expect similar impacts on animals as on humans. The explanation lies in what we value: we value human impacts as changes in the well-being of individuals. Concentrations of air pollutants are generally so small that the incremental mortality is at most a small percentage of the natural rate. Furthermore, most of the deaths from air pollution occur among individuals well beyond reproductive age. If a small percentage of animals die prematurely after having produced and raised offspring, the effect on the ecosystem is negligible. But if any human dies prematurely, we care a great deal.

Similar remarks can be made about the ecosystem impacts of nuclear radiation. Near Chernobyl, there have been some temporary ecosystem impacts in areas where radiation doses have been extremely high, but after ten years there was no firm evidence of long-term ecosystem impacts due to radiation (Dreicer et al., 1995). Subsequent analysis (discussed by Evans (2011)) shows dissent amongst ecologists working in the area, though this could simply reflect differential sensitivity between species. Some species that are threatened elsewhere may do well thanks to the lack of human intervention in the exclusion area around the plant, whilst others do no less well, either due to the radiation or, conceivably, the lack of human presence. Radionuclide exposures due to modern power plants in normal operation are incomparably lower than those in the vicinity of the Chernobyl accident, and any ecosystem impacts are entirely negligible (even the radiation exposure due to a hypothetical all-nuclear scenario would be about two orders of magnitude lower than the dose from cosmic radiation, see Section 13.5.1).

The situation is different for acid rain, which can cause forest damage in some regions, and aquatic impacts: acidification and eutrophication of rivers and lakes has been shown to be detrimental to aquatic life. A river can collect much of the air pollution from a large region, leading to relatively high concentrations in the water. Whereas acidification has been greatly reduced in the EU and North America thanks to desulfurization of power plants, eutrophication remains a serious problem with much of Europe in excess of the critical load for nutrient N (IIASA, 2010). The effect of nutrient N deposition is rather subtle. Unlike acid rain it does not lead to the obvious death of trees or fish, as was observed during the 1970s and 1980s. The tendency is for species adapted to low nitrogen habitats to be out-competed by plants that require higher levels of nitrogen. In some locations it may appear that ecosystems are healthier than before through the presence of lush green grass and other dense vegetation cover. However, the changed ecosystem is essentially an artificial product of human activity, and one that, overall, reduces the variety of types of ecosystems at local through to global levels.

5.5.3   Eutrophication in the Baltic Sea

Bryhn (2009) presents a very interesting analysis of policy measures to reduce eutrophication of the Baltic Sea, with the goal of restoring the clarity of the water to the conditions before 1960. The water clarity is measured in terms of the Secchi depth.[2] At the present time, it is about 5.5 m, and the goal is to increase it to about 8 m. The culprit of this degradation is excessive P input, especially from waste water and from agriculture. Bryhn argues that since N abatement is costly and would currently have very unpredictable effects and may actually favour cyanobacterial competitiveness instead of water quality, policy measures should instead focus on reducing the TP (total P) input. The relation between nutrient input, algal blooms (cyanobacteria) and the resulting Secchi depth is modeled with the CoastMab model of Håkanson and Bryhn (2008). A key modeling result is reproduced here in Fig. 5.2 (the graphs show two scenarios, a reduction of 6650 tonne/yr and one of 10,200

---

[2] Secchi depth is a measure of the transparency of water. A circular disk with black–white pattern is lowered into the water and the Secchi depth is the depth where the pattern is no longer visible from the surface.

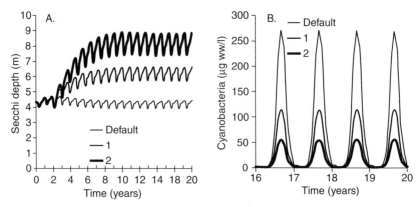

Fig. 5.2 Modelled Secchi depths and cyanobacterial concentrations at late 1990s TP loadings (default), during a 6650 tonne decrease of the annual TP loading after two years (scenario 1) and during a 10,200 tonne decrease of the annual TP loading after two years (scenario 2). TP = total P. From Bryhn (2009), with kind permission.

tonne/yr of TP; here we consider only scenario 2, because it corresponds with the monetary valuation of the benefit).

Bryhn examines in detail the policy options for which cost and performance data are available and identifies those that are most cost-effective for achieving the goal of increasing the Secchi depth from 5.5 m to 8 m. This can be achieved by reducing the input of TP into the Baltic by 10,200 tonne/yr (scenario 2 of Fig. 5.2). He estimates that the total abatement cost would "probably not exceed 0.43 billion euro per year." Finally, Bryhn looks at the valuation literature to find estimates of the monetary benefit of such an improvement of the Baltic ecosystem. As the most relevant study, he finds that by Gren (2001), who estimates 3.6 billion euro per year as the annual basin-wide willingness to pay, for restoring the ecological state of the Baltic Sea to conditions prior to the 1960s (from Gren, 2001, adjusted to 2008 prices).

If these estimates of costs and benefits are correct, the course of action is clear beyond doubt: these measures should be implemented because the benefit is more than eight times larger than the cost. Thus the question for the analysis of the uncertainties becomes: could the uncertainties be large enough to change the conclusion? The benefit/cost ratio could become less than unity if the abatement cost is much larger than 0.43 billion €, or the benefit much smaller than 3.6 billion €. (This is an important general issue: scientists may worry a great deal about the uncertainties in their analysis, and the claim is frequently made that some data set or analysis is too uncertain to be used in decision making. However, if it is the best information available for an active policy debate, the data should be used. In the present case, it is quite possible that the perceived major uncertainty could have no effect on the recommendations arising from the analysis.)

On the side of abatement costs there are mainly three sources of uncertainty:

- Uncertainties in the ecosystem model imply uncertainties in the amount of the required P-load reduction;
- Uncertainties in the performance of the various abatement measures imply uncertainties in the number of measures that are needed;
- And finally the cost of each abatement measure is uncertain.

Indications of the uncertainties in the ecosystem model can be gleaned from comparisons between the model and the data, in particular those of Håkanson and Bryhn (2008), shown here in Fig. 5.3. The graphs in that figure indicate satisfactory agreement. Note that there are comparable uncertainties in the model output and in the data themselves; both contribute to uncertainties in the total abatement cost. Such uncertainties imply that the amount of abatement for the specified goal may have to be increased. The total abatement cost would increase more than proportionally to the TP decrease, because of increasing marginal costs. Visual inspection of Fig. 5.3 suggests, very roughly, that the prediction of Secchi depths may have standard deviations on the order of 1 to 2 m (say 1.75 m). Since scenario 2 corresponds to a reduction of 10,200 tonne/yr and a Secchi depth increase of about 3.5 m, with approximate proportionality of TP reduction and Secchi depth increase, one can estimate the one-standard-deviation upper limit of the required reduction as 10,200 tonne/yr × (1 + 1.75/3.5) = 15,300 tonne/yr, i.e. a 50% increase. The total abatement potential of the cost-effective measures, namely improved sewage treatment and phosphate-free detergents, in Table 2 of Bryhn (reproduced here as Table 5.6), is 15,600 tonne/yr, so the total abatement cost would not increase by much more than about 50% if the required amount is one standard deviation higher, and the required cost would be about 0.43 billion € ×1.5 = 0.65 billion €, still far below the estimated benefit. Assuming approximately normal error distributions (a plausible assumption), the probability of exceeding one standard deviation is about 16%. Even though the uncertainty of the total abatement cost is larger because of questions about the performance of the various measures, these considerations indicate that uncertainties in the abatement cost estimate are very unlikely to reduce the benefit/cost ratio below unity.

Table 5.6 *Possible measures for decreasing the phosphorus input to the Baltic Sea.*
From Bryhn (2009), with permission.

Measure	Area	Tonnes phosphorus/year
Sewage treatment	Poland	5,332
Sewage treatment	Russia	3,844
Sewage treatment	Belarus	1,984
Sewage treatment	Baltic States	992
Sewage treatment	Czech Rep.	372
Phosphate-free detergents[a]	All	3,100
Agriculture	All	5,600
Rural sewage treatment	Sweden	175
Mussel cultivation	Sweden	35
Dams	Sweden	10
Protective zones	Sweden	7.5
Wetlands	Sweden	4.3
Soil drainage	Sweden	2.6

[a] In combination with sewage treatment.

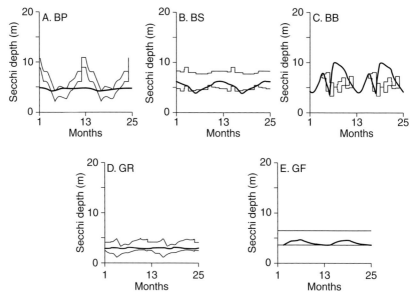

Fig. 5.3 Model results of Håkanson and Bryhn for monthly data on Secchi depths (thick lines) against plus/minus one standard deviation of the mean empirical monthly data (thin lines) for the five sub-basins in the Baltic Sea: Baltic Proper (BP), Bothnian Sea (BS), Bothnian Bay (BB), Gulf of Finland (GF), Gulf of Riga (GR). With thanks to Lars Håkanson for kindly providing the original figures.

By contrast, the uncertainty of the benefit is much greater, in relative terms, than that of the abatement cost, because the benefit is based on a contingent valuation (CV), a method with potentially large uncertainties. The details of the monetary valuation used by Bryhn are described in Gren (2001). Gren reviews the available CV studies and bases her estimates on the two she considers most reliable; one was done in Sweden, the other in Poland. Based on discussions with CV experts, as well as surveys of the CV literature (see Chapter 15), we estimate that typical geometric standard deviations σ_g are about 2 to 3. The probability distributions of the nonzero responses are usually not too far from lognormal. Gren transfers the Swedish CV results to Finland, Germany and Denmark. and the Polish results to Estonia, Latvia, Lithuania and Russia. Such transfer entails an additional uncertainty, in this case with geometric standard deviation of about 1.3 for countries with fairly similar GDP/ capita, but since terms with relatively small σ_{gi} make a negligible contribution to the total geometric standard deviation, the resulting geometric standard deviation is still about 2 to 3.

However, a major additional source of uncertainty lies in the good that was valued in the original CV. Since the valuation question in those studies concerned an overall improvement to bring the Baltic Sea back to pre-1960 condition in all aspects, the clarity of the water by itself was only one component of the good. It is certainly a large component, but not the entirety. Thus the value of water clarity is only a fraction of the 3.6 billion €/year that Bryhn obtains from Gren's paper. That fraction might be in the range of one- to two-thirds, the rest of that amount being for removal of floating wastes, for healthier water for swimming, etc. But even if the benefit were only 1.2 billion €/year and the abatement cost were twice as large as Bryhn's estimate, it would still be sufficient to justify implementation of the policy. The probability of a benefit/cost ratio below unity seems small.

5.5.4   An attempt at valuing loss of species

As part of the NEEDS phase of the ExternE series, an attempt was made to calculate the cost of damage to ecosystems due to the emission of NO_x and SO_2 (Ott et al., 2006). Here we sketch their approach, without going into very much detail, because several of the key references are not peer reviewed. A central concept in this approach is the potentially disappeared fraction (PDF) of species when an ecosystem is degraded, due to land use change or due to pollution. The PDF is the relative difference between number of species S_{ref} under reference conditions and number of species S under the conditions that are to be evaluated,

Table 5.7 *Species number and potentially disappeared fractions (PDF) for different land uses. Adapted from Ott* et al. *(2006).*

CORINE category[a]	Type	Number of species per m^2	PDF relative to Swiss Lowlands[b]
10	Built up land	1	0.97
2112	Integrated arable	7	0.82
142	Sport facilities	7	0.83
111	Continuous urban	8	0.8
2111	Conventional arable	10	0.74
2311	Intensive meadow	17	0.58
322	Heath land	18	0.56
412	Peat bog	19	0.53
2312	Less intensive meadow	19	0.53
112	Discontinuous urban	22	0.45
3112	Semi-natural broad-leafed forest (arid)	23	0.43
2212	Organic orchard	23	0.41
3112	Semi-natural broad-leafed forest (moist)	24	0.41
1212	Industrial area, with vegetation	24	0.39
114	Rural settlement	25	0.38
2113	Organic arable	26	0.35
141	Green urban	29	0.27
134	Mining fallow	38	0.04
321	Natural grassland	39	0.02
125	Industrial fallow	40	−0.01
2115	Agricultural fallow	43	−0.09
325	Hedgerows	44	−0.12
2313	Organic meadow	45	−0.14
314	Forest edge	48	−0.20
245	Agricultural fallow with hedgerows	53	−0.34
	Swiss Lowlands[b]	40	0

[a] CORINE categories with four digit numbers are an extension of the classification introduced by Koellner (2001). Information on CORINE is available at the European Environment Agency www.eea.europa.eu/publications/COR0-landcover.
[b] The area of the reference category Swiss Lowlands consists of 8.2% high intensity forest, 17.6% low intensity forest, 52.8% high intensity agriculture, 9.3% low intensity agriculture, 1.3% lakes, 0.3% non use, 5.8% high intensity artificial and 4.8% low intensity artificial.

$$PDF = 1 - S/S_{ref}. \qquad (5.8)$$

As an illustration, Table 5.7 shows the species number and potentially disappeared fractions (PDF) for different land use types, as derived from various empirical studies for Germany and Switzerland. Only plant species are included. For example, conversion of Swiss Lowland to conventional arable land involves a loss of 74% of the species, conversion of Swiss Lowland to conventional organic meadows, a gain of 14%. To apply such data to calculating the effect of a land use change, one should also take into account that such a change affects not only the converted land itself, but also the surrounding areas. Therefore Goedkoop and Spriensma (2000) recommend the formula,

$$PDF_{use1-use2} = (1 + b)(PDF_2 - PDF_1), \qquad (5.9)$$

where b is the species accumulation factor. Relative to Swiss conditions, Ott *et al.* use a value of b = 0.2.

From surveys of restoration projects in Germany, Ott *et al.* have estimated restoration costs for a wide range of land use conversions. Lacking data for other countries, Ott *et al.* transfer the German values by adjusting for income and purchasing power. An extract of their results is presented in Table 5.8.

Ott *et al.* have also used these restoration costs to calculate the damage cost due to terrestrial acidification and eutrophication caused by NO_x, SO_2 and NH_3 emissions in the EU. That requires, in addition, a model for the physical impacts, i.e. the PDF due to such emissions. Such a model has been developed for the Netherlands. It is called Natuurplanner (Latour *et al.*, 1997) and it has been used by Goedkoop and Spriensma (2000) to calculate PDF changes per kg of such emissions. We will not present any detail because Natuurplanner is limited to the Netherlands

Table 5.8 *Restoration costs per m^2 for different target biotopes for EU25 in 2004, when starting biotope = Built Up Land. From Ott* et al. *(2006).*

Land use category	CORINE category	Restoration costs, $€/m^2$
Integrated arable	2112	0.17
Organic arable	2113	0.42
Organic orchards	2222	4.06
Intensive pasture and meadows	2311	0.34
Less intensive pasture and meadows	2312	0.76
Organic pasture and meadows	2313	1.9
Broad-leafed forest	311	2.66
Plantation forest	312	2.66
Forest edge	314	8.39
Average		1.52

and we do not have confidence in the extrapolation to emissions and affected areas in other countries.

The use of restoration costs is questionable for most applications. The following questions arise:

- What form does restoration take, and is it relevant to the impact under investigation?
- How good is restoration? To what extent are ecosystem services restored to their original state? Is it possible even to identify an "original state"?
- How permanent is restoration? In the case of air pollution effects it is possible that repeated restorations (i.e. continued active management) may be needed.
- Is anyone actually willing to pay for restoration?
- And in relation specifically to the work of Ott *et al.*, to what extent should we rely on their limited evidence base, which is focused on the Swiss lowlands?

However, for some applications it may provide a way forward. A case in point could be Natura 2000 areas, which are legally protected under EU regulations. This could naturally lead to an assumption that there is indeed a willingness to pay for restoration, although whether the implicit assumptions of Ott *et al.* on the form of restoration are relevant remains a barrier to the application.

5.5.5   The ecosystem services approach

As already noted, the ecosystem services approach is attracting much interest at the present time, as offering the potential for a more systematic appraisal of the value of ecosystems. An early example was provided by Costanza *et al.* (1997), who sought to evaluate the total value of the services provided by the world's ecosystems. Based on a systematic review of the literature, Costanza *et al.* estimated that the annual value of these services in 1994 was $16–54 trillion, with an estimated average of $33 trillion; see Table 5.9. That is more than the total world GDP at that time, $18 trillion. Only renewable services were counted; the provision of minerals and other non-renewable resources were excluded. The evaluation did not provide any direct information on external costs, especially since it does not consider any link between marginal costs and pollution, but it did suggest that ecosystem damage due to pollution could entail very high costs.

More recently, the ecosystem services approach has been the focus of concerted research effort and policy analyses that have highlighted and attempted to measure the contribution of ecosystems and the biological diversity contained within them to individual and social well-being. In this

Table 5.9 *Annual value, in $\$_{1994}/yr \times 10^9$, of the world's ecosystem services and functions, as estimated by Costanza et al. (1997). Reprinted by permission from Macmillan Publishers Ltd (Nature, May 15, 1997).*

Ecosystem service	Ecosystem functions	Examples	Value
Gas regulation	Regulation of atmospheric chemical composition.	CO_2/O_2 balance, O_3 for UVB protection, and SO_x levels.	1,341
Climate regulation	Regulation of global temperature, precipitation, and other biologically mediated climatic processes at global or local levels.	Greenhouse gas regulation, DMS production affecting cloud formation.	684
Disturbance regulation	Capacitance, damping and integrity of ecosystem response to environmental fluctuations.	Storm protection, flood control, drought recovery and other aspects of habitat response to environmental variability, mainly controlled by vegetation structure.	1,779
Water regulation	Regulation of hydrological flows.	Provisioning of water for agricultural (such as irrigation) or industrial (such as milling) processes or transportation.	1,115
Water supply	Storage and retention of water.	Provisioning of water by watersheds, reservoirs and aquifers.	1,692
Erosion control and sediment retention	Retention of soil within an ecosystem.	Prevention of loss of soil by wind, run-off, or other removal processes, storage of silt in lakes and wetlands.	576
Soil formation	Soil formation processes.	Weathering of rock and the accumulation of organic material.	53
Nutrient cycling	Storage, internal cycling, processing and acquisition of nutrients.	Nitrogen fixation, N, P and other elemental or nutrient cycles.	17,075
Waste treatment	Recovery of mobile nutrients and removal or breakdown of excess or xenic nutrients and compounds.	Waste treatment, pollution control, detoxification.	2,277
Pollination	Movement of floral gametes.	Provisioning of pollinators for the reproduction of plant populations.	117
Biological control	Trophic-dynamic regulations of populations.	Keystone predator control of prey species, reduction of herbivory by top predators.	417
Refugia	Habitat for resident and transient populations.	Nurseries, habitat for migratory species, regional habitats for locally harvested species, or overwintering grounds.	124

Table 5.9 (*cont.*)

Ecosystem service	Ecosystem functions	Examples	Value
Food production	That portion of gross primary production extractable as food.	Production of fish, game, crops, nuts, fruits by hunting, gathering, subsistence farming or fishing.	1,386
Raw materials	That portion of gross primary production extractable as raw materials.	The production of lumber, fuel or fodder.	721
Genetic resources	Sources of unique biological materials and products.	Medicine, products for materials science, genes for resistance to plant pathogens and crop pests, ornamental species (pets and horticultural varieties of plants).	79
Recreation	Providing opportunities for recreational activities.	Eco-tourism, sport fishing, and other outdoor recreational activities.	815
Cultural	Providing opportunities for non-commercial uses.	Aesthetic, artistic, educational, spiritual, and/or scientific values of ecosystems.	3,015
Total value			**33,268**

sense, the term "ecosystem services approach" has come to describe a basis for analyzing how individuals and human systems are dependent upon the condition of the natural environment. The discussion that follows focuses on a report by Jones *et al.* (2012) for the UK government, which investigated the problem specifically from the perspective of quantifying damage from regional air pollutants. Ecosystem services are allocated to four categories: Provisioning (production), Regulating, Cultural and Supporting.

In practice, there is no single ecosystem services approach (ESA) or framework, and different interpretations of the approach are taken. This is to be expected in part because the ESA is a relatively new technique, and in part because it has attracted a wide range of applications. While numerous research initiatives have been undertaken, it is widely recognized that the key contribution in developing a high profile systematic account of ecosystem services was provided by the UN Millennium Ecosystem Assessment (MEA, 2005). Subsequent studies have sought to improve understanding, refine concepts and develop practical applications of ecosystem services approaches. Recent major initiatives include "The Economics of Ecosystems and Biodiversity" (TEEB: http://www.teeweb.org) and the "UK National Ecosystem Assessment" (UK NEA). A substantial academic literature has also developed, particularly with respect to the role of economic analysis within the ecosystem services approach. The Ecosystem Services Approach (ESA), as defined by Defra (2007), identifies three key steps:

- Identifying which ecosystem services will be affected by a policy decision.
- Prioritizing those ecosystem services, including consideration of environmental limits, designated sites and species and other regulatory factors.
- Valuing the benefits obtained from those ecosystem services, and which will be affected by a policy decision.

This may appear to be an obvious statement of method. However, it presents a clearer structure than is evident from some of the cases referred to above.

From an economic perspective, ecosystem services represent flows of economic value that are generated by stocks of ecological assets, i.e. "natural capital" (Bateman *et al.*, 2011). For example, timber represents a flow of benefits that can be realized from forests. Numerous ecological functions and processes (e.g. nutrient cycling) influence the types of ecosystem services that are derived from ecosystems. Various ecosystem services contribute to the production of market and non-market goods, the consumption of which generates human "well-being". The UK NEA makes a particular distinction between a "good" – items that generate well-being – and a "benefit". In particular, the term benefit is applied to changes in the well-being that are generated by the consumption of goods, the economic value of which can be context-specific and dependent on factors such as spatial location and timing. Benefits derived from ecosystem services are also distinguished by the familiar typology of "total economic value" (TEV), which recognizes that they can be attributed to use and/or non-use values.

Jones *et al.* (2012) placed much emphasis on the use of value transfer, based around the DEFRA value transfer guidelines (eftec, 2010). These set out eight steps, establishing the nature of physical impacts on the environment and identifying and applying appropriate economic valuation evidence. They focused on seven market and non-market goods provided by final ecosystem services, which were selected from the UK NEA (2011) list of final services (Table 5.10).

The analysis necessarily considered the provision of the selected goods across a range of UK habitat types, as appropriate to the ecosystem service: enclosed farmland (arable, horticultural and improved pastures); semi-natural unimproved grassland (acidic, calcareous and neutral grasslands, bracken); woodlands (managed coniferous and broadleaved, and unmanaged); mountains, moors and heathlands (bogs and dwarf shrub heaths) and freshwaters (streams, rivers and lakes). Other habitat types including urban were included where relevant for some services.

The analysis highlighted a number of issues, of which we highlight just two:

(1) Quantification of the selected ecosystem services was possible. As a preliminary analysis the work therefore met its immediate objectives. However, it was noted that some of the estimates provided were of significant or even unknown uncertainty. However, a lesson from ExternE is that once an impact pathway can be implemented, it is possible to identify its strengths and weaknesses and target future research accordingly.

(2) It is easy to develop a sense from a small set of economic damage estimates that analysis is far more complete than it actually is, with respect to the range of impacts considered, the range of ecosystems addressed, and the geographic and temporal scope of the analysis. This requires the development of a framework that conveys equally what has been quantified and what has not. Ideally, this should extend to some at least qualitative assessment of which of the unquantified effects are likely to be significant and which are not.

(3) Following from this (and reinforcing the point), it was noted that some of the easiest effects to quantify were, ironically, those that suggest a benefit from pollution, for example through increased carbon sequestration arising from nitrogen deposition.

Table 5.10 *Final ecosystem services and goods provided by those services. Goods selected by Jones et al. (2012) for assessment are highlighted in grey. (P) Provisioning, (R) Regulating, (C) Cultural, (S) Supporting service.*

Final ecosystem services
(P) Food
Crops
Livestock: meat & dairy
Game (grouse, venison)
Wild food (fungi)
(P) Fibre and fuel
Wool
Timber
Peat extraction
(P) Genetic resources
Genetic diversity of wild species
(P) Water supply
Drinking water
(R) Equable climate
C stocks in vegetation
Net C sequestration
Other GHG emissions
(R) Purification
Clean air
Clean water
(R) Hazard regulation
Reduced flooding (rivers)
Reduced flooding (coastal)
(C) Leisure, recreation, amenity
Recreational fishing
Leisure activities
Aesthetic appreciation of natural environment
Appreciation of biodiversity
(S) Intermediate services
Pollination

The ESA is being taken forward in a number of research studies at the present time. One example of specific interest here concerns the ECLAIRE Project,[3] funded by the European Commission, where one component seeks to provide quantification specifically in relation to effects of air pollution under a changing climate. In addition to working

---

[3] ECLAIRE: effects of climate change on air pollution impacts and response strategies for European ecosystems.

through the impact pathway for various impacts, the work involves development of a framework specifically to ensure that information on what has been quantified and what has been omitted from analysis is communicated. It is also being performed at a European level, to test the extent to which methods are transferable between countries.

References

Baker, C. K., Colls, J. J., Fullwood, A. E. and Seaton, G. G. R. 1986. Depression of growth and yield in winter barley exposed to sulphur dioxide in the field. *New Phytologist* **104**: 233–241.

Bateman, I. J., Mace, G. M., Fezzi, C., Atkinson, G. and Turner, R. K. 2011. Economic analysis for ecosystem service assessments, *Environmental and Resource Economics* **48** (2): 177–218.

Bolsinger, M. and Flukiger, W. 1989. Amino-acids changes by air pollution and aphid infestation. *Environ. Pollut* **56**: 209–216.

Bryhn, A. C. 2009. Sustainable phosphorus loadings from effective, cost-effective and feasible phosphorus management around the Baltic Sea. Uppsala University, Department of Earth Sciences, Villavägen **16**, 75236 Uppsala, Sweden.

Costanza, R., d'Arge, R., de Groot, R. *et al.* 1997. The value of the world's ecosystem services and natural capital. *Nature* **387**: 253–260.

DEFRA. 2007. *Introductory Guide to Valuing Ecosystem Services*, Department for Environment, Food and Rural Affairs, December 2007. http://ec.europa.eu/environment/nature/biodiversity/economics/pdf/valuing_ecosystems.pdf.

DEFRA. 2010. *Fertiliser Manual RB209*, 8th Edition. Section 1: Principles of nutrient management and fertiliser use D2A: Nitrogen (N) for field crops. UK Department for Environment, Food and Rural Affairs. http://adlib.everysite.co.uk/adlib/defra/content.aspx?id=2RRVTHNXTS.88UEOT33RA.

Dreicer, M., Tort, V. and Margerie, H. 1995. *Nuclear fuel cycle: implementation in France*. Final report for ExternE Program, contract EC DG12 JOU2-CT92–0236. CEPN, F-92263 Fontenay-aux-Roses.

EC 1998. Council Directive 98/83/EC of 3 November 1998 on the quality of water intended for human consumption. http://europa.eu.int/eur-lex/en/index.html

eftec 2010. Valuing environmental impacts: practical guidelines for the use of value transfer in policy and project appraisal. Technical report submitted to UK Department for Environment, Food and Rural Affairs.

Evans, P. 2011. Wildlife in the dead zone. *Geographical* www.geographical.co.uk/Magazine/Chernobyl_-_Apr_11.html.

ExternE. 1995. ExternE: *Externalities of Energy Volume 2: Methodology*. European Commission report number EUR16521EN. www.externe.info/externe_d7/?q=node/37

ExternE. 2005. ExternE: Externalities of Energy. Methodology, 2005 update. www.externe.info/externe_d7/?q=node/30.

ExternE. 2008. With this reference we cite the methodology and results of the NEEDS (2004–2008) and CASES (2006–2008) phases of ExternE. For the damage costs per kg of pollutant and per kWh of electricity we cite the numbers of the data CD that is included in the book edited by Markandya, A., Bigano, A.

and Roberto Porchia, R. in 2010: *The Social Cost of Electricity: Scenarios and Policy Implications*. Edward Elgar Publishing Ltd, Cheltenham, UK.

Fantke, P., Charles, R., de Alencastro, L. F., Friedrich, R. and Jolliet, O. 2011a. Plant uptake of pesticides and human health: Dynamic modeling of residues in wheat and ingestion intake. *Chemosphere* **85**: 1639–1647.

Fantke, P., Juraske, R., Antón, A., Friedrich, R. and Jolliet, O. 2011b. Dynamic multicrop model to characterize impacts of pesticides in food. *Environ. Sci. Technol.* **45**: 8842–8849.

Fantke, P., Friedrich, R. and Jolliet, O. 2012a. Health impact and damage cost assessment of pesticides in Europe. *Environ. Int.* **49**: 9–17.

Fantke, P., Wieland, P., Juraske, R., *et al.* 2012b. Parameterization models for pesticide exposure via crop consumption. *Environ. Sci. Technol.* **46** (23): 12864–12872.

Farrar, J. F., Relton, J. and Rutter, J. 1977. Sulphur dioxide and the scarcity of Pinus sylvestris in the industrial Pennines. *Environmental Pollution* **14**: 63–68.

Fuhrer, J. 1996. *The critical level for effects of ozone on crops and the transfer to mapping. Testing and Finalising the Concepts*. UN-ECE Workshop, Department of Ecology and Environmental Science, University of Kuopio, Kuopio, Finland, 15–17 April.

Goedkoop, M. and Spriensma, R. 2000. *The Eco-indicator 99. A damage oriented method for Life Cycle Impact Assessment*. Methodology Report, Pre Consultants, Amersfoort, NL.

Gren, I.-M. 2001. International versus national actions against nitrogen pollution of the Baltic Sea. *Environmental and Resource Economics* **20**: 41–59.

Gruncharov, I. *et al.* 2000. Final report on the environmental impact assessment of NEOCHIM-Plc, Dimitrovgrad. See also www.neochim.bg/.

Gupta, S. K., Gupta, R. C., Seth, A. K. *et al.* 1999. Adaptation of cytochrome-b5 reductase activity and methaemoglobinaemia in areas with a high nitrate concentration in drinking-water. *Bulletin of the World Health Organization* **77** (9).

Håkanson, L. and Bryhn, A. C. 2008. *Eutrophication in the Baltic Sea*. Springer, Berlin/Heidelberg, 261 p.

Harmens, H. and Mills, G. (2012). *Ozone pollution: Impacts on carbon sequestration in Europe*. ICP Vegetation Programme, Coordination Centre. CEH Bangor, UK., ISBN: 978-1-906698-31-7. http://icpvegetation.ceh.ac.uk/publications/documents/Csequestrationreportfinalwebversion.pdf.

Holland, E. A., Braswell, B. H., Sulzman, J. M. and Lamarque, J.-F. 2005. Nitrogen Deposition onto the United States and Western Europe. Data set. Available online [www.daac.ornl.gov] from Oak Ridge National Laboratory Distributed Active Archive Center, Oak Ridge, Tennessee, U.S.A. doi:10.3334/ORNLDAAC/730.

Holland, M. 2013. Cost-benefit Analysis of Policy Scenarios for the Revision of the Thematic Strategy on Air Pollution: Version 1 Corresponding to IIASA TSAP Report #10, Version 1. March 2013. Contract report to European Commission DG Environment.

Huijbregts, M. A. J., Rombouts, L. J. A., Ragas, A. M. J. and van de Meent, D. 2005. Human-toxicological effect and damage factors of carcinogenic and noncarcinogenic chemicals for life cycle impact assessment. *Integr. Environ. Assess. Manage* **1**, 181–244.

IIASA. 2010. *Baseline Emission Projections and Further Cost-effective Reductions of Air Pollution Impacts in Europe – A 2010 Perspective.* International Institute for Applied Systems Analysis, Laxenburg Austria. http://ec.europa.eu/environment/air/pollutants/pdf/nec7.pdf.

IPCC 1995. *Economic and Social Dimensions of Climate Change.* Bruce, J. P., Lee, H. and Haites, E. F. (Eds.) Contribution of Working Group III to the Second Assessment Report of the Intergovernmental Panel on Climate Change. Cambridge University Press, Cambridge, 1995.

Jones, M. L., Provins, A., Harper-Simmonds, L. *et al.* 2012. Using the Ecosystems Services Approach to Value Air Quality. DEFRA Project NE0117. Report to Department for Environment Food and Rural Affairs (DEFRA).

Kahneman, D., and Knetsch, J. L. (1992). Valuing public goods: The purchase of moral satisfaction. *J Environmental Economics and Management* **22**: 57–70.

Karlsson, P. E., Pleijel, H., Belhaj, M. *et al.* 2005. Economic assessment of the negative impacts of ozone on the crop yield and forest production. A case study of the Estate Östads Säteri in southwestern Sweden. *Ambio* **34**: 32–40.

Knabe, W. 1970. Keifernwaldbreitung und Schwefeldioxid Immissionen im Ruhrgebiet. *Staub, Reinhalt. Luft* **39**: 32–35.

Koellner, T. 2001. *Land Use in Product Life Cycles and its Consequences for Ecosystem Quality.* University of St. Gallen, ETH Zürich.

Latour, J. B., Staritsky, I. G., Alkemade, J. R. M. and Wiertz, J. 1997. De Natuurplanner, Decision support system natuur en milieu. RIVM report 71191019.

MEA. 2005. Millennium Ecosystem Assessment. *Ecosystems and Human Wellbeing: Synthesis.* Island Press, Washington, DC. www.maweb.org/en/index.aspx

Mills, G., Holland, M., Buse, A., *et al.* 2003. Introducing response modifying factors into a risk assessment for ozone effects on crops in Europe. In P. E. Karlson, G. Sellden and H. Pleijel: *Establishing Ozone critical levels II*, UNECE Workshop report. IVL Report B 1523. IVL Swedish Environmental Research Institute, Gothenburg.

Mills, G. and Harmens, H. 2011. *Ozone Pollution: A hidden threat to food security.* Report prepared by the ICP Vegetation. September, 2011. ICP Vegetation Programme Coordination Centre, Centre for Ecology and Hydrology, Environment Centre Wales. http://icpvegetation.ceh.ac.uk/publications/documents/ozoneandfoodsecurity-ICPVegetationreport%202011-published.pdf.

Mills, G. and Simpson, D. 2013. Modelling of ozone impacts to crops and forests. Presentation to the 41st meeting of the Task Force on Integrated Assessment Modelling under the UN/ECE Convention on Long Range Transboundary Air Pollution. www.iiasa.ac.at/web/home/research/researchPrograms/Mitigationof AirPollutionandGreenhousegases/7._Mills_Ozone_impacts.pdf.

Mills, G., Wagg, S. and Harmans, H. 2013. Ozone Pollution: Impacts on ecosystem services and biodiversity. Report prepared by the ICP Vegetation. http://icpvegetation.ceh.ac.uk/publications/documents/ICPVegetationozoneecosystemservicesandbiodiversityreport2013.pdf.

NRCC. 1939. *Effects of sulphur dioxide on vegetation*. National Research Council of Canada.

Nunes, P. A. L. D. and van den Bergh, J. C. J. M. 2001. Economic valuation of biodiversity: sense or nonsense? *Ecological Economics* **39**(2): 203–222.

Ott, W., Baur, M., Kaufmann, Y., Frischknecht, R. and Steiner, R. 2006. Assessment of Biodiversity Losses. Deliverable D.4.2.- RS 1b/WP4. Available at www.needs-project.org/RS1b/RS1b_D4.2.pdf

Pretty, J. N., Mason, C. F., Nedwell, D. B. *et al.* 2003. Environmental costs of freshwater eutrophication in England and Wales. *Environmental Science & Technology* **37**(2): 201–208.

Reis, S., Grennfelt, P., Klimont, Z. *et al.* 2012. From acid rain to climate change. *Science* **338**.

Riemer, J. and Whittaker, J. B. 1989. Air pollution and insect herbivores: Observed interactions and possible mechanisms. *Insect Plant Interactions* **1**: 73–105.

Sutton, M. A., Bleeker, A., Howard, C. M. *et al.* 2013. *Our Nutrient World. The challenge to produce more food and energy with less pollution*. Prepared by the Global Partnership on Nutrient Management in collaboration with the International Nitrogen Initiative. http://initrogen.org/uploads/rte/ONW.pdf.

UK NEA 2011. UK National Ecosystem Assessment. Synthesis of key findings and technical report. http://uknea.unep-wcmc.org/Resources/tabid/82/Default.aspx.

van Grinsven, H. J. M., Rabl, A. and de Kok, T. M. 2010. Estimation of incidence and social cost of colon cancer due to nitrate in drinking water in the EU: a tentative cost-benefit assessment. *Environmental Health* **9**: 58 (12 p).

von Blottnitz, H., Rabl, A., Boiadjiev, D., Taylor, T. and Arnold, S. 2006. Damage costs of nitrogen fertilizer in Europe and their internalization. *J. Environmental Planning and Management* **49**(3): 413–433, May 2006.

Wittig, V. E. and Ainsworth, E. A. 2007. To what extent do current and projected increases in surface ozone affect photosynthesis and stomatal conductance of trees? A meta-analytic review of the last 3 decades of experiments. *Plant Cell and Environment* **30**(9): 1150–1162.

Wittig, V. E. and Ainsworth, E. A. 2009. Quantifying the impact of current and future tropospheric ozone on tree biomass, growth, physiology and biochemistry: a quantitative meta-analysis. *Global Change Biology* **15**(2): 396–424.

Wood, S. and Cowie, A. 2004. A review of greenhouse gas emission factors for fertilizer production, *IEA Bioenergy Task* **38**.

6   Other impacts

Summary

In this chapter we address several further impacts sometimes considered in environmental assessments:
- Visibility,
- Noise,
- Traffic congestion,
- Depletion of non-renewable resources,
- Accidents.

Of these only loss of visibility is due to air pollution. Nevertheless we will comment briefly on the other impacts, because they can be relevant as ancillary benefits for the cost–benefit analysis of some pollution abatement measures.

The chapter begins with a brief discussion of the loss of visibility due to air pollution, an impact that has been found to contribute several percent to the cost of air pollution in the USA, but that has not yet been evaluated in Europe. For noise and congestion we cite some typical values. Noise and traffic congestion clearly impose external costs. For some of the other categories it is not clear to what extent they are external costs. We do not discuss the question of externalities related to employment or energy supply security. But we do have a section on depletion of non-renewable resources, because that category is important in most LCA methods.

6.1   Visibility

The most visible impact of air pollution is a loss of visibility. Three indicators are in common use for measuring visibility: standard visual range (SVR), light extinction and deciviews. Light extinction states what fraction of light is lost per meter due to absorption and/or scattering. Standard visual range is inversely proportional to light extinction. Since these two quantities are directly related to the absorption and scattering properties of the constituents of the atmosphere, their change due to air pollution can be calculated from atmospheric models. However, a change in SVR is not what matters for human perception, because perception is a nonlinear function of the light received by the eye. Like the ear, the eye perceives relative changes rather than absolute changes. Thus it is more appropriate to use the logarithm of SVR as an indicator of the importance of visibility changes. That is the deciview, defined to change by one unit for each 10% change of light extinction. It is the visual analog of the decibel. A change of one deciview is considered to be just perceptible by the average person for typical conditions. The same amount of pollution can have vastly different effects on perceived visibility: the resulting loss of deciviews is much greater if the pollutant is added to a clean environment, than if the environment were already highly polluted.

Landrieu (1997) provides a simple model for the contribution of different pollutants to light extinction. It is the following function, adding together extinction from various fractions of ambient air:

$$
\begin{aligned}
b = {}& b_{air} + e_S(SO_4^{--})f(RH) + e_N(NO_3^-)f(RH) \\
& + e_O(organics)g(RH) + e_C(elemental carbon) \\
& + e_D(other fine particles) + e_G(NO_2),
\end{aligned}
\tag{6.1}
$$

where

b = light extinction coefficient of the atmosphere,
b_{air} = scattering of light by molecules in unpolluted air = $0.011\ km^{-1}$ (average value),
SO_4^{--}, NO_3^-, etc. are air concentrations of the pollutants of concern,
$e_{subscript}$ defines the scattering efficiencies of each fraction.
The units are $m^2\ mg^{-1}$, except for NO_2, for which e_G is expressed in $km^{-1}ppm^{-1}$). Values for each are as follows:

e_S = 0.003	e_N = 0.003
e_O = 0.003	e_C = 0.012
e_D = 0.001	e_G = 0.33

$f(RH)$ and $g(RH)$ are ratios of the scattering due to hygroscopic aerosols at a given relative humidity RH to the scattering at 0% RH.

For the interaction between humidity and scattering efficiency, the following two equations can be used for the average annual scattering effect of humidity on nitrates and sulfates:

$$f(RH) = 4.6 - 15(RH) + 19(RH)^2 \qquad (6.2)$$

and on organics,

$$g(RH) = 2.5 - 6(RH) + 5(RH)^2 . \qquad (6.3)$$

A difficulty arises with the application of these equations, because the calculation demands knowledge of all pollutants that affect visible range. It is not sufficient to base the calculations on the level of one pollutant, because of nonlinearities in the calculations. A further problem may arise in the temporal averaging of data.

For the monetary values, many contingent valuation studies have been carried out in the USA to determine the willingness to pay (WTP) for visibility improvements. For a good review see Abt (2000). The valuation questions are based on relative changes and can be stated in terms of $ per % improvement of SVR or $/deciview. The authors of the Methodology Update 2005 report of ExternE (2005) have summarized the valuation numbers of Abt (2000) in terms of Table 6.1.

Maddison (1997) reviewed a number of different studies to produce an alternative estimate of WTP. One of the purposes of the analysis was to test whether the results of certain studies were statistically different from the rest of the literature. The function derived by Maddison, converted to give WTP in $€_{1990}$, and omitting terms that covered studies that Maddison concluded were flawed, was:

$$WTP = 125 \ln \left(\frac{VR_2}{VR_1} \right) €_{1990}, \qquad (6.4)$$

with
VR_1 = initial visual range,
VR_2 = final visual range.

For example, the WTP for a 20% improvement is 23 $€_{1990}$, well in the range for residential WTP in Table 6.1 (the precise value depending on

Table 6.1 *Annual willingness to pay (WTP) per household for visibility improvements in the USA. Adapted from ExternE (2005).*

Visibility improvement (% of visual range)	Recreational US$_{1999}	Residential US$_{1997}
10%	7 to 10	
20%	11 to 19	24 to 278
100%	42 to 69	

the year chosen for the currency conversion and whether one adjusts for inflation before or after the conversion – see Section 9.6.2). Since VR is inversely related to the extinction coefficient b, calculated in Eq. (6.1), the last equation can also be written as,

$$\text{WTP} = 125 \, \text{€}_{1990} \times \ln(b_1/b_2). \qquad (6.5)$$

Studies in the USA have always found the loss of visibility to be a significant contribution to the total cost of air pollution. For example, Muller and Mendelsohn (2007) find that in the USA, the cost of visibility loss contributes $2.7 billion/yr out of the total air pollution damage of $74.3 billion/yr, see Table 12.6 in Chapter 12. This is not only because of the impact on recreational activities in the spectacular landscapes of the national parks, but also because even in large cities such as New York and Chicago people value visibility (Chestnut and Dennis, 1997). The model used by USEPA for air pollution damage costs finds even higher visibility costs than the model of Muller and Mendelsohn, as can be seen from Table 12.7 of Chapter 12 which compares these two models: the cost of visibility is seven times higher in the results of USEPA than in Muller and Mendelsohn.

And yet ExternE has made no attempt to assess this end point. There have been no studies of the value of visibility in Europe and the economists of ExternE decided that this issue was not important. The reason for this lack of interest stems from the fact that most environmental economists of Europe, in particular the ones who made this decision for ExternE, live in the northern parts of the continent with little sunshine and beautiful scenery. A visit to, for example, the Côte d'Azur, would be sufficient to demonstrate the degradation of magnificent vistas by air pollution. It would be strange if the value of visibility were really negligible, at least for the southern parts of Europe.

6.2   Noise

Noise is associated with stress as it disturbs sleep patterns, it interferes with mental activities, in particular learning, and it has harmful health impacts. Noise levels are measured in decibels, a measure based on the logarithm of the physical pressure of sound waves. Such a measure corresponds well with the logarithmic sensitivity of the sense of hearing. But noise is complex. There are many different types of noise and the harm can be very different between, for example, background noise from power plants, street noise in large cities and the noise of planes near airports. The simple decibel measure of noise level does not capture the complexity.

Unfortunately it is usually the only measure available for studies of health effects or amenity costs.

For certain fairly typical types of noise, dose–response functions have been established that make it possible to estimate impacts or costs. As an example we mention in Section 9.2.2 values for the reduction of rent, as determined by hedonic pricing. In that section we also illustrate the use of such values for cost–benefit analysis, using data for the percentage reduction of rent as a function of noise.

Very high sound levels, above 85 dB, can cause permanent hearing damage, but such exposures tend to be limited to special situations, such as discotheques, where the exposure is voluntary, or certain industrial settings. Even though for the general public the exposures are usually well below such levels, they can nevertheless have harmful health impacts. They arise not only because night time noise affects the quality of sleep, but quite generally because there seems to be a link between noise and cardiovascular health. Such a link, well established by numerous epidemiological studies (see e.g. Babisch, 2008), is probably due to noise-induced stress. In particular, Babisch carried out a meta-analysis and derived the following dose–response relationship between daytime transportation noise L_{day} (during 6–22 h, in decibels) and the odds ratio OR for myocardial infarction for noise levels ranging from 55 to approximately 80 dB. Presumably this relation is for chronic exposure but, unfortunately, it is not entirely clear what the relevant exposure duration window is:[1]

$$OR = 1.63 - 0.000613 \times L_{day}^{2} + 0.00000736 \times L_{day}^{3} \qquad (6.6)$$

We will not discuss these topics in detail and only cite in Table 6.2, numbers from a recent assessment of external costs of transport in the EU (CE Delft, 2011). We also recommend a recent review by EEA (2010) as a good source for detailed information on ERFs for annoyance and for health impacts due to noise.

6.3   Traffic congestion

Congestion contributes one of the most important external costs of transport. Here we present a summary of estimates of congestion costs for cars, based on the review by CE Delft (2008). During congested conditions, each additional driver imposes an external cost on other drivers, in

---

[1] As so often, the epidemiological studies limit themselves to demonstrating an effect, without making any attempt to derive relations that can be used for the quantification of impacts.

Table 6.2 *Typical marginal damage cost (€cent$_{2008}$/vkm) due to noise from various transport modes. Adapted from Table 37 of CE Delft (2011).*

Mode	Time of day	Traffic	Urban	Suburban	Rural
Car	Day	Dense	0.90	0.05	0.02
		Thin	2.19	0.14	0.01
	Night	Dense	1.65	0.09	0.01
		Thin	3.99	0.26	0.04
Motorcycle	Day	Dense	1.81	0.11	0.01
		Thin	4.38	0.28	0.04
	Night	Dense	3.29	0.19	0.02
		Thin	7.98	0.52	0.06
Bus	Day	Dense	4.51	0.25	0.04
		Thin	10.96	0.70	0.08
	Night	Dense	8.23	0.46	0.07
		Thin	19.95	1.30	0.15
Light goods vehicle	Day	Dense	4.51	0.25	0.04
		Thin	10.96	0.70	0.08
	Night	Dense	8.23	0.46	0.07
		Thin	19.95	1.30	0.15
Heavy goods vehicle	Day	Dense	8.30	0.46	0.07
		Thin	20.14	1.30	0.15
	Night	Dense	15.14	0.85	0.13
		Thin	36.70	0.24	0.27
Passenger train	Day	Dense	28.01	1.24	1.54
		Thin	55.35	2.44	3.04
	Night		92.38	4.08	5.08
Freight train	Day	Dense	49.67	2.45	3.06
		Thin	119.84	4.75	5.92
	Night		202.62	8.02	10.01

addition to suffering such a cost him/herself. There are several components of the cost of congestion:

- Increased travel time: this is usually the largest cost component, usually more than 90% of the total. To choose the appropriate value of time, one has to distinguish between different purposes for a trip, in particular, business and leisure. In addition to the real travel time increase there is the cost of unreliability because travel times are difficult to foresee.
- Disamenities on crowded roads: the value of time may be increased by about 50% to account for the annoyance and stress.
- Additional operating costs: under stop-and-go conditions, fuel consumption increases and there is extra wear and tear on the vehicle. This category can contribute on the order of 10% of the congestion cost.

Table 6.3 *Typical estimates of marginal external costs due to congestion* *(€$_{2008}$/vkm, all journeys).* *Adapted from Table 38 of CE Delft (2011).*

Area and road type	Min.	Central	Max.
Large urban areas (> 2,000,000)			
Urban motorways	0.33	0.56	1.00
Urban collectors	0.22	0.56	1.33
Local streets centre	1.67	2.22	3.33
Local streets cordon	0.56	0.83	1.11
Small and medium urban areas (< 2,000,000)			
Urban motorways	0.11	0.28	0.44
Urban collectors	0.06	0.33	0.56
Local streets cordon	0.11	0.33	0.56
Rural areas			
Motorways	0.00	0.11	0.22
Trunk roads	0.00	0.06	0.17

Congestion is a highly nonlinear function of traffic flow. As an indication of the magnitude of congestion costs, we cite some typical estimates for European conditions in Table 6.3, based again on the review by CE Delft (2011). These are rough general values. For specific situations specific calculations are called for.

6.4   Depletion of non-renewable resources

There has been much concern about the depletion of non-renewable resources, especially during episodes of rapid price increases. This concern is reflected in the treatment of resources in LCA, where depletion of non-renewable energy and of mineral resources are among the impact categories to be quantified (in units of GJ for energy and kg for each mineral resource). There seems to be a deep-seated fear that we will no longer be able to enjoy our accustomed standard of living when a particular resource is exhausted.

External cost assessments, by contrast with LCA, have not addressed this subject at all. Are there any relevant external costs? Searching the internet, we have found countless studies on sustainability and resources, but none that try to calculate external costs. For example, the final report of the SAUNER (2000) project of the EC, coordinated by the economists of ExternE, does not even mention external costs.

The depletion of a resource is not abrupt, but gradual, reflected in a gradual increase in its price (apart from temporary fluctuations, such as disruptions of oil supply due to political crises). Such price increases have been noticeable for quite a few resources in recent years, because of increasing demand in emerging economies, especially China. As the price of a resource increases, one finds substitutes. Such substitutes can take many forms: geological deposits that were not economically exploitable in the past (e.g. tar sands in Canada for the production of oil), alternative means of producing that resource (e.g. biofuels), more efficient use of the existing resource (e.g. more efficient cars), or substitution of end uses (e.g. electric trains instead of cars). Resource depletion is nothing new. For example, when the supply of wood was no longer sufficient for heating, coal was used instead.

If the marginal cost of production increases due to depletion of a non-renewable resource, the consumption of one unit now increases the cost of all future production, all else being equal. This increase is an external cost if not paid by the consumer of this unit. However, it is merely a so-called pecuniary externality and does not call for internalization, unlike the externalities due to pollution, sometimes called technological externalities. Pecuniary externalities are taken into account in the market through the usual price mechanisms, unlike technological externalities such as pollution that do not carry a price tag.

To explain which externalities should be internalized, we find it helpful to define internalization as a government policy that makes the perpetrator of a burden act as though he/she were also its victim. As an example, consider the supply of land in a given region. The first buyer reduces the remaining supply and drives the price up. This is an external cost imposed on future buyers. Now suppose the first buyer does not have the funds to buy all of the desired land now, but has the intention of buying more land in the future. In this case, the external cost of the first purchase is automatically internalized, but it does not change the actions of the buyer at all. By the same token, the increase of resource prices due to depletion is not in itself a relevant externality that needs to be internalized (e.g. by a tax equal to the external cost) in order to improve economic efficiency.

This does not mean that there is no role for government intervention, but it requires considerations beyond the above argument of a price increase for future generations. One such consideration is the rate of such a price increase. In a classic paper, Hotelling (1931) showed that the optimal rate of the price increase for a finite resource is equal to the discount rate ("Hotelling's rule"), under some simplifying assumptions, including zero extraction cost. The basic idea is that the higher the rate of extraction of the resource, the faster the price increase due to depletion. Consider an impatient owner who has sold too much too soon, causing the price to increase at a higher rate than the discount rate. He can invest the proceeds from the sale, yielding a rate of return equal to the discount rate. But in the meantime, another owner who has left her resource in the ground, has enjoyed a price increase and thus a rate of return that is higher than the discount rate. Conversely, leaving too much in the ground too long reduces the return on the resource left in the ground below the discount rate. Thus the profit obtainable from such a resource is maximized by the extraction rate that causes the rate of the price increase to equal the discount rate.

A difference between social and private perspectives arises because private discount rates are higher than the social discount rate. Thus the extraction rate is higher than socially optimal. However, this consideration is difficult to implement in practice. Hotelling's rule does not tell us what the optimal extraction rate is, because that rate depends on many factors in addition to discount rate, such as the extraction costs and the relation between supply and demand. The rule merely states that whatever the resulting price increase, it should be equal to the discount rate in order for the extraction rate to be optimal. Because of these complications, Hotelling's rule will not yield an estimate of externalities, even if private and social discount rates were exactly known.

To appreciate why these complications are important, it is sufficient to look at the price history of resources: in most cases real prices have been decreasing rather than increasing, contrary to the simple result of Hotelling. In fact, Hotelling's rule does not take into account such effects as: the increase in reserves due to increasing ability to extract resources of lower quality (e.g. tar sands instead of traditional oil wells), the increase in extraction and processing costs for resources of lower quality, the decrease in these costs due to technological progress and economies of scale, and imperfect competition. Livernois (2008) reviews the evidence and finds that Hotelling's rule can indeed explain the price history of resources, if these additional effects are taken into account. However, it is difficult and often impossible to obtain the data needed for an analysis of all these effects, and so the empirical evidence for Hotelling's rule is not clear.

Equity is another consideration for pecuniary externalities: should the ones who are lucky to get a resource at a low price compensate later generations who will have to pay more? Instead of a simple direct cash transfer to future generations, it may be more appropriate to ask how we should use our wealth to assure sustainability. With regard to nonrenewable resources, we are in a sense like a lucky individual who has inherited a pile of gold. What's the best way to use it? We could spend all now to have a good time, or save all for a rainy day, but those are extreme options and usually not the best. Common sense would suggest that we should evaluate the entire portfolio of possible options to find the optimal allocation. This depends on the specifics of the situation. We would use some of this wealth to pay for urgent items, we would invest some of it for future use, and we would consume some of it for present enjoyment. There is, of course, a tradeoff between current and future consumption, and that is where the difficulties arise. If we apply too high a discount rate, we end up not leaving enough for the future. If, on principle and no matter what, we always say "save it all for the future," we don't have any benefit at all – and neither will future generations if they apply the same rigid principle.

As guidance for that tradeoff between current and future consumption, Hartwick's rule is appropriate (Hartwick, 1977). It says that the income from the extraction of a non-renewable resource (the technical term is "resource rent") should be invested in such a way that our standard of living will not decline. If there is substitutability between the resource and capital, and if such investment (in infrastructure, education, research, technology development, etc.) brings returns that compensate the loss of the resource, then our consumption of the resource is sustainable.

There is a role for government intervention because the discount rates of individuals and industry are higher than the social discount rate, implying that not enough of the resource rent is invested for the future. But it is not clear that such intervention should take the form of a tax. For example, if the elasticity of consumption is low, a tax would have insufficient impact on the rate of consumption. Furthermore, if the tax goes into general government revenue, there would be no targeted investment in the sense of Hartwick's rule. Instead of a tax, investment in research is more appropriate, such as for the development of new energy resources.

6.5   Accidents

Quantification of human health damage costs for accidents is often straightforward, particularly when dealing with events that include fatalities, as data tends to be most readily available for these. One takes data from official statistics, normalizes against an appropriate functional unit (e.g. kWh generated or vehicle.km travelled) and then applies to the situation under investigation. Quantification of the environmental consequences of major accidents can be far more challenging because of the diversity of ecosystems that may be at risk. Effects at one location cannot be simply extrapolated to another in the same way that death and injury can.

Passing from damage costs to externalities is also challenging. Accidents are obviously an externality if they impose costs on the general public, for example oil spills. Such costs are internalized to the extent that the perpetrator is made to pay: An obvious case is the Deepwater Horizon Disaster in the Gulf of Mexico, where the company held responsible, BP, has had to pay many billions of dollars in clean-up costs, compensation and fines. However, to continue with the example of oil, even if oil companies know that they will be held accountable in the event of a spill, they may not do enough to minimize the risk of an accident, because their discount rates are too high (higher than the social discount rate), or they underestimate the risk. Or they can afford to pay enough for legal defense to avoid a penalty that would truly compensate the victims.

Many accidents hurt or kill employees of their respective industries. In theory the corresponding costs are internalized, if the employees correctly assessed their risks while negotiating their contracts and could find alternative employment elsewhere. That may be the case if their unions are sufficiently powerful and well informed, but in practice many workers do not have enough power to obtain an equitable deal. However, regardless of the extent to which the accidents of workers are internalized, it is important for the government to evaluate the accident risks of workers, and safety regulations seem more appropriate for internalization than an externality tax.

Major accidents are rare events, and most are limited to the local zone and affect only a small number of individuals. Compared with air pollution, which is emitted continuously and affects a large population, the expectation value of the cost of accidents is small, with the exception of catastrophic accidents such as Chernobyl and Fukushima. But for the affected individuals the effects and costs can be horrendous. Since public policy should be concerned with the safety of individuals, an assessment of accidents is very important for public policy.

A major challenge in assessing the risk of a major accident in complex operations such as refineries or nuclear power plants lies in the difficulty of modeling all of the elements that may fail and estimating their failure probabilities. It comes as no surprise that the results of different analysts can be very different. And as the Fukushima accident demonstrates, the accident risk may be far higher than the predictions of the customary probabilistic safety assessments (PSA), if some factor of the real operations has not been taken into account in the assessment. The historical frequency of catastrophic nuclear accidents has been an order of magnitude higher than the results of PSAs (Rabl and Rabl, 2013).

Oil spills are a highly visible type of accident, attracting a great deal of media attention. The social cost per accident can be enormous, the Exxon Valdez accident being an egregious example. Nonetheless, averaged over the total amount of oil that is consumed, the external cost of oil spills is relatively small. The most recent externality assessment in the USA

(NRC, 2010) has examined this issue and arrives at an average marginal damage cost of $0.0076 per barrel shipped (see p. 229 of the NRC report). Even with more pessimistic assumptions, outlined in a footnote on that p. 229, a high estimate comes to only $0.06 per barrel, very small compared with the market price of oil; around $100 per barrel at the beginning of 2013.

For accident data we refer to the ENSAD (2013) database, maintained by the Paul Scherrer Institut in Switzerland. ENSAD is a very comprehensive database on severe accidents with emphasis on the energy sector. In Section 13.2.2, we show some ENSAD data for accidents in the energy sector. It makes it possible to perform comprehensive analyses of accident risks, covering not only power plants but the full energy chains, including exploration, extraction, processing, storage, transport and waste management. Use of the recent historic record, whilst desirable in a number of ways, may provide an overly conservative view on accidents. For example, the Deepwater Horizon Disaster adds to the number of major accidents in the records and increases the long-run (historic) accident rate. However, given the cost to operators, Deepwater Horizon may cause the future accident rate to fall as oil companies generally recognize the potential consequences of presiding over a disaster.

In summary, the impact pathway for accidents involving death or injury to people is usually straightforward and supportable with evidence from official statistics. This is not the case for assessment of environmental impacts of accidents. In spite of these difficulties, it seems possible to quantify effects with sufficient robustness to enable comparison with other types of effect, and to identify areas where further attention should be paid to mitigation.

Whereas there is a theoretical difficulty in determining to what extent worker accidents are already internalized, this is not relevant in practice because an externality tax is not an appropriate tool for limiting accidents. There is a crucial difference between pollution, for which a tax is appropriate, and worker accidents. An industrial enterprise can evaluate the characteristics of abatement technologies with a fair degree of confidence and when a technology is installed, the reduction in pollution can be measured accurately and immediately. Evaluating the characteristics of safety measures is more difficult and when a safety measure has been implemented, it can take many years until its benefit can be confirmed; for small enterprises, the statistics may never be sufficient to confirm the benefit. Whereas a pollution tax is instantaneous and precise, a safety tax would be slow and vague. To protect workers from death and injury, government regulations are required, with safety standards to reduce the risk of accidents to the social optimum.

References

ABT Associates (2000), *Out of Sight: The Science and Economics of Visibility Impairment*. Report prepared for Clean Air Task Force, Boston, MA. www.catf.us/resources/publications/files/Out_of_Sight2.pdf

Babisch, W. 2008. Road traffic noise and cardiovascular risk. *Noise and Health* **10** (38): 27–33.

CE Delft 2008. *Handbook on estimation of external costs in the transport sector*. Produced within the study Internalisation Measures and Policies for All external Cost of Transport (IMPACT), Version 1.1. CE Delft, February, 2008. www.cedelft.eu/publicatie/deliverables_of_impact_%28internalisation_measures_ and_ policies_for_all_external_cost_of_transport%29/702

CE Delft 2011. External Costs of Transport in Europe: Update Study for 2008. CE Delft, Oude Delft 180, 2611 HH Delft, The Netherlands. www.cedelft.eu/ publicatie/external_costs_of_transport_in_europe/1258

Chestnut, L. G. and Dennis, R. L. 1997. Economic benefits of improvements in visibility: Acid Rain Provisions of the 1990 Clean Air Act Amendments. *Journal of the Air & Waste Management Association* 47(3): 395–402.

EEA 2010. *Good practice guide on noise exposure and potential health effects*. EEA Technical report No 11/2010. European Environment Agency, Kongens Nytorv 6, 1050 Copenhagen K, Denmark

ENSAD 2013. ENSAD – Energy-Related Severe Accident Database. Paul Scherrer Institut, Switzerland. https://gabe.web.psi.ch/research/ra/

ExternE 2005. *ExternE: Externalities of Energy, Methodology 2005 Update*. Edited by P. Bickel and R. Friedrich. Published by the European Commission, Directorate-General for Research, Sustainable Energy Systems. Luxembourg: Office for Official Publications of the European Communities. ISBN 92-79-00429-9.

Hartwick, J. M. 1977. Intergenerational equity and the investment of rents from exhaustible resources. *American Economic Review* **67**, December: 972–974.

Hotelling, H. 1931. The economics of exhaustible resources. *Journal of Political Economy* **39**(2): 137–175.

Landrieu, G. 1997. Visibility impairment by secondary ammonium, sulphates, nitrates and organic particles. Draft note prepared for the meeting of the Task Force on Economic Aspects of Abatement Strategies. UNECE Convention on LRTAP, Copenhagen 9–10 June 1997.

Livernois, J. 2008. The empirical significance of the Hotelling rule. *Review of Environmental Economics and Policy*, 2008: 1–20

Madisson, D. 1997. *The Economic Value of Visibility: A Survey*. CSERGE, unpublished.

Muller, N. Z. and Mendelsohn, R. 2007. Measuring the damages of air pollution in the United States. *Journal of Environmental Economics and Management* 54: 1–14.

NRC 2010. *Hidden Costs of Energy: Unpriced Consequences of Energy Production and Use*. National Research Council of the National Academies Press. National Academies Press, 500 Fifth Street, NW Washington, DC 20001.

Rabl, A. and Rabl, V. A. 2013. External costs of nuclear: Greater or less than the alternatives? *Energy Policy* **57**: 575–584.

SAUNER 2000. *SAUNER: Sustainability And the Use of Non-renewable Resources.* Summary Final Report, November 2000. Prepared by University of Bath, UK, IER, Universität Stuttgart, Germany, and Montanuniversität Leoben, Austria. European Commission DG Environment. Environment and Climate Programme, Contract ENV4-CT97–0692

7 Atmospheric dispersion of pollutants

Summary

Atmospheric dispersion and chemistry is a complex subject, for which this chapter offers only a brief introduction, with focus on a special class of models that are appropriate for damage cost calculations. Such models can be relatively simple, because damage costs involve long-term averages over large areas. Gaussian plume models, suitable for the local zone, are described in some detail and equations are provided for a specific version to allow the reader to carry out calculations. Further from the source, the removal of pollutants from the atmosphere becomes important and is crucial for regional modeling. The removal rates can be expressed in terms of a velocity that we call the depletion velocity, a quantity that accounts for dry and wet deposition and, for reactive pollutants, chemical transformation. To illustrate key features of regional modeling, we develop a simple model and compare it with results from the EMEP model. We present several methods of estimating depletion velocities. We also develop a simple model for an approximate calculation of impacts and damage costs due to air pollution. It is called the "uniform world model" (UWM), because it is exact in the limit where the depletion velocity and the receptor density are uniform. We have validated the model by about 200 comparisons with detailed site-specific calculations using the EcoSense software of the ExternE projects in Europe, Asia and the Americas. For emissions from stacks of 50 m or more, detailed calculations agree with the simplest version of the UWM, within a factor of two in most cases. We provide modifications for site, stack height and receptor distribution that greatly improve the accuracy and applicability of the UWM. The UWM is very relevant for policy applications because it yields representative results for typical situations, rather than for one specific site.

7.1   Introduction – damage function

The goal of an impact pathway analysis (IPA) is to calculate the damage
cost due to a constant source of pollution. Sometimes the damage cost is
called damage function, to express its complicated dependence on all the
relevant input parameters and data. Since the exposure–response functions (ERFs) are usually stated in terms of annual impact due to an annual
exposure, it is convenient to use the year as the basis. Therefore, we
indicate emissions, impacts and damage costs as rates, using a dot above
the symbol according to standard engineering practice, because they are
in fact the time derivative of the cumulative pollutant masses, impacts and
costs. We find such a notation helpful for keeping track of units.
Designating the impact rate due to an emission rate $\dot{m}$ by $\dot{I}(\dot{m})$ and likewise the damage cost rate by $\dot{D}(\dot{m})$, one obtains the unit impact and unit
damage cost as:

$$I_u = \frac{\dot{I}(\dot{m})}{\dot{m}} \text{ and } D_u = \frac{\dot{D}(\dot{m})}{\dot{m}}. \tag{7.1}$$

We write the damage function for air pollution exposure impacts on a set
of receptors (people, buildings, plants etc.) in its most general form as
damage cost rate (e.g., €/yr) due to an emission rate $\dot{m}$ (kg/yr):

$$\dot{D}(\dot{m}) = \iint dx\, dy\, \rho(\mathbf{x}) \sum_i P_i(\mathbf{x}) \left(\frac{1}{t_2 - t_1}\right) \int_{t_1}^{t_2} dt\, \text{ERF}_i\left(\mathbf{x},\, c(\mathbf{x}, t, \dot{m})\right), \tag{7.2}$$

where $c(\mathbf{x}, t, \dot{m})$ is the concentration at time t and location $\mathbf{x}$ (bold face to
indicate the location as a vector with coordinates x and y) for an emission
rate $\dot{m}$, ERF_i $(\mathbf{x}, c(\mathbf{x}, t, \dot{m}))$ is the ERF for impact i at $\mathbf{x}$ due to concentration c with units cases per (year·person·µg/m^3), $P_i(\mathbf{x})$ is the unit cost of
impact i (e.g. cost per case of asthma attack) and $\rho(\mathbf{x})$ is the receptor
density at $\mathbf{x}$ (pers/km^2).

When several sets of receptors are affected, one also has to sum over the
receptor sets to obtain the total damage cost. This very general form
allows for the ERF to be a nonlinear function of the concentration, and
since the concentrations vary with time, an average is taken over the
relevant exposure period, usually one year, although the averaging period
can be different, for instance, for crop losses it is the growing season.

Most of the time one is interested in the marginal (= incremental)
damage for an incremental emission, and, therefore, we interpret $\dot{m}$, c, $\dot{I}$
and $\dot{D}$ as incremental emissions, concentrations, impacts and damage
costs throughout this book (with a minor exception for the Hg impacts

in Section 5.3.4, where we also look at total impacts). For marginal impacts, the ERF can in effect be taken as the ERF slope S_{ERF} multiplied by the marginal concentration, thus:

$$\mathrm{ERF}_i\left(\mathbf{x}, c(\mathbf{x}, t, \dot{m})\right) = S_{ERF}\ c(\mathbf{x}, t, \dot{m}) \text{ for marginal impacts.} \qquad (7.3)$$

The slope could vary with $c(\mathbf{x}, t, \dot{m})$, but in most cases linear ERFs have been assumed, in particular for health impacts (for health impacts of O_3, ExternE (2008) assumes linearity above a no-effect threshold, so the form of Eq. (7.3) still applies, but only for times when the concentration is above the threshold). For linear ERFs, Eq. (7.2) simplifies to:

$$\dot{D}(\dot{m}) = \iint dx\,dy\,\rho(\mathbf{x})\sum_i S_{ERF,i}P_i(\mathbf{x})\ c(\mathbf{x}, \dot{m}),$$

$$\text{where } c(\mathbf{x}, \dot{m}) = \left(\frac{1}{t_2 - t_1}\right)\int_{t_1}^{t_2} dt\ c(\mathbf{x}, t, \dot{m}), \qquad (7.4)$$

where $c(\mathbf{x}, \dot{m})$ is now the annual average concentration increment at $\mathbf{x}$ due to $\dot{m}$. If, furthermore, the unit cost $P_i(\mathbf{x})$ of each impact is assumed to be the same for all locations $\mathbf{x}$ (which is the case for assessments in the EU and USA), both P_i and $S_{ERF,\ i}$ can be taken outside the integral and one obtains:

$$\dot{D}(\dot{m}) = \sum_i S_{ERF,i}P_i \iint dx\,dy\,\rho(\mathbf{x})\ c(\mathbf{x}, \dot{m}). \qquad (7.5)$$

The sum $\sum_i S_{ERF,i}\ P_i$ is the exposure cost. For the set of ERFs for $PM_{2.5}$ and unit costs assumed in ExternE (2008), the sum equals 38.76 €/yr· $(\text{persons·μg/m}^3)^{-1}$ (see Section 12.1.3). For practical calculations, receptor data are available as number of receptors $N_{pers,\ j}$ in each grid cell j, and so one calculates:

$$\dot{D}(\dot{m}) = \sum_i S_{ERF,i}P_i \sum_j c_j(\dot{m})\ N_{pers,\ j}, \qquad (7.6)$$

where $c_j(\dot{m})$ is the annual mean incremental concentration in cell j.

How to calculate $c_j(\dot{m})$ is the subject of this chapter. For the classical air pollutants, only the inhalation exposure matters. Some persistent pollutants, in particular dioxins and toxic metals, can also act through ingestion. How to take ingestion pathways into account will be explained in Chapter 8. For air pollution, we present detailed models as well as a simplified approach, called UWM (uniform world model) for approximate calculations.

7.2   The Gaussian plume model

Note that this is not a textbook on environmental dispersion and chemistry; for that we recommend the book by Seinfeld and Pandis (1998)[1]. Rather the purpose of this chapter is to give a general introduction, with emphasis on a small class of models that are suitable for the calculation of damage costs. There are in fact many, very different environmental problems and accordingly many types of models that may be best for a particular problem. For example, some models are capable of describing local phenomena in great detail, but unmanageable for larger areas, because input data and calculation time would become excessive. Likewise, some models are great for precise modeling of pollution episodes but not suited for long-term averages. For damage costs, the modeling domain must extend over the scale of an entire continent, but the models should also have sufficient resolution at the local scale to account for the population distribution at the scale of a few km, if the source is in or near a large population center. Since only long-term averages are needed, temporal modeling can be greatly simplified.

The reality of environmental pathways is extremely complex. One can already get a hint of the complexity of atmospheric dispersion by looking at the plume emerging from a smoke stack, as sketched in Fig. 7.1 for some typical situations. The details depend on meteorological conditions, in particular wind speed and atmospheric stability class. Needless to say, all models simplify reality. The art, therefore, lies in choosing the appropriate simplifications that allow efficient calculation with sufficient accuracy for the problem of interest. It turns out that for damage cost calculations, the simplifications can be extreme because only long-term averages of regional totals are needed. Thus the result can be reliable as long as errors in spatial or temporal detail cancel out for the averages.

Atmospheric dispersion can be subdivided into "local-range" dispersion, extending up to about 50 kilometers in all directions from the emission source, and "regional-scale" dispersion beyond that. The regional domain extends up to several thousand kilometers. Depending on the scale, different air transport algorithms (models) are employed to estimate the concentration profiles. For the local range, the degree of pollutant dispersion depends mostly on meteorological parameters. Of

---

[1] A third edition of the book was released in 2006.

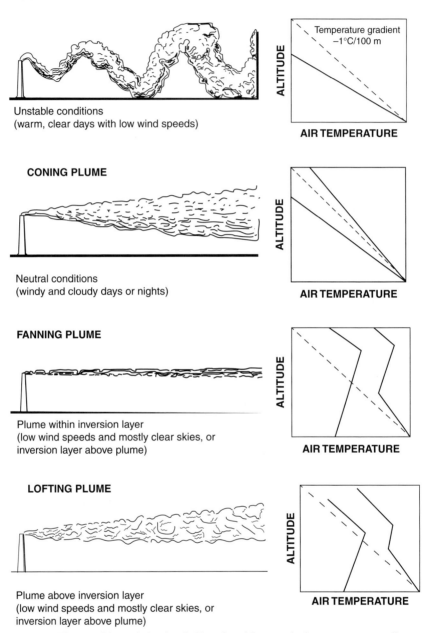

LOOPING PLUME

Unstable conditions
(warm, clear days with low wind speeds)

Temperature gradient
−1°C/100 m

ALTITUDE

AIR TEMPERATURE

CONING PLUME

Neutral conditions
(windy and cloudy days or nights)

ALTITUDE

AIR TEMPERATURE

FANNING PLUME

Plume within inversion layer
(low wind speeds and mostly clear skies, or
inversion layer above plume)

ALTITUDE

AIR TEMPERATURE

LOFTING PLUME

Plume above inversion layer
(low wind speeds and mostly clear skies, or
inversion layer above plume)

ALTITUDE

AIR TEMPERATURE

Fig. 7.1 Plume behavior (left) and ambient vertical temperature gradient (right, solid line). The adiabatic temperature gradient is indicated by the dashed line, it corresponds to a temperature drop of $10°C$ per 1000 m height change. Adapted from Zannetti (1990) with kind permission.

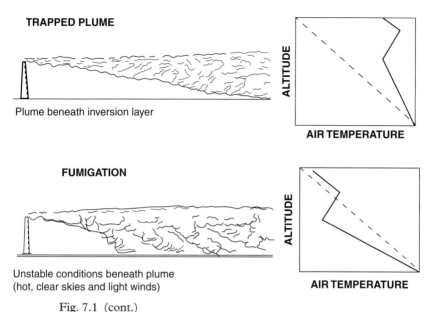

TRAPPED PLUME

Plume beneath inversion layer

FUMIGATION

Unstable conditions beneath plume
(hot, clear skies and light winds)

Fig. 7.1 (cont.)

these, the most important are wind speed, wind direction, ambient temperature, Pasquill dispersion category (atmospheric stability or turbulence intensity) and the mixed layer or planetary boundary layer height (lower portion of atmosphere within which turbulent transport takes place). For the regional domain, chemical interactions between direct emissions and airborne species, pollutant removal through dry (e.g. gravitational settling) and wet deposition (precipitation) are as important as meteorological data in determining concentration levels in the air, and, consequently, the final pollutant fate in the environment through air–water, air–soil, air–vegetation interactions. Figure 7.2 highlights the main features of atmospheric dispersion along the horizontal plane with increasing distance from emission source. For a cross-sectional view along the vertical direction, see Fig. 7.7.

Lagrangian and Eulerian models for long-range transport

Far from the source, dispersion is modeled using a Lagrangian or an Eulerian transport model. In the Lagrangian transport model, atmospheric turbulent diffusion is simulated by focusing on the statistical properties of individual fluid particles and following their random motion through space in time. The mean concentrations or expectation values

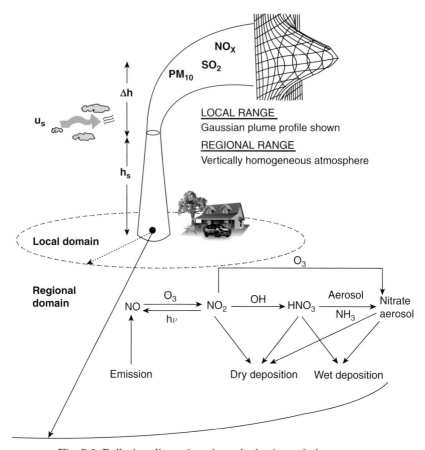

Fig. 7.2 Pollution dispersion along the horizontal plane.

are determined by solving the specie balance equation in integral form. The Lagrangian model is not well suited to situations involving nonlinear chemistry. Because of limited knowledge on turbulence properties of particles, approximate solutions are possible under certain limiting circumstances (statistical theory approach).

An alternative transport scheme is the Eulerian model. Mean concentrations are determined by focusing on the statistical properties of the fluid flow, in particular, velocities, which are measured with respect to a fixed coordinate system. Analysis is based on solving the species conservation equation in differential form. The Eulerian approach is well suited for modeling chemically reactive flows, but a serious mathematical problem arises when the number of unknown variables exceeds the number of mathematical expressions available; namely we are faced with the problem of mathematical closure. Nevertheless, approximate solutions are available by invoking the atmospheric dispersion equation (K theory approach).

In the ExternE project (ExternE, 2005), the Windrose Trajectory Model (WTM) was used for transboundary transport of air pollutants (Krewitt et al., 1995). WTM is a simplified implementation of the more complex Harwell trajectory dispersion model (HTM), developed by Derwent and colleagues at the Harwell Laboratory, AEA Technology in the UK in the mid 1980s (Derwent and Nodop, 1986, Derwent et al., 1988 and Derwent et al., 1989). The HTM algorithm is a Lagrangian plume model. It has already been used extensively in Europe to estimate pollutant concentrations and deposition rates of nitrogen and sulfur, as well as several other chemically related compounds (nitrate and sulfate aerosols, for example). The model uses statistically averaged regional meteorological data. Weather parameters include: annual wind speeds, averaged along 24, equally spaced, wind directions centered at the receptor location, frequency of occurrence or probability arrays of wind direction data, distributed in 15° sectors about the receptor point (windrose summaries), and annual precipitation rates. The Windrose Trajectory Model has been used for modeling both primary and secondary pollutants. It has been implemented in ExternE's EcoSense software, an integrated impact assessment program, combining dispersion results with databases of receptor distributions that cover the whole of Europe with a resolution scale 10 by 10 km (including population statistics, crop production data and building material stocks), ERFs for impacts to human health, agricultural crops and materials, and unit costs for aggregating the various impacts into an overall damage cost. At the local scale, concentrations in EcoSense are calculated with the ISC software of EPA (1995).

Gaussian plume model for modeling near-source ambient concentrations

Near the source, incremental concentrations of primary pollutants (direct emissions) are modeled using a steady state Gaussian plume model (Fig. 7.3). The basic premise is that once the stack gases are emitted into the air, the vertical and horizontal concentration profiles across the plume may be adequately represented as two independent normal distributions, each one characterized by its own standard deviation or sigma parameter. The exhaust gases are hotter than the surrounding air and have considerable vertical momentum, effects that cause the plume to rise well above the chimney height, h_s. Plume growth and trajectory depend on the

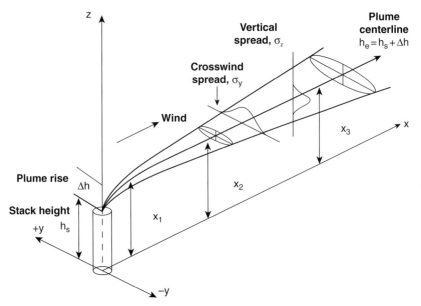

Fig. 7.3 Representation of Gaussian plume model in a wind-oriented Cartesian coordinate system.

local wind speed, wind direction and the surrounding atmospheric turbulence intensity (characterized by the Pasquill stability parameter). While the plume rises across the atmosphere, driven by inertia and buoyancy forces, it continually spreads along the vertical (z) and crosswind (y) directions, because of turbulent mass entrainment. As a consequence of thermal losses to the surrounding air and mass growth, the plume levels off eventually and its centerline defines an effective stack height, h_e, equal to the sum of source height h_s and plume rise Δh. When the plume reaches the ground it may be partially or totally reflected back up into the air. The same may occur at the top of the mixing height (the plume is reflected back towards the ground), and indeed multiple reflections can occur between the two barriers as the plume is transported along the downwind (x) direction. Total reflection at both barriers is generally assumed in most calculations.

The steady-state incremental concentration $c(x, y, z)$ of a pollutant at (x, y, z) for a point source at $(0, 0, h_e)$ with continuous and constant emission rate $\dot{m}$ (kg/s) over flat terrain is given by Eq. (7.7) and for cylindrical coordinates (x, z) by Eq. (7.8) ($x, y, z, h_s, h_e, h_{mix}$ and Δh are all expressed in m):

Gaussian plume in Cartesian coordinates (x, y, z)

$$c(x, y, z) = \frac{\dot{m}}{2\pi\,\sigma_y(x)\,\sigma_z(x) u_s}\,\exp\left[-\frac{1}{2}\left(\frac{y}{\sigma_y}\right)^2\right] S(x, z)\,R(x). \qquad (7.7)$$

Gaussian plume in cylindrical coordinates $(x, z)^2$

$$c(x, z) = \frac{\dot{m}}{(2\pi)^{1.5} x\,\sigma_z(x) u_s}\,S(x, z)\,R(x). \qquad (7.8)$$

The concentration c has units of (kg/m^3 or µg/m^3). σ_y and σ_z are, respectively, the lateral and vertical standard dispersion coefficients (Section 7.2.4) in m. Both parameters only depend on downwind distance x and atmospheric turbulence intensity (Section 7.2.1). Along a particular downwind path, the random fluctuations in the y and z directions are constant, i.e. the atmosphere is homogeneous. u_s is the mean horizontal wind speed at the source stack top h_s in m/s (see Section 7.2.5). The mean wind speed and direction are both assumed to be constant. In fact, all meteorological variables are assumed to be constant along the plume path from the source to the receptor (obviously, this is not the case in the real world). Downwind diffusion in the x direction is negligible compared with downwind advection.

S(x, z) is the vertical plume term (defined in Eqs. (7.9) to (7.13)), which accounts for plume reflections at the ground and at the top of the mixing height h_{mix} (Section 7.2.6). S is a function of x, z, effective stack height h_e ($h_e = h_s + \Delta h$) (Section 7.2.5), h_{mix} and σ_z. R_g and R_m are, respectively, the ground and mixing height reflection coefficients. Generally, both coefficients are set to unity, total reflection. Emissions reflected at the surface or at the top of the mixing height are distributed vertically as though originating from virtual sources located beneath the surface ($z = -h_e$) or above the mixing height ($z = 2 h_{mix} - h_e$) (Zannetti 1990). Virtual concentrations are added to the actual emission source values. There are no barriers for crosswind dispersion. When the air concentration no longer depends upon z (Eq. (7.13c)), the atmosphere is vertically homogeneous or "well" mixed, and the concentration depends solely on x. The distance to vertical homogeneity depends on the level of atmospheric turbulence, landscape and the mixing height (which varies diurnally).

^{2} $\int_{-\infty}^{\infty}\exp\left[-\frac{1}{2}\left(\frac{y}{\sigma_y}\right)^2\right] dy = \frac{\sigma_y}{\sqrt{2\pi}\,x}$

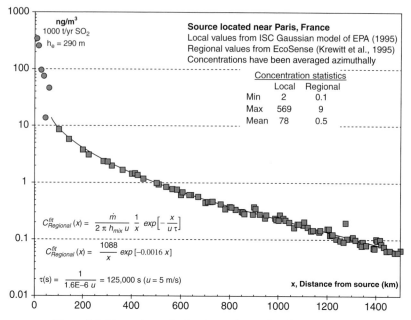

Fig. 7.4 Example of local and regional annual incremental concentrations, averaged horizontally, versus x (km).

R(x) is the pollutant decay term (defined in Eqs. (7.14) and (7.15)). R(x) is a function of x, u_s, h_e, σ_z, dry deposition velocity v_{dry} (m/s) (Section 7.3.1), scavenging or washout coefficient Λ (1/s)[3] and the pollutant chemical transformation rate τ_{ct} (1/s) (Section 7.3.2). τ_R (s) is the atmospheric residence time, and is related to the pollutant removal rate (Section 7.3.2).

At the spatial scale, the Gaussian model is most accurate for predicting pollutant concentrations at downwind distances ranging from 100 m to 10 km, although it is frequently applied outside this interval, up to about 50 km from the source. Local concentrations are typically two to three orders of magnitude higher than regional concentrations (see Fig. 7.4). The accuracy of Gaussian model concentrations depends on meteorological conditions, particularly atmospheric turbulence intensity and dilution rate (wind speed times mixing height), pollutant removal through deposition and transformation, source location (urban versus rural, inland

---

[3] Not to be confused with the washout ratio (see Seinfeld and Pandis, 1998; Underwood, 2001).

versus coastal site), surrounding topography (flat versus complex terrain) and land coverage (nearby buildings and other structures versus vegetative coverage). For short-term comparisons over flat terrain, Gaussian plume model simulations of 1-hour maximum ground-level concentrations, averaged across the study area for multiple hours, are usually accurate to a factor two or three of field measurements, but the residual scatter at individual receptor points is quite large (see, for example, the validation studies by EPA[4] and CERC[5]). For dispersion over complex terrains, the deviation between observed and modeled data for short-term averages can be even larger (factor of 4 to 6 or even more) (Beychok, 1979, Benarie, 1987). However, hourly concentration comparisons show much greater scatter than long-term average concentration comparisons (see comparisons on the EPA website listed below).

Vertical Plume term S(x, z)

Case I. No reflecting boundaries:

$$S(x,z) = \exp\left[-\frac{1}{2}\left(\frac{h_e - z}{\sigma_z}\right)^2\right]. \tag{7.9}$$

Case II. Reflection at ground-level only:

$$S(x,z) = \exp\left[-\frac{1}{2}\left(\frac{h_e - z}{\sigma_z}\right)^2\right] + R_g \exp\left[-\frac{1}{2}\left(\frac{h_e + z}{\sigma_z}\right)^2\right]. \tag{7.10}$$

Case III. Multiple plume reflections along ground and at top of mixing boundary layer height h_{mix}:

$$\begin{aligned}S(x,z) = &\exp\left[-\frac{1}{2}\left(\frac{h_e - z}{\sigma_z}\right)^2\right] + R_g \exp\left[-\frac{1}{2}\left(\frac{h_e + z}{\sigma_z}\right)^2\right] \\ &+ R_m \exp\left[-\frac{1}{2}\left(\frac{h_e + z - 2h_{mix}}{\sigma_z}\right)^2\right].\end{aligned} \tag{7.11}$$

Case IV. Multiple plume reflections along ground surface and at top of the mixing boundary layer height (total or perfect reflection aloft and along the ground is assumed, $R_g = R_m = 1$):

---

[4] USEPA website, *Preferred/Recommended Models*, www.epa.gov/scram001/dispersion_prefrec.htm

[5] Cambridge Environmental Research Consultants (CERC) website, *Validation Studies*, www.cerc.co.uk/environmental-software/model-documentation.html

$$S(x, z) = \sum_{i=0, \pm1, \pm2, \ldots} \left\{ \exp\left[-\frac{1}{2}\left(\frac{h_e - z - 2ih_{mix}}{\sigma_z} \right)^2 \right] \right.$$

$$\left. + \exp\left[-\frac{1}{2}\left(\frac{h_e + z + 2ih_{mix}}{\sigma_z} \right)^2 \right] \right\}. \quad (7.12)$$

Approximations of $S(x, z)$:

- for $\dfrac{\sigma_z}{h_{mix}} \leq 0.63$ (7.13a)

  truncate series at $i = 0, \pm1$

- for $\dfrac{\sigma_z}{h_{mix}} = (0.63, 1.08]$

  $$S(x, z) = \frac{(2\pi)^{0.5}\sigma_z}{h_{mix}}(1 - \varepsilon^2)$$

  $$\times \left[1 + \varepsilon^2 + 2\varepsilon\cos\left(\frac{\pi z}{h_{mix}} \right)\cos\left(\frac{\pi h_e}{h_{mix}} \right) \right].$$

  $$\varepsilon = \exp\left[-\frac{1}{2}\left(\frac{\pi\sigma_z}{h_{mix}} \right)^2 \right]$$

  (7.13b)

- for $\dfrac{\sigma_z}{h_{mix}} > 1.08$

  $$S(x, z) = \frac{(2\pi)^{0.5}\sigma_z}{h_{mix}} \quad \text{(vertically homogeneous}$$

  "well mixed" atmosphere).   (7.13c)

Pollutant decay or total depletion term R(x)

$$R(x) = \exp\left[-\frac{x}{u_s\tau_R} \right]; \quad (7.14)$$

where $\dfrac{x}{u_s}$ is the pollutant travel time to reach x τ_R is a characteristic time scale related to pollutant removal rate (sec. 7.3.2)

$$R(x) = \exp\left\{ -\overbrace{\frac{\tau_{ct} + \Lambda}{u_s}}^{\text{Wet deposition and chemistry}}x - \underbrace{\frac{v_{dry}}{u_s}\sqrt{\frac{2}{\pi}}\int_0^x \frac{1}{\sigma_z}\exp\left[-\frac{1}{2}\left(\frac{h_e}{\sigma_z} \right)^2 \right]dx}_{\text{Mass removal by dry deposition}} \right\} \quad (7.15)$$

The Gaussian plume model is considered most accurate for long-term or annual averages (Zannetti 1990), and therefore we have used it for damage cost calculations. It is easy to see why the average distribution of concentrations downwind of the source is approximately Gaussian. Consider a pollutant puff that is carried along in the direction x of the dominant wind, while being subjected to random vertical and lateral fluctuations. From one time step i to the next i+1 such fluctuations cause successive random lateral and vertical displacements. For a particular path, the total displacement in the y direction after N time steps is simply the sum of the N random fluctuations:

$$\text{Total displacement in } y = y_1 + y_2 + \cdots + y_N.$$

Averaged over a large number of paths, the expectation value of y is 0 and its variance is:

$$\text{Variance of } y, \sigma^2(y) = \sigma^2(y_1) + \sigma^2(y_2) + \cdots + \sigma^2(y_N),$$

since the y_i are uncorrelated with each other and have finite variance. Now one can invoke the central limit theorem of statistics, which implies that in the limit of large N, the distribution of y approaches a Gaussian, even if the individual displacements are not Gaussian. Likewise the vertical distribution approaches a Gaussian.

In our own calculations of long-term averages, we have compared results of the ISC Gaussian model of EPA (1995) with measured SO_2 ambient concentrations near the Donges refinery, located between the cities of St. Nazaire and Nantes along the Atlantic coast of France. For the year 2000, the average SO_2 concentration at a measuring station just 2 km from the refinery was 8.8 $\mu g/m^3$. Additional measurements over the same time period in St. Nazaire and Nantes were, respectively, 3.6 and 4.7 $\mu g/m^3$. Since the refinery is the main source of SO_2 emissions in the area, the increment above background due to the refinery is simply the difference between the nearby measured value 8.8 $\mu g/m^3$ and the regional background, which we take as 4.1 $\mu g/m^3$, the mean of the values for St. Nazaire and Nantes. The excess due to the refinery is 4.7 $\mu g/m^3$. Using detailed refinery emissions data for 2000, along with complete information on source parameters, including the stack heights, the ISC model predicts an annual concentration increase of 3.5 $\mu g/m^3$ at the measuring station, which is quite close to the excess above the regional contribution 4.7 $\mu g/m^3$.

At the regional level, Barrett (1992) has compared long-term averages of regional transport models and found agreement within a factor of two or better.

226 Atmospheric dispersion of pollutants

7.2.1 Pasquill stability categories

Turbulence intensity can be characterized in terms of six Pasquill stability categories (Table 7.1): classes A, B and C apply to very, moderately and slightly unstable conditions, respectively; class D refers to neutral conditions, and classes E and F denote slightly and moderately stable atmospheres. Atmospheric stability is a function of ground-level wind speed, solar radiation (solar altitude angle a_{SOL}, Section 7.2.2) and cloud cover. Surface roughness (the presence of obstacles along the surface) can also enhance horizontal and vertical turbulent flow, although this effect is included through modification of the dispersion parameters σ_y and

Table 7.1 *Pasquill stability categories or classes*

(a) *Relation to daytime insolation (incoming solar radiation), cloud cover and wind speed (m/s, anemometer height 10 m). Adapted from Zannetti, 1990*

Time of day, insolation and cloud cover	≤ 2	2 to ≤ 3	3 to ≤ 5	5 to ≤ 6	> 6
Daytime (cloud cover < 50%)					
Strong insolation ($a_{SOL} > 60°$)	A	A–B	B	C	C
Medium insolation ($35° < a_{SOL} < 60°$)	A–B	B	B–C	C–D	D
Slight insolation ($a_{SOL} < 35°$)	B	C	C	D	D
Daytime (cloud cover > 50%)					
Strong insolation ($a_{SOL} > 60°$)	B	B–C	C	D	D
Medium insolation ($35° < a_{SOL} < 60°$)	B–C	C	C–D	D	D
Slight insolation ($a_{SOL} < 35°$)	C	D	D	D	D
Day or night (overcast)	D	D	D	D	D
Night					
Cloud cover > 50%	F	E	D	D	D
Cloud cover < 50%	F	F	E	D	D

(b) *Relation to ambient temperature gradient, $\frac{dT}{dz}$ [From US NRC, 1972] (potential temperature gradient, $\partial \Gamma / \partial z = \frac{dT}{dz} + \gamma_d$, where $\gamma_d = 0.00986\,°C/m$ is the dry adiabatic lapse rate).*

Pasquill dispersion class	Turner classification	Parcel buoyancy	$\frac{dT}{dz}(°C/m)$
A (very unstable)	1	Positive	< -0.019
B (unstable)	2	Positive	-0.019 to ≤ -0.017
C (slightly unstable)	3	Positive	-0.017 to ≤ -0.015
D (neutral)	4	Neutral	-0.015 to ≤ -0.005
E (slight stable)	5	Negative	-0.005 to ≤ 0.015
F (stable)	6	Negative	> 0.015

σ_z (Zannetti, 1990). Neutral conditions normally occur during daytime–nightime transition periods and on mostly cloudy or windy days (winds greater than 6 m/s). Stable conditions are the norm during clear nights with weak winds; whereas unstable conditions are characteristic during daytime hours when there is a positive heat flux at the earth's surface.

Atmospheric pressure decreases with altitude (about 1200 Pa per 100 m at sea level). Stability classification can be interpreted in terms of the vertical motion of a parcel of air as it expands with increasing altitude, while subjected to an external buoyancy force. If the atmosphere is unstable, an air parcel rises across the atmosphere, since its temperature is always greater than that of its surroundings (*positively* buoyant). For a stable atmosphere, the opposite occurs. Namely, vertical ascent is opposed, since the air parcel has a temperature lower than that of the surrounding atmosphere (*negatively* buoyant). Under these conditions, airborne pollution is trapped close to the ground, which leads to high ground concentrations. Also, depending on atmospheric stratification, the pollutant can be carried far downstream of the source. Finally, for neutral conditions, the net buoyancy force is null, meaning that an air parcel has neither a tendency to rise nor a tendency to fall, but rather its vertical motion depends solely on its vertical momentum.

7.2.2 Solar altitude angle, a_{SOL}

The solar altitude angle or angular position relative to the horizon is evaluated using the relationship (Rabl, 1985),

$$a_{SOL} = \mathrm{Sin}^{-1}(\mathrm{Cos}\,\delta\,\mathrm{Cos}\,L_\theta\,\mathrm{Cos}\,\omega + \mathrm{Sin}\,\delta\,\mathrm{Sin}\,L_\theta), \qquad (7.16)$$

where, a_{SOL} is the solar altitude angle measured in degrees above the horizon ($\alpha = 90° -$ zenith angle), δ is the declination angle (degrees), L_θ is the latitude angle (degrees) and ω is the hour angle based on the solar time, T_{SOL} measured in hours ($\omega = 15° \cdot [T_{SOL} - 12]$, in degrees). ω is the angular displacement of the sun east or west of the local meridian (L_{LOC}) due to the earth's rotation (15°/hour). It is zero at solar noon, when the sun crosses the observer's meridian, negative in the morning (east) and positive in the afternoon (west).

The declination angle is calculated using the formula,

$$\delta = 23.45 \sin\left[\frac{360}{365}(284 + \mathrm{DoY})\right]. \qquad (7.17)$$

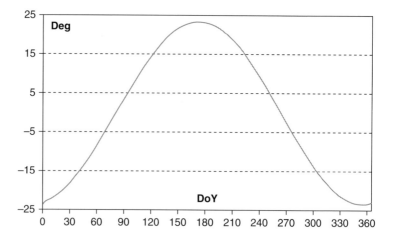

DoY stands for day of year. For example, DoY equals 1 for January 1st, 32 for February 1st, and so on. Solar and local standard time (T_{LOC}) scales are related through Eq. (7.18):

$$T_{SOL} = T_{LOC} + \frac{1}{60}[4\,(L_{STD} - L_{LOC}) + E_T].\qquad(7.18)$$

T_{SOL} is the solar time, T_{LOC} is the observer's local time, L_{STD} is the standard meridian corresponding to the local time zone (in degrees) and E_T is the *equation of time* (in units of minutes), it accounts for perturbations due to the earth's orbit (Eq. (7.19)):

$$E_T = 19.74\,\mathrm{Sin}\,\psi\,\mathrm{Cos}\,\psi - 7.53\,\mathrm{Cos}\,\psi - 1.5\,\mathrm{Sin}\,\psi$$
$$\psi = \frac{360}{364}(\mathrm{DoY} - 81)\qquad(7.19)$$

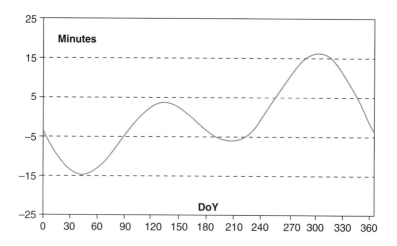

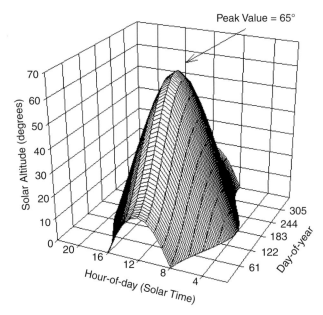

Fig. 7.5 Solar altitude annual variation for Paris, France ($L_\theta = 48.7°$ north). Solar radiation at earth's surface $= 1376 \text{ W/m}^2 \cdot \text{Sin}(a_{SOL}) \cdot (1 - \text{albedo}) \cdot [0.2 + 0.8 (1 - \text{cloud cover})]$. For most land areas, the albedo ranges between 0.1 (forest) and 0.4 (desert sand). Values for vegetated surfaces range between 0.1 and 0.25, while fresh snow can reach 0.9. Earth's average albedo is 0.3.

Figure 7.5 shows the annual variation of a_{SOL} as a function of solar time and day of year for the city of Paris in France ($L_\theta = 48.7°$ north).

7.2.3   Surface roughness length, z_o

The surface roughness is a characteristic length scale indicating the size of obstacles encountered along the ground surface as the plume moves away from the source – higher values correspond to larger objects. Values of z_o are shown in Fig. 7.6.

7.2.4   Dispersion parameters, σ_y and σ_z

Concentrations calculated with Eq. (7.7) depend on several key input parameters, in particular the choice for the lateral (σ_y) and vertical (σ_z) standard dispersion coefficients. Modeling the evolution of these

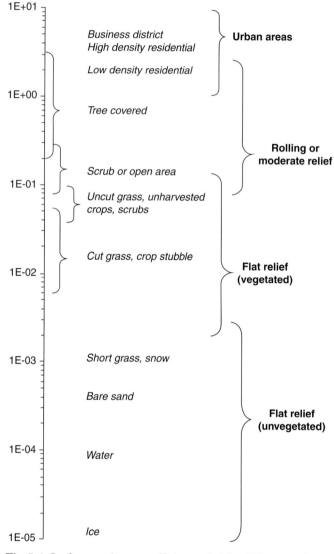

Fig. 7.6 Surface roughness coefficient z_o (m) for different surfaces (From Zannetti, 1990). Typically z_o in about 1/30 of mean physical height of obstacles in area.

parameters with increasing distance from the source is a major challenge for all Gaussian models. Dispersion coefficients can be calculated using a formal mathematical treatment based on similarity theory, which incorporates turbulence intensity measurements linked to wind component fluctuations along the horizontal and vertical directions (Hanna *et al.*, 1977, Draxler, 1976 and Irwin, 1979), or using semi-empirical relations derived from field experiments (Pasquill, 1961, Gifford, 1961 and 1975, Smith, 1968, McElroy and Pooler, 1968, McElroy, 1969, Turner, 1970, Briggs, 1973, McMullen, 1975, Vogt, 1977 and Green *et al.*, 1980).

In Table 7.2, we report several possible functional relations for the Gaussian dispersion coefficients. Two sets of results are presented, one for urban and another for rural locations. Plume widths for urban areas are larger than for rural dispersion, because surface obstacles and generally warmer temperatures enhance atmospheric mixing (turbulence intensity). The surrounding air is less stable, and consequently the plume spread in urban or industrial zones is greater than in rural areas. Furthermore, the peak ground-level concentration of an urban plume lies closer to the source and its value is greater than for a rural plume. For a comprehensive discussion and historical perspective, we recommend Zannetti (1990) or Seinfeld and Pandis (1998).

Table 7.2 *Gaussian dispersion parameters*

(a) *Briggs sigmas (recommended)*
The Briggs curves represent an interpolation of the Pasquill–Gifford and Brookhaven schemes mentioned below. The urban sigmas were derived from previous experimental work conducted by McElroy and Pooler (1968 and 1969). Range of validity is between 100 m and 10,000 m.

Briggs (1973) *and Gifford (1975)*

$$\sigma(x) = \kappa_1 x \left(1 + \kappa_2 x\right)^{\kappa_3} \qquad (x \text{ and } \sigma \text{ in meters}). \qquad (7.20)$$

Constants κ_1 through κ_3 depend on Pasquill stability class
RURAL *dispersion*

	σ_y			σ_z		
	κ_1	κ_2	κ_3	κ_1	κ_2	κ_3
Stability A	0.22	0.0001	−0.5	0.20	0	0
Stability B	0.16	0.0001	−0.5	0.12	0	0
Stability C	0.11	0.0001	−0.5	0.08	0.0002	−0.5
Stability D	0.08	0.0001	−0.5	0.06	0.0015	−0.5
Stability E	0.06	0.0001	−0.5	0.03	0.0003	−1
Stability F	0.04	0.0001	−0.5	0.016	0.0003	−1

Table 7.2 (*cont.*)

URBAN *dispersion*

	σ_y			σ_z		
	κ_1	κ_2	κ_3	κ_1	κ_2	κ_3
Stability A	0.32	0.0004	−0.5	0.24	0.001	0.5
Stability B	0.32	0.0004	−0.5	0.24	0.001	0.5
Stability C	0.22	0.0004	−0.5	0.20	0	0
Stability D	0.16	0.0004	−0.5	0.14	0.0003	−0.5
Stability E	0.11	0.0004	−0.5	0.08	0.0015	−0.5
Stability F	0.11	0.0004	−0.5	0.08	0.0015	−0.5

(b) *Pasquill-Gifford sigmas*
Curves were derived from field experiments conducted over flat, smooth
terrain in a rural setting, *characterized by a surface roughness coefficient* z_o
equal to 3 cm, for a near-surface, non-buoyant emission source using a
sampling rate of 3 to 10 minutes up to a distance x *of 800 m from the*
source.

Green et al. (1980)

$$\sigma_y(x) = \frac{\kappa_1 x}{[1 + (x/\kappa_2)]^{\kappa_3}}; \; \sigma_y(x) = \frac{\kappa_4 x}{[1 + (x/\kappa_2)]^{\kappa_5}} \quad (\text{x and } \sigma \text{ in meters}). \quad (7.21)$$

Constants κ_1 *through* κ_5 *depend on Pasquill stability class*

	κ_1	κ_2	κ_3	κ_4	κ_5
Stability A	0.250	927	0.189	0.1020	−1.918
Stability B	0.202	370	0.162	0.0962	−0.101
Stability C	0.134	283	0.134	0.0722	0.102
Stability D	0.0787	707	0.135	0.0475	0.465
Stability E	0.0566	1070	0.137	0.0335	0.624
Stability F	0.0370	1170	0.134	0.0220	0.700

McMullen (1975)

$$\sigma(x) = \exp\left[\kappa_1 + \kappa_2 \ln x + \kappa_3 \ln^2 x\right] \quad (\text{x in km and } \sigma \text{ in meters}). \quad (7.22)$$

Constants κ_1 *through* κ_3 *depend on Pasquill stability class (7.22)*

Table 7.2 (*cont.*)

	σ_y			σ_z		
	κ_1	κ_2	κ_3	κ_1	κ_2	κ_3
Stability A	5.357	0.8828	−0.0076	6.035	2.1097	0.2770
Stability B	5.058	0.9024	−0.0096	4.694	1.0629	0.0136
Stability C	4.651	0.9181	−0.0076	4.110	0.9201	−0.0020
Stability D	4.230	0.9222	−0.0087	3.414	0.7371	−0.0316
Stability E	3.922	0.9222	−0.0064	3.057	0.6794	−0.0450
Stability F	3.533	0.9191	−0.0070	2.621	0.6564	−0.0540

(c) *Brookhaven sigmas*
The Brookhaven sigmas were developed from field experimental data at the Brookhaven National Laboratory in the United States and are appropriate for elevated gas releases (108 m) dispersed over forested areas ($z_o \sim 1$ m, rough surface). The sampling rate was 1 hour, and concentration measurements were recorded up to several kilometers from the source.

<u>Smith (1968)</u>

$$\sigma(x) = \kappa_1 x^{\kappa_2} \qquad \text{(x and } \sigma \text{ in meters)}. \tag{7.23}$$

Constants κ_1 and κ_2 depend on Pasquill stability class

	σ_y		σ_z	
	κ_1	κ_2	κ_1	κ_2
Stability A (gustiness category B2)	0.40	0.91	0.41	0.91
Stability B or C (gustiness category B1)	0.36	0.86	0.33	0.86
Stability D (gustiness category C)	0.32	0.78	0.22	0.78
Stability E	Take average of gustiness categories C and D			
Stability F (gustiness category D)	0.31	0.71	0.06	0.71

7.2.5 *Effective stack height (plume centerline), h_e*

The exhaust gases emitted from the elevated stacks of power plants and industrial facilities usually rise well above the stack due to the plume's considerable initial stack exit velocity and to buoyancy momentum because the combustion flue gases are typically much hotter than the surrounding ambient air temperature. Plume growth and trajectory are affected by the prevailing ambient conditions, specifically wind speed and direction at plume height and atmospheric stability (Pasquill class).

Turbulent air entrainment, due in part to the prevailing atmospheric turbulence intensity and in part self-induced by the initial motion of the plume, contributes to plume growth, increasing its mass and decelerating its vertical ascent. Meanwhile, wind acts on the developing plume by bending its shape along the wind direction. Typically, after a few hundred m, the plume height levels off and its trajectory is determined solely by the prevailing wind direction (Figure 7.3).

From there on it looks as though the pollution originates from an imaginary (virtual) point located some distance above the physical stack height. The plume centerline defines what is called the effective stack height h_e; it is a function of downwind distance from the source. The incremental height above the chimney is defined as the plume rise Δh:

$$h_e(x) = h_s + \Delta h(x), \qquad (7.24)$$

where x is the downwind distance from the source. The maximum ground concentration (y and z = 0) varies roughly as the inverse square of the effective stack height, which is why plume uplift is so critical to pollutant dilution (see Chapter 18 in Seinfeld and Pandis, 1998). Equations for estimating Δh in a thermally stratified atmosphere are summarized below in Table 7.3 (Briggs 1972 and 1975). A plume is buoyancy-driven if the temperature difference between the stack exhaust gases T_s and the ambient air T_a is greater than 20°C. F_B is the Briggs buoyancy flux parameter (in m^4/s^3); it is proportional to the flue gas sensible heat emission rate $\dot{Q}_s$ (J/s) relative to the surrounding air temperature:

$$F_B = g\, w_s\, d_s{}^2 \left(\frac{T_s - T_a}{4T_s}\right) = \frac{g}{\pi\, c_{pa} \varrho_a T_a} \dot{Q}_s, \qquad (7.25)$$

g is the gravitational acceleration (9.81 m/s^2), w_s and d_s are, respectively, the flue gas exhaust speed and stack exit diameter, c_{pa} is the specific heat capacity at constant pressure measured at T_a (in K) and ϱ_a is the ambient air density. The characteristic downwind distance from source to location where the plume centerline has leveled off is identified by x_f (i.e. the distance to maximum plume rise, in m). The value of x_f is shortest for stable atmospheric conditions; it can range from several 100 m for stable ambient dispersion to several km for unstable conditions.

Wind profile exponent

Wind speed at source height h_s is calculated using a power law function of height (Eq. (7.26)):

Table 7.3 *Plume rise equations (F_B is defined in Eq. (7.25))*
[From Briggs (1972 and 1975) and EPA (1995)]

Stability class	Formulae
Unstable (Pasquill A, B, C) or Neutral atmosphere (Pasquill D)	Distance to final plume rise, x_f (in m):

$$x_f = 49\, F_B^{5/8} \quad (F_B < 55)$$

$$x_f = 119\, F_B^{2/5} \quad (F_B \geq 55)$$

Transitional phase plume rise (in m; $x < x_f$):

$$\Delta h\,(x) = 1.6\frac{F_B^{1/3}\, x^{2/3}}{u_s} \leq \Delta h_{max} \ (\text{all } F_B) \tag{7.29}$$

Final or maximum plume rise (in m; $x \geq x_f$):

$$\Delta h_{max} = 21.425\,\frac{F_B^{3/4}}{u_s} \quad (F_B < 55)$$

$$\Delta h_{max} = 38.71\,\frac{F_B^{3/5}}{u_s} \quad (F_B \geq 55) \tag{7.30}$$

Stable atmosphere (Pasquill E and F class)

Distance to final plume rise, x_f (in m):

$$x_f = 2.0715\,\frac{u_s}{\sqrt{\xi}} \ (\text{all } F_B) \tag{7.31}$$

Stability parameter, ξ (in $1/s^2$)

$$\xi = \frac{g}{T_a}\frac{\partial\Gamma}{\partial z} = \frac{g}{T_a}\left[\frac{\partial T_a}{\partial z} + \gamma_d\right]$$

As default approximations, EPA (1995) recommends $\partial\Gamma/\partial z = 0.020$ °C/m for E class and $\partial\Gamma/\partial z = 0.035$ °C/m for stability class F. See also Table 7.1b for other choices. γ_d is the dry adiabatic lapse rate (0.00986 °C/m)

Transitional phase plume rise (in m; $x < x_f$):

$$\Delta h(x) = 1.6\frac{F_B^{1/3}\, x^{2/3}}{u_s} \leq \Delta h_{max} \ (\text{all } F_B) \tag{7.32}$$

Final or maximum plume rise (in m; $x \geq x_f$):

$$\Delta h_{max} = 2.6\left(\frac{F_B}{u_s\xi}\right)^{1/3} \ (\text{all } F_B) \tag{7.33}$$

Table 7.3 nomenclature
g 9.81 m/s^2
w_s, T_s and d_s Flue gas exhaust speed (m/s), temperature (K) and stack exit diameter (m)
u_s Ambient wind speed at stack or release height (m/s); see Table 7.4
T_a Ambient temperature (K)
$\partial\Gamma/\partial z$ Potential temperature gradient (K/m)

Table 7.4 *Wind profile exponent η for Eq. (7.26)*

Reference	A	B	C	D	E	F
			Exponent for Pasquill stability category			
EPA (1995) (recommended)						
Rural	0.07	0.07	0.10	0.15	0.35	0.55
Urban	0.15	0.15	0.20	0.25	0.30	0.30
Smedman-Högström *et al.* (1978), $\eta = \kappa_0 + \kappa_1 \log(z_0) + \kappa_2 \log^2(z_0)$ (sensitivity analysis)						
		$\kappa_0 = 0.18$	$\kappa_0 = 0.18$	$\kappa_0 = 0.30$	$\kappa_0 = 0.52$	$\kappa_0 = 0.80$
		$\kappa_1 = 0.13$	$\kappa_1 = 0.13$	$\kappa_1 = 0.17$	$\kappa_1 = 0.20$	$\kappa_1 = 0.25$
		$\kappa_2 = 0.03$	$\kappa_2 = 0.03$	$\kappa_2 = 0.03$	$\kappa_2 = 0.03$	$\kappa_2 = 0.03$
Touma (1977)						
Rural (range)	0.10–0.14	0.09–0.18	0.08–0.17	0.12–0.21	0.20–0.33	0.41–0.56
Rural (average)	0.11	0.12	0.12	0.17	0.29	0.45
Hsu (1982)						
Tropical coast	< 0.165	0.165	0.19	0.235	0.40	> 0.40

$$\frac{u_s}{u_{ref}} = \left(\frac{h_s}{z_{ref}}\right)^\eta, \tag{7.26}$$

where u_s is the wind speed (m/s) at stack height h_s, u_{ref} is the wind speed at ground station height z_{ref} (ground wind speeds are typically recorded at a reference height of 10 m). The wind profile exponent η is a function of z_0 and Pasquill stability class (Table 7.4). For near-neutral (adiabatic) conditions in rural areas, η is often chosen as $1/7$.

For neutral conditions and relatively flat terrain, the vertical wind profile can be evaluated using the well-known and widely used logarithmic wind profile function (Eq. (7.27), Panofsky and Dutton, 1984):

$$\frac{u_2(z_2)}{u_1(z_1)} = \frac{\ln(z_2/z_0)}{\ln(z_1/z_0)} \tag{7.27}$$

Equation (7.27) should be applied up to a height of about 200 m; beyond this height, wind speed can be assumed constant with increasing altitude. For rough terrain conditions (and neutral stability), Eq. (7.28) can be used instead (Zannetti, 1990):

$$\frac{u_2(z_2)}{u_1(z_1)} = \frac{\ln\left(\frac{z_2-\epsilon}{z_0}\right)}{\ln\left(\frac{z_2-\epsilon}{z_0}\right)}. \tag{7.28}$$

The displacement distance ∈ equals 70% to 80% of the height of the large roughness elements found in the study area. A comparative assessment of different approaches of estimating η and methods commonly used to extrapolate measured wind speeds has been carried out by Gualtieri and Secci (2011) for coastal areas.

7.2.6   The mixed layer height, h_{mix}

The planetary (or atmospheric) boundary layer (PBL) is the lowest part of the troposphere where pollutant transport and mixing is influenced by local surface forces, meteorology and terrain landscape. Surface effects contribute to vertical exchanges of mass and energy. The PBL can be subdivided into several layers, each one characterized by different dispersion properties (Fig. 7.7). The mixing height (mixing boundary layer or mixing depth) h_{mix} refers to that portion of the PBL where turbulent flow is fully developed and dominates air pollution phenomena. The presence of strong eddy currents enhances the mixing of nearby air parcels, which greatly influences the pollutant dilution rate compared with the case of only molecular motion.

The turbulence intensity of the atmosphere is characterized by the Pasquill stability class (Table 7.1). For unstable or neutral conditions, the temperature of the PBL decreases with increasing height z until a temperature inversion layer is reached (z_{INV}), at which point the air temperature increases with increasing altitude. After some distance,

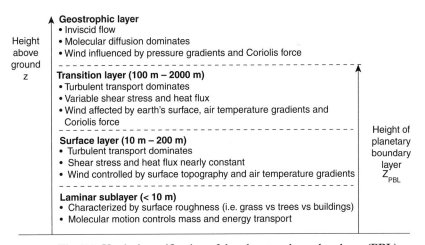

Fig. 7.7 Vertical stratification of the planetary boundary layer (PBL).

a second inversion point is reached,[6] and the air temperature once again starts to decrease with increasing height (Fig. 7.1). Under stable conditions, which normally occur during clear nights with calm winds, a ground-based inversion layer is present and the air temperature increases with vertical height up to z_{PBL}. A stable mixing layer height is formed because of mechanical turbulence. For unstable or neutral conditions, h_{mix} is often set equal to the planetary boundary layer height z_{PBL} unless measurements of an elevated inversion layer are available, in which case h_{mix} is set equal to the inversion height (i.e. $h_{mix} = z_{INV} < z_{PBL}$). For a stable atmosphere, h_{mix} is always less than z_{PBL}. Typically, z_{PBL} varies between 100 m (stable conditions) to 2000 m (very unstable atmosphere); much higher values are possible, but only during unusual weather such as thunderstorms. h_{mix} values can be calculated using the scheme summarized in Table 7.5. As an illustration, we show in Fig. 7.8 calculations using the model by Luhar (1998).

7.2.7   Calculation of concentration profiles and local population exposure

The information presented in Sections 7.2.1 through 7.2.6 serves as input to calculate the concentrations in the proximity of the emission source,

Table 7.5 *Proposed scheme for estimating the mixed layer height h_{mix} (in meters).* *(From EPA (1989), Zannetti (1990), Cenedese et al. (1997), Seinfeld and Pandis 1998)*

Stability	Formulae	
Unstable atmosphere, or convective boundary layer (Pasquill class A, B and C)	$h_{mix}(t) = \left\{ \dfrac{2\int_{t_0}^{t} H_f(t)dt}{c_{pa}\, \varrho_a\, [\gamma_d - \gamma(t=t_0)]} \right\}^{0.5}$	(7.34)
	$c_{pa} = 1006\ J/(^{\circ}C \cdot kg_{dry\text{-}air})$ $\varrho_a = 1.2\ kg/m^3$ $\gamma_d = 0.00986\ ^{\circ}C/m$ $\gamma = -\partial T/\partial z$ at sunrise $t = t_0$ Note, $h_{mix} \propto t$	
Neutral atmosphere (Pasquill class D)	$h_{mix} = \kappa_1 \dfrac{u_*}{f}, \quad u_* = \kappa_2 \dfrac{u(z)}{\ln\left(\frac{z}{z_0}\right)}, \quad f = 2\Omega \sin L_\theta$	(7.35)
	$\kappa_1 = 0.3$ (EPA, 1989), 0.15 to 0.25 (Zannetti, 1990) $\kappa_2 \approx 0.4$ (von Karman constant) $z_0 =$ see Fig. 7.6 $\Omega = 0.0000729/s$	

[6] It is possible to have multiple inversion layers.

Table 7.5 *(cont.)*

Stability	Formulae
Stable atmosphere (Pasquill class E and F)	$h_{mix} = \kappa_3 \left(\dfrac{u_*}{f} L_{MO}\right)^{0.5}$, $\quad \dfrac{1}{L_{MO}} = \kappa_4 z_0{}^{\kappa_5}$ (7.36)

$$u_* = \kappa_2 \frac{u(z)}{\ln\left(\frac{z}{z_0}\right) + \frac{4.7(z - z_0)}{L_{MO}}}$$

$\kappa_3 \approx 0.4$

$\kappa_4 = 0.00807$ (E class) and 0.03849 (F class)

$\kappa_5 = -0.3049$ (E class) and -0.1714 (F class)

Table 7.5 nomenclature

$H_f(t)$ Surface heat flux at time t (t_0 is sunrise time)

c_{pa} Specific heat capacity at constant pressure of dry air

ϱ_a Air density

γ Ambient lapse rate (γ_d = dry adiabatic lapse rate)

u_* Friction velocity

$u(z)$ Wind speed at height z

z_0 Surface roughness coefficient (z = vertical height above ground)

f Coriolis parameter (Ω = earth's rotation rate; L_0 = latitude)

L_{MO} Monin–Obukhov length ($L_{MO} > 0$ for stable atmosphere and $L_{MO} < 0$ for unstable conditions)

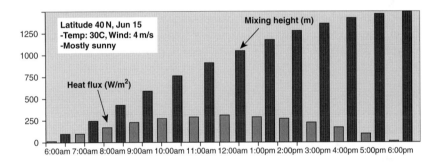

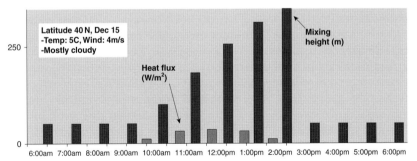

Fig. 7.8 Example of surface heat flux (W/m^2) and daytime atmospheric mixing height (m) for a rural site. Graph generated by the authors using the CSIRO model of Luhar (1998) as implemented at www.cmar.csiro. au/airquality/mixheight/index.html.

and, therefore, to estimate the exposure to a pollutant across the affected population and to then calculate the resulting damage cost. This information will also be needed to prepare the input meteorological file to run the RiskPoll model.

The exposure $E(x, y, z)$ at point (x, y, z) is the product of the concentration at that location multiplied by the population that is affected (N_{pers}). The collective exposure is the sum across the local domain:

$$\text{Local Exposure} = \sum_{\text{Local domain}} c(x, y, z)$$

$$\times N_{pers}(x, y, z) \left(\text{with units} \frac{\mu g}{m^3} \cdot \text{persons} \right).$$

Gaussian concentration plumes for different dispersion assumptions, atmospheric conditions and source parameters, including stack height and exhaust gas conditions, are presented in Fig. 7.9. Ground concentration profiles $(z = 0)$ as a function of distance from source x have been averaged along the y-axis (crosswind direction). Mean estimates were calculated with Eq. (7.8). Values for the reflection term $S(x, z = 0)$ were calculated using Eq. (7.13) p. 224 when the source height h_s or the effective stack height h_e, as determined by the plume rise equations, summarized in Table 7.3, was less than the mixing height h_{mix}. Otherwise ground concentrations are zero. The mixing layer height is a function of surface roughness and Pasquill stability category. Equations in Table 7.5 were used to estimate the mixing height h_{mix}, based on the assumption that $z_o = 0.1$ m for the rural site and $z_o = 1.5$ m for the urban location. Dispersion coefficients were calculated using the empirical formulae in Table 7.2. The pollutant emission rate (continuous source) is 1000 metric tonnes per year (31.7 g/s), while the other stack parameters are identified in the graph inserts. The decay coefficient $R(x)$ is set to unity; that is to say, the plume mass is conserved for all values of x, or total reflection occurs along the ground and at the top of the mixing layer.

Maximum ground-level concentrations (Fig. 7.9g) occur beneath the plume centerline $(z = y = 0)$. These values have been calculated with Eq. (7.7). The graphs in Fig. 7.9h show the vertical concentrations at different downwind distances x for two release heights under different assumptions about atmospheric turbulence intensity. When the vertical concentration profile no longer depends upon z, the atmosphere is said to be uniformly mixed. The downwind distance to a vertically mixed atmosphere depends on the plume growth rate, and therefore, on the Pasquill stability class. The final series of plots in Fig. 7.9j illustrate the effect of stack parameters on the estimation of the collective population exposure for a source located near Paris and Albi (rural site) in France. Calculations cover the local domain only, which consists of a circular

(a) Base case (curves for urban source in black and for rural source in gray).

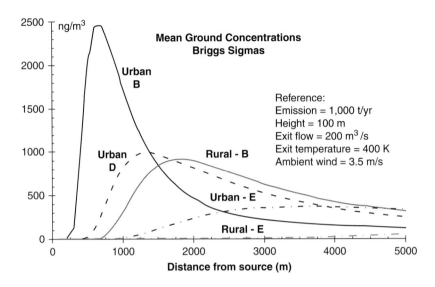

(b) Effect of source height (h_s).

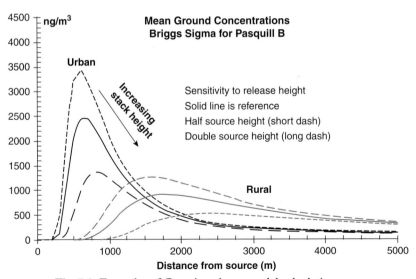

Fig. 7.9 Examples of Gaussian plume model calculations

(c) Influence of effective source height (h_e).

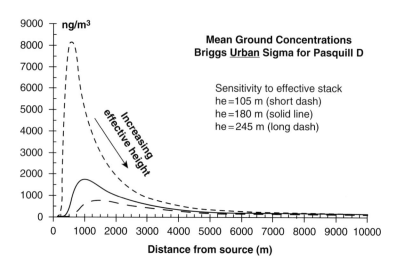

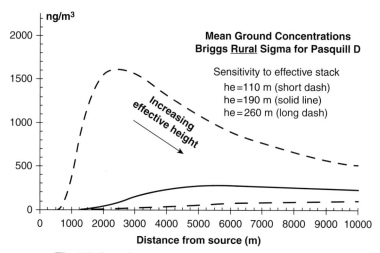

Fig. 7.9 (cont.)

(d) Sensitivity to surface wind speed (u).

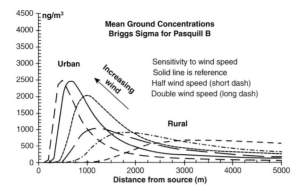

(e) Sensitivity to exhaust flue gas flow rate $\left(\frac{\pi}{4} d_s^2 w_s\right)$

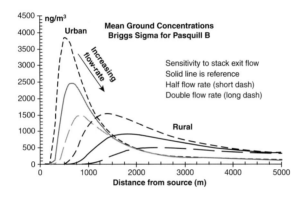

(f) Sensitivity to exhaust temperature (T_s).

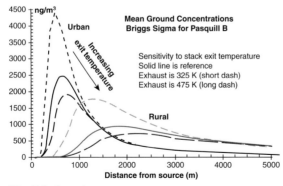

Fig. 7.9 (cont.)

(g) Maximum concentrations at ground level versus downwind distance x for three Pasquill stability classes. For Class B, the long-term mean concentration is also shown (as a solid line).

Urban source

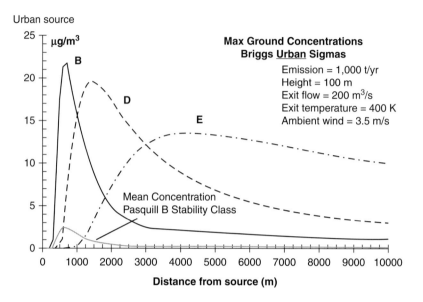

Rural source

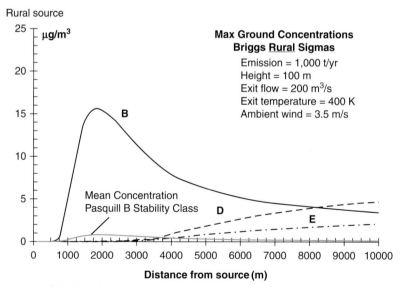

Fig. 7.9 (cont.)

(h) Vertical concentration profiles as a function of stack height (h_s) and Pasquill stability category for a continuous source located near Paris, France (1000 t/yr).

Plume can be considered uniformly mixed when concentration variation with vertical height (z) becomes negligible. Results calculated using the ISC software of US EPA (1995).

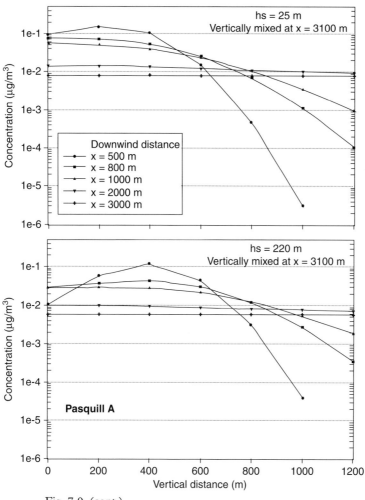

Fig. 7.9 (cont.)

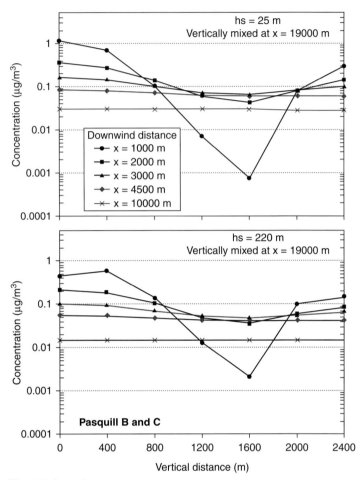

Fig. 7.9 (cont.)

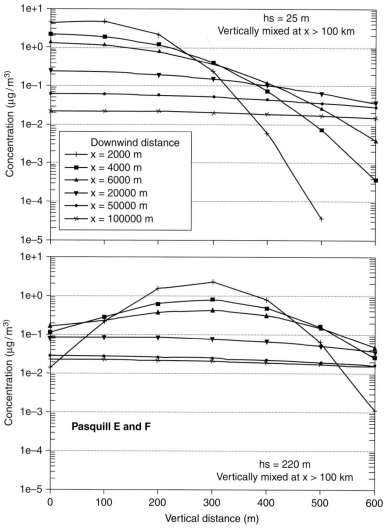

Fig. 7.9 (cont.)

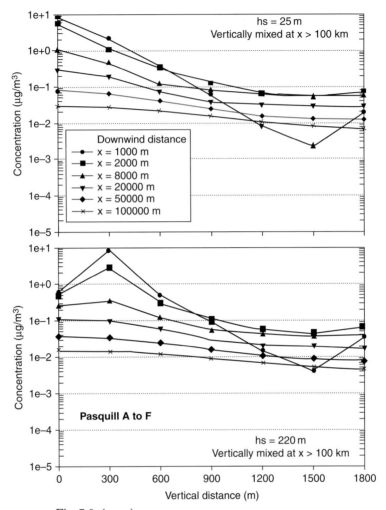

Fig. 7.9 (cont.)

(i) Consequence of the choice of dispersion parameters, σ_y and σ_z

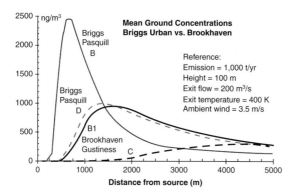

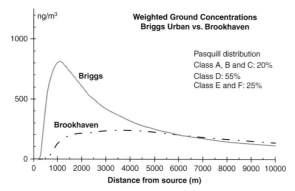

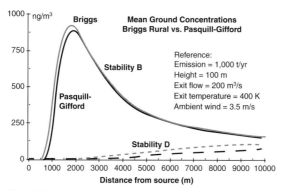

Fig. 7.9 (cont.)

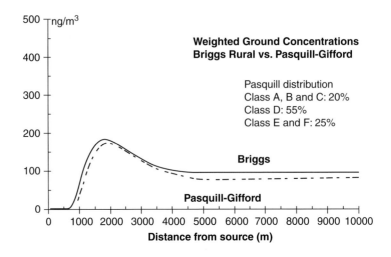

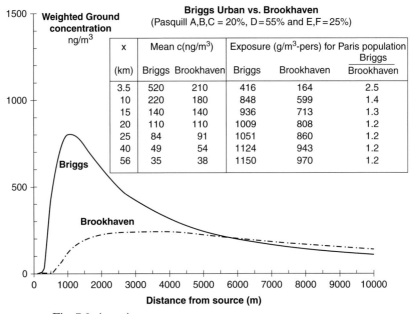

| x | Mean c(ng/m³) | | Exposure (g/m³-pers) for Paris population | | Briggs |
	Briggs	Brookhaven	Briggs	Brookhaven	Brookhaven
(km)					
3.5	520	210	416	164	2.5
10	220	180	848	599	1.4
15	140	140	936	713	1.3
20	110	110	1009	808	1.2
25	84	91	1051	860	1.2
40	49	54	1124	943	1.2
56	35	38	1150	970	1.2

Fig. 7.9 (cont.)

(j) Effect of stack parameters and wind speed in local population exposure calculations within 56 km from the source (population-averaged concentration for local domain size = 10,000 km^2).

Source located near Paris, France (local population of 10.1 million)

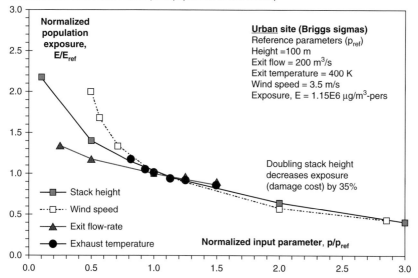

Source located near Albi, France (rural site with local population of 560,000)

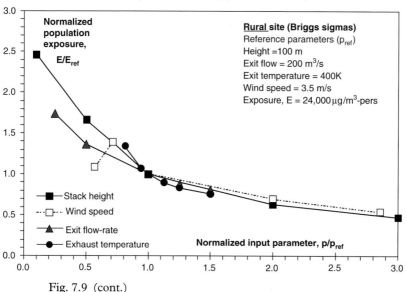

Fig. 7.9 (cont.)

area centered at the source with radius 56 km. Results are presented in a format which is well suited for sensitivity studies. Each input parameter p and the associated exposure are first normalized by their reference values, which are indicated in the figure insert, and the pair is then plotted on the graph. All variations are relative to the reference state. Thus, for example, for a doubling of the reference stack height, the reference exposure (damage cost) would decrease by about 40%. A decrease in the population exposure is certainly expected, since the plume travels further downstream before touching the ground. Greater dilution results in lower ground-level concentrations.

Several conclusions can be drawn from the results illustrated in Fig. 7.9:

- Compared with the rural location, urban dispersion results reveal higher peak concentrations that are closer to the source as a consequence of enhanced vertical mixing. But further downstream, rural concentrations exceed urban values (Fig. 7.9a).
- High turbulence and mixing result in greater maximum concentrations as compared with stable conditions, but unstable profiles have a narrower width (Fig. 7.9a).
- As commented on earlier, tall stacks result in lower ground concentrations and lower exposures of the population in the vicinity of the source (Fig. 7.9b). In addition, the peak concentration is shifted farther downstream with increasing height. Over the range of stack heights considered here, 10 m to 300 m, the local exposure decreased by a factor of five (Fig. 7.9j).
- Unlike the other input variables, the wind speed simultaneously affects the dilution rate and the plume rise, with opposite effects on the concentration. Higher wind speeds result in lower concentration due to greater pollutant dilution, and the peak concentration lies closer to the source (Fig. 7.9d). At low wind speeds ("calm" conditions), the concentration increases at first because of the reduced dilution rate, but further reductions lead to lower concentration as a result of the larger plume rise. Under certain conditions, the effective height may exceed the mixing height, which would then significantly reduce population exposure near the source (Fig. 7.9j).
- Concentrations decrease when either the exhaust flow rate or temperature, or both, increase, since the plume rises higher in the air and travels a longer distance before reaching ground, thus reducing concentrations in the proximity of the source (Fig. 7.9e–f). The sensitivity to temperature change (buoyancy-driven plume) is more pronounced than the effect of flow rate (mechanical momentum).

- Maximum ground concentrations (y = 0) exceed mean estimates (averaged along y direction) by an order of magnitude. Both max and mean concentration profiles peak at about the same downwind distance x (Fig. 7.9g).
- The downwind distance when the vertical concentration gradient no longer depends upon z varies with stability category. The distance from the source when the atmosphere is vertically homogeneous is reduced with increasing turbulence intensity. Values range between a few kilometers from the source for very unstable ambient conditions and tens of km for neutral atmospheres (Fig. 7.9h). For stable conditions, the distance can be quite large, reaching hundreds of km or even more, which is certainly well beyond the range of validity of the Gaussian plume model. The distance to vertically mixed x_{mixed} can be approximated by Eq. (7.7)c ($x_{mixed} = \sigma_z^{-1}[1.08h_{mix}]$, where $\sigma_z^{-1}[...]$ is the inverse function $\sigma_z[x_{mixed}]$).
- As demonstrated in Fig. 7.9i, the choice of the sigma parameters has a significant impact on both magnitude and location of the peak concentration and also on the shape of the concentration profile. This can have a significant effect on the calculated population exposure if the population distribution is very non-uniform. The last part of Fig. 7.9i shows an example where close to the source, the Brookhaven model underestimates the exposure of the population calculated by the Briggs model by a factor of 2.5. Over the entire local domain, however, the discrepancy is a modest underestimation of just 20%.

7.3   Atmospheric dispersion far from the source

7.3.1   Pollutant removal

Far from the source, the vertical concentration profile becomes nearly uniform (see Fig. 7.9h), as a consequence of plume spread and reflections at the ground and at the top of the lowest, elevated temperature inversion (mixed boundary layer height). At downwind distances greater than about 50 km (the typical "extended" range assumed in most Gaussian plume models), dry and wet deposition and chemical transformation contribute in a significant way to pollutant removal. If there were no removal mechanisms, in fact, the pollutant concentration in the atmosphere would increase indefinitely.

Dry deposition occurs at the earth's surface due to adsorption onto soil particles or vegetation, or absorption into water. Wet deposition involves pollutant absorption into suspended water droplets which are then removed by precipitation. Transformation occurs through chemical interactions with other pollutants present in the surrounding air (see Fig. 7.10) or because of radioactive decay. Wet and dry processes can be characterized in terms of a deposition velocity which can be interpreted as the average velocity at which a pollutant is removed at the ground. Dry deposition velocity is dependent on pollutant characteristics, such as particle aerodynamic size and terrain features, such as surface roughness (smooth or complex terrain) and surface coverage type (vegetated or not, for instance). Values exhibit diurnal and seasonal fluctuations. Typical ranges of dry deposition velocities are indicated below (Seinfeld and Pandis, 1998, Feliciano et al., 2001, Underwood, 2001). Deposition velocities of gases can be up to an order of magnitude smaller than those of particulate matter. Field measurements have revealed a wide range of possible values, with a high to low ratio of velocities spanning more than two orders of magnitude (Sehmel, 1980). The smaller the deposition velocity the farther the pollutant is transported in the atmosphere.

Pollutant	Typical range of dry deposition velocity (cm/s)
PM_{10}	Theory – urban: 0.4 to 0.6; rural: 0.1 to 0.4; observed – 0.06 to 1
$PM_{2.5}$	0.05 to 0.8 (observed)
SO_2	0.2 to 3 (observed)
NO_2	0.2 to 1 (observed)

Wet deposition velocity is proportional to the precipitation rate and the washout ratio, the latter representing the ratio of pollutant concentration in surface-level precipitation to pollutant concentration in surface-level air. The washout ratio depends on location, time, rainfall intensity, particle size, rain droplet size and scavenging efficiency for below- or in-cloud scavenging. Wet deposition velocities for PM_{10} can range between 3.5 cm/s (for a rainfall intensity of 0.5 mm/h, drizzle) to 65 cm/s (for 25 mm/h, heavy rain). For $PM_{2.5}$, a representative range is 1.8 to 3.3 cm/s (Seinfeld and Pandis, 1998, Underwood, 2001).

For chemical transformation, the relevant timescale is the pollutant transformation rate. For SO_2, a typical value is 1% per hour (Zannetti, 1990, Luria et al., 2001, Khoder, 2002; see also Section 7.3.2).

Accounting for pollutant removal requires more complex dispersion algorithms than Gaussian plumes, in order to capture more faithfully atmospheric variations in time and space (horizontal and vertical

(a) NO to nitrate aerosols.

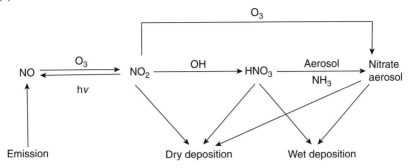

(b) SO$_2$ to sulfate aersols.

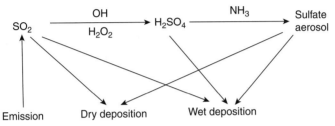

(c) NH$_3$ to ammonium aerosols.

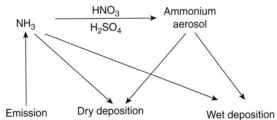

(d) Ozone formation.

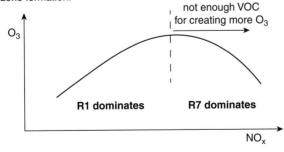

Fig 7.10 (*cont.*)

Scheme for ozone O_3 formation (Sillman 1999)

Photolysis of NO_2 (hv = light, 420 nm) and additional ozone

$$NO_2 + hv \rightarrow NO + O \tag{R1}$$

$$O + O_2 \rightarrow O_3 \tag{R2}$$

Oxidation of NO without destruction of O_3

$$CO + OH \rightarrow HO_2 + CO_2 \tag{R3}$$

$$HO_2 + NO \rightarrow NO_2 + OH \tag{R4}$$

$$RH + OH \rightarrow RO_2 + H_2O \tag{R5}$$

$$RO_2 + NO \rightarrow NO_2 + RO + HO_2 \tag{R6}$$

(RH – directly emitted hydrocarbon radicals, RO – intermediate organic pollutant and RO_2 – alkyl peroxy radicals are collectively referred to as Volatile Organic Compounds, VOC)

Daytime creation
(NO_x-VOC-CO)

Oxidation of NO and destruction of O_3

$$NO + O_3 + NO_2 \tag{R7}$$

This reaction counterbalances O_3 formation by NO_2 photolysis (R1). The equilibrium ozone concentration depends on the ratio of NO_2 and NO concentrations. Increasing the presence of VOCs in the atmosphere favors oxidation of NO according to reactions (R1 – R6), and consequently there is a net increase in ozone concentration. O_3 formation is highly non-linearly with changes in VOC and NO_x emissions. At low ambient NO_x concentrations, ozone increases almost linearly with NO_x emission (relatively independent of VOC changes) up to a local maximum concentration. Beyond this point, ozone formation is inhibited by further increases of NO_x, but O_3 increases with increasing VOC emissions.

Ozone reduction

Typically, ozone concentrations are lowest at nighttime because there is no formation via photolysis of NO_2. Concentrations are also reduced in the vicinity of large sources of NO emissions, which is typically the case of stack emissions from power plants. This process is referred to as NO_x titration, and may occur during daytime and nighttime hours.

Fig. 7.10   Formation of secondary pollutants from precursor SO_2, NO and NH_3 emissions. From ExternE (2005).

directions). Weather variability, whether occurring on a daily or seasonal time-scale, can have a profound impact on air concentrations. A rainstorm can wash out the bulk of a pollutant in the plume, while NO_x transformation is strongly influenced by ambient temperature, sunlight and the presence of other compounds in the air, especially ammonia (Zhou et al., 2003). Some pollutants remain in the air for several days and can expose a large number of receptors, while others have an environmental residence time on the order of years, decades (or even longer), and thus have a global reach (carbon dioxide, methane, mercury are a few examples), or even an inter-generational impact.

There are numerous long-range transport models, including the Windrose Trajectory Model of ExternE (Krewitt et al., 1995), EMEP[7] (the official dispersion model for policy decisions on trans-boundary air pollution in Europe) and CALPUFF (Scire et al., 2000). Developed by the US Environmental Protection Agency (EPA), CALPUFF is a Lagrangian puff model (Zannetti, 1990). It simulates mass transport as pollutant puffs that are released into the air at regular intervals. Transport is based on Gaussian dispersion, along with pollutant removal by dry and wet deposition and chemical transformation. The ambient wind flow carries the puffs downwind from the source. The prevailing wind direction and wind speed vary with time and space, unlike the Gaussian model, which assumes a horizontally homogeneous wind field. At a particular receptor site, the pollutant concentration is a weighted mean of all puffs crossing that point. CALPUFF can be used to model both primary (direct emissions) and secondary pollutants formed in the air via chemical transformation.

The USEPA has compared concentration estimates from CALPUFF with ISC, EPA's widely used default model for regulatory applications until it was replaced by AERMOD (EPA, 1998a). The EPA has also compared CALPUFF with tracer gas concentrations from two short-term field experiments (EPA, 1998b). The conclusions from these studies are summarized below. The EPA findings suggest that a factor of two between modeled and measured concentrations is to be expected. This conclusion agrees with the population-total exposure uncertainty analysis carried out by Spadaro and Rabl (2005), who in their analysis recommended a geometric standard deviation σ_g of 1.2 for a large city, 1.9 for a rural site and a value of 1.5 for average sites (see Section 11.3.1.2 in Chapter 11).

---

[7] www.emep.int

EPA (1998a) A comparison of CALPUFF with ISC3

"*Overall trends have been noted in the percentage difference comparisons in simulated concentration values between CALPUFF and ISC3. For taller point sources, there is a trend toward higher concentrations being simulated by CALPUFF in comparison to ISC3. For annual averages, the closer a receptor is to the source and the taller the stack, the greater the chance that the CALPUFF concentration values will be higher than those simulated by ISC3. At the more distant downwind receptor rings, the bias changes direction from CALPUFF yielding higher concentrations, to CALPUFF yielding relatively lower concentrations and sometimes these concentrations are lower than their respective ISC3 counterpart.*"

EPA (1998b) A comparison of CALPUFF modeling results to two tracer field experiments

"*The performance of the CALPUFF atmospheric dispersion model for two field tracer experiments is summarized. The first tracer experiment was in 1975 at Savannah River Laboratory and the second was in 1980 in the central United States. Both experiments examined long-range transport of an inert tracer material. The results generally were encouraging, with the simulated results within a factor of two of the observed data for the statistical measures presented in the report. However, there is not a consistent pattern of over- or under-estimation relative to the observations.*"

7.3.2   A simple model for regional dispersion

7.3.2.1 Primary pollutants

Primary pollutants are those emitted directly by the source. The concentration relationships presented here for primary pollutants, and later on for secondary pollutants, are appropriate for distances beyond about 50 km from the source. At these distances, the source can be treated as though it were a vertical line stretching from the ground to the top of the mixing height, h_{mix}. We assume a continuous and constant source at the origin of our coordinate system, under steady-state conditions with uniform atmospheric characteristics and a constant horizontal wind speed u in all directions (uniform windrose). At a distance r from the source, the mass balance for the control volume shown in Fig. 7.11 yields in the limit of small Δr:

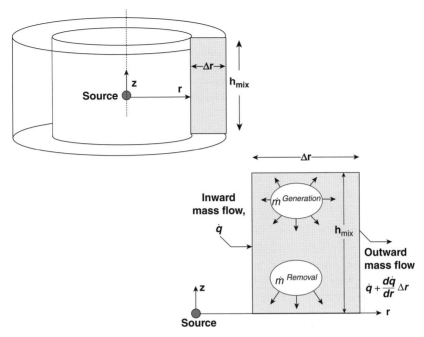

Fig. 7.11 Control volume for mass balance calculations of primary and secondary pollutants (volume of cylindrical annulus is $2\pi\,r\,\Delta r\,h_{mix}$).

$$\frac{d\dot{q}_p(r)}{dr} + \frac{\dot{m}_p^{Removal}}{\Delta r} = \frac{\dot{m}_p^{Generation}}{\Delta r}$$

with $\dot{q}_p = 2\pi\,r\,u\,h_{mix}c_p$; $\dot{m}_p^{Generation} = 0$ and $\dot{m}_p^{Removal} = 2\,\pi\,r\,\Delta r\,k_p\,c_p$.

$$\text{hence,}\quad \frac{d\dot{q}_p(r)}{dr} + \frac{k_p}{u\,h_{mix}}\,\dot{q}_p(r) = 0 \quad \text{with } \dot{q}_p(0) = \dot{m}_p$$

$$(7.37)$$

$\dot{m}_p$ is the primary pollutant p emission rate, $\dot{m}_p^{Removal}$ is its surface removal rate between r and r + Δr, $\dot{m}_p^{Generation}$ is the pollutant generation rate within the control volume, $\dot{q}_p(r)$ is the mass flow rate across a cylindrical surface of radius r and height h_{mix} and k_p is the pollutant depletion velocity (m/s). For the mean wind speed u, typical values over land are around 5 to 8 m/s (wind speed maps can readily be found on the web).[8] The mixing height varies with time of day and year, and depends on solar radiation and atmospheric turbulence (see Fig. 7.8). An annual mean range is 600 to 800 m.

---

[8] e.g. http://www.3tier.com/en/support/resource-maps/
http://www.nrel.gov/wind/international_wind_resources.html

The solution for the primary pollutant concentration c_p is:

$$c_p(\dot{m}_p, r) = \frac{\dot{m}_p}{2\pi \, u \, h_{mix}} \frac{1}{r} e^{-\frac{k_p}{u \, h_{mix}} r} = \frac{\alpha_p}{r} e^{-\beta_p \, r}, \qquad (7.38)$$

with coefficients $\alpha_p = \dfrac{\dot{m}_p}{2\,\pi\,u\,h_{mix}}$ and $\beta_p = \dfrac{k_p}{u\,h_{mix}}$,

and $k_p = \dfrac{\dot{m}_p\,\beta_p}{2\,\pi\,\alpha_p}$.
$$(7.39)$$

The units of c_p depend on the units chosen for $\dot{m}_p$ (usually, c_p has units $\mu g/m^3$). The term $\frac{\alpha_p}{r}$ is the solution to the special case when there is no pollutant removal, i.e. $k_p = 0$. The substance is simply diluted in the atmosphere ($\dot{q}_p = \dot{m}_p$). Pollutant removal occurs through the exponential factor whose exponent is the ratio of the plume transit time $\frac{r}{u}$ to reach position r, and the atmospheric residence (or decay) time τ_R, ($\tau_R = h_{mix}/k_P$). β_p is the pollutant decay factor, it represents the rate at which the pollutant is removed with distance (% loss per m), and the product $\dot{m}_p \, exp(-\beta_p r)$ is the pollutant mass remaining in the plume after traveling a distance r from the source.

The depletion velocity k_p depends on the ratio $\frac{\beta_p}{\alpha_p}$, which can be determined through a regression analysis of either measured or simulated concentration data as a function of source distance. We have calculated depletion velocities of various pollutants by fitting concentrations of the Windrose Trajectory Model (WTM) (Krewitt et al., 1995) and of EMEP (Barrett, 1992; Sandness, 1993). An example for SO_2 has been presented above in Fig. 7.4. Table 7.6 lists the results for a source located in the middle of France. The value of α_p has been expressed as a function of the

Table 7.6 *Fit coefficients to Eq. (7.39) and Eq. (7.42) and depletion velocities of primary and secondary pollutants for an emission source located in the middle of France.*

Pollutant	α_p (kg/m^2)	β_p (1/m)	k_p (cm/s)	α_s (kg/m^2)	β_s (1/m)	k_{ps} (cm/s)
PM$_{10}$	3.567E-5 $\dot{m}_p$	1.502E-6	0.68			
SO$_2$	3.188E-5 $\dot{m}_p$	1.470E-6	0.73			
NOx	2.971E-5 $\dot{m}_p$	2.741E-6	1.47			
Sulfates				2.078E-5 $\dot{m}_p$	4.198E-6	1.73
Nitrates				−2.272E-5 $\dot{m}_p$	2.159E-6	0.71

$\dot{m}_p$ is the primary or precursor pollutant emission rate, expressed in kg/s. The precursor pollutant for sulfates is SO_2 and for nitrates it is NO_2.

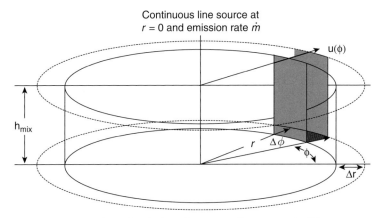

Continuous line source at
$r = 0$ and emission rate $\dot{m}$

Fig. 7.12 Dispersion of emissions from a continuous vertical line source.

emission rate $\dot{m}_p$, assuming units of kg/s. Additional depletion velocities are summarized in Table 7.8 (in Section 7.4).

The model can be generalized to consider (Fig. 7.12):

(i) The case of variable wind speeds in each direction with distribution $f(u(\phi), \phi)$, which represents the fraction of the time the wind blows in the direction ϕ at speed $u(\phi)$.

(ii) The case when plume trajectories meander instead of being straight lines; in this case, we replace the term $\exp(-\beta\, r)$ by $\exp(-k\, t_r\, /\, h_{mix})$, where t_r is the transit time to reach location r.

7.3.2.2 Secondary pollutants

Secondary pollutants are formed in the air through chemical change of precursor pollutants. As indicated in Fig. 7.10, the presence and abundance of ammonia (NH_3) particles in the atmosphere plays a key role in the formation of sulfate and nitrate aerosols, through neutralization of sulfuric and nitric acids. Ammonia ambient concentrations are strongly dependent on local farming activity and, to a lesser degree, on regional industrial production or electricity generation. Cars fitted with catalytic converters are another important source of ammonia emissions. Both sulfur and nitrogen compete for the available ammonia in the air, but sulfur tends to be neutralized first. Unlike sulfates, nitrate formation is reversible, the equilibrium between nitric acid, ammonium nitrate and ammonia can shift, depending on ambient temperature and ammonia availability. During summer months, formation of ammonium nitrates is inhibited, and concentrations tend to be several times lower than during the wintertime; in fact, concentrations can be up to an order of magnitude lower (West *et al.*, 1999, Zhou *et al.*, 2003). But, Khoder (2002) noted the opposite effect in

North Africa, where summer concentrations were higher than in the wintertime, with corresponding higher transformation rates, because of greater availability of ammonia in the air in the summertime. If transformation of nitrogen is limited primarily to the formation of nitric acid, the removal of nitrogen could be quite fast, as nitric acid has a high deposition velocity. Saturation of atmospheric ammonia levels is another possibility, as reported by Krewitt et al. (1999). If saturation occurs, the incremental exposure to sulfates (nitrates) due to an incremental tonne of SO_2 (NO_x) would be zero.

Unlike primary pollutants, secondary pollutants are less sensitive to local meteorology and source parameters (e.g. height) as chemical transformation is a relatively slow process that takes place over tens to hundreds of km downwind of the source. In the meantime, the plume has spread uniformly across the mixing boundary layer. Following the same procedure as carried out for primary pollutants, let us derive an expression for the concentration profile of a secondary pollutant s, based on mass conservation for the control volume shown in Fig. 7.11. Assume both primary and secondary pollutants are vertically well mixed, and the transformation from primary to secondary is proportional to the concentration of the precursor pollutant, c_p. The proportionality constant is the chemical transformation velocity k_{ct}, given as the product of the mixing height and the chemical transformation rate τ_{ct} ($k_{ct} = \tau_{ct} h_{mix}$). The concentration of the secondary pollutant at the origin is zero. Thus,

$$\frac{d\dot{q}_s(r)}{dr} + \frac{\dot{m}_s^{Removal}}{\Delta r} = \frac{\dot{m}_s^{Generation}}{\Delta r}$$

with $\dot{q}_s = 2\pi r u h_{mix} c_s$; $\dot{m}_s^{Generation} = 2\pi r \Delta r k_{ct} c_p$; $\dot{m}_s^{Removal} = 2\pi r \Delta r k_s c_s$

hence,
$$\frac{d\dot{q}_s(r)}{dr} + \frac{k_s}{u h_{mix}}\dot{q}_s(r) = \frac{k_{ct}}{u h_{mix}}\left(\dot{m}_p e^{-\frac{k_p}{u h_{mix}}r}\right) \text{ with } \dot{q}_s(0) = 0,$$

$$(7.40)$$

where $\dot{m}_s^{Generation}$ is the creation rate of the secondary pollutant from transformation of the precursor pollutant and $\dot{m}_s^{Removal}$ is its removal rate at the surface.

The solution to Eq. (7.40) is:

$$c_s(\dot{m}_p, r) = \frac{\dot{m}_p}{2\pi u h_{mix}}\frac{1}{r}\left(\frac{k_{ct}}{k_s - k_p}\right)\left[e^{-\frac{k_p}{u h_{mix}}r} - e^{-\frac{k_s}{u h_{mix}}r}\right]$$

$$= \frac{\alpha_s}{r}\left(e^{-\beta_p r} - e^{-\beta_s r}\right),$$

$$(7.41)$$

$$\text{with coefficients } \alpha_s = \frac{\alpha_p k_{ct}}{(k_s - k_p)} \text{ and } \beta_s = \frac{k_s}{u\, h_{mix}}$$

$$\text{and } k_{ps} = \frac{k_p k_s}{k_{ct}} = \frac{\dot{m}_p}{2\pi\, \alpha_s} \frac{\beta_s}{(\beta_s - \beta_p)} \frac{\beta_p}{}$$

$$(7.42)$$

The first exponential term inside the brackets in Eq. (7.41) represents the secondary pollutant creation rate (which is proportional to the amount of the precursor pollutant), while the second exponential is its removal rate (deposition and possibly further chemical transformation). Coefficients α_s, β_s and k_{ps} are obtained through regression analysis of secondary concentrations with source distance using WTM or EMEP; values for France are shown in Table 7.6. k_{ps} can be interpreted as the secondary pollutant "effective" depletion velocity. The ratio $k_{ct}/k_p = \tau_R\, \tau_{ct}$ with τ_R = atmospheric residence time and τ_{ct} = transformation rate.

7.3.2.3 Alternative methods to determine the depletion velocity

Value based on atmospheric removal rates
Consider a well-mixed column of polluted air of height h_{mix} and cross-sectional area ΔA. A mass balance for primary and secondary pollutants yields the following relationships:

Primary pollutants

$$-\frac{d(\Delta A\, h_{mix}\, c_p)}{dt} = \Delta A\, k_p\, c_p = (\Delta A\, v_{dry,p}\, c_p + \Delta A\, v_{wet,p}\, c_p + \Delta A\, k_{ct,p}\, c_p)$$

$$= \left(\overbrace{\Delta A\, v_{dry,p}\, c_p}^{\text{Dry deposition}} + \overbrace{\underbrace{\Delta A\, \Lambda_p h_{mix}\, c_p}_{v_{wet,\, p}}}^{\text{Wet deposition}} + \overbrace{\underbrace{\Delta A\, \tau_{ct,p} h_{mix}\, c_p}_{k_{ct,p}}}^{\text{Chemical transformation}} \right).$$

$$(7.43)$$

Secondary pollutants

$$\frac{d(\Delta A\, h_{mix}\, c_s)}{dt} = \left(\overbrace{\underbrace{\Delta A\, \tau_{ct,p}\, h_{mix}\, c_p}_{k_{ct,p}}}^{\text{Generation}} \right)$$

$$- \left(\overbrace{\Delta A\, v_{dry,s}\, c_s}^{\text{Dry deposition}} + \overbrace{\underbrace{\Delta A\, \Lambda_s\, h_{mix}\, c_s}_{v_{wet,\, s}}}^{\text{Wet deposition}} + \overbrace{\underbrace{\Delta A\, \tau_{ct,s}\, h_{mix}\, c_s}_{k_{ct,s}}}^{\text{Chemical transformation}} \right)$$

$$\underbrace{\qquad\qquad\qquad\qquad\qquad\qquad\qquad\qquad}_{\text{Removal rate} = \Delta A\, k_s\, C_s}$$

$$(7.44)$$

where,

Parameter	Description
t	Time
v_{dry}, v_{wet}	Dry and wet deposition velocities for primary (subscript p) and secondary (subscript s) pollutants in m/s. In general, distributions of dry deposition velocities can span up to two orders of magnitude (Nicholson, 1988, Sehmel, 1980), although typical values usually lie in the range 0.1 to 2 cm/s.
Λ	Washout (scavenging) coefficient, which determines the wet deposition velocity, the rate at which a pollutant is removed from the air because of local "scavenging" by water droplets and removal in the form of precipitation ($v_{wet} = \Lambda h_{mix}$).
	Λ is a function of droplet size, particle size and local precipitation rate. Λ has units 1/s. For a 1 mm/h precipitation rate, its value can vary between 4×10^{-5} 1/s for 1 μm particles and 4×10^{-4} 1/s for 10 μm particles (Seinfeld and Pandis, 1998, Underwood, 2001). For PM_{10}, the scavenging coefficient is a size-weighted average estimate.
$\tau_{ct,p}$ $\tau_{ct,s}$	Chemical conversion rates for primary-to-secondary and secondary-to-other pollutant, respectively ($k_{ct} = \tau_{ct}h_{mix}$). Values are reported as % of precursor pollutant transformed per hour. For SO_2 to sulfates transformation, literature values range between 0.2% and 7% per hour, depending on atmospheric photochemical activity, air temperature and humidity, and time of day or day of year (Luria *et al.*, 2001, Khoder, 2002, Miyakawa *et al.*, 2007). Lee and Watkiss (1998) recommend a value of 1%/h for Europe. For NO_x to nitrates, the transformation rate is usually lower, about a quarter to half of the rate of SO_2 to sulfates transformation (Khoder, 2002), but there can be considerable regional and temporal variation. Transformation rates have a lognormal distribution with geometric standard deviation 1.5 to 2.

Solving for k_p in Eq. (7.43) and then using Eq. (7.42) to calculate the effective depletion velocity for secondary pollutants one obtains:

$$\frac{k_p}{h_{mix}} = \left(\frac{v_{dry,p}}{h_{mix}} + \Lambda_p + \tau_{ct,p} \right) \tag{7.45}$$

$$\frac{k_{ps}}{h_{mix}} = \frac{1}{h_{mix}} \frac{k_p k_s}{k_{ct,p}} = \frac{1}{\tau_{ct,p}} \left(\frac{v_{dry,p}}{h_{mix}} + \Lambda_p + \tau_{ct,p} \right) \left(\frac{v_{dry,s}}{h_{mix}} + \Lambda_s + \tau_{ct,s} \right). \tag{7.46}$$

A value based on atmospheric residence time

Another method to calculate depletion velocities is based on the pollutant atmospheric residence time τ_R. Using data in Jolliet and Crettaz (1997), the depletion velocity[9] of primary pollutants can be related to τ_R according to Eq. (7.47):

---

[9] k is the reciprocal of the airborne fate factor used in the Life Cycle Impact Assessment methodology.

$$k_p = 1.2\,\tau_R^{-0.4};\ k_p \text{ in cm/s and } \tau_R \text{ in days } (\tau_R < 75 \text{ days}). \qquad (7.47)$$

For typical European conditions, atmospheric residence times for NO_x, SO_2 and particulates are, respectively, 1.1, 4.4 and 7.3 days (Dinkel *et al.*, 1996). Estimates of removal velocities given by Eqs. (7.39), (7.42), (7.45), (7.46) and (7.47) are compared in Table 7.7. As noted, all three approaches yield similar results across all pollutants (within a factor of two). In any case, whenever possible, estimates obtained from Eqs. (7.39) and (7.42) are preferred.

What to do in the case of lack of input data

In the absence of empirical or simulated input data to determine depletion velocities using any of the methods discussed above, the results from Table 7.8 can serve as a guide. These reference values can be "transferred" to other locations (study area) on the basis of similarity between the reference and target sites. As a preliminary assessment, one may also use the mean estimates of the data presented in Table 7.8, and then carry out sensitivity analyses to assess the variability of the answer.

Means of the depletion velocities in Table 7.8 are:

- $PM_{2.5}$ 0.68 cm/s
- PM_{10} 1.03 cm/s
- SO_2 0.97 cm/s
- NO_x 1.45 cm/s
- Sulfates 1.98 cm/s
- Nitrates 1.06 cm/s

7.4   The Uniform World Model (UWM)

7.4.1   Methodology

Almost all policy applications concern pollution sources in an entire country or region, rather than sources at a particular location. For example, regulations for vehicle emissions affect all vehicles of the region. Power plant regulations affect all plants of a given type or new plants that will be built in locations that are not yet known. Therefore, one needs typical damage estimates for an entire region rather than installation-specific results. Unless the sources are systematically associated with certain characteristics or particular sites (such as cars, most of whose impacts are imposed in urban regions), an average over all possible emission sites is a good estimate of typical values. Averaging over source sites is equivalent to making the receptor distribution uniform.

Table 7.7 *Comparison of depletion velocities (cm/s) using different methods of estimation.*

Pollutant	Estimates based on regression analysis Values from Table 7.6 for a source in France	Estimates based on atmospheric decay rate Eqs. (7.45) and (7.46)	Estimates based on atmospheric residence Eq. (7.47)
PM_{10}	0.68	0.73 to 2.16[1]	0.54
	0.86 (EU-27, Table 7.8)	1.33 (best guess)	
SO_2	0.73	0.44 to 3.08[2]	0.66
	0.88 (EU-27, Table 7.8)	0.81 (best guess)	
NO_x	1.47	0.27 to 2.2[3]	1.16
	1.36 (EU-27, Table 7.8)	0.73 (best guess)	
Sulfates	1.73	1.8 to 2.41[4]	
	1.85 (EU-27, Table 7.8)	1.86 (best guess)	
Nitrates	0.71	3.45, 4.42[5]	
	1.00 (EU-27, Table 7.8)	1.51 (best guess)	

1. PM_{10} input data
   - Low estimate: v_{dry} = 0.02 cm/s, Λ = 1.41E-5 s^{-1} (rain = 500 mm/yr) and h_{mix} = 500 m
   - High estimate: v_{dry} = 0.48 cm/s, Λ = 2.10E-5 s^{-1} (rain = 850 mm/yr) and h_{mix} = 800 m
   - Best guess: v_{dry} = 0.36 cm/s, Λ = 1.62E-5 s^{-1} (rain = 600 mm/yr) and h_{mix} = 600 m
2. SO_2 input data
   - Low estimate: v_{dry} = 0.35 cm/s, Λ = 0.127E-5 s^{-1} (rain = 500 mm/yr), h_{mix} = 500 m, τ_{ct} = 0.2%/hour
   - High estimate: v_{dry} = 1.1 cm/s, Λ = 0.526E-5 s^{-1} (rain = 850 mm/yr), h_{mix} = 800 m, τ_{ct} = 7%/hour
   - Best guess: v_{dry} = 0.5 cm/s, Λ = 0.238E-5 s^{-1} (rain = 600 mm/yr), h_{mix} = 600 m, τ_{ct} = 1%/hour
3. NO_x input data
   - Low estimate: v_{dry} = 0.2 cm/s, Λ = 0.127E-5 s^{-1} (rain = 500 mm/yr), h_{mix} = 500 m, τ_{ct} = 0.05%/hour
   - High estimate: v_{dry} = 1 cm/s, Λ = 0.526E-5 s^{-1} (rain = 850 mm/yr), h_{mix} = 800 m, τ_{ct} = 3.5%/hour
   - Best guess: v_{dry} = 0.5 cm/s, Λ = 0.238E-5 s^{-1} (rain = 600 mm/yr), h_{mix} = 600 m, τ_{ct} = 0.5%/hour
4. Sulfates input data
   - Low estimate: v_{dry} = 0.05 cm/s, Λ = 0.127E-5 s^{-1} (rain = 500 mm/yr) and h_{mix} = 500 m
   - High estimate: v_{dry} = 0.8 cm/s, Λ = 0.526E-5 s^{-1} (rain = 850 mm/yr) and h_{mix} = 800 m
   - Best guess: v_{dry} = 0.24 cm/s, Λ = 0.238E-5 s^{-1} (rain = 600 mm/yr) and h_{mix} = 600 m
5. Nitrates input data
   - Estimate #1: v_{dry} = 0.05 cm/s, Λ = 0.127E-5 s^{-1} (rain = 500 mm/yr) and h_{mix} = 500 m
   - Estimate #2: v_{dry} = 0.8 cm/s, Λ = 0.526E-5 s^{-1} (rain = 850 mm/yr) and h_{mix} = 800 m
   - Best guess: v_{dry} = 0.24 cm/s, Λ = 0.151E-5 s^{-1} (rain = 600 mm/yr) and h_{mix} = 600 m
     For NO_x, v_{dry} = 0.5 cm/s, Λ = 0.151E-5 s^{-1}, τ_{ct} = 1%/hour
Input data from the following sources: Carmichael (1994), Eliassen (1978), Sehmel (1980), Seinfeld and Pandis (1998) and Underwood (2001)

Table 7.8 *Depletion velocities (cm/s), determined through regression analysis of EcoSense concentrations (Krewitt* et al., *1995) with distance from source (Eqs. (7.38) and (7.41)); the authors of EcoSense also developed versions for China and South America. Values for North America and Asia are from literature that used the UWM methodology. Results for NH$_3$, sulfates and nitrates are effective depletion velocities of Eq. (7.42).*

Location/Region	PM$_{2.5}$	PM$_{10}$	SO$_2$	NO$_x$	NH$_3$	Sulfates	Nitrates
EU-27	0.57	0.86	0.88	1.36	1.07	1.85	1.00
Austria	0.56	0.84	0.85	1.19	1.05	1.95	1.03
Belgium	0.66	0.99	1.01	1.19	1.24	1.80	1.01
Bulgaria	0.49	0.74	0.88	1.51	0.91	1.85	0.88
Cyprus	0.43	0.65	0.77	1.27	0.81	1.36	0.84
Czech Repl.	0.59	0.89	0.87	1.04	1.11	2.15	1.26
Denmark	0.86	1.29	1.27	1.83	1.61	2.05	1.26
Estonia	0.62	0.93	1.00	1.67	1.16	1.35	1.29
Finland	0.62	0.93	1.00	1.67	1.16	1.35	1.29
France	0.45	0.68	0.73	1.47	0.84	1.73	0.71
Germany	0.52	0.78	0.73	1.01	0.98	1.94	0.83
Greece	0.49	0.74	0.88	1.51	0.91	1.85	0.88
Hungary	0.57	0.86	0.94	1.53	1.06	1.77	1.01
Ireland	0.59	0.89	0.94	1.18	1.11	2.03	1.28
Italy	0.71	1.07	0.99	1.38	1.33	1.86	1.04
Latvia	0.62	0.93	1.00	1.67	1.16	1.35	1.29
Lithuania	0.62	0.93	1.00	1.67	1.16	1.35	1.29
Luxembourg	0.59	0.89	0.73	1.24	1.11	1.84	0.77
Malta	0.45	0.68	0.73	1.36	0.84	1.40	0.73
Netherlands	0.66	0.99	1.01	1.19	1.24	1.80	1.01
Poland	0.57	0.86	0.90	0.96	1.08	2.00	1.23
Portugal	0.54	0.81	0.89	1.40	1.01	1.65	0.91
Romania	0.57	0.86	0.94	1.53	1.06	1.77	1.01
Slovakia	0.58	0.87	0.94	1.53	1.09	1.77	1.01
Slovenia	0.57	0.86	0.94	1.53	1.06	1.77	1.01
Spain	0.50	0.75	0.80	2.16	0.94	1.65	0.91
Sweden	0.86	1.29	1.27	1.83	1.61	2.05	1.26
UK	0.59	0.89	0.94	1.18	1.11	2.03	1.28
China & provinces	0.67	1.01	0.87	1.21	1.26	2.14	0.96
Gansu	0.66	0.99	1.06	2.35	1.24	2.06	0.67
Ningxia	0.72	1.08			1.35	2.24	0.91
Qinghai	0.71	1.06			1.33	2.61	1.17
Shaanxi	0.85	1.27			1.58	2.31	1.11
Xinjiang	0.68	1.02			1.23	2.32	0.89
Beijing	0.43	0.64	0.84	1.44	0.80	1.77	0.82
Hebei	0.55	0.83			1.04	2.05	0.84
Inner Mongolia	0.67	1.01			1.26	2.36	1.00
Shanxi	0.69	1.03			1.29	2.18	0.97

Table 7.8 (*cont.*)

Location/Region	$PM_{2.5}$	PM_{10}	SO_2	NO_x	NH_3	Sulfates	Nitrates
Tianjin	0.43	0.64			0.80	1.77	0.82
Chongqing	0.89	1.34			1.68	2.32	0.93
Guizhou	1.11	1.66			2.08	2.72	1.05
Sichuan	0.90	1.35			1.69	2.39	1.02
Tibet	0.68	1.02			1.28	2.32	0.89
Yunnan	1.22	1.83	1.16	0.90	2.29	2.26	0.81
Heilongjiang	0.42	0.63			0.79	1.69	0.79
Jilin	0.42	0.63			0.79	1.69	0.79
Liaoning	0.42	0.63			0.79	1.73	0.80
Guangdong	0.85	1.28			1.60	2.09	1.04
Guangxi	1.02	1.53			1.91	2.17	0.92
Hainan	0.85	1.28			1.60	2.09	1.04
Henan	0.63	0.94			1.18	1.99	0.92
Hong Kong	0.85	1.28			1.60	2.09	1.04
Hubei	0.59	0.88			1.10	1.70	0.72
Hunan	1.01	1.51			1.89	3.27	1.36
Macau	0.85	1.28			1.60	2.09	1.04
Anhui	0.61	0.92			1.15	1.96	0.86
Fujian	0.90	1.35			1.69	2.37	1.16
Jiangsu	0.56	0.84			1.05	2.14	0.99
Jiangxi	1.34	2.01			2.51	2.80	1.48
Shandong	0.49	0.74	0.66	0.96	0.93	0.90	0.97
Qingdao	*0.62*	*0.93*	*0.84*	*1.23*	*1.16*	*0.97*	*1.17*
Yantai	*0.56*	*0.84*	*0.74*	*1.33*	*1.05*	*1.00*	*1.04*
Jinan	*0.35*	*0.53*	*0.49*	*0.73*	*0.66*	*0.90*	*1.00*
Shanghai	0.64	0.96	1.06	1.40	1.20	2.27	1.02
Zhejiang	0.90	1.35			1.69	2.37	1.16
Balkans	0.49	0.74	0.88	1.51	0.91	1.85	0.88
Norway	0.89	1.34	1.27	1.83	1.68	2.05	1.26
Serbia	0.49	0.73	0.88	1.51	0.91	1.85	0.88
Switzerland	0.55	0.83	0.84	1.24	1.04	1.89	0.96
Tunisia	0.48	0.72	0.74	1.21		1.57	0.83
Indonesia							
• Jakarta	0.65	0.97	0.92	1.69		2.61	1.02
Malaysia	0.58	0.65	0.87	1.53		0.76	1.82
Pakistan							
• Kabirwala (Punjab)	1.41	2.11	2.00	2.24		3.49	1.31
• Karachi	1.09	1.64	1.55	1.74		2.71	1.02
Thailand	0.43	0.65	0.87	1.53		1.82	0.76
USA	0.37	0.55	0.83	0.40	0.69	1.96	0.99
Mexico		1.20					
• Salamanca			1.09	0.81		1.19	2.85
• Tuxpan (Atlantic coast)			1.10	2.48		1.15	1.73

Table 7.8 (*cont.*)

Location/Region	$PM_{2.5}$	PM_{10}	SO_2	NO_x	NH_3	Sulfates	Nitrates
• Manzanillo (Pacific Coast)			1.01	2.22		0.55	0.73
Argentina							
• Buenos Aires	1.46	2.19	2.08	0.40		3.63	1.59
Brazil							
• Belo Horizonte	0.84	1.26	0.84	1.49		3.11	1.33
• North Amazonas	1.91	2.86	1.38	2.26		4.76	3.00
• Rio Grande do Sul	1.05	1.58	1.17	2.05		3.38	1.36
Paraguay	0.75	1.13	1.05	2.13		3.13	1.04

A simple and convenient tool for estimating *typical* damages is the Uniform World Model (UWM), first presented by Curtiss and Rabl (1996) and further developed, with detailed validation studies, by Spadaro (1999), Spadaro and Rabl (1999) and Spadaro and Rabl (2002). In Spadaro and Rabl (2004), the model was extended to toxic metals and their pathways through the food chain (see Chapter 8). We begin with the simple UWM, which is a product of a few factors. It is instructive because it shows, at a glance, the role of the most influential parameters of the impact pathway analysis. As the following derivation shows, it is exact in the limit where the source is a vertical line, the distribution of the receptors is uniform and the atmospheric parameters are the same everywhere. For source heights above about 50 m it agrees with detailed site-specific calculations within a factor of two to three.

To improve the model, we then examine how the total population exposure changes when the population distribution, source height and/or the atmospheric parameters are changed. This allows us to derive modifications of the simple UWM to make it applicable even for ground-level sources and population distributions that are quite non-uniform. We also provide appropriate values of the atmospheric parameters for different countries of the world. In Section 7.4.2 we present validation studies. The model is implemented in the RiskPoll software (brief description presented in Appendix B).[10]

---

[10] The RiskPoll software has been written by J. Spadaro and is available for free at www.arirabl.org

7.4.1.1 Derivation of the simple UWM

We begin with the damage function of Eq. (7.5) for the emission of a primary pollutant at a rate $\dot{m}_p$,

$$\dot{D}_p(\dot{m}_p) = \sum_i S_{ERF,p,i} P_{p,i} \int\int r \, dr \, d\phi \, \rho(\bar{r}) c_p(\bar{r}, \dot{m}_p), \qquad (7.5)$$

but rewritten in cylindrical coordinates $\bar{r} = (r, \phi)$. $S_{ERF,p,i}$ and $P_{p,i}$ are for the impacts of the primary pollutant. Then we use the relation,

$$M_p = (\dot{m}_p, \bar{r}) = k_p(\bar{r}) c_p(\dot{m}_p, \bar{r}) \text{ with } k_p(\bar{r}) = v_{dry,p} + v_{wet,p} + k_{ct,p} \quad (7.48)$$

to replace the concentration by the removal flux $M_p(\dot{m}, \bar{r})$ to obtain,

$$\dot{D}_p(\dot{m}_p) = \sum_i S_{ERF,p,i} P_{p,i} \int\int r \, dr \, d\phi \, \frac{\rho(\bar{r}) M_p(\dot{m}_p, \bar{r})}{k_p(\bar{r})}. \qquad (7.49)$$

In a world where the receptor density and the depletion velocity are the same everywhere, one can replace $\rho(\bar{r})$ by ρ_{uni} and $k_p(\bar{r})$ by a constant k_p and take them outside the integral. That leaves an integral over the removal flux which is equal to the emission rate by conservation of matter,

$$\int\int r \, dr \, d\phi \, M_p(\dot{m}_p, \bar{r}) = \dot{m}_p. \qquad (7.50)$$

Thus we obtain the simple UWM for the damage cost rate of a primary pollutant,

$$\dot{D}_{UWM,p} = \left(\frac{\dot{m}_p \, \rho_{uni}}{k_p} \right) \sum_i \left[S_{ERF,p,i} P_{p,i} \right]. \qquad (7.51)$$

For the damage cost of a secondary pollutant due to the emission of a primary pollutant at a rate $\dot{m}_p$, the damage function is

$$\dot{D}_s(\dot{m}_p) = \sum_i S_{ERF,s,i} P_{s,i} \int\int r \, dr \, d\phi \, \rho(\bar{r}) c_s(\bar{r}, \dot{m}_p), \qquad (7.52)$$

where now $S_{ERF,s,i}$ and $P_{s,i}$ are for the impacts of the secondary pollutant. Analogous to the relation between removal flux and concentration for primary pollutants, we can use

$$M_s = (\dot{m}_p, \bar{r}) = k_s(\bar{r}) c_s(\dot{m}_p, \bar{r}) \text{ with } k_s(\bar{r}) = v_{dry,s} + v_{wet,s} + k_{ct,s} \quad (7.53)$$

to replace the concentration $c_s(\dot{m}_p, \bar{r})$ by the removal flux $M_s(\dot{m}_p, \bar{r})$,

$$\dot{D}_s(\dot{m}_p) = \sum_i S_{ERF,s,i}\, P_{s,i} \iint r\, dr\, d\phi\, \frac{\rho(\bar{r}) M_s(\dot{m}_p, \bar{r})}{k_s(\bar{r})}. \qquad (7.54)$$

Replacing $\rho(\bar{r})$ by ρ_{uni} and $k_s(\bar{r})$ by a constant k_s leaves an integral over the removal flux which is equal to the production rate of the secondary pollutant by conservation of matter,

$$\iint r\, dr\, d\phi\, M_s(\dot{m}_p, \bar{r}) = \dot{m}_s, \qquad (7.55)$$

and so we obtain,

$$\dot{D}_{UWM,s} = \left(\frac{\dot{m}_s \rho_{uni}}{k_s}\right) \sum_i \left[S_{ERF,s,i} P_{s,i}\right]. \qquad (7.56)$$

Since $\dot{m}_s$ is also equal to the integral of the creation flux of the secondary from the primary pollutant,

$$\dot{m}_s = \iint r\, dr\, d\phi\, M_{ct,p}(\dot{m}_p, \bar{r}), \qquad (7.57)$$

and the creation flux can be replaced by the depletion flux of the primary,

$$M_{ct,p}(\dot{m}_p, \bar{r}) = k_{ct,p}(\bar{r}) c_p(\dot{m}_p, \bar{r}) = \frac{k_{ct,p}(\bar{r}) M_p(\dot{m}_p, \bar{r})}{k_p(\bar{r})}, \qquad (7.58)$$

one obtains after replacing $k_{ct,p}(\bar{r})$ and $k_p(\bar{r})$ by constants $k_{ct,p}$ and k_p,

$$\dot{m}_s = \frac{k_{ct,p}}{k_p}\, \dot{m}_p \qquad (7.59)$$

So we finally obtain the simple UWM for the damage cost rate of a secondary pollutant due to the emission rate $\dot{m}_p$ of the primary pollutant,

$$\dot{D}_{UWM,s} = \left(\frac{\dot{m}_p\, \rho_{uni}}{k_{ps}}\right) \sum_i \left[S_{ERF,s,i} P_{s,i}\right] \text{ with } k_{ps} = \frac{k_s k_p}{k_{ct,p}}. \qquad (7.60)$$

The quantity k_{ps} is the effective depletion velocity for the secondary pollutant which accounts for the creation velocity of the secondary and for the depletion velocities of the primary and secondary pollutants.

The derivation of the simple UWM involves replacing the average of the product $\frac{\rho(\bar{r}) M(\dot{m}_p, \bar{r})}{k(\bar{r})}$ by the product of their averages. This is exact only if the factors are uncorrelated. That is the case in the limit of vertical line sources in a world of uniform receptor density and uniform terrain and weather, because there the depletion velocity k_p does not vary at all. For primary pollutants emitted at ground level in large population centers, ρ and M vary strongly and together. Therefore the simple UWM is good for primary pollutants only if the source is not too close to a large city and its height is more than about 50 m. For the secondary pollutants nitrates and sulfates, the simple UWM is better than for primary pollutants because their creation takes place sufficiently far from the source, where variations in receptor density and M do not matter very much.

Of course in the real world neither receptor density nor depletion velocities are constants. To improve the accuracy of the UWM we take a closer look at the relation between receptor distribution, analysis range and concentrations, using the regional dispersion model of Section 7.3.

7.4.1.2 Size of analysis area and cumulative impact

Figs. 7.13 and 7.14 show some examples of the cumulative impact distribution as a function of the radius R_0 of the analysis area A. It should be large enough to capture at least 95% of the total damage cost. In Fig. 7.13 the receptor density is uniform and R_0 has to be at least 1200 km for PM_{10} and SO_2. For secondary pollutants (nitrates and sulfates), the range has to be larger, about 1600 km. Figure 7.14 takes into account real population

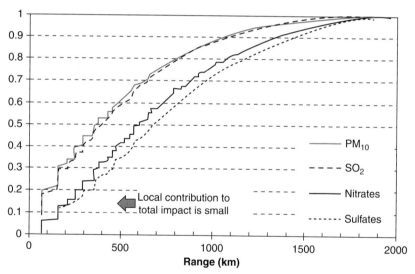

Fig. 7.13 Cumulative impact distribution (fraction of total impact) for primary and secondary pollutants versus radius of analysis area, as calculated by the EcoSense software with *uniform* population distribution.

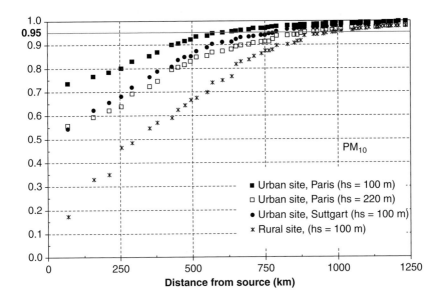

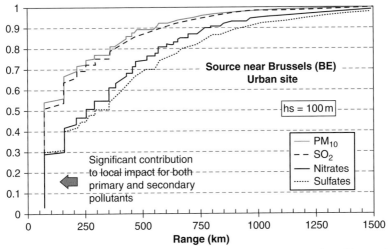

Fig. 7.14 Cumulative impact distribution (fraction of total impact) for primary and secondary pollutants versus radius of analysis area, as calculated by the EcoSense software with *actual* population distributions.

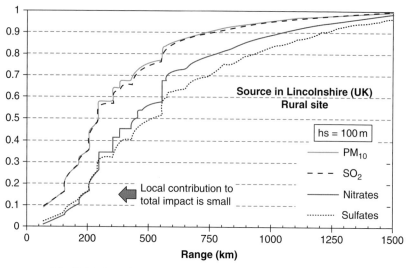

Fig. 7.14 (cont.)

distributions. For primary pollutants R_0 depends on source height and location. A range of 500 km is sufficient for source height 100 m in a large city, but for a rural site it has to be at least 1,000 km. For secondary pollutants, such as particulate aerosols, the analysis range should be 1000 to 1500 km, and is relatively independent of source location, stack height and exhaust gas conditions.

7.4.1.3 Accounting for stack height, local population and nonlinear chemistry

Since the formation of secondary pollutants takes place tens to hundreds of km from the source, their impacts are not very sensitive to source parameters, local weather and receptor distribution. This is illustrated in Figs. 7.15 and 7.16. The variability is a factor of about 1.4 for sulfates, a factor of 2 for nitrates. The only exception is Forssa in Finland where most of the affected areas have a low or no population. Results for secondary pollutants are normalized per kg of precursor pollutant, which for nitrate and sulfate aerosols are NO_x and SO_2, respectively.

Figure 7.16 illustrates how the impact of primary pollutants varies with local receptor distribution and stack height. For primary pollutants there is a very strong variation. Paris has the highest damage cost per kg of

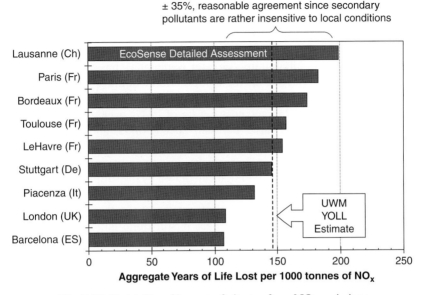

Fig. 7.15 Variability of impact of nitrates from NO_x emissions.

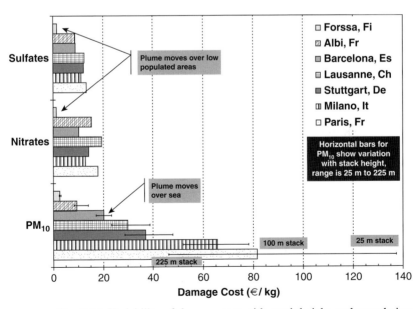

Fig. 7.16 Variability of damage cost with stack height and population density.

pollutant, whereas Forssa, a rural site in Finland, has the lowest cost, about 1/50 of the cost determined for Paris. Damage estimates for Milano (1.4 million), Stuttgart (0.6 million) and Lausanne (130,000), lie in between. Barcelona, a city whose metropolitan population numbers around 3.2 million inhabitants (Greater Barcelona), has a lower damage cost than Stuttgart despite having a population five times larger, because the wind carries the plume out to sea, part of the time. Depending on local receptor density, the damage cost per kg of PM_{10} over the stack height range 25 to 225 meters varies between 20% (Forssa) and 300% (Paris) (horizontal bars in graph).

As shown in Fig. 7.17, for short stacks close to or inside large cities, the impact can be much larger than that calculated by the simple UWM, since the high receptor density is correlated with the high concentrations near the source. Although Porcheville is a small town with a population of around 3000 people, the total impact is still quite large, because it is located about 40 km north west of Paris. For rural locations, the agreement with the simple UWM is best for short stacks, because multiple

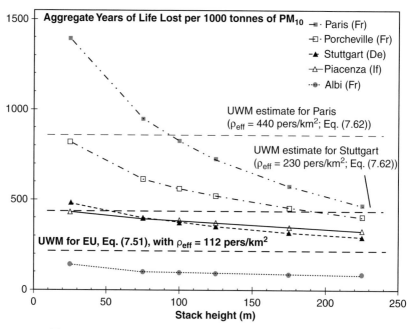

Fig. 7.17 Comparison of UWM results for PM_{10} (Eq (7.51)) with detailed site-specific results computed with the EcoSense model of ExternE (Krewitt *et al.*, 1995).

Table 7.9 S_{sh} and S_{ct} coefficients for Eqs. (7.63) and (7.64) (to be applied after replacing ρ_{uni} by ρ_{eff}, Eq.7.61); h_s is the source height in m. ρ_{loc} = receptor density within about 50 km of source.

Pollutant	Site characterization	S_{sh}	S_{ct}
PM$_{10}$, SO$_2$ and NO$_x$	Rural, $\left(\frac{\rho_{loc}}{\rho_{uni}}\right) < 2$	• 1.5 for h_s = 25 m • ≈ 1 for h_s = 100 m • 0.9 for h_s = 225 m • 1.5–2 for $h_s \rightarrow 0$	1
	Small city, $\left(\frac{\rho_{loc}}{\rho_{uni}}\right) < 6$	• 1.3 for h_s = 25 m • ≈ 1 for h_s = 100 m • 0.8 for h_s = 225 m • 3–5 for $h_s \rightarrow 0$	
	Medium city, $\left(\frac{\rho_{loc}}{\rho_{uni}}\right) < 10$	• 1.4 for h_s = 25 m • ≈ 1 for h_s = 100 m • 0.7 for h_s = 225 m • 5–8 for $h_s \rightarrow 0$	
	Large city, $\left(\frac{\rho_{loc}}{\rho_{uni}}\right) > 10$	• 1.6 for h_s = 25 m • ≈ 1 for h_s = 100 m • 0.6 for h_s = 225 m • Up to 15 for $h_s \rightarrow 0$	
Sulfates	Weak dependence on location and source parameters	• 0.75 to 1.25	• Marginal emissions: ≈1 • For non-marginal emissions: 0.75 to 1.5 (1 recommended)
Nitrates	Weak dependence on location and source parameters	• 0.6 to 1.4	• Marginal emissions: ≈1 • For non-marginal emissions: 0.25 to 0.5 (0.5 recommended)

reflections at the surface enhance vertical mixing, which tends to smooth out variations in concentration with height above ground, and, therefore, the atmosphere becomes more homogeneous. The UWM result using the mean EU receptor density lies in the middle between a rural area and a typical city in Europe. For this reason, we recommend the use of UWM when an average estimate of the impact or damage cost is necessary in environmental policy applications at the regional scale. At the local scale, the multipliers in Table 7.9 can be used to improve the accuracy of UWM calculations.

In order to improve the simple UWM, we have carried out numerous comparisons of UWM with detailed results from several computer models, including EMEP and the Windrose Trajectory Model (Krewitt *et al.*, 1995) of ExternE. The first step is to distinguish between local ρ_{loc} (within

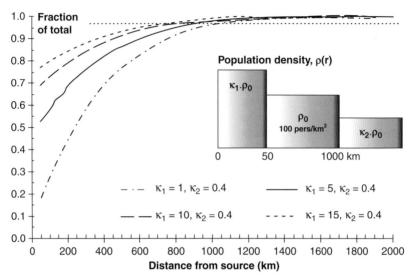

Fig. 7.18 Cumulative population exposure distribution (dotted line indicates range of analysis to account for 95% of total; population distribution is indicated in the insert). A value of $\kappa_1 = 5$ is typical for a small city in Europe, $\kappa_1 = 10$ for a medium-sized city and $\kappa_1 = 15$ would be the case for Paris or London (a very large metropolis).

about 50 km of source) and background ρ_{uni} receptor densities and to replace ρ_{uni} of the simple UWM by

$$\rho_{eff} = \left[f_{loc} \left(\frac{\rho_{loc}}{\rho_{uni}} \right) + (1 - f_{loc}) \right] \rho_{uni}, \qquad (7.61)$$

where f_{loc} is the local share of the damage cost. For primary pollutant, f_{loc} is 5% to 20%, with 15% representing a typical estimate. The role of different population densities in different zones is illustrated in Fig. 7.18. An empirical expression for computing ρ_{eff} (for a 100 m stack height) is

$$\rho_{eff} = 2.273(\rho_{loc})^{0.7617}. \qquad (7.62)$$

Results from these two equations can be averaged to obtain a third estimate. For EU-27, the mean value of ρ_{uni} is 112 pers/km^2; for China it is 231 pers/km^2.

For sulfates and nitrates, the local impact is small because chemical transformations occur over tens to hundreds of km from the source;

hence, $f_{loc} \approx 0$ and $\rho_{eff} \approx \rho_{uni}$, unless a source is located in a region of very high local receptor density (the case for Brussels in Fig. 7.14).

We also introduce modifying factors, S_{sh}, for stack height and source location and S_{ct} for chemical transformation to obtain,

$$\dot{D}_{UWM,p}^{Improved} = S_{sh}\left(\frac{\dot{m}_p\,\rho_{eff}}{k_p}\right) \sum_i \left[S_{ERF,p,i}P_{p,i}\right] \qquad (7.63)$$

for primary pollutants and

$$\dot{D}_{UWM,s}^{Improved} = S_{sh}S_{ct}\left(\frac{\dot{m}_p\,\rho_{eff}}{k_{ps}}\right) \sum_i \left[S_{ERF,s,i}P_{s,i}\right] \qquad (7.64)$$

for secondary pollutants.

Recommendations for S_{sh} and S_{ct} are listed in Table 7.9, for short and tall stack heights. S_{ct} is a correction factor that accounts for nonlinearities in chemical reactions, an important issue that comes up in the transformation of non-marginal emissions of nitrogen oxides (Zhou *et al.*, 2003). Both nitrogen and sulfur compete for the ammonia that is present in the atmosphere, to neutralize nitric and sulfuric acid (Fig. 7.10), but ammonia preferentially neutralizes sulfur over nitrogen. Moreover, unlike sulfates, nitrate formation is reversible; the equilibrium between nitric acid, ammonium nitrate and ammonia can shift depending on atmospheric ammonia availability and ambient temperature. During summer months, for instance, higher temperatures limit nitrate formation considerably (see West *et al.*, 1999 and Table 2 in Zhou *et al.*, 2003, for example). Hence, for non-marginal NO_x emissions, S_{ct} is assigned a value between 0.25 and 0.5 (with 0.5 recommended). For all other pollutants and marginal NO_x emissions, generally, a change in background ammonia has no appreciable effect on other pollutant concentrations, and S_{ct} is unity.

7.4.1.4 Concentrations for elevated sources

For a homogeneous (k constant) and well-mixed atmosphere with uniform wind speed in all directions, the steady-state incremental concentration c at $\bar{r}$ due to an emission rate $\dot{m}$ has been derived in Section 7.3.2 (Eqs. (7.38) and 7.41) and is repeated here for use in the UWM.

Primary pollutant, p ($\dot{m}_p$ emission rate)

$$c_p(r) = \frac{\dot{m}_p}{2\,\pi\,u\,h_{mix}}\frac{1}{r}e^{-\frac{k_p}{u\,h_{mix}}r}. \qquad (7.65)$$

Secondary pollutant, s (equivalent emission rate $\dot{m}_s = \frac{k_{ct}}{k_p} \dot{m}_p$)

$$c_s(r) = \frac{\dot{m}_p}{2\,\pi\,u\,h_{mix}} \left(\frac{k_{ct}}{k_s - k_p}\right) \frac{1}{r} \left[e^{-\frac{k_p}{u\,h_{mix}}r} - e^{-\frac{k_s}{u\,h_{mix}}r}\right] \tag{7.66}$$

where h_{mix} is the mixing layer height which varies with time of day/year and as a function of solar radiation and atmospheric turbulence, increasing from morning to late afternoon (Figure 7.8). A typical annual mean estimate is between 600 m and 800 m (less for stable Pasquill classes and more for unstable atmospheric conditions). u is the mean wind speed (m/s) for which a typical value over land is 5 to 8 m/s.

7.4.1.5 Average concentration for elevated sources

For primary pollutants, the average concentration $\bar{c}_p(R_o)$ over the range $r = 0$ and $r = R_o$ is obtained by integrating Eq. (7.65) over the analysis area $A = \pi R_o^2$:

$$\bar{c}_p(R_0) = \frac{1}{A} \int_0^{R_o} 2\,\pi\,r\,c_p(r)dr = \frac{\dot{m}_p}{\pi\,k_p R_0^2} \left[1 - e^{-\frac{k_p R_0}{u\,h_{mix}}}\right] \tag{7.67}$$

The average concentration decreases towards 0 as $R_o \to \infty$. In some of the following we will find it convenient to look at the average in an analysis area A, and in most cases A will be chosen so that 95% of the total impact will be captured. Then the radius R_o, for elevated sources, should be between 1000 and 1500 km for primary and 1500 to 2000 km for secondary pollutants (Fig. 7.13). In this case, one can neglect the exponential and approximate the average concentration in A as

$$\bar{c}_p(R_0) \approx \frac{\dot{m}_p}{k_p A}. \tag{7.68}$$

In some cases we will look at the local zone separately, taking a circular local domain of area $A = 10,000$ km^2 with $R_o = 56$ km. For secondary pollutants, k_p is replaced by the effective depletion velocity k_{ps}.

The exponential inside the square brackets is the fraction of the pollutant mass remaining in the plume at R_o. Equation (7.67) is plotted in Fig. 7.19, as a function of downwind distance r for different dilution rates q_R and depletion velocities k_p. The dilution rate for elevated sources is defined as the product of the average values of wind speed and mixing height:

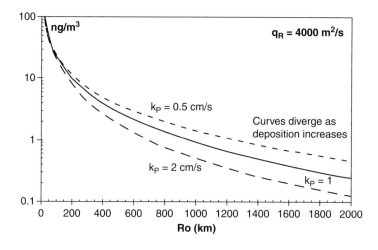

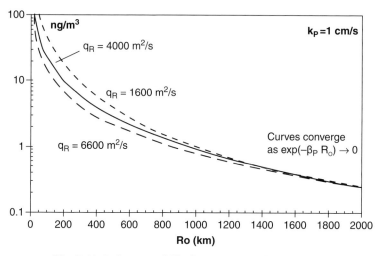

Fig. 7.19 Influence of dilution rate q_R and depletion velocity k_p on the average concentration for a primary pollutant (Eq. (7.67)). The emission rate is 1000 t/yr.

$q_R = uh_{mix}$ (with units m^2/s) = dilution rate for elevated sources.

As can be seen in these figures, near the source and for mid-range distances, the dilution rate is the contributing factor for decreasing pollutant concentrations. The depletion velocity influence on concentration is negligible near the source, but increases significantly with downwind distance; eventually, pollutant removal becomes the leading cause for changes in concentration.

7.4.1.6 Inhalation intake fraction

The pollutant intake fraction iF (dimensionless) is a metric that is often used in life cycle analyses to determine human exposure to chemicals. The intake fraction is the fraction of the primary pollutant that is emitted into the environment and then enters a human body via different routes of exposure, primarily by inhalation, ingestion and dermal contact. In the case of air pollution, inhalation is the exposure pathway for a chemical to enter a human body. The intake fraction for a population of N_{pers} individuals (men, women and children) over an analysis area A is calculated using Eq. (7.69) (with p and s indices added to identify primary and secondary pollutants):

$$iF = S_{sh} S_{ct} \frac{\bar{c}(R_o) N_{pers} B_R}{\dot{m}_p} = S_{sh} S_{ct} \frac{N_{pers} B_R}{A k} = S_{sh} S_{ct} \frac{\rho_{eff} B_R}{k} \quad (7.69)$$

$\bar{c}(R_o)$ is the mean incremental concentration (Eq. (7.68)). B_R is the population weighted daily mean breathing rate. It varies with age, sex and activity level. The choice of B_R from different studies, range between 9 and 20 m^3/day per person. We will assume a value of 13 m^3/day, determined on the basis of the information presented in the *Exposure Factors Handbook* of the EPA (2011). For secondary pollutants, k is the effective depletion velocity k_{ps}. In the last equality, ρ_{eff} has replaced N_{pers}/A, following Eq. (7.61).

7.4.1.7 Damage and cumulative damage distributions
for elevated sources

For primary pollutants, $c_p(r)$ of Eq. (7.65) can be used to determine the distribution function $DF_{UWM,p}(r)$ of the UWM damage cost at downwind distance r, whereas the average concentration $\bar{c}_p(R_o)$ of Eq. (7.67) can be used to derive the cumulative distribution function $CDF_{UWM, p}(R_o)$ of the damage cost, assuming a *uniform population* distribution (Eq. (7.71)).

$$DF_{UWM,p}(r) = N_{pers}(r) c_p(r) \sum_i [S_{ERF,i} P_i] = \left[\frac{\Delta A}{2 \pi r u} \frac{k_p}{h_{mix}} e^{-\frac{k_p r}{u\, h_{mix}}} \right] \dot{D}_{UWM,p}$$

$$(7.70)$$

$N_{pers}(r)$ is the exposed population in area ΔA at a distance r from the source ($N_{pers}(r) = \rho_{uni} \times \Delta A$). For an annular region, $\frac{\Delta A}{2\pi r} = \Delta R$, the annulus thickness (r is the half-way point between edges).

$$CDF_{UWM,p}(R_o) = N_{pers}(R_o)\bar{c}_p(R_o)\sum_i \left[S_{ERF,i}P_i \right]$$

$$= \left[1 - e^{-\frac{k_p R_o}{u\, h_{mix}}} \right]\dot{D}_{UWM,p}. \qquad (7.71)$$

$N_{pers}(R_o)$ is the number of people integrated over the interval $r \in [0, Ro]$ ($N_{pers}(Ro) = \rho_{uni} \times \pi R_o^2$). Equation (7.71) is the integral of Eq. (7.70) over the range $r = 0$ to R_o, with ΔA replaced by $2\pi r\, dr$. The exponential term in Eq. (7.71) is the regional share, beyond R_o, of the total damage cost,

$$\begin{array}{c}\text{damage cost}\\\text{from } r > R_0\end{array} = \frac{1}{\Delta R}\int_{R_0}^{\infty} DF_{UWM,p}(r)dr = \dot{D}_{UWM,p} - CDF_{UWM,p}(R_o)$$

$$= e^{-\frac{k_p R_o}{u\, h_{mix}}}\dot{D}_{UWM,p}$$

$$(7.72)$$

Equations (7.70) and (7.71) are plotted in dimensionless form in Fig. 7.20. Similar equations can be derived for secondary pollutants starting from Eq. (7.66).

To improve on Eq. (7.71), we assume different, but still uniformly distributed, local and background receptor densities and obtain the following relations for UWM cumulative damage cost distribution:

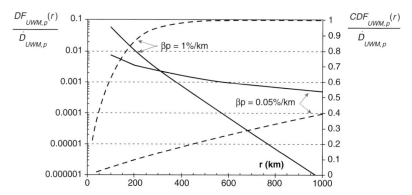

Fig. 7.20 Normalized UWM damage cost distribution (solid lines, left axis; Eq. (7.70)) and cumulative damage distribution (dashed lines, right axis; Eq. (7.71)) as a function of distance from an elevated source for different pollutant decay constants (β_p) and uniform population. $\Delta A = 10,000$ km^2.

$$\frac{CDF_{UWM,p}(R_o)}{\dot{D}_{UWM,p}}\bigg|_{Improved} = 1 - \frac{\rho_{uni}\,e^{-\beta_p R_o}}{\left[\rho_{loc}(1 - e^{-\beta_p R_{loc}}) + \rho_{uni}e^{-\beta_p R_{loc}}\right]} \quad \langle R_o \geq R_{loc} \rangle$$

(7.73a)

$$\frac{CDF_{UWM,p}(R_o)}{\dot{D}_{UWM,p}}\bigg|_{Improved} = 1 - e^{-\beta_p R_o} \quad \langle R_o < R_{loc}; \rho_{uni} = \rho_{loc} \rangle \quad (7.73b)$$

R_{loc} is the local domain radius. The expression in the denominator is an improved assessment of the total UWM damage cost, where the local and regional damage cost shares are weighted according to their respective receptor densities. Typical European values of the ratio $\left(\dfrac{\rho_{loc}}{\rho_{uni}}\right)$ are 0.5 to 2 for rural sites, 6 to 10 for urban locations and up to 15 for very large cities, such as London and Paris, assuming $R_{loc} = 56$ km. Equation (7.73a) is plotted in Fig. 7.21 for various ratios of local-to-background receptor densities. Curves are similar to those presented in Fig. 7.14. When $R_o < R_{loc}$, Eq. (7.73b) is used with ρ_{uni} replaced by ρ_{loc}.

7.4.1.8 Concentrations for ground-level sources in cities
For ground-level sources of primary pollutants in cities most of the impact occurs within a few tens of km of the source, and further modifications are needed for UWM. We begin by presenting a suitable equation for the resulting concentration in the local zone:

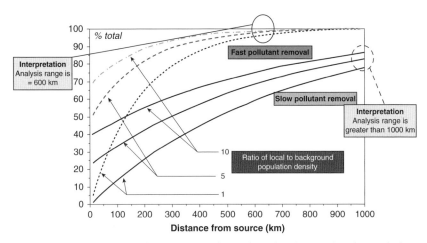

Fig. 7.21 Plot of Eq. (7.73a) for various local-to-regional population density ratios ($R_{loc} = 56$ km).

Table 7.10 *Country-averaged dilution rates DR (m²/s) for ground-level urban sources. Based on data from Apte* et al. *(2012).*

Country	DR	Country	DR	Country	DR
Argentina	570	Austria	400	Australia	690
Belgium	540	Brazil	630	Bulgaria	420
Canada	650	Chile	500	China	450
Czech Repl.	480	Denmark	900	Egypt	620
Estonia	770	Finland	530	France	520
Germany	510	Greece	690	Hungary	390
India	530	Indonesia	460	Ireland	1050
Italy	380	Japan	610	Korea	710
Latvia	700	Lithuania	450	Malaysia	850
Mexico	410	Netherlands	590	Norway	270
Philippines	590	Poland	520	Portugal	910
Puerto Rico	3390	Romania	370	Russia	420
Slovakia	410	Slovenia	230	South Africa	590
Spain	590	Sweden	400	Switzerland	420
Thailand	510	Turkey	480	UK	680
Ukraine	520	USA	580		

Concentration due to ground-level source in cities ($\dot{m}_p$ emission rate)

$$c_p^{\text{ground}}(r) = \frac{\dot{m}_p}{2\pi\,DR}\frac{1}{r}e^{-\frac{k_p}{DR}r} \qquad (7.74)$$

DR is the pollutant dilution rate (m²/s), the product of the ground wind speed (m/s) and a characteristic mixing depth (m) for ground-level sources, similar to the planetary boundary layer for an elevated source. Site-specific and country averaged DR estimates are summarized in Tables 7.10 and 7.11 (based on data from Apte *et al.*, 2012). Compared with the dilution rate q_R for elevated sources, which is on the order of several thousand m²/s, DR is smaller by a factor of 5 to 10, since the plume lies near the ground. The value of DR is proportional to the vertical dispersion coefficient σ_z, which for a neutral atmosphere is on the order of hundreds of meters at a downwind distance of several kilometers (this is the analysis range for ground-level concentration calculations).

The average concentration within a circle of R_0 is calculated as:

Average concentration over range r = 0 to R_0 ($k_p R_0 < DR$)

$$\bar{c}_p^{\text{ground}}(R_0) = \frac{1}{A}\int_0^{R_0} 2\pi r\, c_p^{\text{ground}}(r)\,dr = \frac{\dot{m}_p}{\pi\,k_p R_0^2}\left[1 - e^{-\frac{k_p R_0}{DR}}\right] \approx \frac{\dot{m}_p}{\pi\,R_0 DR}$$

$$(7.75)$$

Table 7.11 *Ground-level dilution rates DR (m^2/s) for select worldwide cities. From Apte et al. (2012) with permission from American Chemical Society.*

Country	City	DR	Country	City	DR	Country	City	DR	Country	City	DR
Argentina	Buenos Aires	600	China	Xinhua	390	Italy	Turin	380	South Africa	Pretoria	300
Argentina	Cordoba	360	China	Yantai	400	Italy	Venice	760	Spain	Barcelona	310
Argentha	Rosario	560	China	Yinan	660	Japan	Fukushima	400	Spain	Bilbao	510
Australia	Canberra	430	China	Zaoyang	440	Japan	Matsuyama	370	Spain	Madrid	770
Australia	Melbourne	350	China	Zhaoqing	440	Japan	Nagasaki	420	Spain	Sevilla	350
Australia	Perth	900	Colombia	Bogota	410	Japan	Osaka	380	Spain	Valencia	1150
Australia	Sydney	340	Costa Rica	San Jose	480	Japan	Tokyo	140	Sweden	Goteborg	490
Austria	Salzburg	270	Croatia	Zagreb	330	Korea, Rep.	Pusan	320	Sweden	Stockholm	1620
Austria	Vienna	440	Cyprus	Nicosia	260	Korea, Rep.	Seoul	370	Sweden	Upsala	580
Belgium	Brugge	550	Czech Rep.	Brno	450	Korea, Rep	Taejon	460	Switzerland	Bern	410
Belgium	Brussels	550	Czech Rep.	Prague	380	Latvia	Riga	490	Switzerland	Geneva	760
Belgium	Charleroi	510	Denmark	Copenhagen	400	Lithuania	Vilnius	970	Switzerland	Zurich	470
Bolivia	La Paz	560	Ecuador	Quito	380	Malaysia	Kuala Lumpur	540	Thailand	Bangkok	850
Brazil	Belo Horizonte	430	Egypt	Alexandria	400	Mexico	Guadalajara	670	Turkey	Ankara	300
Brazil	Parnaiba	570	Egypt	Cairo	570	Mexico	Mexico	560	Turkey	Istanbul	220
Brazil	Rio de Janeiro	670	Finland	Helsinki	480	Mexico	Puerto Vallarta	650	Turkey	Izmir	210
Brazil	Sao Paolo	470	France	La Rochelle	670	Morocco	Marrakech	1140	UK	Belfast	490
Bulgaria	Sofia	380	France	Le Havre	700	Morocco	Rabat	930	UK	Bristol	500
Canada	Calgary	360	France	Lille	720	Netherlands	Amsterdam	530	UK	Edinburgh	610
Canada	Montreal	500	France	Lyon	640	Netherlands	Utrecht	420	UK	Glasgow	580
Canada	Ottawa	300	France	Marseille	640	New Zealand	Auckland	530	UK	London	1770
Canada	Quebec	320	France	Nantes	590	New Zealand	Wellington	540	UK	Milton Keynes	2040
Canada	Toronto	1270	France	Paris	700	Nigeria	Lagos	520	UK	Newport	340
Canada	Vancouver	220	France	Toulouse	720	Norway	Oslo	390	UK	Plymouth	210
Chile	Santiago	510	Germany	Berlin	650	Paraguay	Asuncion	520	UK	Southampton	520
China	Anshun	510	Germany	Essen	640	Peru	Lima	530	UK	York	550

Country	City		Country	City		Country	City		Country	City	
China	Beijing	380	Germany	Frankfurt	460	Philippines	Cebu	1380	Ukraine	Kiev	500
China	Chongqing	340	Germany	Leipzig	500	Philippines	Metro Manila	480	Uruguay	Montevideo	1040
China	Funan	380	Germany	Munich	430	Poland	Gdansk	900	USA	Atlanta	320
China	Fuzhou	510	Germany	Nuremberg	450	Poland	Krakow	470	USA	Boston	320
China	Guangzhou	480	Germany	Stuttgart	440	Poland	Warszawa	480	USA	Chicago	610
China	Jinan	420	Greece	Athens	740	Portugal	Funchal	4230	USA	Cleveland	420
China	Jining	620	Greece	Thessaloniki	530	Portugal	Lisboa	1130	USA	Dallas	610
China	Kunming	660	Hungary	Budapest	400	Portugal	Porto	470	USA	Detroit	570
China	Laizhou	450	India	Bangalore	550	Puerto Rico	San Juan	3820	USA	Houston	460
China	Leizhou	550	India	Chennai	480	Romania	Bucharest	350	USA	Los Angeles	270
China	Liaoyuan	460	India	Delhi	330	Russia	Moscow	470	USA	Milwaukee	830
China	Liaozhong	450	India	Jaipur	340	Russia	Saint Petersburg	290	USA	New Orleans	1290
China	Qinzhou	480	India	Kolkota	380	Russia	Vladivostok	520	USA	New York	320
China	Quanzhou	580	India	Mumbai	1200	Russia	Volgograd	520	USA	Norfolk	1080
China	Quzhou	380	Indonesia	Denpasar	230	Saudi Arabia	Riyadh	570	USA	Philadelphia	450
China	Shanghai	530	Indonesia	Jakarta	310	Serbia	Beograd	360	USA	Seattle	260
China	Taipei	670	Italy	Bari	500	Singapore	Singapore	390	Venezuela	Caracas	300
China	Tianjin	480	Italy	Florence	340	Slovakia	Bratislava	440	Vietnam	Ho Chi Minh	400
China	Weinan	330	Italy	Milano	280	Slovenia	Ljubljana	230	Zimbabwe	Harare	530
China	Wuhai	430	Italy	Naples	350	South Africa	Cape Town	800			
China	Wuhan	430	Italy	Palermo	660	South Africa	Durban	700			
China	Xiantao	380	Italy	Rome	330	South Africa	Johannesburg	390			

The appropriate value for R_0 is taken as the radius of a circle with area equal to the urban footprint (urban conurbation area). Typical values of R_o for cities less than half a million, between 0.5 and 1 million, between 1 and 3 million and greater than 3 million are, respectively, 4, 6, 9 and 18 km (Atlas of urban expansion, www.Lincoln.inst.edu).

Figure 7.22 compares Gaussian plume concentrations (ISC model, EPA, 1995) with our calculations of concentrations for elevated (Eq. (7.65)) and ground-level sources (Eq. (7.74)). Our concentration estimates do not account for stack height or exit gas flow rate, and, consequently, the concentration comparisons for short stacks and for

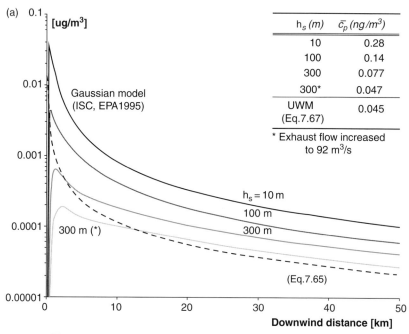

Fig. 7.22 Concentration comparisons of Gaussian model and Eq. (7.65) for an elevated and Eq. (7.74) for a ground source located near Stuttgart (Germany). $PM_{2.5}$ annual emission of 1000 tonnes. (Source parameters: d_s = stack exit diameter, T_s = exhaust gas temperature and w_s = exhaust speed.) $\bar{c}_p$ = average concentration for R_0 = 50 km.
   (a) Elevated source: d_s = 4 m, T_s = 373 K, w_s = 1.3 m/s (16.3 m³/s);
         UWM input parameters: k_p = 0.52 cm/s, q_R = 4400 m²/s
         (u = 5.5 m/s and h_{mix} = 800 m)
   (b) Ground-level source: d_s = 0.3 m, T_s = 373 K, w_s = 1.3 m/s (0.1 m³/s)
         UWM input parameters: k_p = 0.52 cm/s, DR = 440 m²/s
         (Table 7.11)

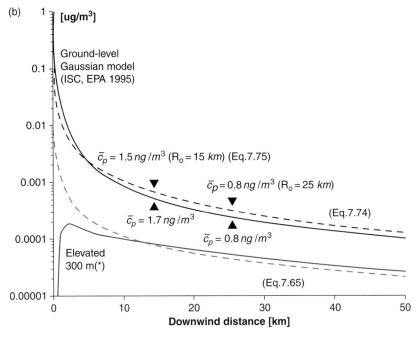

Fig. 7.22 (cont.)

distances near the source fall short of the Gaussian model results, the largest deviation being about a factor of 7. But, for tall chimneys and downwind distances greater than 5 km, Eq. (7.65) is quite satisfactory, with deviations in the range ±30%. For ground emissions, Eq. (7.74) results closely follow the Gaussian derived concentrations over almost the entire range of r, with deviations within ±35% when r > 2 km. The difference between our results and those calculated by ISC for the locally averaged concentrations is quite small (< 10%).

7.4.1.9 Damage costs of ground-level sources in cities

For primary pollutants emitted by such sources, most of the impact occurs within a small area extending a few tens of km from the point of emissions, in contrast with sulfates and nitrates for which the UWM does not need modifications for ground-level sources. Ozone formation, on the other hand, occurs more rapidly, over distances from several km to tens of km, but we have not yet developed a UWM for ozone. Intake fractions and damage costs for primary pollutants emitted by ground-level sources in cities are calculated by the following equations:

Intake fraction

$$iF = \frac{\overline{c}_p^{ground}(R_o)N_{pers}B_R}{\dot{m}_p} = \left(\frac{N_{pers}}{\sqrt{\pi}\,R_o}\right)\frac{1}{DR}\left(\frac{B_R}{\sqrt{\pi}}\right) = \left(\frac{N_{pers}}{\sqrt{A_o}}\right)\frac{1}{DR}\left(\frac{B_R}{\sqrt{\pi}}\right)$$

$$= \left(\frac{LPD}{DR}\right)\left(\frac{B_R}{\sqrt{\pi}}\right)$$

where the linear population density $LPD = \left(\dfrac{N_{pers}}{\sqrt{A_o}}\right)$ (7.76)

Damage cost

$$\dot{D}_{UWM,p}^{ground} = \overline{c}_p^{ground}(R_o)\,N_{pers}\sum_i\left[S_{ERF,i}P_i\right]$$

$$= \left[\frac{iF}{B_R}\frac{k_p}{\rho_{uni}}\right]\dot{D}_{UWM,p} = S_{eg}\dot{D}_{UWM,p}$$

where $S_{eg} = \left[\dfrac{iF}{B_R}\dfrac{k_p}{\rho_{uni}}\right] = \left[\left(\dfrac{LPD}{\sqrt{\pi}DR}\right)\dfrac{k_p}{\rho_{uni}}\right]$ (7.77)

$\overline{c}_p^{ground}(R_o)$ is the average concentration of Eq. (7.75) in the urban conurbation area A_o, characterized as a circle with radius R_o. DR is the dilution rate for ground-level emissions (see Tables 7.10 and 7.11). N_{pers} is the number of exposed individuals across area A_o. B_R is the daily mean breathing rate (for example, 13 m^3/day). LPD is the linear-population density, with units persons per m. Its value is computed by dividing the population by the square root of A_o. $\dot{D}_{UWM,p}$ is the simple UWM of Eq. (7.51). The multiplier S_{eg} is the elevated-to-ground-damage scaling factor, which varies linearly with conurbation population. Values of S_{eg} for the cities of Paris (France) and Stuttgart (Germany) are calculated below.

Paris

N_{pers} = 10.4 million people[11], A_o = 2845 km^2, DR = 520 m^2/s (from Table 7.11), k_p = 0.45 cm/s ($PM_{2.5}$ for France from Table 7.8), ρ_{uni} = 112 pers/km^2 (EU-27 mean)

$$S_{eg} = \left[\left(\frac{LPD}{\sqrt{\pi}DR}\right)\frac{k_p}{\rho_{uni}}\right] = \frac{1}{\sqrt{\pi}\times 520}\frac{10.4\times 10^6}{\sqrt{2845\times 10^6}}\frac{0.0045}{112\times 10^{-6}} = 8.5$$

---

[11] Institut National de la Statistique et des Études Économiques (http://www.recensement.insee.fr)

Stuttgart

N_{pers} = 2.7 million people[12], A_o = 3654 km^2, DR = 440 m^2/s (from Table 7.11)

k_p = 0.52 cm/s (PM$_{2.5}$ for Germany from Table 7.8), ρ_{uni} = 112 pers/km^2 (EU-27 mean)

$$S_{eg} = \left[\left(\frac{LPD}{\sqrt{\pi}DR}\right)\frac{k_p}{\rho_{uni}}\right] = \frac{1}{\sqrt{\pi} \times 440}\frac{2.7 \times 10^6}{\sqrt{3654 \times 10^6}}\frac{0.0052}{112 \times 10^{-6}} = 2.7$$

S_{eg} calculations are based on the background population density ρ_{uni}. UWM results correspond to the extrapolated values of the curves for Paris and Stuttgart in Figure 7.17 in the limit as $h_s \rightarrow 0$. The agreement between UWM damage costs (Eq. (7.77)) and results based on detailed dispersion models of transport emissions is between a factor of 2 to 3, a significant achievement given the simplicity of Eq. (7.77). Accuracy of the UWM estimate can be improved by calculating the population exposure using Eq. (7.74), rather than assuming the mean incremental concentration determined by Eq. (7.75), which treats the city as a single spatial grid point. The accuracy of the UWM is also improved by using the urban built-up area (www.Lincoln.inst.edu) rather than the city footprint.

As a validation of our approach, Eq. (7.76) is compared with intake fractions computed in the assessment by Apte *et al.* (2012) in Fig. 7.23. Deviations are within ±30% between the two approaches, with the consistent tendency of a negative bias on the part of UWM. For radial dispersion, the case for the UWM, the cross-wind area is larger than the constant cross-sectional area assumed by Apte and colleagues. In other words, UWM assumes a circular urban footprint, rather than a square box shape. Pollutant dilution is greater in the UWM case, and consequently, the concentrations (and intake fractions) are lower.

7.4.2 *Validation of the UWM methodology*

7.4.2.1 **Comparisons of PM$_{2.5}$ marginal damage costs with ExternE (2008)**

Objective: Compare country-specific unit damage costs of PM$_{2.5}$ (€/kg) for elevated sources calculated by UWM (Eq. (7.63)) with ExternE[13] (2008) results.

---

[12] Statistische Berichte Baden-Württemberg (http://www.statistik.baden-wuerttemberg.de)
[13] http://www.needs-project.org/

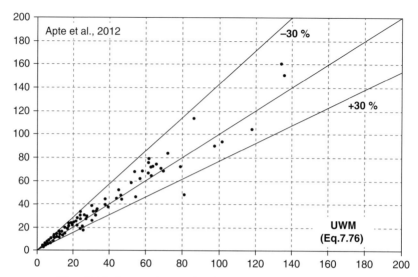

Fig. 7.23 Comparison of urban intake fractions for ground-level particulate emissions.

Input data and results: Country-specific input data (ρ_{eff} and k_p) are listed in Table 7.12. ρ_{eff} is the effective receptor density, computed according to Eq. (7.61), with $f_{loc} = 15\%$ and ρ_{loc} equal to the country-level population density. The analysis area is a circular domain centered in the middle of the country of interest with a radius of 1000 km. Based on the set of ERFs and unit costs of ExternE (2008), the exposure cost is 38.76 (€/yr) per (person·μg/m³).

The UWM results are compared with ExternE in Fig. 7.24. The maximum deviation at the country-level is 26%. The mean deviation (bias) is −0.4; UWM underestimates ExternE by 2% (mean relative difference). The standard error and coefficient of variability are, respectively, 2.2 and 0.1. The average marginal damage cost for EU-27 is 25.7 € per kg of $PM_{2.5}$ based on the UWM, this value is 5% higher than the ExternE result.

Example Calculate the marginal damage cost of $PM_{2.5}$ for an emission source in France.

Solution: We use Eq. (7.63) of UWM with input data listed in Table 7.12. For France, the effective population density ρ_{eff} is 105 pers/km² and the depletion velocity k_p is 0.45 cm/s. The exposure cost $\sum_i S_{ERF,i} P_i$ is 38.76 (€/yr) (person·μg/m³)⁻¹. We assume unity for S_{sh} (Table 7.9). Plugging these data into Eq. (7.63) we obtain:

Table 7.12 *Country-specific input data for PM$_{2.5}$ and NH$_3$ marginal damage cost calculations.*

Country	Population (millions of pers)	ρ_{eff} pers/km^2	$k_{PM2.5}$ cm/s	k_{NH3} cm/s
Austria	8.3	110	0.56	1.05
Balkans	23.6	73	0.49	0.91
Belgium	10.6	214	0.66	1.24
Bulgaria	7.6	53	0.49	0.91
Cyprus	1.0	56	0.43	
Czech Rep.	10.3	116	0.59	1.11
Denmark	5.5	83	0.86	1.61
Estonia	1.3	33	0.62	1.16
Finland	5.3	36	0.62	1.16
France	61.7	105	0.45	0.84
Germany	82.2	152	0.52	0.98
Greece	11.2	55	0.49	0.91
Hungary	10.0	106	0.57	1.06
Ireland	4.3	59	0.59	1.11
Italy	59.1	150	0.71	1.33
Latvia	2.3	40	0.62	1.16
Lithuania	3.4	52	0.62	1.16
Luxembourg	0.5	138	0.59	1.11
Malta	0.4	33	0.45	0.84
Netherlands	16.4	228	0.66	1.24
Norway	4.7	43	0.89	1.68
Poland	38.1	97	0.57	1.08
Portugal	10.6	62	0.54	1.01
Romania	21.5	73	0.57	1.06
Slovakia	5.4	106	0.58	1.09
Slovenia	2.0	110	0.57	1.06
Spain	45.3	55	0.50	0.94
Sweden	9.1	75	0.86	1.61
Switzerland	7.6	139	0.55	1.04
UK	60.9	122	0.59	1.11
Population weighted mean		112	0.57	1.07

$$\frac{\dot{D}_{UWM}^{improved}}{\dot{m}_p} = S_{sh} \left(\frac{\rho_{eff}}{k_p} \right) \sum_i [S_{ERF,i} P_i] \tag{7.63}$$

$$= 1 \times \frac{105 \frac{pers}{km^2} \times 10^{-6} \frac{km^2}{m^2}}{0.45 \frac{cm}{s} \times 10^{-2} \frac{m}{cm}} \times 38.76 \frac{\text{€}}{yr} \times \frac{yr}{3.154 \times 10^7 s} \times \frac{10^9 \frac{\mu g}{kg}}{\frac{\mu g}{m^3} pers}$$

Marginal damage cost for France $= 28.7 \frac{\text{€}}{kg}$

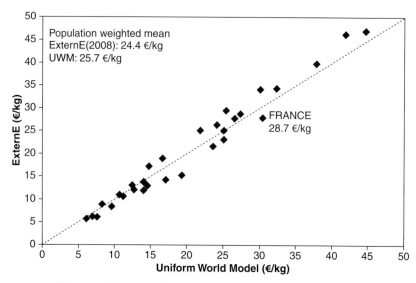

Fig. 7.24 Marginal damage costs of $PM_{2.5}$ emissions, ExternE UWM (Eq. (7.63)) versus ExternE (2008).

7.4.2.2 Comparisons with EC4MACS project for ammonia emissions in Europe

Objective: Assess the health costs of ammonia emissions in Europe using UWM (Eq. (7.64)) and compare with results from the EC4MACS[14] project of the EU-LIFE program.

Input data and result: Input data are summarized in Table 7.12. The exposure cost is 41.47 (€/yr) per (person·µg/m³), a slightly higher value than ExternE (2008) because ERFs and unit costs chosen in EC4MACS are different. The UWM damage costs are compared with EC4MACS results in Fig. 7.25. For most countries, deviations are within ±30%, but residuals scatter is greater than that observed for primary $PM_{2.5}$ emissions. The residuals standard deviation is 3.4 and the mean relative difference is 5%. The larger scatter arises because of variability in the chemical transformation rates. For EU-27, UWM overestimates EC4MACS by 7.5%.

The highest unit damage costs are calculated for the Benelux countries since the local population density is largest for this region.

---

[14] http://www.ec4macs.eu/

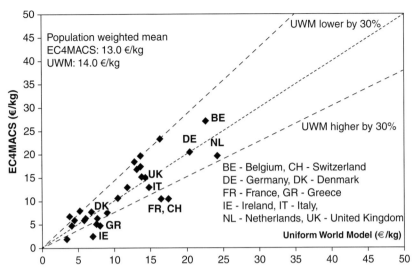

Fig. 7.25 Marginal damage costs of ammonia emissions, EC4MACS versus UWM (Eq. (7.64)).

The damage cost for Greece, Ireland and Spain is lower than UWM, because the wind carries the plume out to sea, some of the time. To improve the UWM calculation, one can subtract from the total exposure the contribution over water; this fraction can be determined by integration of the population-weighted concentration (Eq. (7.66)) multiplied by ρ_{eff}). For the UK, the local contribution is significant because the population density is high, and this compensates for the lower exposure when the plume is carried out to sea.

7.4.2.3 Comparison with EcoSense for China for PM$_{10}$, SO$_2$ and NO$_x$

Objective: Calculate the mortality impact of emissions from elevated sources for different sites across Shandong province and compare UWM with results of EcoSense for China. This version of EcoSense was used for the China Electricity Technology Project (CETP), and the results were published by Hirschberg *et al.* (2004).

Input data and results: Table 7.13 shows effective population density and the mortality results (presented in YOLLs per kg of precursor pollutant) for EcoSense and UWM as well as the ratios UWM-to-EcoSense. For these calculations we assume S_{sh} and S_{ct} are both unity. For Weihai, Yantai and Longkou the population density is

Table 7.13 *Mortality impacts of emissions in the Shandong Province of China: Comparison of UWM calculations with EcoSense for China results.*

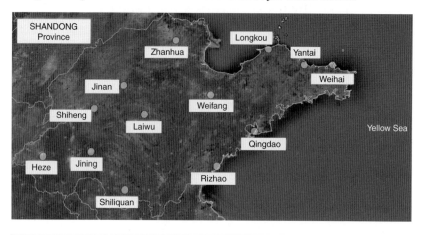

Site	ρ_{eff} [pers/km²]	PM₁₀ (YOLL/kg)				Sulfates (YOLL/kg)				Nitrates (YOLL/kg)			
		UWM	Ecosense		UWM/Ecos	UWM	Ecosense		UWM/Ecos	UWM	Ecosense		UWM/Ecos
Weifang	314	0.21	0.18	<	1.18	0.29	0.25	<	1.16	0.16	0.14	<	1.12
Zhanhua	308	0.21	0.15	<	1.38	0.28	0.26	<	1.08	0.16	0.16	>	0.98
Jinan	323	0.22	0.26	>	0.83	0.30	0.36	>	0.84	0.17	0.18	>	0.91
Laiwu	329	0.22	0.24	>	0.93	0.30	0.32	>	0.94	0.17	0.16	<	1.06
Shiheng	334	0.22	0.27	>	0.82	0.31	0.36	>	0.85	0.17	0.16	<	1.11
Jining	354	0.24	0.28	>	0.85	0.33	0.38	>	0.85	0.18	0.16	<	1.14
Heze	359	0.24	0.31	>	0.78	0.33	0.45	>	0.73	0.18	0.28	>	0.65
Shiliquan	357	0.24	0.25	>	0.95	0.33	0.35	>	0.95	0.18	0.16	<	1.17
Qingdao	213	0.14	0.15	>	0.94	0.20	0.20	<	1.00	0.11	0.10	<	1.13
Rizhao	220	0.15	0.17	>	0.87	0.20	0.27	>	0.76	0.11	0.16	>	0.70
Huangdao	212	0.14	0.10	<	1.48	0.20	0.19	<	1.02	0.11	0.10	<	1.10
Weihai	97	0.07	0.04	<	1.63	0.09	0.09	>	0.95	0.05	0.04	<	1.31
Yantai	96	0.06	0.08	>	0.81	0.09	0.14	>	0.64	0.05	0.07	>	0.75
Longkou	101	0.07	0.10	>	0.66	0.09	0.18	>	0.51	0.05	0.10	>	0.55
ALL sites	258	**0.17**	**0.18**	>	**0.94**	**0.24**	**0.27**	>	**0.88**	**0.13**	**0.14**	>	**0.95**

Legend: < UWM over-estimates EcoSense; > UWM under-estimates EcoSense

calculated taking into account that the plume will be carried out to sea, away from the local population and in the direction of the Korean peninsula. Depletion velocities for the Shandong province are listed in Table 7.8. As seen at the bottom of the table, differences at the site level are typically within ± 50%, with the largest deviations at coastal sites (e.g. Weihai). For the population-weighted average, the deviation between the UWM and EcoSense is less than about 10%.

7.4.2.4 Comparisons with GAINS-China model for PM_{10}, SO_2, NO_x and NH_3

Objective: Use the UWM to calculate the loss of life expectancy (LLE) that is attributable to 2005 anthropogenic emissions in China and compare against the GAINS[15]-China model, as implemented in the study by Amman *et al.* (2008).

Input data and results: For simplicity, we assume the emissions and the mortality risk remain constant over an individual's entire lifetime. Then, the reduction in life expectancy (also known as Years of Life Lost, YOLL) can be calculated as

$$LLE = \bar{c}\, S_{ERF}\, LE = \frac{\dot{m}_p\, S_{ERF}\, LE}{A\, k} \tag{7.78}$$

where, the mean concentration $\bar{c}$ is determined by Eq. (7.68) and includes only anthropogenic emissions of primary particulates ($PM_{2.5}$) and secondary inorganic aerosols formed from precursor emissions of SO_2, NO_x and NH_3. S_{ERF} is the chronic mortality ERF slope (here we assume a value of 6.51E-4 YOLL per [yr-pers-$\mu g_{PM2.5}/m^3$]), LE is the Chinese life expectancy (74 years), A is the analysis area (3.1 million km^2), which includes Central, Southeast and East China (based on Figure 2.5, p. 13 in Amman *et al.*, 2008), and k is the pollutant depletion velocity from Table 7.8.

UWM input data and results, along with GAINS-China values for comparison, are presented in Table 7.14. Additional details of the calculation can be found in the footnotes at the end of the table. The UWM total concentration (60 $\mu g/m^3$) is consistent with the annual mean concentrations of $PM_{2.5}$ (incl. aerosols) in ambient air as indicated in Figure 2.5 in Amman *et al.* (2008). The UWM and GAINS-China LLE central estimates are close; the difference being only 11%.

7.4.2.5 Comparison of concentrations in Beijing with results of the Integrated Environmental Strategies program (IES, 2005)

Objective: Use the UWM to determine the impact on urban air quality due to PM_{10} emissions from the power sector in Beijing. We calculate the mean urban concentration using Eq. (7.67) and compare with values from the Integrated Environmental Strategies program (IES,[16] 2005).

---

[15] http://gains.iiasa.ac.at/gains/EAS/index.login?logout=1
[16] http://en.openei.org/wiki/EPA-Integrated_Environmental_Strategies

Table 7.14 *Comparison of UWM (Eq. (7.78)) and GAINS-China results for the loss of life expectancy from 2005 anthropogenic emissions in China,*

| Pollutant | Emissions kt/year[1] | $k^{(2)}$ cm/s | $\bar{c}$ μg/m³ Eq. (7.68) | LLE (months) UWM[4] | | | |
				Eq. (7.78)	low[5]	high[5]	GAINS[6]
$PM_{2.5}$	12,725	0.60	21.9	12.7	8.5	19.1	
SO_2	31,528	1.99	16.4	9.5	4.8	9.5	**Mean: 39**
NO_x	16,926	0.90	9.7[3]	5.6	2.8	11.2	**Min: 21**
NH_3	12,844	1.13	11.8	6.8	3.4	13.6	**Max: 53**
Sum[5]			59.8	34.6	19.5	53.4	

[1] Anthropogenic emissions data taken from Table 2.3, p. 10 in Amann *et al.* (2008).
[2] Depletion velocities are population-weighted averages using data from Table 7.8 and assuming a circular area centered on the province of Hubei with a radius extending up to 1000 km to cover central, southeast and east China (analysis area, A = 3.1 million km²).
[3] Estimates for nitrates have been adjusted by S_{ct} = 0.5 (Table 7.9).
[4] Input data for Eq. (7.78): LE = 74 years and S_{ERF} = 6.51E-4 YOLL per yr-pers-μg/m³
[5] Low and high UWM estimates correspond to the 68% confidence interval CI calculated by Eq. (11.28) (see Section 11.2.4) with σ_g values of 1.5 for $PM_{2.5}$ and 2 for the remaining pollutants. For the sum, 68%CI was calculated according to Section 11.4 (σ_g of sum is 1.35).
[6] GAINS results for loss of life expectancy from Table 2.5, p. 16 in Amann *et al.* (2008).

Input data and results: Input data and comparison of UWM output with results from the IES program are shown in Table 7.15. For the UWM calculations, we treat the combined power plant emissions for the business-as-usual scenario (BAU) as a single point source. The most damaging power plants are located close to the outer edges of the city, shaded area in Fig. 7.26. We calculate two estimates for the mean incremental concentration, as indicated in Table 7.15 below. In the first estimate, we assume the sources are located near the city center and the averaging distance for the mean concentration is equal to the urban footprint ($R_o = R_{loc}$ = 23.8 km). In the second case, we assume the sources lie near the outskirts of the metropolitan area of Beijing ($R_o = 2 \cdot R_{loc}$ = 47.6 km). The second scenario, clearly, is more realistic of the actual locations of the power plants, and, therefore, more representative of the true population exposure. A weighted average of the two UWM estimates, assuming a 2:1 weighting factor in favor of the second scenario gives a concentration estimate of 0.61 μg/m³. This result is about 20% higher than the value computed by IES (2005) using the ISC Gaussian model of EPA (1995). According to the methodology of Chapter 11, we compute the UWM 68%

Fig. 7.26 Locations of fossil fuel power plants in Beijing, China. The Beijing metropolitan area is distinguished from the rest of the Beijing Municipality by the gray shaded area. Combined emissions from Jingfeng, Datang and Jingneng plants contributed to more than 70% of total PM_{10} power emissions in 2000, while Huadian's contribution was only 5%. (From Hao *et al.* (2007)).

confidence intervals assuming lognormal concentration distributions with geometric standard deviation σ_g of 1.5.

Example Calculate (i) the PM_{10} mean concentration assuming the power plants are located at the edge of the city of Beijing and (ii) the regional share of the damage cost.

Solution: (i) We use Eq. (7.67) of the UWM to calculate the mean concentration across the city of Beijing. The value of R_o is 47.6 km. The remaining input data are listed in Table 7.15.

$$\bar{c}_p(R_o) = \frac{m_p}{\pi\, k_p\, R_o^2}\left[1 - e^{-\frac{k_p R_0}{u\,h_{mix}}}\right] = \frac{m_p}{\pi\, k_p\, R_o^2}\left[1 - e^{-\beta_p R_o}\right] \qquad (7.67)$$

$$\beta_p = \frac{\text{depletion velocity}}{\text{dilution rate for elevated sources}}$$

$$= \frac{k_p}{u\,h_{mix}} = \frac{k_p}{q_R} = \frac{0.0064\,\dfrac{m}{s}}{4200\,\dfrac{m^2}{s}} = 1.5\times10^{-6}\,m^{-1}$$

Table 7.15 *Influence of power sector emissions (PM$_{10}$) on urban air quality in Beijing, China.*

IES (2005) results for BAU 2010 (Table 5.4 and Figure 5.6 in IES report) Mean urban concentration calculated with the ISC Gaussian model of USEPA (1995) $\bar{c}_p = 0.5 \frac{\mu g}{m^3}$; uncertainty: $\pm 10 - 15\%$		

UWM calculations (Eq. (7.67))		
Input data Beijing inner city zone		
Urban area	1,782 km^2 (p.38 in IES report)	R_{loc} = 23.8 km
Population	11,488,000 persons (p.83, IES report)	ρ_{loc} = 6,447 pers/km^2
Power sector emissions	9,000 t/yr (Table 7.8 in IES report)	$\dot{m}_p$= 2.854E8 µg/s
Depletion velocity	0.64 cm/s	Table 7.8
Dilution rate	4,200 m^2/s (u = 6 m/s, h$_{mix}$ = 700 m)	β_p = 1.5E-6 m^{-1}
Mean concentration		
Plants along city edge	0.43 µg/m^3 (68% CI: 0.29, 0.65)	R_o = 47.6 km, σ_g = 1.5
Power plants inside city	0.88 µg/m^3 (68% CI: 0.59, 1.3)	R_o = 23.8 km, σ_g = 1.5
Average of above estimates for a 2:1 weighting factor	**0.61 µg/m^3 (68% CI: 0.46, 0.82)**	σ_g = 1.33

$$\bar{c}_p(R_o = 47.6 \text{ km}) = \cfrac{2.85 \times 10^8 \dfrac{\mu g}{s}}{\begin{array}{l}\pi \times 0.0064 \dfrac{m}{s} \times (47.6 \times 10^3)^2 m^2 \\[6pt] \times \left[1 - \exp(-1.5 \times 10^{-6} m^{-1} \times 47.6 \times 10^3 \text{ m})\right]\end{array}}$$

$$\bar{c}_p = 0.431 \frac{\mu g}{m^3}$$

(ii) To calculate the regional share of the total damage cost we use Eq. (7.73a). The background population density ρ_{uni} is 213 pers/km^2 and the local receptor density ρ_{loc} is 6447 pers/km^2. The value of β_p is $1.5 \cdot 10^{-6}$ m^{-1} (calculated in part (i)).

$$R_{loc} = \sqrt{\frac{\text{Urban area}}{\pi}} = \sqrt{\frac{1782 \times 10^6 m^2}{\pi}} = 23,817 \text{ m}$$

$$\left. \frac{CDF_{UWM,p}(R_o)}{\dot{D}_{UWM,P}} \right|_{Improved} = 1 - \frac{\rho_{uni} \, e^{-\beta_p R_o}}{\left[\rho_{loc}(1 - e^{-\beta_p R_{loc}}) + \rho_{uni} \, e^{-\beta_p R_{loc}}\right]} \quad \langle R_o \geq R_{loc} \rangle$$

(7.73a)

$$\text{Regional share of total damage} = \frac{\rho_{\text{uni}}\, e^{-\beta_p R_o}}{\left[\rho_{\text{loc}}(1 - e^{-\beta_p R_{\text{loc}}}) + \rho_{\text{uni}}\, e^{-\beta_p R_{\text{loc}}}\right]} ;$$

with $R_o = R_{\text{loc}}$

$$= \frac{1}{\left[\left(\dfrac{\rho_{\text{loc}}}{\rho_{\text{uni}}}\right)\left(e^{\beta_p R_{\text{loc}}} - 1\right) + 1\right]}$$

$$= \frac{1}{\left[\dfrac{6447}{213}\left(\exp\left(1.5 \times 10^{-6}\,\text{m}^{-1} \times 23817\,\text{m}\right) - 1\right) + 1\right]}$$

Regional share of total damage $= 48\%$

7.4.2.6 Comparison of intake fractions for $PM_{2.5}$, sulfates and nitrates with published results

Objective: Compare the inhalation intake fraction iF using Eq. (7.69) of the UWM with values reported in the literature for elevated sources in the United States and China.

Input data and results: Table 7.16 summarizes the input data and our iF comparisons. In the original studies, the spatial distribution of concentrations was simulated by the CALPUFF long-range air transport model (EPA 1998b), and a breathing rate of 20 m³/day was assumed. The UWM uncertainty intervals are calculated assuming σ_g equals 1.5 and 2 for primary and secondary pollutants, respectively, using the methodology of Chapter 11. Further details and commentary can be found in the table footnotes.

The UWM calculations compare quite favorably with literature data. Whether the range of the analysis is local, regional or national, most deviations are within ±50%, and in only two cases the difference is a factor of two. UWM estimates can be improved by integrating the population weighted concentration (Eq. (7.65) for primary and Eq. (7.66) for secondary pollutants) between r = 0 and $R_o \approx 1500$ km, rather than applying a mean concentration across the entire analysis area (Eq. (7.68)). Indeed, the improvement can be significant, as illustrated in the example below in the comparison with Zhou *et al.* (2003) for $PM_{2.5}$. iF increases from 11.5 ppm (Eq. (7.69)) to 15.4 ppm (value obtained after splitting the impact domain into separate local and regional areas and then integrating Eq. (7.65) across each area), which is quite close to the 15 ppm calculated by Zhou and colleagues.

Table 7.16 *Intake fractions (in parts per million, ppm) for elevated sources in the US and China.*

Pollutant	k (cm/s)	UWM (Eq.7.69)		CALPUFF[5] (iF, ppm)	
		iF (ppm)	68% CI	mean	Low-High
Study[1]: Beijing, China (Zhou *et al.*, 2003); ρ_{eff} = 213 pers/km^2, S_{ct} = 1 (all pollutants)					
PM$_{2.5}$	0.43	11.5	7.7–17	**15**	9–25
		(15.4)	(10–23)		
Sulfates	1.77	**2.8**	1.4–5.6	**6.0**	3–11
Nitrates	0.82	**6.0**	3.0–12	**6.5**	2–15
Study[2]: 29 sites in China (Zhou *et al.*, 2006); ρ_{eff} = 248 pers/km^2, S_{ct} = 0.5 (nitrates)					
PM$_{2.5}$	0.60	**9.6**	6.4–14	**7.1**	2–14
Sulfates	1.99	**2.9**	1.5–5.8	**4.4**	0.73–7.3
Nitrates	0.90	**3.2**	1.6–6.4	**3.5**	0.80–7.1
Study[3]: 9 sites in Illinois, USA (Levy, Spengler *et al.*, 2002); ρ_{eff} = 60 pers/km^2, S_{ct} = 0.5 (nitrates)					
PM$_{2.5}$	0.37	**2.6**	1.7–3.9	**2.3**	0.6–4
Sulfates	1.96	**0.32**	0.16–0.64	**0.2**	0.1–0.3
Nitrates	0.99	**0.32**	0.16–0.64	**0.3**	0.2–0.5
Study[4]: 40 sites in the USA (Levy, Wolff *et al.*, 2002); ρ_{eff} = 60 pers/km^2, S_{ct} = 0.5 (nitrates)					
PM$_{2.5}$	0.37	**2.6**	1.7–3.9	**2.2**	0.25–6.3
Sulfates	1.96	**0.32**	0.16–0.64	**0.22**	0.083–0.3
Nitrates	0.99	**0.079**	0.04–0.16	**0.035**	0.0096–0.075

[1] For the Beijing case study, two PM$_{2.5}$ iF estimates have been calculated: (i) using Eq. (7.69) with default data shown in the table and (ii) an improved estimate in which the modeling domain has been subdivided into local and regional areas. At the local-level (Beijing municipality), the mean concentration was evaluated using Eq. (7.67), where R$_o$ (73 km) is the effective radius of the municipality area (16,800 km^2) and 4000 m^2/s is assumed for the dilution rate. The incremental concentration was multiplied by the population (19.6 million inhabitants) to obtain the collective exposure and by the breathing rate (20 m^3/day) to determine the local intake fraction (4.75 ppm). Equation (7.65) was then used to determine the regional intake, integrating from r = R$_o$ to ∞, which contributed 10.6 ppm. Total intake is the sum of local and regional parts, 4.75 + 10.6 = 15.4 ppm. This method can be applied to any of the other calculations in this table.

[2] For the Chinese national-level calculations, the depletion velocities are population weighted averages for a domain area excluding the provinces of Gansu, Qinghai, Xinjiang, Tibet and Yunnan, as no emission sources were located in those regions in the original study. These values are slightly different from those reported in Table 7.8 because of differences in normalized areas. Estimates for nitrates have been adjusted by S_{ct} = 0.5 (Table 7.9).

[3] In the Levy, Spengler *et al.* (2002) study, the intake fraction was evaluated over an impact radius of 400 to 500 km. To account for the smaller impact range, the UWM calculation (Eq. (7.69)) has been reduced by 30% for PM$_{2.5}$ and by 55% for secondary pollutants based on the cumulative damage curves presented in Fig. 7.14. This choice is also corroborated by the cumulative profiles shown in Figure 2 in Zhou *et al.* (2003).

Example   Calculate the $PM_{2.5}$ intake fraction for an elevated source in Beijing assuming the mean breathing rate B_R is 20 m^3/day. The municipality of Beijing covers a surface area of 16,800 km^2 and the population is 19.6 million. The dilution rate (q_R) is 4000 m^2/s (u = 5 m/s and h_{mix} = 800 m).

Solution: We calculate the intake fraction iF as the sum of two contributions: (i) the population intake at the local scale (Beijing municipality) and (ii) the intake across the regional domain. For part (i) we use Eq. (7.67) and for the second part Eq. (7.65).

Parameter	Symbol or equation	Value
Depletion velocity (Table 7.8)	k_p	0.0043 m/s
Pollutant decay factor	$\beta_p = \frac{k_p}{q_R}$	$\frac{0.0043\text{m/s}}{4000\text{m}^2/\text{s}} = 1.075 \cdot 10^{-6}\text{m}^{-1}$
Equivalent radius of local area	$R_o = \left(\frac{\text{municipality area}}{\pi}\right)^{0.5}$	$\sqrt{\frac{16800 \times 10^6\text{m}^2}{\pi}} = 73,127\text{m}$
Municipality population	N_{pers}	19.6 million
Background population	ρ_{eff}	213 pers/km^2
Population breathing rate	B_R	20 m^3/day per person

Table 7.16 (*cont.*)

[4] UWM results for nitrates have been divided by four (in addition to the correction S_{ct} = 0.5) to be consistent with the assumption in Levy, Wolff *et al.* (2002) that nitrate formation only happens during the wintertime. Although nitrate formation is significantly limited during the warmer months of the year, as shown by Zhou *et al.* (2003) and Tarrasón *et al.* (2004), dividing the annual concentration by four may lead to underestimation of the annual intake fraction (see, for example, Table 2 in Zhou *et al.* 2003 and Figure 5.8 in Tarrasón *et al.*, 2004). To account for the smaller local impact domain size in the original analysis (≈ 500 km), we use the same correction factors as in comment 3 above.

[5] To test the influence of parameter uncertainty and variability on the resulting CALPUFF intake fraction estimates for emission sources in Beijing, Zhou *et al.* (2003) carried out sensitivity analyses. They found their results varied by as much as a factor of three, depending on which input parameter was changed. Likewise, Hao *et al.* (2007) calculated intake fractions based on CALPUFF concentration simulations for power plant emissions in the metropolitan area of Beijing using a 4 by 4 km resolution grid, 1/50 of the size used by Zhou and colleagues. Hao *et al.* reported a factor of two difference between their own iF calculations and the results published by Zhou *et al.* In explanation of the difference between studies, Hao *et al.* proposed two likely causes: choice of power plants included in the exposure assessment (impact of pollution mix) and grid size differences (impact of source–receptor distance). These studies clearly demonstrate that when comparing UWM with other models, one has to take into account their uncertainty and variability.

(i) Local contribution

$$\frac{\bar{c}_p(R_o)}{\dot{m}_p} = \frac{1}{\pi\, k_p R_o^2}\left[1 - e^{-\beta_p R_o}\right]$$

$$= \frac{1}{\pi \times 0.0043\,\frac{m}{s} \times 86400\,\frac{s}{day} \times (73127)^2 m^2}$$
$$\times \left[1 - \exp(-1.075 \times 10^{-6}\,m^{-1} \times 73127\,m)\right]$$

$$= 1.2113 \times 10^{-14}\,\frac{\mu g/m^3}{\mu g/day}\quad \begin{array}{l}\text{(mean concentration increase per}\\ \text{unit emission)}\end{array}$$

$$(7.67)$$

$$iF_{local} = \frac{\bar{c}_p(R_o)}{\dot{m}_p}\,N_{pers}\,B_R = 1.2113 \times 10^{-14}\,\frac{day}{m^3} \times 19.6 \times 10^6\,pers$$

$$\times 20\,\frac{m^3}{day \times pers}$$

$$iF_{local} = 4.75 \times 10^{-6} = 4.76\,ppm$$

(ii) Regional contribution
We start by integrating the concentration (Eq. (7.65)) from $r = R_o$ to ∞ to obtain the mean concentration increase across the regional domain $\bar{c}_{p,back}(R_o)$. The collective exposure is the product of the mean concentration increase and the population at risk $N_{pers,back}$. Finally, the exposure is multiplied by the mean breathing rate to obtain the regional intake fraction $iF_{regional}$.

$$\frac{\bar{c}_{p,back}(R_o)}{\dot{m}_p} = \frac{1}{A}\int_{R_o}^{\infty} c_p(r)2\,\pi\,r\,dr = \frac{1}{A}\int_{R_o}^{\infty} \frac{1}{2\pi u\,h_{mix}}\frac{1}{r}e^{-\beta_p\,r}2\,\pi\,r\,dr = \frac{e^{-\beta_p R_o}}{k_p\,A}$$

The collective population exposure per unit emission across the regional domain is given by,

$$\text{Regional exposure} = \frac{\bar{c}_{p,back}(R_o)}{\dot{m}_p}N_{pers,back} = \frac{e^{-\beta_p R_o}}{k_p A}N_{pers,back} = \frac{\rho_{uni}e^{-\beta_p R_o}}{k_p}$$

$$= 213\cdot 10^{-6}\,\frac{pers}{m^2} \times \frac{1}{0.0043\,\frac{m}{s}\times 86400\,\frac{s}{day}}\,\cdot$$

$$\times \exp(-1.075 \times 10^{-6}\,m^{-1} \times 73127\,m)$$

$$= 5.300\cdot 10^{-7}\,\frac{\mu g \cdot day/m^3}{\mu g/day}$$

$$iF_{regional} = \frac{\bar{c}_{p,back}(R_o)}{\dot{m}_p} N_{pers,back} \, B_R = 5.300 \times 10^{-7} \frac{pers \cdot day}{m^3} \times 20 \frac{m^3}{day \cdot pers}$$

$$iF_{regional} = 10.6 \times 10^{-6} = 10.6 \, ppm$$

The total intake fraction is the sum of local and regional contributions:

$$iF = iF_{local} + iF_{regional} = (4.76 + 10.6)ppm = 15.4ppm$$

Comment: Alternatively, we could have calculated the total intake fraction using Eq. (7.69), although less precisely (with $S_{sh} = S_{ct} = 1$) :

$$iF = \frac{\rho_{eff} \, B_R}{k_p} \tag{7.69}$$

$$= 213 \cdot 10^{-6} \frac{pers}{m^2} \times \frac{1}{0.0043 \frac{m}{s} 86400 \frac{s}{day}} \times 20 \frac{m^3}{pers \times day}$$

$$= 11.5 \cdot 10^{-6} = 11.5 \, ppm$$

7.4.2.7 Comparison of concentrations with EMEP source–receptor relationships

Objective: Use the UWM to calculate the change in ambient concentration due to emission reductions of primary particulate matter and from changes in secondary inorganic aerosols formed from precursor emissions of SO_2, NO_x and NH_3 and compare with source–receptor relationships derived by the European Monitoring and Evaluation Programme (EMEP) model. For continental scale changes, such as across the EU-27, we use Eq. (7.68) and for changes across individual countries we use Eq. (7.67).

Input data and results: An example of a source–receptor matrix is given in Table 7.17; in this case, for reductions of $PM_{2.5}$ primary emissions in 2010. The country at the top of the column is the source or emitter country, while each row identifies the receptor country. The numbers in the table indicate the reduction in background $PM_{2.5}$ concentration in the receptor country (ng/m^3) for a 15% reduction in $PM_{2.5}$ emissions in the emitter country. Moving down a column shows where a pollutant ends up once it is released into the air from a specified country. Whereas, moving across a row identifies where a pollutant comes from, namely, the contribution of own emissions versus emissions transported from other countries. The improvement in air quality for a 15% decrease in emissions from all countries is the sum of the values along a row.

Input data and comparisons of the UWM results with EMEP source–receptor matrices are shown in Table 7.18. Results are reported as a decrease in $PM_{2.5}$ ambient concentration (incl. aerosols) across the

EU-27 (= receptor, last row of Table 7.17 for $PM_{2.5}$ reductions) for a 15% decrease in precursor emissions in the emitter country. We consider reductions in the emissions of $PM_{2.5}$, SO_2, NO_x and NH_3 in various emitter (= source) countries and emission reductions for the entire EU-27 for the year 2010. Several interesting conclusions can be drawn from these comparisons. The UWM consistently underestimates the benefit of sulfur reductions by about a factor of 2, while overestimating the reduction potential of nitrogen by the same amount. This result suggests that the rate of transformation of sulfur to sulfates is too low in the UWM and it is too high for the conversion of nitrogen to secondary aerosols. The calculation routine in EMEP for the production of nitric acid was updated in 2007 to correct a coding error, and as a result of the changes there was a significant reduction in the reaction rate leading to the formation of nitrates. This would explain the deviation of the UWM and EMEP results. It would also explain the higher SO_2 transformation rate. For 2004 comparisons (see Figure 7.27), the deviation between UWM and EMEP results for reductions in SO_2 and NO_x emissions is ±25% for all emitter countries considered in the analysis.

For ammonia reductions, the results of the UWM are within ±50% of EMEP, with Spain being the only notable exception because the plume is transported out to sea, part of the time. Since changes in ammonia emissions affect the formation of both sulfates and nitrates, and UWM underestimates the effect of sulfates and over-predicts the effect of nitrates, the deviation is smaller because of partial cancellation of errors. For changes in primary PM emissions, the relative deviation of UWM to EMEP ranges between -24% and +53%, and for uniform emission reductions at the EU-27 level, UWM overestimates by about 10%.

We have also compared UWM calculations with EMEP for the years 2004 and 2006. These results, along with data from 2010, are plotted in Fig. 7.27. There is significant annual and site variability. UWM estimates can be adjusted using the factors in Table 7.9, or integrating Eq. (7.65) for primary pollutants or Eq. (7.66) for secondary pollutants over the appropriate averaging areas. In any case, we can say with confidence that the deviation of the results calculated by UWM and those calculated using the very detailed dispersion algorithms of EMEP are within a factor of two (actually, 70% of comparisons shown in Fig. 7.27 are within ±50%).

Table 7.19 and Figure 7.28 show the improvement in air quality in the emitter country for a 15% reduction of its own $PM_{2.5}$ emissions. We calculate changes in the mean country-level concentration using Eq.7.67, assuming the dilution rate q_R for elevated sources is 4,000 m^2/s ($u = 5$ m/s and $h_{mix} = 800$ m). Depletion velocities are listed in Table 7.8.

Table 7.17 *2010 EMEP source–receptor relationships ("blame matrices") for primary $PM_{2.5}$ emissions. Numbers indicate change in mean background concentration (ng/m^3) in receiver (receptor) country (table row) for a 15% emission reduction in source (emitter) country (table column).*

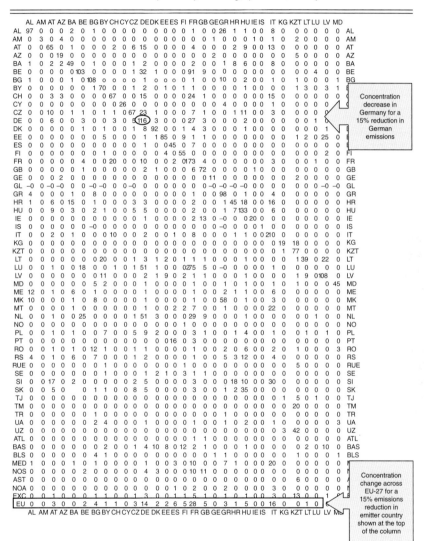

Concentration decrease in Germany for a 15% reduction in German emissions

Concentration change across EU-27 for a 15% emissions reduction in emitter country shown at the top of the column

Table 7.17 (*cont.*)

	ME	MK	MT	NL	NO	PL	PT	RO	RS	RUE	SE	SI	SK	TJ	TM	TR	UA	UZ	ATL	BAS	BLS	MED	NOS	AST	NOA	BIC	DMS	VOL	EXC	EU	
AL	4	10	0	0	0	1	0	3	18	0	0	0	0	0	0	1	0	0	0	0	0	6	0	0	0	0	0	7	175	42	AL
AM	0	0	0	0	0	0	0	0	0	1	0	0	0	1	13	0	0	0	0	0	0	0	7	0	0	0	0	0	24	0	AM
AT	0	0	0	0	0	5	0	2	1	0	0	18	8	0	0	0	0	0	0	0	0	1	0	0	0	0	0	12	154	147	AT
AZ	0	0	0	0	0	0	0	0	0	6	0	0	0	0	2	3	0	2	0	0	0	0	0	15	0	0	0	1	39	0	AZ
BA	4	0	0	0	0	3	0	5	12	1	0	2	2	0	0	0	0	0	0	0	0	2	0	0	0	0	0	7	110	35	BA
BE	0	0	0	23	1	2	0	0	0	1	1	0	0	0	0	0	1	0	0	0	14	0	0	0	0	0	0	22	271	269	BE
BG	0	2	0	0	0	1	0	34	9	4	0	1	0	0	7	3	0	0	0	2	2	0	0	0	0	0	0	5	190	162	BG
BY	0	0	0	0	0	17	0	6	1	21	1	0	1	0	0	0	10	0	0	0								3	144	40	BY
CH	0	0	0	0	0	1	0	0	0	0	0	0	0	0	0	0	0	0	0	0								13	128	61	CH
CY	0	0	0	0	0	0	0	1	0	2	0	0	0	0	73	1	0	0	0	0									110	33	CY
CZ	0	0	1	0	35	0	3	1	1	0	3	20	0	0	1	0	0	0	0									94	187		CZ
DE	0	4	1	15	0	1	0	1	1	2	0	0	0	0	1	0	0	2												192	DE
DK	0	2	6	11	0	0	0	1	7	0	1	0	0	0	1	0	0	11										41	152		DK
EE	0	0	1	5	0	1	0	16	3	0	0	0	0	1	0	0	4											158	135		EE
ES	0	0	0	0	0	7	0	0	0	0	0	0	0	0	0	2	0											63	62		ES
FI	0	0	0	1	1	0	0	0	8	3	0	0	0	0	0	0	0	1										2	78	67	FI
FR	0	0	0	2	0	1	0	0	0	0	0	0	0	0	0	0	0	2	0	0	2	3	0	0	0	0	0	19	204	201	FR
GB	0	0	0	1	1	1	0	0	0	0	0	0	0	0	0	0	4	0	0	0	6	0	0	0	0	0	20	87	85		GB
GE	0	0	0	0	0	0	0	0	0	3	0	0	0	0	8	0	0	0	0	2	0	0	0	0	1	23	0				GE
GL	0	0	0	0	0	-0	0	-0	-0	0	0	-0	-0	0	0	-0	-0	0	0	-0	0	0	0	0	0	0	0	0	0	0	GL
GR	0	5	0	0	0	1	0	4	4	2	0	0	0	0	6	1	0	0	0	0	13	0	0	1	0	0	7	141	117		GR
HR	1	0	0	0	0	5	0	7	12	1	0	19	5	0	0	0	0	0	0	0	5	0	0	0	0	0	10	162	87		HR
HU	0	0	0	0	0	16	0	38	14	2	0	7	37	0	0	3	0	0	0	0	1	0	0	0	0	0	12	295	254		HU
IE	0	0	0	0	0	0	0	0	0	0	0	0	0	0	0	0	0	5	0	0	0	1	0	0	0	0	0	35	38	37	IE
IS	0	0	0	0	0	0	0	0	0	0	0	0	0	0	0	0	0	0	1	0	0	0	0	0	0	0	0	211	2	1	IS
IT	0	0	0	0	0	1	0	1	1	0	0	6	0	0	0	0	0	0	0	0	0	12	0	0	0	0	0	5	239	234	IT
KG	0	0	0	0	0	0	0	0	0	0	0	0	4	0	0	0	42	0	0	0	0	0	1	0	0	0	1	84	0		KG
KZT	0	0	0	0	0	0	0	0	0	21	0	0	0	1	1	0	9	0	0	0	0	1	0	0	0	4	111	1			KZT
LT	0	0	0	0	1	22	0	3	0	15	2	0	1	0	0	3	0	0	2	0	0	1	0	0	0	0	0	4	142	101	LT
LU	0	0	0	4	0	3	0	0	1	0	0	0	0	0	0	0	0	1	0	0	3	0	0	0	0	0	26	436	434		LU
LV	0	0	0	1	10	0	2	0	14	2	0	1	0	0	2	0	0	2	0	0	1	0	0	0	0	0	3	179	150		LV
MD	0	0	0	0	0	7	0	88	1	11	0	0	1	0	4	22	0	0	0	2	1	0	0	0	0	0	5	193	109		MD
ME	46	1	0	0	0	1	0	3	18	1	0	0	0	1	0	0	0	0	0	3	0	0	0	0	0	0	7	103	19		ME
MK	1	48	0	0	0	1	0	5	23	1	0	0	1	0	0	1	1	0	0	0	2	0	0	0	0	0	7	166	79		MK
MT	0	0	45	0	0	0	0	0	0	0	0	1	0	0	0	0	0	0	0	74	0	0	2	0	0	0	6	84	82		MT
NL	0	0	0	83	1	4	0	0	0	1	1	0	1	0	0	0	0	0	1	1	0	28	0	0	0	0	0	12	214	211	NL
NO	0	0	0	0	43	0	0	0	0	1	2	0	0	0	0	0	0	0	0	1	0	0	0	0	0	0	2	49	5		NO
PL	0	0	0	0	1	174	0	5	1	6	1	1	9	0	0	5	0	0	1	0	0	1	0	0	0	0	0	5	244	222	PL
PT	0	0	0	0	0	0	127	0	0	0	0	0	0	0	0	0	0	13	0	0	1	0	0	0	0	0	0	3	147	147	PT
RO	0	0	0	0	0	4	0	242	6	4	0	0	3	0	0	2	5	0	0	0	1	0	0	0	0	0	0	6	299	274	RO
RS	3	4	0	0	0	3	0	30	114	1	0	1	4	0	0	1	1	0	0	0	1	0	0	0	0	0	0	7	211	73	RS
RUE	0	0	0	0	0	0	0	0	0	41	0	0	0	0	0	1	0	0	0	0	0	0	0	0	0	0	0	2	51	2	RUE
SE	0	0	0	0	8	2	0	0	0	2	23	0	0	0	0	0	0	0	2	0	0	1	0	0	0	0	0	2	45	35	SE
SI	0	0	0	0	0	5	0	4	2	1	0	212	3	0	0	0	0	0	0	0	3	0	0	0	0	0	0	10	317	293	SI
SK	0	0	0	0	0	39	0	14	4	2	3	136	0	0	3	0	0	0	0	0	0	0	0	0	0	0	0	11	253	255	SK
TJ	0	0	0	0	0	0	0	0	0	0	0	0	0	64	1	0	0	35	0	0	0	0	11	0	0	0	0	107	0		TJ
TM	0	0	0	0	0	0	0	0	0	4	0	0	0	3	46	1	0	37	0	0	0	0	14	0	0	0	1	111	0		TM
TR	0	0	0	0	0	0	0	0	0	2	0	0	0	0	0	111	1	0	0	1	3	3	0	0	0	0	0	2	119	4	TR
UA	0	0	0	0	0	9	0	17	1	22	0	0	2	0	4	42	0	0	0	1	0	0	0	0	0	0	0	4	115	37	UA
UZ	0	0	0	0	0	0	0	0	0	6	0	0	0	11	8	0	0	144	0	0	0	0	5	0	0	0	0	2	214	1	UZ
ATL	0	0	0	0	1	0	1	0	0	0	0	0	0	0	0	0	0	2	0	0	0	0	0	0	0	0	13	5	4		ATL
BAS	0	0	0	3	12	0	1	0	7	11	0	1	0	0	0	1	0	0	12	0	0	2	0	0	0	0	0	3	90	76	BAS
BLS	0	0	0	0	0	1	0	11	1	16	0	0	0	0	35	8	0	0	0	11	0	0	0	0	0	0	4	85	19		BLS
MED	0	0	0	0	0	0	1	0	1	1	0	1	0	0	13	1	0	0	0	45	0	1	3	0	0	0	5	66	47		MED
NOS	0	0	0	3	8	2	0	0	0	0	1	0	0	0	0	0	1	1	0	0	15	0	0	0	0	0	13	43	39		NOS
AST	0	0	0	0	0	0	0	0	0	2	0	0	0	1	2	7	0	3	0	0	0	1	0	23	0	0	0	1	23	1	AST
NOA	0	0	0	0	0	0	0	0	0	0	0	0	0	0	0	0	0	0	0	0	0	0	0	0	7	0	0	3	11	9	NOA
EXC	0	0	0	0	1	4	1	4	1	24	1	0	1	1	1	4	2	5	0	0	0	1	0	1	0	0	0	4	85	32	EXC
EU	0	0	0	2	1	17	4	16	1	2	3	2	4	0	0	1	1	0	1	1	0	2	2	0	0	0	0	9	159	149	EU
	ME	MK	MT	NL	NO	PL	PT	RO	RS	RUE	SE	SI	SK	TJ	TM	TR	UA	UZ	ATL	BAS	BLS	MED	NOS	AST	NOA	BIC	DMS	VOL	EXC	EU	

Concentration decrease in Germany for a 15% reduction in EU-27 emissions (sum of numbers in row)

Source: EMEP country-to-country source-receptor matrices www.emep.int/SR_data/sr_tables.html

Table 7.18 $PM_{2.5}$ concentration decrease (incl. aerosols) in the EU-27 for 15% emission reduction in emitter country, UWM versus 2010 EMEP source–receptor matrices.

Emitter country				UWM (Eq.7.68)		$\Delta \bar{c}_{15\%}$ (ng/m^3)		
Precursor Pollutant	$\dot{m}_p^{(1)}$(kt)	k(cm/s)	$A^{(2)}$(km²)	$\bar{c}^{(3)}$ ng/m³	UWM	EMEP	UWM/ EMEP
FRANCE							
$PM_{2.5}$	255	0.45	7.1E+06	254	38	28	**1.36**
SO_2	262	1.73	9.6E+06	50	7	16	**0.47**
NO_x	1,080	0.71	9.6E+06	251	38	16	**2.35**
NH_3	645	0.84	9.6E+06	252	38	25	**1.51**
GERMANY							
$PM_{2.5}$	111	0.52	7.1E+06	96	14	14	**1.03**
SO_2	449	1.94	9.6E+06	76	11	34	**0.34**
NO_x	1,323	0.83	9.6E+06	263	39	16	**2.46**
NH_3	548	0.98	9.6E+06	185	28	45	**0.62**
HUNGARY							
$PM_{2.5}$	32	0.57	7.1E+06	25	3.8	5.0	**0.76**
SO_2	32	1.77	9.6E+06	6	0.9	2.0	**0.45**
NO_x	162	1.01	9.6E+06	26	4.0	2.4	**1.65**
NH_3	65	1.06	9.6E+06	20	3.0	7.0	**0.43**
ITALY							
$PM_{2.5}$	173	0.71	7.1E+06	109	16	16	**1.02**
SO_2	210	1.86	9.6E+06	37	6	11	**0.51**
NO_x	963	1.04	9.6E+06	153	23	17	**1.35**
NH_3	379	1.33	9.6E+06	94	14	18	**0.78**
POLAND							
$PM_{2.5}$	137	0.54	7.1E+06	114	17	17	**1.00**
SO_2	974	1.98	9.6E+06	162	24	42	**0.58**
NO_x	867	1.27	9.6E+06	113	17	4.1	**4.12**
NH_3	271	1.01	9.6E+06	88	13	33	**0.40**
UK							
$PM_{2.5}$	67	0.59	7.1E+06	51	8	5	**1.53**
SO_2	406	2.03	9.6E+06	66	10	22	**0.45**
NO_x	1,106	1.28	9.6E+06	142	21	9	**2.37**
NH_3	284	1.11	9.6E+06	85	13	15	**0.85**
EU-27							
$PM_{2.5}$	1,397	0.57	7.1E+06	1099	165	149	**1.11**
SO_2	4,526	1.85	9.6E+06	806	121	223	**0.54**
NO_x	9,069	1.00	9.6E+06	1495	224	99	**2.26**
NH_3	3,588	1.07	9.6E+06	1106	166	213	**0.78**
AMMONIA							
Denmark	75	1.61	9.6E+06	15	2.3	3.1	**0.74**
Spain	368	0.94	9.6E+06	129	19	7.0	**2.76**
Switzerland	63	1.04	9.6E+06	20	3.0	1.9	**1.58**
NO_x							
Spain	881	0.91	9.6E+06	160	24	8	**2.95**

[1] Precursor emissions for 2010 (*Source*: http://www.emep.int/mscw/mscw_publications.html)
[2] A is the averaging surface area, a circular domain of radius R_o centered on the middle of the region of interest. The value of R_o is 1500 km for $PM_{2.5}$ and 1750 km for all other pollutants.
[3] Nitrates concentrations have been scaled by $S_{ct} = 0.5$ (Table 7.9).

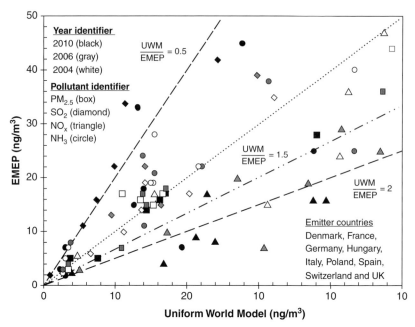

Fig. 7.27 Comparison of $PM_{2.5}$ concentration changes in the EU-27 for a 15% decrease in precursor emission in various emitter countries using UWM versus EMEP source–receptor matrices for years 2004, 2006 and 2010 (note: since 2007, EMEP made changes to the routine for calculation of nitrate concentrations; details can be found in the EMEP Status Report 2008).

Annual emissions by country and source–receptor relationships for the year 2010 can be downloaded for free from the EMEP website indicated at the bottom of the table. For large countries with a surface area greater than 100,000 km^2, R_o is the radius of a circle of area A equal to the geographical surface area of the country $\left(R_o = \sqrt{\frac{A}{\pi}}\right)$. For small countries, R_o is estimated as $\sqrt{A}$. This is necessary to avoid overestimating the pollutant dilution rate, which under the assumption of a circular cross-wind area would be artificially greater than the dilution for a constant rectangular area of width R_o and height h_{mix}. The difference is negligible for large countries, and there is no need for modifications. As indicated in the figure at the bottom of the table, the UWM values are within ±35% of the EMEP results. The mean relative difference is −12% (UWM tends to underestimate, as most data points in the figure lie to the left of the diagonal line). The scatter of residuals is due to variability in local

Table 7.19 Improvement in air quality in the emitter country for a 15% reduction of its own $PM_{2.5}$ emissions (emitter = receiver country), comparison UWM with 2010 EMEP source-receptor matrices.

Emitter country	Geographical area 1000's km²	UWM (Eq.7.67)				$\Delta\bar{c}_{15\%}$ (ng/m³)[3]		
		$\dot{m}_p$ kt	$k_{PM2.5}$ cm/s	$R_0^{(1)}$ km	$\bar{c}_p(Ro)^{(2)}$	UWM	EMEP	$\dfrac{UWM}{EMEP}$
EU-27		**1397**	**0.57**	**1500**	**1099**	**165**	**149**	**1.11**
France	552	255	0.45	419	1224	184	173	1.06
Spain	506	74	0.50	401	366	55	46	1.19
Germany	357	111	0.52	337	673	101	116	0.87
Finland	338	41	0.62	328	247	37	55	0.67
Poland	313	137	0.57	316	882	132	174	0.76
Italy	301	173	0.71	310	1085	163	210	0.78
UK	243	67	0.52	278	510	77	72	1.06
Romania	238	118	0.57	275	895	134	242	0.55
Hungary	93	32	0.57	305	675	101	133	0.76
Czech Rep.	79	20	0.59	281	462	69	67	1.03
Denmark	43	26	0.86	207	802	120	92	1.31
Netherlands	42	15	0.66	205	492	74	83	0.89
Switzerland	41	10	0.55	202	342	51	67	0.77
Luxembourg	2.6	2	0.59	51	300	45	72	0.62

[1] $R_o = \sqrt{\frac{A}{\pi}}$ for large countries, and $\sqrt{A}$ for small ones ($<10^5\ km^2$); A=geographical area

[2] Dilution rate $q_R = 4000$ m²/s (u = 5 m/s and h_{mix} = 800 m)

[3] $\Delta\bar{c}_{15\%}$ is the $PM_{2.5}$ concentration reduction in the emitter country for a 15% emission reduction in that country (diagonal elements in Table 7.17). Emissions data, and other useful information, can be found in the documentation accessible from http://emep.int/publ/reports/2012/status_report_1_2012.pdf

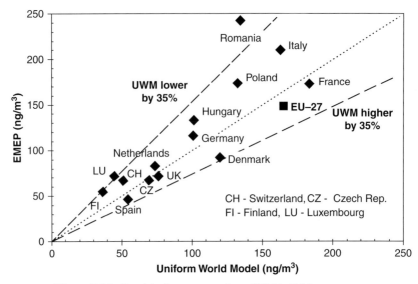

Figure 7.28 Graphical representation of Table 7.19.

meteorology and terrain cover. Another source of variability is the appropriate choice of R_o.

7.4.3   Final remarks

The scope of Section 7.4 has been to present a simple and convenient environmental impact assessment tool, the Uniform World Model (UWM). It yields the damage cost as a product of a few factors; it is simple and transparent, showing at a glance, the role of the most important parameters of the impact pathway analysis. For uniform receptor distributions and homogeneous (k = constant) and well-mixed atmospheres (vertical concentration gradients are negligible), the UWM is an exact solution of the damage function. The methodology also yields results for concentrations and intake factors.

Key conclusions of the validation studies:

- UWM damage costs have been compared with results from detailed impact assessments that have been carried out in Europe, China, India and the USA, and found to be reasonably robust and reliable, the observed differences are usually in the ±50% range.
- Rural and regional (continental) estimates are more accurate than results for site-specific case studies involving large urban areas, but suitable correction factors have been determined to improve the

agreement with results from detailed models, even in the case of ground-level sources (Table 7.9).

- For urban emissions from elevated stacks greater than 25 m, the UWM damage costs, as computed by Eq. (7.63), are usually within a factor of two.
- For ground-level emissions, damage costs calculated by Eq. (7.77) are expected to be within a factor of 2 to 3 of detailed modeling results.
- UWM estimates for average concentrations have been compared with the detailed assessment carried out by EMEP for the EU-27 and at individual country level. Differences are usually no more than a factor of 2.
- For stacks in excess of 150 meters, even in the proximity of large cities, the ground-level mean concentration (< 50 km) is within a factor of 2 (Fig. 7.22).

The reason why the UWM is such a good representation of typical results is that averaging over many sites is equivalent to averaging over different distributions of population, thus rendering the distribution more uniform. The UWM involves replacement of the average of a product by the product of the averages, an approximation that is justified to the extent that the factors are not correlated with each other and do not vary too much. In practice, the concentration varies the most, being high near the source and decreasing with downwind distance x as $1/x \exp(-\beta x)$. For sources close to or inside large cities, this variation is correlated with the population density and so the UWM, understandably, underestimates the impact. For sources far from large cities, the spatial variation of population density occurs in a region where the concentration varies slowly, consequently, taking the mean population density is adequate and the UWM prediction is acceptable.

References

Amann, M., Kejun, J., Jiming, H. and others. 2008. *GAINS-ASIA, Scenarios for Cost-Effective Control of Air Pollution and Greenhouse Gases in China.* IIASA, Schlossplatz 1, Laxenburg, 2361, Austria.

Apte, J. S., Bombrun, E., Marshall, J. D. and Nazaroff, W. N. 2012. Intraurban intake fractions for primary air pollutants from vehicles and other distributed sources. *Environmental Science & Technology* 46: 3415–3423.

Barrett, K. 1992. *Dispersion of Nitrogen and Sulfur across Europe from Individual Grid Elements: Marine and Terrestrial Deposition.* EMEP/MSC-W Note 3/92. August 1992. Norwegian Meteorological Institute, P.O.Box 43, Blindern, N-0313 Oslo 3.

Benarie, M. M. 1987. The limits of air pollution modeling (Editorial). *Atmospheric Environment* 21: 1–5.

Beychok, M. R. 1979. How accurate are dispersion estimates? *Hydrocarbon Processing* October 1979.

Briggs, G. A. 1972. Discussion on chimney plumes in neutral and stable surroundings. *Atmospheric Environment* 6: 507–510.

Briggs, G. A. 1975. *Plume Rise Predictions*. Lectures on Air Pollution and Environmental Impact Analysis, American Meteorological Society, Boston, MA, USA.

Briggs, G. A. 1973. *Diffusion Estimation for Small Emissions*. Air Resources Atmospheric Turbulence and Diffusion Laboratory, NOAA, ADTL-106, Oak Ridge, TN, USA.

Carmichael, G. R. and Ardnt, R. 1994. *Long Range Transport and Deposition of Sulfur in Asia* in RAINS-ASIA: An Assessment Model for Acid Rain in Asia, Report from the World Bank sponsored project, Acid Rain and Emissions Reduction in Asia, Technical report (51 pp).

Cenedese, A., Cosemans, G., Erbrink, H. and Stubi, R. 1997. *Vertical Profiles of Wind, Temperature and Turbulence*. COST Action 710, Preprocessing of Meteorological Data for Dispersion Modelling, Report of Working Group 3, October 1997.

Curtiss, P. and Rabl, A. 1996. Impacts of air pollution: General relationships and site dependence. *Atmospheric Environment* 30: 3331–3347.

Derwent, R. G. and Nodop, K. 1986. Long-range transport and deposition of acidic nitrogen species in north-west Europe. *Nature* 324: 356–358.

Derwent, R. G., Hov, Ø., Asman, W.A.H., van Jaarsveld, J. A. and de Leeuw, F.A. A.M. 1989. An intercomparison of long-term atmospheric transport models; The budgets of acidifying species for the Netherlands. *Atmospheric Environment* 23(9): 1893–1909.

Derwent, R. G., Dollard G. J. and Metcalfe, S. E. 1988. On the nitrogen budget for the United Kingdom and North-west Europe. *Quarterly Journal of the Royal Meteorological Society* 114: 1127–1152.

Dinkel, F., Pohl, C. H., Matjaz, R. and Waldeck, B. 1996. Okologische Bewertung mit der wirkungsorientierten Methode, Buwal.

Draxler, R. R. 1976. Determination of atmospheric diffusion parameters. *Atmospheric Environment* 10: 99–105.

Eliassen, A. 1978. The OECD Study of Long-Range Transport of Air Pollutants: Long-Range Transport Modeling. *Atmospheric Environment* 12: 479–487.

ExternE 2005. *ExternE: Externalities of Energy, Methodology 2005 Update*. Edited by P. Bickel and R. Friedrich. Published by the European Commission, Directorate-General for Research, Sustainable Energy Systems. Luxembourg: Office for Official Publications of the European Communities. ISBN 92-79-00429-9.

ExternE 2008. With this reference we cite the methodology and results of the NEEDS (2004–2008) and CASES (2006–2008) phases of ExternE. For the damage costs per kg of pollutant and per kWh of electricity we cite the numbers of the data CD that is included in the book edited by Markandya, A., Bigano, A. and Porchia, R. in 2010: *The Social Cost of Electricity: Scenarios and Policy Implications*. Edward Elgar Publishing Ltd, Cheltenham, UK. They can also be downloaded from http://www.feem-project.net/cases/ (although in the latter some numbers have changed since the data CD in the book).

Feliciano, M. S., Pio, C. A. and Vermeulen, A. T. 2001. Evaluation of SO_2 dry deposition over short vegetation in Portugal. *Atmospheric Environment* 35: 3633–3643.

Gifford, F. A. 1975. *Lectures on Air Pollution and Environmental Impact Analyses*, Haugen, D. A. (Editor), American Meteorological Society, September 1975.

Gifford, F. A. 1961. Use of routine meteorological observations for estimating the atmospheric dispersion. *Nuclear Safety* 2(4): 47–57.

Green, A. E., Singhal, R. P. and Venkateswar, R. 1980. Analytic extensions of the Gaussian Plume Model. *Journal of the Air Pollution Control Association (JAPCA)* 30(7): 773–776.

Gualtieri, G. and Secci, S. 2011. Comparing methods to calculate atmospheric stability-dependent wind speed profiles: A Case Study on Coastal Location. *Renewable Energy* 36(8): 2189–2204.

Hao, J, Wang, L., Shen, M. Li, L. and Hu, J. 2007. Air quality impacts of power plant emissions in Beijing. *Environmental Pollution* 147: 401–408.

Hanna, S. R. *et al.* 1977. AMS Workshop on stability classification schemes and sigma curves – Summary of Recommendations. *Journal of Climate and Applied Meteorology* 58(12):1305–1309.

Hirschberg, S., Heck, T., Gantner, U. *et al.* 2004. Health and Environmental Impacts of China's Current and Future Electricity Supply, with Associated External Costs. Special Issue on China's Energy Economics and Sustainable Development in the 21st Century, *International Journal of Global Energy Issues*, Y. M. Wei, H. T. Tsai, C. H. Chen, Guest Editors, Volume 22 (2/3/4), InderScience Publishers 2004.

Hsu, S. A. 1982. Determination of the power-law wind profile exponent on a tropical coast. *Journal of Applied Meteorology* 21: 1187–1190.

IES 2005. Integrated Environmental Strategies, *Energy Options and Health Benefit – Beijing Case Study*. Report by NREL, USA, Department of Environmental Science, Tsinghua University, China, School of Public Health, Peking University, China and School of Public Health, Yale University, USA, Nov 2005. IES Program www.epa.gov/ies/china/national_assessment.html

Irwin, J. S. 1979. Estimating Plume Dispersion – A Recommended Generalized Scheme. Presented at the 4th AMS Symposium on Turbulence and Diffusion, Reno, Nevada, USA.

Jolliet O. and Crettaz, P. 1997. Fate coefficients for the toxicity assessment of air pollutant. *International Journal of Life Cycle Assessment* 2(2): 104–110.

Khoder, M. I. 2002. Atmospheric conversion of sulfur dioxide to particulate sulfate and nitrogen dioxide to particulate nitrate and gaseous nitric acid in an urban area. *Chemosphere* 49: 675–684.

Krewitt, W., Heck, T. and Friedrich, R. 1999. Environmental damage costs from fossil electricity generation in Germany and Europe. *Energy Policy* 27(4): 173–183.

Krewitt, W. Trukenmueller, A., Mayerhofer, P. and Friedrich, R. 1995. ECOSENSE – An Integrated Tool for Environmental Impact Analysis, in Kremers, H. and W. Pillmann (Ed.), *Space and Time in Environmental Information Systems*, Umwelt-Informatik aktuell, Band 7, Metropolis-Verlag, Marburg.

Lee and Watkiss. 1998. Working Paper for the ExternE Project of the European Commission.

Levy, J., Spengler J. D., Hlinka, D., Sullivan, D. and Moon, D. 2002. Using CALPUFF to evaluate the impacts of power plant emissions in Illinois: Model Sensitivity and Implications. *Atmospheric Environment* 36: 1063–1075.

Levy, J., Wolff, S. K. and Evans, J. S. 2002. A regression-based approach for estimating primary and secondary particulate matter intake fractions. *Risk Analysis* 22(5): 895–904.

Luhar, A. K. 1998. An analytical slab model for the growth of the coastal thermal internal boundary layer under near-neutral onshore flow conditions. *Boundary-Layer Meteorology* 88: 102–120.

Luria, M, Imhoff, R. E., Valente, R. J., Parkhurst, W. J. and Tanner, R. L. 2001. Rates of conversion of sulfur dioxide to sulfate in a scrubbed power plant plume. *Journal Air Waste Management Association* 51: 1408–1413.

McElroy, J. L. 1969. A comparative study of urban and rural dispersion. *Journal of Applied Meteorology* 8(1): 19.

McElroy, J. L. and Pooler, F. 1968. *The St. Louis Dispersion Study, Vol. II-Analysis*, US EPA Publication AP-53, December 1968.

McMullen, R. W. 1975. The change of concentration standard deviations with distance. *Journal of the Air Pollution Control Association (JAPCA)*, October 1975.

Miyakawa, T., Takegawa, N. and Kondo, Y. 2007. Removal of sulfur dioxide and formation of sulfate aerosol in Tokyo. *Journal of Geographical Research* 112, D13209.

Nicholson, K. W. 1988. The dry deposition of small particles: A review of experimental measurements. *Atmospheric Environment* 22: 2653–2666.

Panofsky, H. A. and Dutton, J. A. 1984. *Atmospheric Turbulence*. John Wiley & Sons, Inc. New York.

Pasquill, F. 1961. The estimation of the dispersion on windborne material. *Meteorological Magazine* 90: 33–49.

Rabl, A. 1985. *Active Solar Collectors and Their Applications*. Oxford University Press, New York.

Sandness, H. 1993. *Calculated Budgets for Airborne Acidifying Components in Europe*, EMEP/MSC-W Report 1/93 (July 1993), Norwegian Meteorological Institute, P.O. Box 43, Blindern, N-0313, Oslo 3.

Scire J. S., Strimaitis, D. G. and Yamartino, R. J. 2000. *A user's guide for the CALPUFF dispersion model (Ver. 5)*, Earth Tech Inc.

Sehmel, G. 1980. Particle and gas dry deposition: a review. *Atmospheric Environment* 14: 983.

Seinfeld, J. H. and Pandis, S. N. 1998. *Atmospheric Chemistry and Physics: from Air Pollution to Climate Change*, John Wiley & Sons, Inc., New York.

Smedman-Högström, A. S. and Högström, U. 1978. A practical method for determining wind frequency distributions for the lowest 200 m from routine meteorological data. *Journal of Applied Meteorology*, 17: 942–54.

Smith, M. E. 1968. *Recommended Guide for the Prediction of the Dispersion of Airborne Effluents, 1st Edition*. American Society of Mechanical Engineers, New York, USA.

Spadaro, J. V. 1999. Quantifying the Damages of Airborne Pollution: Impact Models, Sensitivity Analyses and Applications. Ph.D. Doctoral Thesis, Ecole des Mines de Paris, Boulevard St. Michel, 60, Paris Cedex 06, F75272.

Spadaro, J. V. and Rabl, A. 2005. *Dispersion Models for Time-Averaged Collective Air Pollution Exposure: An Estimation of Uncertainties*. Centre Energétique et Procédés, Ecole des Mines (ARMINES), 60 Boulevard St. Michel, Paris, France.

Spadaro, J. V. and Rabl, A. 2004. Pathway analysis for population-total health impacts of toxic metal emissions. *Risk Analysis* 24(5): 1121–1141.

Spadaro, J. V. and Rabl, A. 2002. *Assessing the Health Impacts due to Airborne Emissions: The AirPacts Model*. Probability Safety Assessment and Management Conference, San Juan, Puerto Rico, June 23–28, 2002.

Spadaro, J. V. and Rabl, A. 1999. Estimates of real damage from air pollution: site dependence and simple impact indices for LCA. *International J. of Life Cycle Assessment* 4(4): 229–243.

Tarrasón, L, Fagerli, H., Jonson, J. E. *et al.* 2004. *Transboundary Acidification, Eutrophication and Ground Level Ozone in Europe*. EMEP/MSC-W, Norwegian Meteorological Institute, EMEP Status Report 2004, ISSN 0806-4520.

Touma, J. S. 1977. Dependence of the wind profile law on stability for various locations. *Journal of the Air Pollution Control Association (JAPCA)*, September 1977.

Turner, D. B. 1970. *Workbook of Atmospheric Dispersion Estimates*, US EPA, Research Triangle Park, NC, USA. Publication AP-26 (NTIS PB191-482).

Underwood, B. 2001. *Review of deposition velocity and washout coefficient*. Technical report, AEA Technology. www.hpa.org.uk/webc/HPAwebFile/HPAweb_C/1194947314056

US Environmental Protection Agency (EPA). 2011. *Exposure Factors Handbook*: 2011 Edition. Office of Research and Development, Washington, DC 20460, USA. EPA/600/R-090/052F.

US Environmental Protection Agency (EPA). 1998a. *A Comparison of CALPUFF with ISC3*. Office of Air Quality, Planning and Standards, Research Triangle Park, NC, USA. EPA-454/R-98-020, Dec 1998.

US Environmental Protection Agency (EPA). 1998b. *A Comparison of CALPUFF modeling Results to Two Tracer Field Experiments*. Office of Air Quality, Planning and Standards, Research Triangle Park, NC, USA. EPA-454/R-98-009, Jun 1998.

US Environmental Protection Agency (EPA) 1995. *User's Guide for the Industrial Source Complex (ISC3) Dispersion Models, Volume II – Description of Model Algorithms*, Office of Air Quality Planning and Standards, Emissions, Monitoring and Analysis Division, Research Triangle Park, NC, USA, EPA-454/B-95-003b.

US Environmental Protection Agency (EPA) 1989. *User's Guide to the CTDM Meteorological Preprocessor Program*, Atmospheric Research and Exposure Assessment Laboratory, Research Triangle Park, NC, USA, EPA/600/8-88/004.

US Nuclear Regulatory Commission (US NRC) 1972. *On-Site Meteorological Programs*, Regulatory Guide 1.23 (Safety Guide 23), February 1972.

Vogt, K. J. 1977. Empirical investigations of the diffusion of waste air plumes in the atmosphere, *Nuclear Technology* 34: 43–57 (June 1977).

West, J. J., Ansari, A. S. and Pandis, S. N. 1999. Marginal $PM_{2.5}$: Nonlinear aerosol mass response to sulfate reductions in the eastern United States. *Journal of the Air and Waste Management Association* 49: 1415–1424.

Zannetti, P. 1990. *Air Pollution Modeling. Theories, Computational Methods and Available Software*, Van Nostrand-Reinhold.

Zhou, Y, Levy, J. I., Evans, J. S. and Hammitt, J. K. 2006. The influence of geographic location on population exposure to emissions from power plants throughout China. *Environment International* 32: 365–373.

Zhou, Y., Levy, J. I., Hammitt, J. K. and Evans, J. S. 2003. Estimating population exposure to power plant emissions using CALPUFF: A case study in Beijing, China. *Atmospheric Environment* 37: 815–826.

8   Multimedia pathways

Summary

Whereas the classical air pollutants are harmful only via inhalation, persistent pollutants such as toxic metals are also harmful after entering the food chain. This is an important pathway, because the total population dose due to ingestion can easily be an order of magnitude larger than the dose via inhalation. Note that the geographic range of the analysis must be even larger than for atmospheric dispersion, because most food is transported over large distances, often worldwide. This chapter describes several approaches for estimating ingestion doses. Even though the detail about dispersion in the environment is exceedingly complex and difficult to model, some shortcuts are possible if one can find the right data for the relation between total emissions and total population dose under steady-state conditions. Thus one can carry out calculations of total population dose that are far simpler and probably more reliable than detailed site-specific models. That is the case for dioxins, the subject of Section 8.2. The pathways for mercury are complex, and because of its long residence time in the atmosphere it is dispersed over the entire hemisphere. But with data for the global emissions and the global ingestion dose, one can again obtain a simple model for the global health impact, as described in Section 8.3. In Section 8.4 we present a more detailed model, based on transfer factors between different environmental compartments that have been published by USEPA. The key result is the intake fraction, defined as the fraction of the emitted pollutant mass that will be inhaled or ingested by a human being. Results for impacts and damage costs of toxic metals can be found in Section 8.5.

8.1   General considerations

Whereas the classical air pollutants affect human health only through inhalation, many micropollutants such as toxic metals and dioxins are persistent and can be harmful by ingestion of food or water (dermal contact is another possible pathway but not of concern for pollutants emitted to air). It turns out that for many persistent pollutants, the dose by ingestion is larger than the dose by inhalation, roughly by one to two orders of magnitude. Analysis of multimedia pathways through soil, water, plants and animals is also needed for the assessment of ecosystem impacts. A simple schematic of the transfers through the main environmental compartments can be seen in Fig. 8.1. Of course, this is a very simplified picture. Countless details are not shown, for example, that residue of agricultural vegetation can get back into the soil or that some food for fish farms comes from agriculture. Most such details have little or no effect on the modeling and can safely be ignored in order to make the models manageable.

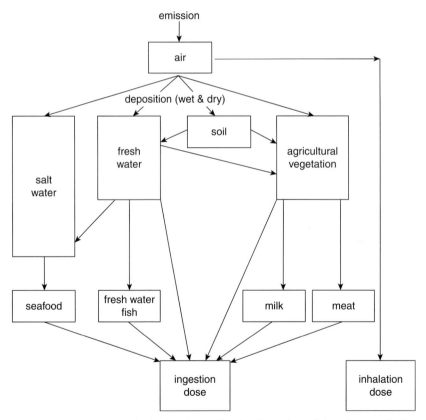

Fig. 8.1 The principal pathways for the dispersion of air pollutants in the environment.

A wide variety of different models are available for analyzing the pathways. For the calculation of damage costs, a model should calculate the expectation value of the collective doses, for the entire population that is affected. For linear ERFs (exposure–response functions) without threshold, one needs the total dose, for ERFs with a no-effect threshold, the dose above the threshold. Models should evaluate both local and regional impacts (and in the case of Hg, even global impacts). The requirement to determine the dose for the entire population, rather than just the most highly exposed individuals near the source, rules out numerous models that only evaluate local impacts.

Among the many models that have been developed we mention the following:

- The models of MSC-East (Meteorological Synthesizing Centre – East) (www.msceast.org/),
- Multimedia models of the Mackay-type (Mackay, 2001) that have been used for LCA (life cycle assessment) or external cost assessments, such as CalTox (McKone and Enoch 2002) and WATSON (Bachmann, 2006),
- IMPACT2002+ (Jolliet et al., 2003),
- USES-LCA (Huijbregts et al., 2005),
- RiskPoll (www.arirabl.org).

WATSON is a multizonal model that links the regional air quality model of EcoSense to a soil and water multimedia model of the Mackay level III/IV type. WATSON is part of the EcoSense software developed by the ExternE project series.

Detailed models such as CalTox, WATSON and those of MCS-East require a great deal of site-specific data. Of course, the developers of the models have these data for their region, but for other regions it is usually difficult to obtain all of the required data. And even with the best input data the uncertainties are large because of modeling uncertainties. Therefore it is sometimes interesting to use simplified approaches, based on generic data for transfer factors. A transfer factor is the ratio of the concentration (or emission) in one compartment to that in another compartment under steady state conditions. If the relevant transfer factors are known, calculation of the exposure is straightforward. Even in the age of massive data and powerful personal computers, such approaches can be valuable; their accuracy can be comparable to detailed models, given the overall uncertainties. Therefore, we present a simple and transparent model, RiskPoll for trace pollutants. It is a multimedia version of the "uniform world model" (UWM), based on the transfer factors published by EPA (1998). It is part of the RiskPoll package and can be downloaded for free at www.arirabl.org.

As further justification of the value of our simple models, note that most micropollutants are emitted from stacks of sufficient height (usually 40 m or more) that the impact in the immediate vicinity (usually with low population density) is only a small percentage of the total. Local detail is important only for local impact assessments, but not for the calculation of total damage (which is significant over hundreds to thousands of km). Furthermore, most estimates of total damage are needed for applications where the precise site of an installation is either unknown (future installations) or not relevant (regulations that make no site-specific distinctions). Thus estimates for typical sites are more relevant than precise values for specific sites.

In this section we present the RiskPoll results for doses that we published in Spadaro and Rabl (2004). One of the parameters in that calculation is the inhalation rate, which we took as 20.6 m^3/(person.day). More recent values for this rate recommended by the EPA are lower. For the depletion velocity we took 0.0049 m/s. These values are also in the current version of RiskPoll (version 1.052), but the user can change them after unprotecting the spreadsheet.

Our starting point is the UWM (Uniform World Model) for the collective inhalation dose rate $\dot{d}_{inhal}(\dot{m})$ (kg/yr or g/s) of a primary pollutant, which is readily obtained from Eq. (7.51) of Section 7.4 by replacing the sum of the S_{ERF} P by $\dot{V}_{inhal}$,

$$\dot{d}_{inhal}(\dot{m}) = \dot{V}_{inhal}\rho_{eff}\dot{m}/k_{dep}, \qquad (8.1)$$

where

$\qquad$ $\dot{m}$ = rate at which pollutant is emitted to the air (kg/yr),

$\qquad$ $\dot{V}_{inhal}$ = population-averaged inhalation rate for which we take 20.6 m^3/(pers·day),

$\qquad$ ρ_{eff} = average population density (pers/km^2) within 1000 km of source, and

$\qquad$ k_{dep} = depletion velocity of pollutant (dry + wet) (m/s).

If the emissions have been sufficiently stable for a long enough time (long compared to the time constants of the environmental dispersion), the ratio of measured dose and emission data can provide a good estimate of a transfer factor which can then be used in other situations. This idea has been applied by Rabl et al. (1998) for dioxins and by Spadaro and Rabl (2008a) for Hg. Here we present the argument for dioxins in Section 8.2, followed by an analogous argument for Hg in Section 8.3. The latter is a little more involved, because we also consider the possibility of a no-effect threshold. For both of these substances we proceed immediately to the calculation of the health impacts and costs, rather than postponing it past

Chapter 9. Finally, in Section 8.4, we describe the multimedia model of RiskPoll and its results for As, Cd, Cr, Ni and Pb.

8.2   Simplified approach for dioxins

Data for typical doses, both due to inhalation and to ingestion, can be found in Fig. II-5, p. 37 of the report "Estimating exposure to dioxin-like compounds" Vol. I Executive Summary (EPA, 1994). For the ratio of total dose over ingestion dose this figure indicates a typical value of

$$\frac{d_{tot}}{d_{inhal}} = \frac{119}{2.2} = 54. \tag{8.2}$$

These dose data are based on conditions in the USA in around 1980 when dioxin emissions were relatively stable before declining to the present low levels. This fact is relevant since the time constants for inhalation and ingestion doses are very different, inhalation being essentially simultaneous with emission, while ingestion occurs later, after passage through the food chain (dioxins stay in the environment for several years). Of course, the precise value can vary strongly with diet and local conditions, but since the total population impact involves an average over such local variations, one can also use Eq. (8.2) as a good approximation for other countries and times, if the diet is not too different.

In any case the ingestion dose involves dispersion over large distances (e.g. by transport of cattle feed, followed by transport of meat). This has the effect of making the receptor distribution more uniform. Therefore we recommend estimating the total collective dioxin dose by calculating the inhalation dose rate $\dot{d}_{inhal}(\dot{m})$ and multiplying it by 54. Using Eqs. (8.1) and (8.2) the total dose rate due to an emission rate $\dot{m}$ is

$$\dot{d}_{tot}(\dot{m}) = 54 \, \dot{V}_{inhal} \, \rho_{eff} \, \dot{m}/k_{dep}, \tag{8.3}$$

where
$\dot{d}_{tot}(\dot{m})$ = total dose rate,
V_{inhal} = inhalation rate,
ρ_{eff} = average population density of the region,
$\dot{m}$ = emission rate, in g/s, and
k_{dep} = mean depletion velocity.

Example 8.1   Calculate the total population dose rate due to a dioxin emission rate of 1 kg/yr.

Solution: See this table. For the depletion velocity we take k_{dep} = 0.005 m/s for particles. The left part of the table shows the quantities in customary units, the right half in consistent SI units as used for the calculation. The total dose turns out to be 0.705 mg/day.

Quantity	Customary units		Consistent units	
$\dot{V}_{inhal}$	20.6	m^3/(person·day)	0.000238	m^3/(person·s)
ρ_{eff}	1.00E+02	persons/km^2	1.00E−04	persons/m^2
$\dot{m}$	1	kg/yr	3.17E−05	g/s
k_{dep}	0.5	cm/s	0.005	m/s
$\dot{d}_{inhal}(\dot{m})$	1.31E−05	g/day	1.51E−10	g/s
d_{tot}/d_{inhal}	54		54	
$\dot{d}_{tot}(\dot{m})$	7.05E−01	mg/day	8.16E−09	g/s

The EPA has said that dioxins (2,3,7,8-TCDD and HxCDD) are "the most potent carcinogen(s) evaluated by the EPA's Carcinogen Assessment Group". Here we only consider cancers. Cancer risks are usually stated in terms of the slope factor, defined as probability of getting a cancer as result of a constant daily intake during 70 years. According to EPA (2000) the slope factor for cancers due to ingestion of dioxins, expressed as TEQ (=toxic equivalent of 2,3,7,8-TCDD) is 1×10^6 per mg/kg_{body}/day. However, this is very uncertain, and Searle (2005) suggests that the true cancer risk may be five times smaller.

Example 8.2 Estimate the damage cost of dioxin.

Solution: Start with $\dot{d}_{tot}(\dot{m})$ = 0.705 mg/day for an emission rate of 1 kg/yr TEQ, as per Example 8.1. For the average body weight assume 64 kg, and for the ERF assume a straight line corresponding to the slope factor 1×10^6 per mg/kg_{body}/day of EPA (2000). The resulting cancer rate is 1.10E+04 cancers/70 yr or 1.57E+02 cancers/yr. Now both emission rate and damage rate refer to the same duration and their ratio is the number of cancers per kg of emission. If cancers are valued at 2 M€, the damage cost turns out to be 3.15×10^8 €/kg TEQ. But that is probably an overestimate. Reducing S_{ERF} by a factor of 5 and the average cost of the resulting cancers by 2 (if 50% of these cancers can be cured), the damage cost is only 31.5 M€/kg TEQ.

Emission rate $\dot{m}$	1	kg/yr
$\dot{d}_{tot}(\dot{m})$	7.05E−01	mg/day
Average body weight	64	kg
D_{tot} per body weight	1.10E−02	mg/(kg$_{body}$-day)
Slope factor	1.00E+06	per mg/kg$_{body}$/day
Cancers/70yr	1.10E+04	cancers/70yr
Cancers/yr	1.57E+02	cancers/yr
Cancers per emitted kg	1.57E+02	cancers/kg
Cost of cancer	2.00E+06	€/cancer
Damage cost per kg dioxin	**3.15E+08**	**€/kg**
Damage cost, if S_{ERF} 5 times smaller and average cancer cost two times smaller	**3.15E+07**	**€/kg**

8.3   Global health impacts and costs due to mercury emissions

8.3.1   *Why a simplified approach is necessary and appropriate*

Estimating the impacts of Hg is challenging because the modeling of the environmental pathways is extremely complex and uncertain. A further complication arises from the fact that Hg is a global pollutant (Lamborg *et al.*, 2002) and therefore the benefits should be evaluated at the global scale, something that has not been attempted before. More precisely, it is Hg(0), the gaseous metallic part of Hg emissions, that is dispersed globally because its effective atmospheric residence time (including re-suspension) is about 1 to 2 yr, long enough for the distribution in the hemisphere to become quite uniform. The other main species are RGM (reactive gaseous mercury Hg^{++}) and Hg$_p$ (particulate mercury), and they deposit regionally rather than globally. The speciation of Hg emissions depends on the source, but the contribution of Hg(0) to the total is large.

We argue that a global estimate is relatively simple, because the dominant pathway for health impacts is the ingestion of fish and seafood, and the transfer factor from emission to ingestion can be estimated by comparing the global average dose and emissions, both of which have been measured in a number of studies. Such a comprehensive transfer factor is much easier to determine and less uncertain than a detailed calculation of all the individual transfer factors, in particular the fraction transformed to methyl-Hg (MeHg) and the bioconcentration factor in seafood. The most troubling and currently best understood health impacts of Hg are neurotoxic (see Axelrad *et al.* (2007) and references therein), and we evaluate them using as proxy dose–response function data for IQ decrement due to MeHg ingestion. We neglect other forms of Hg, because they are far less toxic, despite relatively high doses, for example from dental amalgam (see e.g. Bellinger *et al.*, 2006).

The main assumptions of our analysis (Spadaro and Rabl, 2008a) are the following.

(1) For an assessment of global impacts of Hg emissions to air the dependence on emission site can be neglected, because the residence time of metallic Hg in the atmosphere is sufficiently large to imply a fairly uniform hemispherical distribution of the ingestion dose. Even though the actual distribution of ambient total Hg is not very uniform (because of the contribution of RGM and Hg_p), the ingestion dose of MeHg becomes far more uniform because of the wide international trading of fish.

(2) A comprehensive transfer factor, defined as incremental average dose due to an incremental emission, can be estimated as the ratio of the average dose and total (natural + anthropogenic) global emission at steady state conditions.

(3) The worldwide average dose from fish and seafood is about 2.4 µg/day-person of MeHg as reported by UNEP (2002).

(4) The worldwide emission rate is about 6000 tonnes/yr, as estimated by the UN study (UNEP 2002); about one-third of that is from natural sources.

(5) The ERF for IQ points lost can be approximated by a straight line with two possibilities for a threshold, either no threshold or a threshold corresponding to the reference dose RfD of EPA (2001); as slope of the ERF we take the value found by the integrative analysis of Axelrad *et al.* (2007) and we convert it to an ingestion dose using conversion factors from the literature.

(6) The neurotoxic impact on a population can be estimated by applying the ERF of Axelrad *et al.* to women of childbearing age.

(7) The average ingestion dose of women of childbearing age is equal to the average dose reported by UNEP (2002); for the threshold case, we assume a distribution of worldwide doses similar to that in the USA (NCHS 2005).

(8) The social cost of an IQ point lost in each country is calculated by modifying the value in the USA (for which we take $\$_{2005}18,000$/IQ point) in proportion to the GDP/capita, adjusted for purchase power parity (PPP).

We believe that in view of the evidence, these assumptions are the most plausible choice for calculating the global average Hg damage cost, given the limited data currently available. An examination of the uncertainties suggests that the result estimates the global damage cost within a factor of about 4 (see Section 11.3.5.1).

The analysis is quite simple if the ERF is linear without threshold. But consideration of a possible no-effect threshold introduces a major complication.

8.3.2   The comprehensive transfer factor

At the present state of knowledge a detailed modeling of global impacts would be difficult and very uncertain. The difficulties begin with the speciation during emission. The dominant species emitted from coal fired power plants are Hg(0), RGM and Hg_p, and typical percentages in the USA are around 58% for Hg(0), 40% for RGM and 2% for Hg_p, although highly variable from one plant to another (RGM from utility boilers in the USA can vary from 10% to 90%) (Sullivan et al., 2003). Pirrone et al. (2001) report Hg speciation estimates for all anthropogenic sources as 64% Hg(0), 28.5% RGM and 7.5% Hg_p. Natural emissions are mostly elemental mercury, according to EMEP (www.msceast.org). Because of the high percentage of Hg(0) a global analysis is necessary.

RGM and Hg_p are water soluble and deposit by wet and dry deposition (Rea et al., 2000 and 2001; Vette et al., 2002; Landis et al., 2002). Several papers (Rossler, 2002, and EPA, 1997) indicate that, in the USA, about 60% of the Hg deposition is from local sources, the global reservoir contributing the rest. Hg(0) becomes part of the global mercury cycle. The effective lifetime of Hg(0) in the atmosphere is in the range 1 to 2 years, long enough to cause mixing in the entire hemisphere (Lamborg et al., 2002). Note that this lifetime is much longer than what would be implied by a simple consideration of the wet and dry deposition velocities, because much of the deposited Hg(0) evaporates again for another deposition–volatization cycle.

The dose from fish and seafood involves the transformation of Hg into MeHg by aquatic micro-organisms. Even though only a small percentage of the Hg is thus transformed, MeHg is much more toxic than the other forms. Furthermore, the bioconcentration factor of MeHg is very large, leading to relatively high concentrations in seafood, especially among predatory fish. For these reasons the ingestion of fish and seafood is by far the most important pathway for human health impacts.

As an alternative to detailed pathway modeling, we use a comprehensive transfer factor T_{av}, defined as the ratio of the global average ingestion dose d_{av} per person and the global emission rate E

$$T_{av} = d_{av}/E; \qquad (8.4)$$

it has units of $(\mu g_{MeHg}/yr)/(kg_{Hg}/yr)$. Estimates of global emissions and global ingestion dose have been reviewed in a major study by UNEP (2002). Based on this study, we take a global emission rate of E = 6000 tonnes/year (uncertainty range, 3000 to 9000 tonnes/yr). A large part of that, about one-third, is due to natural emissions; the anthropogenic part has been increasing steadily since the industrial revolution (Lamborg et al., 2002). One of the difficulties of estimating global emissions lies in the fact that a significant fraction of the deposited Hg is later, possibly much later, re-emitted to the atmosphere.

For the global average ingestion dose of MeHg, we take d_{av} = 2.4 μg_{MeHg}/day per person, based on a report of the WHO (World Health Organization) as cited in Table 4.3 of UNEP (2002). For comparison, we note that in the USA the ATSDR dose estimate is 50 ng_{MeHg}/kg_{body}/day, implying 3.0 μg_{MeHg}/day for a body weight of 70 kg_{body} (www.atsdr.cdc. gov/toxprofiles/phs46.html), quite similar in view of the uncertainties. Data for the UK (2002) indicate intakes of about 3 μg/day for adults.

With d_{av} = 2.4 μg_{MeHg}/day per person we obtain a transfer factor of

$$T_{av} = 4.0E{-}07\,(\mu g_{MeHg}/yr)/(kg_{Hg}/yr). \qquad (8.5)$$

The product of T_{av} and the world population, 6.4 billion, (times the ratio of molecular weights Hg/MeHg) is the intake fraction, i.e. the fraction of the emitted Hg that passes through a human body as MeHg on its way to the ultimate environmental sink, mostly ocean sediment (Lamborg et al., 2002). Our result for intake fraction, 0.9E-03, is in the range of values found for other toxic metals, see Section 8.4. In spite of the low proportion of Hg that is transformed to MeHg, we find such an intake fraction plausible because of the large bioconcentration factor in fish and seafood.

8.3.3   Calculation of lifetime impact due to current emissions

We have discussed the health impacts of Hg in Section 3.6.7, where we conclude that, at the present time, only the neurotoxic impacts can be quantified with sufficient reliability. In particular, we obtained the ERF for loss of IQ points per ingestion dose of MeHg, Eq. (3.42), reproduced again here:

$$S_{ERF} = 0.18 \text{ IQpoints/ppm}_{hair} \times 0.2 \text{ ppm}_{hair}/(\mu g/L_{cord})$$
$$\times 1.65 \mu g/L_{cord}/(\mu g/L_{mat}) \times 0.61 \mu g/L_{mat}/(\mu g_{MeHg}/\text{day})$$
$$= 0.036 \text{ IQpoints}/(\mu g_{MeHg}/\text{day}). \tag{8.6}$$

To apply this ERF, it might appear necessary to consider the time window during which the brain is affected by Hg. The sensitivity of the brain to Hg is greatest during the early development of the body, but the precise time distribution of the damage is not known. Whereas the damage is incurred only during early development, including pregnancy, it is assumed to be permanent and measurable at the ages reported in the epidemiological studies. Since the ERF of Axelrad *et al.* is based on correlations between the maternal hair concentration and the IQ of the children, implicitly it also includes the effect of diet during early infancy, before the IQ of the children was measured, if one assumes that the diet of the infants is strongly correlated with that of the mothers. Thus the ERF slope of Eq. (8.6) describes the total lifetime impact on children whose mothers are exposed to a specified steady-state ingestion dose, and the detailed time distribution of the sensitivity to Hg does not matter for the calculation of impacts.

If a particular mother, i, has had an ingestion dose d_i, the lifetime impact I_i on the offspring is an IQ loss of

$$I_i = \begin{cases} 0 & \text{if } d_i < d_{th} \\ S_{ERF}(d_i - d_{th}) & \text{if } d_i \geq d_{th} \end{cases}, \tag{8.7}$$

where d_{th} is the threshold dose. The total impact in a population of p individuals is obtained by summing Eq. (8.7) over the doses,

$$I_{tot} = S_{ERF} \sum_{i=p_{th}}^{p} (d_i - d_{th}), \tag{8.8}$$

where p_{th} is the number of individuals with maternal dose below d_{th} and the sum covers only the individuals with maternal dose above d_{th}.[1] We

---

[1] One might think of calculating the impact by simply multiplying the no-threshold dose $d_{av}(0)$ by the fraction $f(d_{th})$ of the population that is above the threshold. That is not a good approximation because it does not take into account how the doses above d_{th} are distributed: if some of them are very high the impact is higher than if they are just a little above d_{th}. Fig. 8.2 shows that $\Delta d_{av}(d_{th})$ is more than twice as large as $f(d_{th}) \Delta d_{av}(0)$ for $d_{th} = 6.7 \mu g_{MeHg}/\text{day}$, and $d_{av}(d_{th})$, needed for the marginal impact, is more than 4 times as large as $f(d_{th})d_{av}(0)$.

find it convenient to express everything as impact per person, averaged over the entire population (men, women, all ages), rather than just the sensitive individuals. Using the notation $\Delta d_{av}(d_{th})$ for the positive differences $d_i - d_{th}$, averaged over the number of people p in the entire population,

$$\Delta d_{av}(d_{th}) = \frac{1}{p} \sum_{i=p_{th}}^{p} (d_i - d_{th}), \qquad (8.9)$$

one can write the average lifetime IQ loss per person $I_{av} = I_{tot}/p$ as,

$$I_{av} = S_{ERF} \Delta d_{av}(d_{th}). \qquad (8.10)$$

To calculate this quantity one needs data for the distribution of doses for the world population. For most countries this is unfortunately not available. The best data are those of NCHS (2005) for the fraction of women of child bearing age in the USA that have various levels of MeHg concentration, measured as µg per L of maternal blood (see also Mahaffey (2005)). This distribution is to a good approximation lognormal with geometric mean $\mu_g = 0.7\ \mu g_{MeHg}/L_{mat}$ and geometric standard deviation $\sigma_g = 3.5$ (a distribution that is fairly similar to data from the UK (2002)). Using this lognormal distribution, one can readily calculate the population averaged dose above d_{th} of Eq. (8.9). Since the distribution is not weighted by the number of children, our calculations assume implicitly that the number of children per mother is independent of her dose.

The resulting mean dose in the USA is about 2.5 μg_{MeHg}/day and very close to the UNEP (2002) estimate of the average dose of the entire world, 2.4 μg_{MeHg}/day, the dose that we use as the basis for our damage cost estimates. Lacking data for the dose distributions in other countries, we use the distribution of NCHS (2005) for the world and scale it down by a factor 2.4/2.5, assuming that the UNEP data also apply to women of childbearing age.

Figure 8.2 shows the resulting relation between threshold dose d_{th} and $\Delta d_{av}(d_{th})$ of Eq. (8.9). The fraction $f(d_{th})$ of the population that is above threshold is indicated by the dashed line and the right-hand scale. The threshold RfD of EPA is 0.1 μg_{MeHg}/day/kg_{body}, implying $d_{th} = 6.7$ μg_{MeHg}/day for an average weight of 67 kg_{body}; this is indicated by the diamonds. The graph also shows the quantity $d_{av}(d_{th})$ of Eq. (8.12) in the next section, which will be needed for the marginal damage cost. Multiplying the dose $\Delta d_{av}(d_{th})$ by S_{ERF} of Eq. (8.6), one finds the corresponding average IQ loss is 0.020 IQpoints if $d_{th} = 6.7$ μg_{MeHg}/day. For zero threshold the loss is 0.087 IQpoints.

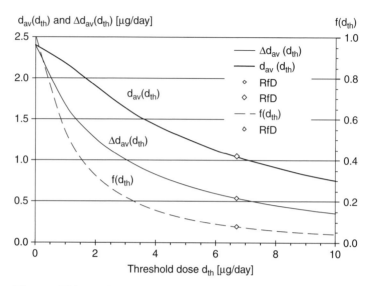

Fig. 8.2 Thin solid line shows the relation between threshold dose d_{th} and dose $\Delta d_{av}(d_{th})$ above d_{th} per person, averaged over the entire population, Eq. (8.9); this is needed for the total lifetime impact.

Thick gray line shows the dose $d_{av}(d_{th})$ of Eq. (8.12), needed for the marginal damage cost.

Dashed line and right-hand scale shows the fraction $f(d_{th})$ of the population that is above threshold.

Diamonds mark the EPA reference dose of 6.7 μg_{MeHg}/day.

8.3.4 Marginal impacts and costs per kg of emitted Hg

By contrast with the lifetime impact, the calculation of impact and cost of an additional kg of emitted Hg (also called marginal or incremental cost) requires consideration of the rate at which new individuals are affected, hence the birth rate b enters into the calculation. For the allocation of an additional impact δI to an additional emission δm, it may be helpful to assume that emission and impact are simultaneous and occur during a certain time interval for which one can arbitrarily take $\Delta t = 1$ year; in reality, the impact occurs later, of course, but the delay does not change the magnitude of the impact. Only the $\Delta p = p\, b\, \Delta t$ individuals born during this time interval are affected, and the incremental impact on the total population is

$$\delta I_{tot} = p\, b\, \Delta t\, S_{ERF}\, \delta \Delta d_{av}(d_{th}), \qquad (8.11)$$

where $\delta \Delta d_{av}(d_{th})$ is the increment of $\Delta d_{av}(d_{th})$ of Eq. (8.9) due to δm. To find how an increase in the emission rate $\delta E = \delta m/\Delta t$ changes the dose $\Delta d_{av}(d_{th})$, note that in the limit of small changes, the number p_{th} of individuals above threshold does not change (a change in p_{th} would yield a term of order δE^2 in the incremental impact and is, by definition, neglected in the calculation of marginal costs). Whereas the doses d_i are proportional to the emission rate E, the quantity $\Delta d_{av}(d_{th})$ is not, because it contains only the contributions above threshold. To find how $\Delta d_{av}(d_{th})$ changes, let us define the population average of the total dose of the individuals who are above d_{th} as,

$$d_{av}(d_{th}) = \frac{1}{p} \sum_{i=p_{th}}^{p} d_i, \qquad (8.12)$$

and rewrite Eq. (8.9) in the form,

$$\Delta d_{av}(d_{th}) = d_{av}(d_{th}) - f(d_{th})d_{th} \text{ with } f(d_{th}) = (1 - p_{th}/p), \qquad (8.13)$$

where $f(d_{th})$ is the fraction of the population that is above threshold.

To relate this to the emission rate E, we multiply $d_{av}(d_{th})$ by a factor ET_{av}/d_{av} which is equal to unity by its definition in Eq. (8.4),

$$\Delta d_{av}(d_{th}) = ET_{av}d_{av}(d_{th})/d_{av} - f(d_{th})d_{th}. \qquad (8.14)$$

Since a small change δE in emission rate changes neither the transfer factor T_{av} nor the shape of the probability distribution of doses, the quantities $T_{av} d_{av}(d_{th})/d_{av}$ and $f(d_{th}) d_{th}$ are independent of E. Hence the increment $\delta \Delta d_{av}(d_{th})$ equals

$$\delta \Delta d_{av}(d_{th}) = \delta E T_{av}d_{av}(d_{th})/d_{av}. \qquad (8.15)$$

Inserting this into Eq. (8.11) and replacing the incremental emission rate by $\delta E = \delta m/\Delta t$, we obtain,

$$\delta I_{tot} = pb\Delta t S_{ERF} (\delta m/\Delta t) T_{av}d_{av}(d_{th})/d_{av}. \qquad (8.16)$$

The time interval Δt is indeed arbitrary because it cancels out. The ratio of δI_{tot} and incremental emission δm is the incremental impact per emitted Hg,

$$\text{incremental impact per emitted Hg} = S_{ERF}[T_{av}d_{av}(d_{th})/d_{av}] \sum_k p_k b_k,$$
$$(8.17)$$

where $d_{av}(d_{th})$ is given by Eq. (8.12), and we have replaced the product of population p and birth rate b by a sum over countries k to allow for the fact that birth rates in different countries are different. Multiplying by the cost P_k per IQpoint in country k, we obtain the marginal damage cost D in €/kg emitted,

$$D = S_{ERF} [T_{av} d_{av}(d_{th})/d_{av}] \sum_k p_k b_k P_k f_{dis}. \tag{8.18}$$

We have included a factor f_{dis} to account for the time lag between a change in emissions and the impact,

$$f_{dis} = (1 + r_{dis})^{-N}, \tag{8.19}$$

where r_{dis} is the discount rate and N the time lag; following Griffiths *et al.* (2007), we take r_{dis} = 3% and N = 15 yr for the central value, resulting in f_{disc} = 0.64. Note that we apply this discount factor only to the marginal cost, not to the lifetime cost, since the latter corresponds to current exposure.

For the cost associated with the loss of an IQ point, there is a range of estimates in the USA, mostly based on lost earnings or remedial education. In particular we cite

- Muir and Zegarac (2001) $\$_{1999}$14,700/IQpoint,
- Grosse *et al.* (2002) $\$_{2000}$14,500/IQ point,
- Rice and Hammitt (2005) $\$_{2000}$16,500/IQpoint,
- Trasande *et al.* (2005) $\$_{2000}$22,200/IQpoint, and
- Griffiths *et al.* (2007) $\$_{2000}$11,245/IQpoint.

Adjusting these figures to $\$_{2005}$ by means of the CPI we obtain a mean of about,

$$P_{USA} = \$_{2005}18,000/\text{IQpoint in the USA.}$$

To apply this cost in different countries, we modify the cost in proportion to the GDP_{PPP}/capita, the per capita GDP adjusted for purchase power parity (www.cia.gov/cia/publications/factbook/). So the cost of an IQ point in country k is

$$P_k = P_{USA} \frac{(GDP_{PPP}/capita)_k}{(GDP_{PPP}/capita)_{USA}}. \tag{8.20}$$

The worldwide average cost is

$$P_{av} = \sum_k P_k \frac{p_k}{p_{world}} = \$3,890/\text{IQpoint}, \qquad (8.21)$$

where p_k = population of country k.

Let us define a fraction f_{world} of the world population p_{world} that is affected per year by Hg, weighted by GDP/capita, as,

$$f_{world} = \sum_k b_k \frac{p_k}{p_{world}} \frac{(GDP_{PPP}/capita)_k}{(GDP_{PPP}/capita)_{USA}}, \qquad (8.22)$$

its numerical value is 0.00315. Thus we can write D in the form,

$$D = S_{ERF}\left[T_{av}d_{av}(d_{th})/d_{av}\right]f_{world}p_{world}P_{USA}f_{dis} \quad \text{in \$/kg emitted.} \quad (8.23)$$

As shown in Table 8.1, the marginal damage cost D per kg of emitted Hg is about \$1500/kg for a threshold dose of 6.7 μg/day and \$3400/kg for zero threshold. It is also interesting to look at the total average worldwide IQloss per person due to current emissions (0.02 IQpoints/person with and 0.087 IQpoints/person without threshold) and the corresponding costs (\$78/person with and \$344/person without threshold). We note that the marginal damage cost for the case with threshold is appropriate only for small changes; for large changes the fraction of the population above threshold will also change.

In Section 11.3.5.1 we analyze the uncertainties of this calculation and conclude by saying that the total uncertainty can be characterized by a geometric mean σ_g of about 4, which means that the 68% confidence interval extends from about 1/4 to 4 times the values shown in Table 8.1.

Table 8.1 *Our estimates of the total worldwide lifetime impact and damage cost (Eqs. (8.10) and (8.21)), and of the marginal damage cost D (Eq. (8.23)) of Hg emissions, for two assumptions about the threshold dose d_{th}. $f(d_{th})$ is the fraction of the total population that is above threshold.*

			Lifetime impact and damage cost		Marginal damage cost	
d_{th}	$f(d_{th})$	$\Delta d_{av}(d_{th})$	Lifetime impact	Lifetime cost	$d_{av}(d_{th})$	D
(μg/day)	=1-p_{th}/p	(μg/day)	(IQpoints/person)	(\$/person)	(μg/day)	(\$/kg)
6.7	0.08	0.54	0.020	78	1.05	1,500
0	1	2.40	0.087	344	2.40	3,400

The damage cost of 8000 €/kg of Hg of ExternE (2008) was based on an earlier version of our calculations, not the value of $4300 per kg in Spadaro and Rabl (2008). After a review of the evidence for cardiovascular effects of Hg, we now believe that they should be added, making the damage cost significantly higher, probably even more than 8000 €/kg.

8.3.5   Discussion

Some might argue that regional and local variability could invalidate our global approach. As far as environmental pathways are concerned, such variability arises only from the Hg_p and RGM component of the emissions, since Hg(0) disperses globally. Our model would be exact for Hg(0) emissions if we had exact data for the Hg(0) component of global emissions and for its contribution to the global average dose. Instead our transfer factor includes the other components, and thus our model combines global and regional impacts, the latter as global average. The other source of variability arises from the distribution of ingestion doses in different parts of the world. We believe that our uncertainty analysis of Section 11.3.5.1 adequately accounts for the possible effects of such variability by including a wide range of possible doses and dose distributions, and that therefore our results are valid within the uncertainty bounds we have estimated. Another possible criticism concerns our use of a steady-state analysis, whereas in reality emissions and doses have been changing. Here, too, our uncertainty analysis accounts for it, because the range of emissions and doses is wide enough to account for changes during the past one or two decades (i.e. the time constant for the environmental pathways).

In terms of absolute magnitude the impacts and costs are small: 0.02 IQpoints/person with and 0.087 IQpoints/person without threshold, and $78/person with and $344/person without threshold; also one should note that about two-thirds of the current total emissions are anthropogenic. However, because of the small quantity of Hg that is involved, the marginal damage cost per kg of Hg is high compared with other pollutants that have been evaluated.

An examination of the uncertainties leads us to a geometric standard deviation of about 4, in other words, the damage cost could be a factor of 4 smaller or larger than our estimate. However, the uncertainty about a possible threshold is difficult to capture because of the lack of information. As an indication, one can consider the range of results between zero threshold and the RfD threshold of the EPA, but the true threshold could be even higher: the RfD is a guideline for protecting public health and derived by including a large margin of safety (see also the discussion of thresholds by Lipfert *et al.* (1996)). Thus the impact and cost could be smaller than our estimate with threshold.

It is interesting to compare our marginal damage cost D with a study by Rice and Hammitt (2005), who estimate the benefit of the new regulation for Hg emissions by power plants in the USA. Of course, their numbers are not very comparable with ours, because they consider only impacts within the USA, whereas our estimate is a simple global average. Due to local and regional variations in Hg speciation, dispersion, population, dietary habits etc., a regional calculation such as that of Rice and Hammitt can find very different results. In particular the speciation for the power plants considered by Rice and Hammitt has a higher fraction of RGM and Hg_p than the global emissions we have assumed; thus they find a higher fraction of Hg deposited in the USA where the cost of an IQ point is higher than the global average. However, the threshold assumptions are the same.

To estimate what our model implies with the scope and assumptions of Rice and Hammitt, we multiply our result by the ratio of the USA cost (for IQ loss) and the global cost in our global calculation; that ratio is 0.207, relatively high because of the high cost of an IQ point in the USA compared with the rest of the world (in other words, the result of Rice and Hammitt would be a factor 1/0.207 larger if they had included global impacts and if the damage of USA power plant emissions were equal to our very simple global model). We also adjust for different assumptions about

cost of IQ point ($16,500/IQpoint for Rice and Hammitt versus $18,000/IQpoint for us),

and ERF slope S_{ERF} (0.12 IQpoint/(μg_{MeHg}/day) for Rice and Hammitt versus 0.0362 IQpoint/(μg_{MeHg}/day) for us).

Finally, about 60% of USA emissions are deposited on US soil (EPA 1997) because of the large fraction of RGM in the speciation of power plant emissions in the USA; to convert from the typical speciation that we assume to power plant emissions in the USA we therefore have to multiply by a factor of 3. Combining these adjustment factors, we multiply our global damage cost D (but without discount factor f_{disc} for consistency with Rice and Hammitt) by

$$0.207 \times 3.0 \times (0.12/0.0362) \times (16.5/18) = 1.89$$

to obtain our estimate for the USA. The result, $4380/kg with and $9993/kg without threshold, agrees with the corresponding numbers of Rice and Hammitt, $4300/kg and $11200/kg, within 12%, far closer than one could expect in view of the radically different approach.

Abatement of Hg emissions is cost effective if the marginal abatement cost is smaller than the marginal damage cost. Abatement costs are highly variable, depending on technologies and specific conditions. For example, Jones *et al.* (2007) evaluate the cost of Hg removal by activated carbon injection for six power plants, with several variations, and find a range of about $8400 to $365,000/kg$_{Hg}$; in most cases the abatement would exceed by far the benefits we have calculated. However, the benefits may turn out to be much larger if other impacts, in particular cardiovascular, are included (Rice and Hammitt, 2005, Virtanen *et al.*, 2007).

But let us emphasize that for the formulation of new regulations there are other options besides reducing emissions. In particular, a worldwide educational campaign to prevent pregnant women and infants from eating fish and seafood with high Hg content may be far more cost effective in many regions, and it is all the more relevant in view of the fact that a third of the emissions are of natural origin. However, careful analysis of all relevant costs and benefits is required before formulating any such advisory, because the large health benefits of seafood must not be overlooked (see Cohen *et al.*, 2005; and Jorgensen *et al.* 2007).

Example 8.3   To illustrate the use of our results, consider the damage cost of compact fluorescent light (CFL) bulbs if they are thrown in the trash and break at the end of their life. According to a fact sheet by Energy Star (www.energystar.gov/ia/partners/promotions/change_light/downloads/Fact_Sheet_Mercury.pdf) a typical CFL contains about 4 mg of Hg, and since most of it becomes bound to the inside of the light bulb as it is used, only about 11% would be released if the bulb were to break at the end of its life. Compare the damage cost with the value of the energy savings and with the avoided Hg emissions by power plants, if the average plant in the US emits 0.012 mg/kWh (according to the same fact sheet).

Solution: Using our global damage cost estimate of $3,400/kg (if no threshold), the damage cost is

$$\$3,400/\text{kg} \times 0.11 \times 4\text{mg} = \$0.0034 \times 0.11 \times 4 = 0.15\text{cents}.$$

A typical 15 W CFL produces about as much as a conventional incandescent bulb of 60 W, and it lasts about 10,000 hr. Thus the energy savings are 10000 hr $\times$ 45 W = 450 kWh, worth $49.5 at an energy price of $0.11/kWh. The avoided power plant emissions of Hg are 450 kWh $\times$ 0.012 mg/kWh = 5.4 mg, compared with the 0.11 $\times$ 4 mg = 0.44 mg if the CFL is thrown into the trash. Obviously the CFL is a winner on all counts, but nonetheless one should dispose of it properly.

8.4   The uniform world model for ingestion doses

8.4.1   *Assumptions*

1. Our starting point is the model developed for the assessment of multi-media pathways by EPA (1998). It involves transfer factors, i.e. factors that indicate how much pollutant is transferred from one environmental compartment to another under steady-state conditions (Spadaro and Rabl, 2004). A similar model has been developed by IAEA (1994 and 2001) for the assessment of radiological impacts, but its list of parameter values has many gaps. We rely mostly on the EPA model, supplemented in some cases by data from IAEA. Some of our equations are different from those of EPA or IAEA, because we include integration over all time (in the case of very long time constants we also show the results for a time horizon of 100 years). We provide explicit input parameters and results for the most toxic metals: As, Cd, Cr, Ni and Pb; the model can readily be extended to other pollutants, including organics.

As for the validity of a steady-state approach, we show in Section 8.4.2, that if the parameters of a dynamic model can be approximated by time-independent constants (the case for the USEPA methodology), the collective dose depends only on the total emission, regardless of its distribution over time. Therefore the total damage per kg of a pollutant can be calculated by a simple steady-state analysis, assuming a constant emission rate and steady state conditions. The results for the total collective dose should be equal to a calculation with a level IV model in the terminology of the life cycle assessment community (Mackay, 2001), for example the CalTOX model (McKone and Enoch, 2002); however, there are major differences in the detailed modeling of the pathways.

We account for the pathways in Fig. 8.1, except for seafood. Even if the concentration increment in the sea is very small, the collective dose from seafood could be significant for some pollutants if there is bioconcentration and the removal processes (sedimentation) are slow. But detailed modeling would be difficult because one would need compartment models of all the oceans, coupled with data on fish production. Furthermore, the consideration of seafood is crucial only for Hg, discussed above in Section 8.3; for the other metals it does not seem to be very important, as suggested by typical dose data. Like the underlying EPA model, we do not consider ground water, assuming that on average inflow and outflow are equal.

We have implemented this approach in the RiskPoll software, available at www.arirabl.org. Within the uncertainties, our results are consistent with data reported by the World Health Organization (WHO 1988 – 2001) and with results of CalTOX (McKone and Enoch 2002).

8.4.2   Justification for steady state models

If the concentration of a pollutant in a compartment is not uniform, one can subdivide it into smaller compartments, for example, a parcel of air above a city or a layer of soil in a cornfield. With the approximation of uniformity a single variable, the mass of the pollutant in the compartment, characterizes the state of the compartment. In any compartment with first-order processes, the mass m_j in compartment j of the environment can be described by a first-order differential equation in time:

$$\kappa_j m_j + \frac{dm_j}{dt} = \dot{m}_{j,in}, \tag{8.24}$$

where,

κ_j = rate parameter,

$\dot{m}_{j,in}$ = inflow of pollutant into the compartment (e.g. the emission by a smoke stack into the surrounding air column) and

$\kappa_j m_j$ is the pollutant outflow.

m_j and $\dot{m}_{j,in}$ are functions of time t. Henceforth, we assume that the rate parameters can be approximated by time-independent constants, i.e. κ_j = rate constant (= 1/time constant).

Without loss of generality, one can assume that $\dot{m}_{j,in}(t)$ and $m_j(t)$ are nonzero only for t>0. Then the solution is

$$m_j(t) = \exp(-\kappa_j t) \int_0^t \exp(\kappa_j t') \dot{m}_{j,in}(t') dt'. \tag{8.25}$$

Since the dose obtained in this compartment at time t is proportional to the pollutant mass $m_j(t)$ inside, the total (collective) dose d_j, integrated over all time, in this compartment is proportional to the integral,

$$d_j = K \int_0^\infty dt\, m_j(t), \tag{8.26}$$

where K is the proportionality constant between mass and dose. Inserting Eq. (8.25) and changing the order of integration, one can show that the total dose is the integral of the total net flow into the compartment,

$$d_j = (K/\kappa) \int_0^\infty dt\, \dot{m}_{j,in}(t). \tag{8.27}$$

Since with linear dose–response functions only the collective dose matters for the total impact (irrespective of how it is distributed in time or among individuals), a dynamic model consisting of compartments with first-order processes and constant rate parameters, yields exactly the same result as a steady state model with the same compartments, regardless of any detail of the time history of the inflow $\dot{m}_{j,in}(t)$. Therefore a steady state model is sufficient for calculating the total dose, even though the real environment is never in steady state. It is easy to see why the argument breaks down if the rate parameters vary with time. Suppose for example that the deposition from the atmosphere is nonzero only outside the growing season; then there is no direct deposition on crops, even though a calculation with time-averaged parameters would yield such a term.

Under steady-state conditions, $\dot{m}_{j,in}$ = constant and the time derivative of m_j in Eq. (8.24) is zero; thus one finds that the mass m_j inside the compartment is

$$m_j = \dot{m}_{j,in}/\kappa_j. \tag{8.28}$$

In passing, we note that if there is no removal process, the time constant and the total dose are infinite even if the total inflow is finite.

For practical calculations, we find it convenient to work in terms of rates by allocating emissions and impacts on an annual basis; in other words we take the emission rate equal to the average during the year and calculate the steady state dose resulting from a permanent emission at this rate. For example, if the emission rate is x = 1 kg/yr and the resulting steady-state dose is y mg/yr, the dose per emitted kg is y/x, in mg/kg. The actual time distribution of the dose does not matter for the total. Thus the typical periodicity of emissions and of agricultural production is automatically taken into account. Likewise, we need not worry about the time distribution of the transfers into and out of the soil (except for the option of a cut-off time, discussed in the next paragraph).

For some pollutants the time constants of some of the removal processes can be very long, and significant contributions to the total impact may come even centuries after the emission. The models of EPA and IAEA, being conceived in the spirit of calculating peak impacts rather than total impacts, do not address this problem; their equations in effect limit the time horizon over which impacts are counted to the duration of the emissions, on the order of 20 to 40 years for most installations. Therefore we modify the EPA equations so that they correspond with steady-state conditions. Because of the large uncertainties of impacts in the far future, we propose two estimates of such impacts: an upper limit corresponding to true steady state, and a lower limit corresponding to the doses that will be reached at the end of a cut-off for which we choose t_{cut} = 100 yr as an arbitrary but reasonable number; for the purpose of discounting we also show results for a 30 yr cut-off. For a cut-off this means replacing the time constant $1/\kappa$ of a slow process by

$$1/\kappa \to \text{Min}[1/\kappa, (1 - \exp(-\kappa t_{cut}))/\kappa], \qquad (8.29)$$

since the dose rate from a compartment is proportional to the time constant, and the dose rate at time t is $(1 - \exp(-\kappa t))$ times its asymptotic steady-state limit.

8.4.3   Dose and impacts from ingestion

We start with the UWM for the collective inhalation dose rate (readily obtained from Eq. (7.51) of Section 7.4 by replacing the sum of the $S_{ERF,i} P_i$ by $\dot{V}_{inhal}$),

$$\dot{d}_{inhal}(\dot{m}) = \dot{V}_{inhal} \, \rho_{eff} \, \dot{m}/k_{dep} \qquad (8.30)$$

which yields the inhalation health impact rate (cases/yr) in the form

$$\dot{I}_{inhal}(\dot{m}) = S_{ERF,inhal} \, \rho_{eff} \, \dot{m}/k_{dep}, \qquad (8.31)$$

where,

$S_{ERF,inhal}$ = slope of ERF for inhalation $((cases/yr)/(person.\mu g/m^3))$,

$\dot{m}$ = rate at which pollutant is emitted to the air (kg/yr),

$\dot{V}_{inhal}$ = population-averaged inhalation rate for which we take $20.6 \text{ m}^3/(\text{pers·day})$,

ρ_{eff} = population density $(pers/km^2)$, and

k_{dep} = depletion velocity of pollutant (m/s).

$c_{air}(\mathbf{x}, \dot{m})$ = concentration increment $(\mu g/m^3)$ at $\mathbf{x}$ due to emission rate $\dot{m}$.

Since these metals are usually emitted as components of industrial particulate matter we take $k_{dep} = 0.0049$ m/s as deposition velocity.

Let $\dot{d}_{food,j}(\mathbf{x}, \dot{m})$ designate the individual dose rate (kg/(pers·yr)) due to ingestion of a food product j for a person living at $\mathbf{x} = (x,y)$,

$$\dot{d}_{food,j}(\mathbf{x}, \dot{m}) = c_{food,j}(\mathbf{x}, \dot{m}) \, \dot{Q}_{food,j}, \qquad (8.32)$$

where

$c_{food,j}(\mathbf{x}, \dot{m})$ = pollutant concentration in food product j (kg/kg_{food}) at $\mathbf{x} = (x,y)$ due to emission $\dot{m}$, and

Table 8.2 Default annual consumption rates $\dot{Q}_{food,j}$. Adapted from Eurobarometer data.

Water	0.600	$m_{wat}^3/(\text{pers·yr})$
Milk + milk products	250	$kg_{milk}/(\text{pers·yr})$
Meat	100	$kg_{FW}/(\text{pers·yr})$
Above ground fruit & vegetables (moisture 86%)	26.5	$kg_{DW}/(\text{pers·yr})$
Below ground vegetables (moisture 80%)	20.0	$kg_{DW}/(\text{pers·yr})$
Cereals (moisture 14%)	100.6	$kg_{DW}/(\text{pers·yr})$
Freshwater fish	3	$kg_{FW}/(\text{pers·yr})$
Marine fish	6	$kg_{FW}/(\text{pers·yr})$
Shellfish	1	$kg_{FW}/(\text{pers·yr})$

Notes: DW = dry weight, FW = fresh weight.

$\dot{Q}_{food,j}$ = consumption rate of food product j per person ($kg_{food}/(\text{pers·yr})$).

Data for $\dot{Q}_{food,j}$ are shown in Table 8.2. The impact rate $\dot{I}_{food,j}$ (cases/yr) due to this dose rate is

$$\dot{I}_{food,j}(\dot{m}) = S_{ERF,ingest}\, \dot{Q}_{food,j} \int dx \int dy\ \rho(\mathbf{x})\ c_{food,j}(\mathbf{x}, \dot{m}), \qquad (8.33)$$

where

$S_{ERF,ingest}$ = slope of ERF for ingestion (cases/kg), and
$\rho(\mathbf{x})$ = population density at $\mathbf{x}$ (persons/m^2).

Now we define a quantity $X_{food,j}(\mathbf{x})$ as the ratio of the concentration in food product j and the concentration in air, at a point $\mathbf{x} = (x,y)$,

$$X_{food,j}(\mathbf{x}) = c_{food,j}(\mathbf{x}, \dot{m})/c_{air}(\mathbf{x}, \dot{m}); \qquad (8.34)$$

it is in effect a transfer factor from air to food, in units of $(kg/kg_{food})/(\mu g/m^3)$. $X_{food,j}(\mathbf{x})$ does not depend on $\dot{m}$ because all of the incremental concentrations are proportional to $\dot{m}$. The equations for $X_{food,j}(\mathbf{x})$ are developed in the following section and in Appendix C. For pathways that pass through the soil, the contribution to the concentration in food produced at $\mathbf{x}$ is proportional to the local deposition and hence to $c_{air}(\mathbf{x}, \dot{m})$. Using the default parameters of the EPA for the soil pathways the ratio $X_{food,j}(\mathbf{x})$ is the same for all points $\mathbf{x}$ in a watershed. Since for the present version of the model we use the same default values for all $\mathbf{x}$, we assume $X_{food,j}$ to be independent of $\mathbf{x}$.

Replacing the $X_{food,j}(\mathbf{x})$ by site-independent constants $X_{food,j}$, we obtain, analogous to the derivation of the UWM for inhalation impacts, the UWM for the collective ingestion dose rate,

$$\dot{d}_{food,j}(\dot{m}) = X_{food,j}\, Q_{food,j}\, \rho_{eff}\, \dot{m}/k_{dep}. \tag{8.35}$$

Finally, the total impact rate $\dot{I}(\dot{m})$ (cases/yr) of a pollutant is the sum over inhalation and food products,

$$\dot{I}(\dot{m}) = \left(S_{ERF,inhal} + S_{ERF,ingest}\sum_{j} X_{food,j}Q_{food,j}\right) \rho_{eff}\, \dot{m}/k_{dep}. \tag{8.36}$$

Because most food is transported over large distances, tens of thousands of km, the conditions for the validity of the UWM are even better satisfied for ingestion than for inhalation. In fact, precise modeling of all the respective transport details would be difficult or impossible because of the lack of data. In particular, the total ingestion dose is quite insensitive to variations in the site where the pollutant is emitted into the air.

This separation of inhalation and ingestion impacts corresponds to the frequent practice of defining the dose–response functions S_{ERF} for inhalation in terms of ambient concentrations, whereas the dose–response functions S_{ERF} for ingestion are defined in terms of the ingested quantity. Therefore the units of $S_{ERF,inhal}$ and $S_{ERF,ingest}$ are different. Note that all $S_{ERF,inhal}$ and $S_{ERF,ingest}$ in this book are defined relative to the average population; if only a certain group, e.g. people over 60, are affected, we include the respective fraction of the population in $S_{ERF,inhal}$ and $S_{ERF,ingest}$.

The ratio of the rates $\dot{I}(\dot{m})$ and $\dot{m}$ is the unit impact I_u (cases/kg), i.e. the number of cases attributable to the emission of a kg of the pollutant,

$$I_u = \dot{I}(\dot{m})/\dot{m}. \tag{8.37}$$

8.4.4   Concentration in food

The calculation of pollutant concentrations in food begins with an analysis of the concentrations in soil and water. In the EPA model, the following pathways are taken into account for pollutants that enter the soil:

- On cropland, the pollutants in the top 20 cm (typical root depth) can enter the food grown;
- On pasture, the pollutants in the top 10 cm (typical root depth) can enter the feed for farm animals;
- Soil from the top 1 cm can be ingested by farm animals or children.

Like EPA (1998) we do not consider pollutant exchange with deeper soil layers or with ground water (except for leaching, which in our model is an attenuation term, implying a one-way transport to deeper soil layers or ground water for part of the pollutant mass whose impacts we do not consider). In other words we assume steady state conditions in the sense that, averaged over time, the pollutant flow to deep soil or ground water is equal to the outflow from deep soil or ground water to the top 20 cm.

Consider an area A where the deposition flux is $F_{dep} = k_{dep} \, c_{air}$. According to Eq. (8.28) for steady-state conditions, the pollutant mass $m = A \, h_{soil} \, \rho_{soil} \, c_{soil}$ in a layer of depth h_{soil} with density ρ_{soil} and average soil concentration c_{soil} (kg/kg$_{soil}$) equals the ratio of inflow $\dot{m} = A \, F_{dep}$ and decay rate (here called loss constant) κ_{soil}. This yields the concentration,

$$c_{soil} = \frac{F_{dep}}{h_{soil} \, \rho_{soil} \, \kappa_{soil}}. \tag{8.38}$$

We calculate values for c_{soil} for three values of the soil depth h_{soil}, 1 cm, 10 cm and 20 cm. The ratio,

$$\frac{c_{soil}}{c_{air}} = \frac{k_{dep}}{h_{soil} \, \rho_{soil} \, \kappa_{soil}} \tag{8.39}$$

is uniform in the entire region if, as in this chapter, a single set of parameter values is chosen. On irrigated cropland there is an additional deposition term for the pollutant influx from irrigation water. Since this term involves both soil and water concentrations, the equations for c_{soil} and c_{water} would have to be combined to solve for the resulting concentrations. But in Europe, the fraction of cropland that is irrigated is rather small, only 12% (WRI, 1994, p. 295), and so we neglect this complication, as does EPA (1998).

For the soil loss constant κ_{soil} (yr^{-1}) we take into account leaching, run-off and erosion:

$$\kappa_{soil} = \kappa_{soil,leach} + \kappa_{soil,ro} + \kappa_{soil,er}. \tag{8.40}$$

In general, there can also be biotic and abiotic degradation, and volatilization, but for metals these contributions can be neglected. The equations for the soil loss constants are listed in Table C.1 of Appendix C, including the default parameter values used in this chapter.

Example 8.4 Calculate the soil loss constants κ_{soil} for erosion, run-off and leaching, for each of the respective soil depths (ingestion by cattle, pasture, crops), as well as their sum. The inverse of the latter is the time

Solution for Example 8.4

Variable	Description	Value	Units	As	Cd	Cr	Ni	Pb
L_{soil}	Unit soil loss	1.61	$kg_{soil}/(m^2 \cdot yr)$					
R_{sed}	Sedimentary delivery ratio	0.1065						
R_{en}	Soil enrichment ratio	1						
ρ_{soil}	Soil bulk density	1500	kg_{soil}/m^3					
h_{soil}	Soil depth, ingestion	1	cm					
	Soil depth, pasture	10	cm					
	Soil depth, crops	20	cm					
θ_{sw}	Soil volumetric water content	0.2	m_{wat}^3/m_{soil}^3					
K_{sw}	Soil–water partition coefficient		m_{wat}^3/kg_{soil}	0.029	0.075	0.019	0.065	0.9
$K_{soil,er}$	Erosion soil loss constant, ingestion		1/s	3.61E-10	3.62E-10	3.60E-10	3.62E-10	3.62E-10
	Erosion soil loss constant, pasture		1/s	3.61E-11	3.62E-11	3.60E-11	3.62E-11	3.62E-11
	Erosion soil loss constant, crops		1/s	1.80E-11	1.81E-11	1.80E-11	1.81E-11	1.81E-11
v_{ro}	Run-off from pervious area	10	cm/yr					
$K_{soil,ro}$	Run-off soil loss constant, ingestion		1/s	7.26E-09	2.81E-09	1.10E-08	3.25E-09	2.35E-10
	Run-off soil loss constant, pasture		1/s	7.26E-10	2.81E-10	1.10E-09	3.25E-10	2.35E-11
	Run-off soil loss constant, crops		1/s	3.63E-10	1.41E-10	5.52E-10	1.62E-10	1.17E-11
v_{irr}	Irrigation rate	11	cm/yr					
v_{evap}	Evapotranspiration rate	30	cm/yr					
v_{prec}	Precipitation rate	75	cm/yr					
$K_{soil,leach}$	Leaching soil loss constant, ingestion		1/s	3.34E-08	1.29E-08	5.08E-08	1.49E-08	1.08E-09
	Leaching soil loss constant, pasture		1/s	3.34E-09	1.29E-09	5.08E-09	1.49E-09	1.08E-10
	Leaching soil loss constant, crops		1/s	1.67E-09	6.47E-10	2.54E-09	7.46E-10	5.40E-11
$K_{soil,total}$	Total soil loss constant, ingestion		1/s	4.10E-08	1.61E-08	6.22E-08	1.85E-08	1.68E-08
	Total soil loss constant, pasture		1/s	4.10E-09	1.61E-09	6.22E-09	1.85E-09	1.68E-09
	Total soil loss constant, crops		1/s	2.05E-09	8.06E-10	3.11E-09	9.27E-10	8.39E-11
$\boldsymbol{\tau_{soil,total}}$	**Soil time constant, ingestion**		**yr**	**7.73E-01**	**1.97E+00**	**5.10E-01**	**1.71E+00**	**1.89E+01**
	Soil time constant, pasture		**yr**	**7.73E+00**	**1.97E+01**	**5.10E+00**	**1.71E+01**	**1.89E+02**
	Soil time constant, crops		**yr**	**1.55E+01**	**3.93E+01**	**1.02E+01**	**3.42E+01**	**3.78E+02**
t_{cut}	Cut-off time for calculation, Eq. (8.29)	100	yr					
k_{dep}	depletion velocity	0.49	cm/s					
$\boldsymbol{X_{soil}}$	**Soil transfer factor, ingestion**		$\boldsymbol{(kg/kg_{soil})/(kg/m_{air}^3)}$	**7.97E+03**	**2.03E+04**	**5.25E+03**	**1.76E+04**	**1.94E+05**
	Soil transfer factor, pasture		$\boldsymbol{(kg/kg_{soil})/(kg/m_{air}^3)}$	**7.97E+03**	**2.01E+04**	**5.25E+03**	**1.76E+04**	**8.00E+04**
	Soil transfer factor, crops		$\boldsymbol{(kg/kg_{soil})/(kg/m_{air}^3)}$	**7.96E+03**	**1.87E+04**	**5.25E+03**	**1.67E+04**	**4.53E+04**

constant for removal of the pollutant from the respective soil layer. Also calculate the transfer factor X_{soil} from air to soil. Use the equations and default values in Table C.1.

Solution: Here is the output of the spreadsheet for the calculations. It is convenient to convert all inputs to SI units before doing the calculations, and at the end, convert the results to customary units. In this simple model, the time constants are proportional to the depth of the respective soil layers.

The calculation of the water concentration involves the flow rate of the pollutant through the rivers and lakes of a watershed. It has four terms: direct deposition into rivers and lakes, pollutant in the run-off from impervious surfaces, pollutant in the run-off from pervious surfaces, and pollutant in the contribution from erosion. The equations are described in Table C.2 of Appendix C. For agricultural uses, we suppose that the water is not filtered, and so we take the concentration $c_{wc,tot}$ in the water column (which includes suspended sediments). For drinking water, we take the dissolved phase water concentration $c_{wat,d}$ (without suspended sediments). Note that some water utilities employ special treatment for the removal of toxic metals, although lacking general data we do not take this into account in our calculations.

EPA also gives a formula for the concentration in the benthic sediment, but since the consumption of food from this compartment for fresh water is negligible, we do not use it (for seafood, by contrast, the benthic compartment would be important because of the large consumption of shellfish).

Substances are assimilated into vegetation due to foliar absorption of external deposits (dry and wet) and uptake from the soil via roots. The contribution of pollutants in irrigation water is easy to include in foliar absorption (unlike root absorption for which the equations for c_{soil} and c_{wat} would have to be coupled), and so we modify Eq. 5–14 of EPA (1998) by adding to the wet deposition flux on plants the term $c_{wc,tot} v_{irr}$, where v_{irr} represents irrigation rates, averaged over time and over all cropland.

The total plant concentration is the sum of the contributions from foliar absorption and from uptake through the roots. The equations and default parameter values are listed in Table C.3 of Appendix C.

Example 8.5 Sewage sludge is an excellent fertilizer. However, it also contains pollutants, including Pb. For instance, a concentration of 0.044 mg of Pb per kg of soil was measured after the use of sludge as fertilizer on a field in France. How large is the resulting concentration of Pb in wheat?

Solution: Table C.3 lists a value of 0.009 $(kg/kg_{DW})/(kg/kg_{soil})$ for the plant–soil transfer factor for grains B_{root}. Thus the concentration in the

wheat is 0.009 $(kg_{Pb}/kg_{DW})/(kg_{Pb}/kg_{soil}) \times 0.044\ mg_{Pb}/kg_{soil}$
$= 0.396\ mg_{Pb}/kg_{DW}$ per dry weight.

Accumulation in milk and meat occurs when cattle ingest soil, contaminated water and feed. The intake depends on a number of factors, including animal species, mass, age and growth rate, feed digestibility, and the milk yield for lactating animals. For simplicity, we consider only cattle, as though all meat consumed were equivalent to beef. The concentrations in meat and in milk are obtained by multiplying the intake of soil, water and feed by appropriate biotransfer factors B_m. For fish, the dissolved phase concentration in water is multiplied by the bioaccumulation factor B_{fish}. The equations and default parameter values are listed in Table C.4 of Appendix C.

8.4.5   Direct emissions to soil or water

If a pollutant is emitted directly into a river, lake or the sea, the resulting concentration near the source (first few km) can be calculated with a Gaussian plume model. Beyond a few km of flow in a river, one can assume that the pollutant is uniformly mixed; that is the case of interest here. By conservation of mass, the emission rate $\dot{m}$ (kg/s) equals the rate at which the substance flows through a cross-section A_{cross} (m^2) of the river,

$$\dot{m} = c_{river} A_{cross} v, \qquad (8.41)$$

where

v = flow velocity of river (m/s), and
c_{river} = concentration of pollutant in river.
This implies that the concentration is

$$c_{river} = \dot{m}/(A_{cross} v) \quad \text{after a few km from the source.} \qquad (8.42)$$

If this water is used for drinking or irrigation, the resulting health impacts can now be calculated with the method described in the previous section.

For direct emissions to soil, one omits the term for deposition on the vegetation, and one replaces the deposition flux to the soil F_{dep} by F_{soil}, the emission rate per unit soil surface over which the pollutant is emitted,

$$F_{soil} = \dot{m}/A_{soil}. \qquad (8.43)$$

8.4.6   Results for doses

The above equations are for collective dose rates resulting from a constant emission rate. To get the dose per emitted kg, take an emission rate of

Table 8.3 *Collective doses for central European conditions, by exposure pathway as intake fraction in mg per emitted kg. Doses from seafood are not included. These doses are the intakes, not the absorbed doses. Doses from drinking water are lower if the water utilities remove toxic metals.*

Pathway	Arsenic	Cadmium	Chromium VI	Nickel	Lead	*Lead, infants*
Inhalation	3.9	3.9	3.9	3.9	3.9	*1.6*
Water 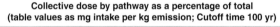	31.1	31.7	31.0	31.5	35.5	*17.8*
Cattle milk	156.2	0.3	38.0	27.2	10.8	*5.4*
Cattle meat	13.8	1.4	36.8	43.9	4.1	*0.6*
Freshwater fish	15.6	31.7	1.6	31.6	9.0	*0.0*
Grains	60.8	119.5	60.4	64.4	80.4	*15.1*
Root vegetables	12.4	24.1	12.0	13.1	16.0	*4.1*
Green vegetables	16.2	47.5	15.9	17.7	24.0	*3.1*
TOTAL	**310**	**260**	**200**	**233**	**184**	*47.6*

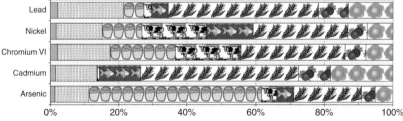

Collective dose by pathway as a percentage of total
(table values as mg intake per kg emission; Cutoff time 100 yr)

Fig. 8.3 Collective doses for central European conditions, by exposure pathway as a percentage of the total.

1 kg/yr; then the corresponding dose rate is the dose per year, and the ratio of dose rate and emission rate is the dose per kg.

Figure 8.3 and Table 8.3 show the collective dose in mg due to the atmospheric emission of 1 kg by industrial installations under typical central European conditions. Here the average population density has been taken as 80 pers/km^2 (land and water). The entries in Table 8.3, taken as dimensionless numbers and multiplied by 10^{-6}, are the fraction of the emitted pollutant that passes through human bodies; this is sometimes called intake fraction.

The doses shown are the total ingested or inhaled quantities, without regard to the fraction that is actually absorbed. If the absorption rates are less than 100%, they have to be included before applying ERFs that are based on absorbed dose. Inhalation and ingestion can be associated with very different ERFs. For example, arsenic (As) is more carcinogenic, per mass, if inhaled than if ingested.

For Pb, Table 8.3 also shows the doses for infants aged 0 to 3 yr, because that information is needed to calculate the IQ loss in the next section. We have estimated the doses for infants by scaling population average doses by the respective ratios of the food intake in Table 8.2 and the food intake for infants in the Exposure Factor Handbook of the EPA (2009).

The total dose can be much larger than the inhalation dose, by about two orders of magnitude. A simple back-of-the-envelope calculation may be instructive to explain why ingestion can be so much more important than inhalation. Consider an average person exposed to air with concentration c_{air}. The annual inhalation dose is $d_{inhal} = c_{air} V_{inhal}$ with an annual inhalation volume $V_{inhal} = 365$ days $\times$ 20 m^3/(pers·day) = 7300 m^3/(pers·yr) (we take a period of a year but that choice has no effect on the result, since the argument involves time-averaged values). If above ground food crops are exposed to the same concentration, the ingestion dose due to direct deposition on the plants is

$$d_{food} = c_{air} k_{dep} A_{crop} \times 1\,yr,$$

where A_{crop} is the horizontal area of the crops intercepting the deposition flux. The plant yield of 2.24 kg$_{DW}$/m^2 in Table C.3 of Appendix C, together with the consumption rate of 127 kg$_{DW}$/(pers·yr) for above ground crops implies an area of 127/2.24 = 56.7 m^2 per person; however, this number has to be reduced by the fraction of the year the plants are grown (say two months/yr) and by the ratio intercepting area/ground area ($f_{int} = 0.39$ in Table C.3). Thus we take,

$$A_{crop} = 56.7\,m^2 \times (2/12) \times 0.39 = 3.7\,m^2.$$

The resulting ratio of ingestion and inhalation doses is

$$d_{food}/d_{inhal} = k_{dep} A_{crop} \times 1\,yr/V_{inhal} = 58,$$

for a k_{dep} of 0.0049 m/s = 1.55E5 m/yr and after multiplying by a factor of 0.75 to account for various losses, in particular losses during retail and the extent that the pollutant is not absorbed by the edible portions of the plant, but it is increased by the contributions of below ground crops, milk and meat. In any case, for pollutants that are absorbed by plants, ingestion can indeed be much more important than inhalation.

Table 8.4 *Ratio of total doses calculated by our model (UWM) and by CalTOX, after multiplying the CalTOX results by the ratio 80/29 of population densities in central Europe and the USA.*

Ratio of doses	As	Cd	Cr	Ni	Pb
UWM/CalTOX	0.61	0.07	0.39	0.60	0.05

We have performed a sensitivity analysis to evaluate how much the intake fractions vary with changes in the input parameters; the results are shown in Table C.5 of Appendix C. The most critical parameters are the yield per planted area, and in some cases the soil–plant bioconcentration factors and the biotransfer factors for meat and milk. The choice of deposition velocities is not very critical because of the relative smallness of inhalation. The last lines of Table C.5 show how the doses change if the cut-off time t_{cut} in Eq. (8.29) is changed to 30 yr and to infinity: this doubles the impact of Pb because of the long soil loss time constants.

The results for the ratio ingestion/inhalation are consistent with data reported by WHO (1988–2001). Among models that should give comparable results, we have found the CalTOX model of McKone and Enoch (2002), a level IV model in the terminology of MacKay (2001). It analyzes essentially the same pathways as our model; in particular, ingestion of seafood and exchanges with ground water are not considered. We have run CalTOX for the most comparable scenario, i.e. in the continuous emission mode with the settings landscape = US, start of exposure = 30 yr, exposure duration = 70 yr, and exposure factors = LCIA, to calculate the intake fractions for inhalation and for ingestion. Our inhalation doses tend to be higher than those of CalTOX, our ingestion doses, lower. Table 8.4 shows the ratios of the total doses calculated by our model and by CalTOX, after multiplying the CalTOX results by the ratio 80/29 of population densities in the EU and the USA. For As, Cr and Ni, the results are not too different, considering the uncertainties; for Cd and Pb, our numbers are 14 to 20 times lower.

There are two principal differences between CalTOX and our model: one lies in the modeling of atmospheric dispersion, the other in the transfer between compartments. Whereas we use empirically determined transfer factors, mostly of the EPA, CalTOX calculates the transfer by means of fugacity data. For dispersion in the atmosphere, we assume that As, Cd, Cr and Ni and Pb are emitted as part of PM_{10} (for industrial and power plant emissions), being dispersed and deposited on the ground like other particulate matter of its size. CalTOX, by contrast, assumes emission in pure metallic form, the metals then attaching themselves to other particulate matter in the atmosphere, according to the fugacity between the metal under consideration and the particles that are already in the atmosphere (their concentration is one of the input parameters). Since the transfer from metallic phase to particles occurs at different rates for different metals, the atmospheric residence time and hence the inhalation dose in CalTOX are different for different metals. The atmosphere of CalTOX is modeled as a homogeneous perfectly mixed compartment with volume equal to the height of the atmosphere times the impact area under consideration (land area of the USA for the setting landscape = US). We believe that our treatment of the atmosphere is more realistic, because it has been explicitly validated by numerous comparisons with detailed atmospheric models. For the transfers between the other compartments, we do not know which approach is more reliable.

Example 8.6   How do the doses change if the depletion velocity is different, all else being the same?

Solution: The inhalation dose of the UWM is proportional to $1/k_{dep}$. To see how the ingestion doses change, note that all the concentrations for the other media (soil, water and food) in Tables C.1 to C.4 are proportional to $c_{air} k_{dep}$. Therefore all the ingestion doses are proportional to the inhalation dose of the UWM times k_{dep}, in other words, they do not change with k_{dep}. Since the UWM with its multimedia extension of Sections 8.4.3 and 8.4.4 is an appropriate approximation of reality, the real ingestion doses cannot vary much if k_{dep} changes.

8.5   Results for impacts and damage costs

In Table 8.5, we evaluate the damage costs due to cancers, using the respective ERFs of Section 3.5. For cancers due to inhalation, the ERFs are stated as unit risk, defined as the probability of getting a cancer due to an exposure to 1 $\mu g/m^3$ for an entire life (taken as 70 yr). Therefore we need to express the inhalation dose as an equivalent concentration c_{eq}. Recall Eq. (8.30) for the collective inhalation dose rate $\dot{d}_{inhal}(\dot{m}) = V_{inhal} \rho_{eff} \dot{m}/k_{dep}$. This implies that the collective inhalation dose rate can be calculated as though a single person were exposed to a concentration (line 7 of Table 8.5),

$$\dot{d}_{inhal}(\dot{m})/\dot{V}_{inhal} = \rho_{eff} \dot{m}/k_{dep.} \qquad (8.44)$$

Table 8.5 *ERFs and impacts, per kg emitted, for the carcinogenic metals. The top lines show the input and the collective inhalation exposure* $\rho_{eff}\,\dot{m}/k_{dep}$ *of the UWM.*

	As	Cd	Cr-VI[a]	Ni
Inhalation				
ρ_{eff} (/m^2)	0.00008	0.00008	0.00008	0.00008
$\dot{m}$ (kg/yr)	1	1	1	1
k_{dep} (m/s)	0.0049	0.0049	0.0049	0.0049
s/yr	31536000	31536000	31536000	31536000
$\rho_{eff}\,\dot{m}/k_{dep}$ (μg/m^3)	0.5177	0.5177	0.5177	0.5177
unit risk (cancers/(70yr·μg/m^3))	4.30E−03	1.80E−03	1.20E−02	2.40E−04
$S_{ERF;inhal}$ ((cancers/yr)/(μg/m^3))	6.14E−05	2.57E−05	1.71E−04	3.43E−06
$S_{ERF;inhal}$ ((cancers/yr)/(kg/m^3))	6.14E+04	2.57E+04	1.71E+05	3.43E+03
Cancers/kg, inhalation, UWM	**3.18E−05**	**1.33E−05**	**8.85E−05**	**1.78E−06**
Ingestion				
Dose (mg$_{ingested}$/kg$_{emitted}$)	31.1[b]			
slope factor (cancers/(mg/(kg$_{body}$·day)))	1.50			
$S_{ERF\ ingest}$ (cancers/kg)	1.07			
Cancers/kg, ingestion	**3.32E−05[b]**			
Total cancers/kg	**6.5E−05**	**1.33E−05**	**8.85E−05**	**1.78E−06**
Cost/kg (€/kg) at 2 M€/cancer	**130**	**27**	**177**	**3.6**

[a] if only total Cr emission is known, one must estimate the fraction in the VI oxidation state; typical numbers are 11% for coal-fired and 18% for oil-fired power plants, according to EPA (www.epa.gov/ttncaaa1/t3/meta/m28497.html and www.epa.gov/ttncaaa1/t3/meta/m27812. html).
[b] Dose from drinking water only, but impact and cost can be much lower if drinking water is filtered.

To combine this concentration with the unit risk, one has to use consistent time intervals. Because of the linearity of the ERF, any exposure interval can be used: the probability of getting a cancer is proportional to the length of the interval. It is convenient to take an emission rate of 1 kg/yr and a cancer risk for exposure during one year, i.e. an ERF slope $S_{ERF;inhal}$ ((cancers/yr)/(kg/m^3)) which is the unit risk divided by 70 yr. Of course one has to do the calculation in consistent units, and so we add in Table 8.5 a line with $S_{ERF;inhal}$ in units of (cancers/yr)/(kg/m^3). For the unit cost, let us assume 2 M€ per cancer. The damage costs are different from the ones of ExternE (2008) shown in Section 12.2, because of different monetary valuation and some of the dispersion modeling was done using WATSON.

The only metal for which EPA indicates an ERF due to ingestion is As. Only the inorganic form is considered carcinogenic. Since most As in food is organic, whereas As in water is inorganic, we take only the dose from drinking water. EPA states cancer risks due to ingestion in terms of a so-called slope factor, defined as the probability of a cancer per dose in $mg/(kg_{body} \cdot day)$.

Finally, we calculate the impact and damage cost of IQ decrement due to Pb, a cost that can be quantified with present knowledge and that is probably the dominant part of the total damage cost of Pb. We use the ERF of Eq. (3.39) in Section 3.6.6,

$$S_{ERF} = 1/3 \times f_{0-3yr} \times 1.14E-04 \, IQpoints/(\mu g_{Pb}/yr),$$
to be applied to total population,

where f_{0-3yr} is the fraction of the population between 0 and 1 yr of age (in the EU $f_{0-3yr} = 0.033$). We have to combine this with the intake fraction for infants, 47.6 mg/kg according to Table 8.3, with due attention to the units, of course. Consider steady-state conditions with an emission rate of 1 kg_{Pb}/yr; then the intake rate by infants is 47.6 mg_{Pb}/yr. Each year a new cohort of infants is born, at a rate of 0.011 of the total population per year and exposed to this Pb. With $1/3 \times f_{0-3yr} = 0.011$ of the total population that is affected per yr, we thus obtain an IQ loss of

$$0.011/yr \times 1.14E-04 \, IQpoints/(\mu g_{Pb}/yr) \times 47.6 \, (mg_{Pb}/yr)/(kg_{Pb}/yr)$$
$$= 0.06 \, IQpoints/kg_{Pb}.$$

At 15,000€/IQpoint the damage cost is 896€/kg.

In this chapter we have calculated intake fractions only for metals. For other pollutants, we note that a database of intake fractions has been assembled by a team at the National Institute for Health and Welfare of Finland and is available at www.thl.fi/expoplatform/if_database_ui/

References

Axelrad, D. A., Bellinger, D. C., Ryan, L. M. and Woodruff, T. J.. 2007. Dose-response relationship of prenatal mercury exposure and IQ: an integrative analysis of epidemiologic data. *Environ Health Perspect* **115**(4): 609–615.

Bachmann, T. M. 2006. *Hazardous substances and human health: exposure, impact and external cost assessment at the European scale*. Elsevier, Amsterdam.

Bellinger, D. C., Trachtenberg, F., Barregard, L. *et al.* 2006. Neuropsychological and renal effects of dental amalgam in children. *JAMA* **295**(15): 1775–1783.

Cohen, J. T., Connor, W. E., Kris-Etherton, P. M. *et al.* A quantitative risk-benefit analysis of changes in population fish consumption. *American Journal of Preventive Medicine* **29**(4): 325–334.

EPA 1994. *Estimating exposure to dioxin-like compounds* and *Health Assessment Document for 2,3,7,8-Tetrachlorodibenzo-p-Dioxin (TCDD) and Related Compounds* Report EPA/600/BP-92/001a, b and c. United States Environmental Protection Agency. Washington, DC 20460.

EPA 1997. *Locating and estimating air emissions from sources of mercury and mercury compounds.* Office of Air Quality Planning and Standards and Office of Air and Radiation. Research Triangle Park, NC. EPA-454/R-97–012. Washington, DC: U.S. Environmental Protection Agency.

EPA 1998. Human Health Risk Assessment Protocol for Hazardous Waste Combustion Facilities, Support Materials. U.S. EPA Region 6 U.S. EPA Multimedia Planning and Permitting Division Office of Solid Waste, Center for Combustion Science and Engineering. Available at www.epa.gov/epaoswer/hazwaste/combust/riskvol.htm#volume1

EPA 2000. *Exposure and Human Health Reassessment of 2,3,7,8-Tetrachlorodibenzo-p-Dioxin (TCDD) and Related Compounds: Part III: Integrated Summary and Risk Characterization for 2,3,7,8-Tetrachlorodibenzo-p-Dioxin (TCDD) and Related Compounds.* Report EPA/600/P-00/001Bg, September 2000. United States Environmental Protection Agency. Washington, DC 20460.

EPA 2001. *Oral Reference Dose for Methylmercury.* United States Environmental Protection Agency. Integrated Risk Information System (IRIS). Office of Research and Development, National Center for Environmental Assessment, Washington, DC.

EPA 2009. *Exposure Factors Handbook.* External Review Draft, July 2009. EPA/600/R-09/052A. Office of Research and Development, National Center for Environmental Assessment, U.S. Environmental Protection Agency, Washington, DC 20460.

ExternE 2008. With this reference we cite the methodology and results of the NEEDS (2004–2008) and CASES (2006–2008) phases of ExternE. For the damage costs per kg of pollutant and per kWh of electricity we cite the numbers of the data CD that is included in the book edited by Markandya, A., Bigano, A. and Porchia, R. in 2010: *The Social Cost of Electricity: Scenarios and Policy Implications.* Edward Elgar Publishing Ltd, Cheltenham, UK. They can also be downloaded from http://www.feem-project.net/cases/ (although in the latter some numbers have changed since the data CD in the book).

Griffiths, C., McGartland, A. and Miller, M. 2007. A comparison of the monetized impact of IQ decrements from mercury emissions. *Environ Health Perspect.* **115**(6): 841–847.

Grosse, S. D., Matte, T. D., Schwartz, J. and Jackson, R. 2002. Economic gains resulting from the reduction in children's exposure to lead in the United States. *Environmental Health Perspectives* **110**(6): 563–569.

Huijbregts, M. A. J., Struijs, J., Goedkoop, M. *et al.* 2005. Human population intake fractions and environmental fate factors of toxic pollutants in life cycle impact assessment. *Chemosphere Chemosphere* **61**: 1495–1504.

IAEA 1994. *Handbook of Parameter Values for the Prediction of Radionuclide Transfer in Temperate Environments,* Technical Report Series No. 364, produced in collaboration with the International Union of Radioecologists, International Atomic Energy Agency, Vienna, Austria.

IAEA 2001. *Generic Models for Use in Assessing the Impact of Discharges of Radioactive Substances to the Environment,* Safety Reports Series No. 19, International Atomic Energy Agency, Vienna, Austria.

Jolliet, O., Margni, M., Charles, R. *et al.* IMPACT 2002+: A new life cycle impact assessment methodology. *International Journal of LCA* **8**(6): 324–330.

Jorgensen, E. B, Grandjean, P. and Weihe, P. 2007. Separation of risks and benefits of seafood intake. *Environmental Health Perspectives* **115**(3): 323–327.

Lamborg, C. H., Fitzgerald, W. F., O'Donnell, J. and Torgersen, T. 2002. A non-steady-state compartmental model of global-scale mercury biogeochemistry with interhemispheric atmospheric gradients. *Geochimica et Cosmochimica Acta* **66**(7): 1105–1118.

Landis, M. S., Vette, A. F. and Keeler, G. J. 2002. Atmospheric Deposition to Lake Michigan During the Lake Michigan Mass Balance Study, submitted to *Environmental Science and Technology*.

Lipfert, F. W., Moskowitz, P. D., Fthenakis, V. and Saroff, L. 1996. Probabilistic assessment of health risks of methylmercury from burning coal. *Neurotoxicology* **17**(1): 197–211.

Mackay, D. 2001. *Multimedia Environmental Models: The Fugacity Approach.* 2nd ed. Lewis Publishers. Boca Raton, Florida.

Mahaffey, K. R. 2005. *NHANES 1999–2002, Update on Mercury.* Presented at: 2005 National Forum on Contaminants in Fish; 2005 Sep 18–21, Baltimore, MD.

McKone, T. E. and Enoch K. G. 2002. CalTOX™, A Multimedia Total Exposure Model. Report LBNL – 47399. Lawrence Berkeley National Laboratory, Berkeley, CA. Available at http://eetd.lbl.gov/ied/ERA.

Muir, T. and Zegarac, M. 2001. Societal costs of exposure to toxic substances: Economic and health costs of four case studies that are candidates for environmental causation. *Environmental Health Perspectives* **109**, Supplement 6 (December): 885–903.

NCHS 2005. NHANES 1999–2000 and NHANES 2001–2002 Public Use Data Files. Hyattsville, MD: National Center for Health Statistics. Available: www.cdc.gov/nchs/about/major/nhanes/nhanes01–02.htm.

Rabl, A., Spadaro, J. V. and McGavran, P. D. 1998. Health risks of air pollution from incinerators: a perspective. *Waste Management & Research* **16**: 365–388.

Rea, A. W., Lindberg, S. E. and Keeler, G. J. 2000. Assessment of dry deposition and foliar leaching of mercury and selected trace elements based on washed foliar and surrogate surfaces. *Environmental Science and Technology* **34**: 2418–2425.

Rea, A. W., Lindberg, S. E. and Keeler, G. J. 2001. Dry deposition and foliar leaching of mercury and selected trace elements in deciduous forest throughfall. *Atmospheric Environment* **35**: 3453–3462.

Rice, G. and Hammitt, J. K. 2005. *Economic Valuation of Human Health Benefits of Controlling Mercury Emissions from US Coal-Fired Power Plants.* Northeast States for Coordinated Air Use Management (NESCAUM). Boston, MA. February 2005.

Rossler, M. T. 2002. The electric power industry and mercury regulation: Protective, cost-effective, and market-based solutions. *EM.* April: 15–21.

Searle A. 2005. *Exposure-Response Relationships for Human Health Effects Associated With Exposure to Metals.* Report Institute of Occupational Medicine, Edinburgh.

Spadaro, J. V. and Rabl, A. 2004. Pathway analysis for population-total health impacts of toxic metal emissions. *Risk Analysis* **24**(5): 1121–1141.

Spadaro, J. V. and Rabl, A. 2008a. Global health impacts and costs due to mercury emissions. *Risk Analysis* **28** (3): 603–613.

Sullivan, T. M., Lipfert, F. W., Morris, S. C. and Moskowitz, P. D. 2003. *Potential health risk reduction arising from reduced mercury emissions from coal-fired power plants*. Report for United States Department of Energy, September 2001. Energy, Environment and National Security Directorate. Brookhaven National Laboratory, Upton, New York 11973–5000.

Trasande, L., Landrigan, P. J. and Schechter, C. 2005. Public health and economic consequences of methyl mercury toxicity to the developing brain. *Environmental Health Perspectives* **113**(5): 590–596.

UK 2002. Committee on toxicity of chemicals in food, consumer products and the environment. "Updated COT statement on a survey of mercury in fish and shellfish". www.food.gov.uk/multimedia/pdfs/cotstatementmercuryfish.PDF, accessed 6 Oct. 2007.

UNEP 2002. United Nations Environment Programme. Global Mercury Assessment. UNEP Chemicals, Geneva, Switzerland.

Vette, A. F., Landis, M. S. and Keeler, G. J. 2002. Deposition and Emission of Gaseous Mercury to and from Lake Michigan During the Lake Michigan Mass Balance Study (July, 1994 – October, 1995), submitted to *Environmental Science and Technology*.

Virtanen, J. K., Rissanen, T. H., Voutilainen, S. and Tuomainen, T-P. 2007. Mercury as a risk factor for cardiovascular diseases. *Journal of Nutritional Biochemistry* **18**: 75–85.

WHO 1988–2001. WHO 1988, *Chromium*. Environmental Health Criteria 61. WHO 1990, *Methyl Mercury*. Environmental Health Criteria 101. WHO 1991, *Inorganic Mercury*. Environmental Health Criteria 118. WHO 1991, *Nickel*. Environmental Health Criteria 108. WHO 1992, *Cadmium*. Environmental Health Criteria 134. WHO 1995, *Inorganic Lead*. Environmental Health Criteria 165. WHO 2001, *Arsenic and arsenic compounds*. Environmental Health Criteria 224. World Health Organization, Geneva, Switzerland.

WRI 1994. *World Resources: a Guide to the Global Environment*. World Resources Institute. Oxford University Press.

9   Monetary valuation

Summary

The chapter on monetary valuation begins with a discussion of discounting, a tool that is necessary for the correct accounting of costs that occur at different times. A particularly important and controversial issue is the intergenerational discount rate, in Section 9.1.3. This is followed, in Section 9.2, by an overview of valuation methods, especially for non-market goods. Section 9.3 addresses the important case of the valuation of mortality, especially the loss of life expectancy due to air pollution. Morbidity valuation follows in Section 9.4, including a discussion of DALY and QALY scores. Section 9.5 addresses the valuation of neurotoxic impacts (value of an IQ point). Section 9.6 discusses the transfer of values to situations that are different from the original valuation studies.

Note that the valuation of some impact categories has been discussed in other chapters: Chapter 4 for buildings, Chapter 5 for agricultural losses and ecosystems, Chapter 6 for noise and traffic congestion (plus brief comments on visibility, non-renewable resources, accidents, employment, and security of energy supply), and Chapter 10 for global warming. A summary of the monetary values for health impacts will be provided in Table 12.3 of Chapter 12.

9.1   Comparing present and future costs

9.1.1   *The effect of time on the value of money*

It may be appropriate to begin this chapter with a tool that is needed whenever there are costs that occur at different times. Such a cost must be adjusted to a common time basis because a dollar (or any other currency) unit to be paid in the future does not have the same value as a dollar available today. This time dependence of money is due to two, totally different, causes. The first is inflation, the well-known and ever present erosion of the value of our currency. The second reflects the fact that a dollar today can buy goods to be enjoyed immediately or it can be invested to increase its value by profit or interest. Thus a dollar that becomes available in the future is less desirable than a dollar today; its value must be discounted. This is true even if there is no inflation. Both inflation and discounting are usually characterized in terms of annual rates.

Let us begin with inflation. To avoid confusion it is advisable to add subscripts to the currency signs, indicating the year in which the currency is specified. For example, throughout the past two decades the inflation rate r_{inf} in the EU and in the USA has generally been around $r_{inf} = 2$ to 3%. At 2.5% a dollar bill in 2011 is worth only $1/(1+0.025)$ as much as the same dollar bill in 2010,

$$\$_{2011}\, 1.00 = \$_{2010}\, \frac{1}{1 + r_{inf}} = \$_{2010}/(1 + 0.025) = \$_{2010}\, 0.976. \quad (9.1)$$

The definition and measure of the inflation rate are not without ambiguities, since different prices escalate at different rates and the inflation rate depends on the mix of goods assumed. Therefore a standard basket of goods is used as the basis. In the USA it is the Consumer Price Index (CPI), in Europe, the Harmonized Index of Consumer Prices. In terms of the CPI, the average inflation rate from year ref to year ref+n is given by,*

$$(1 + r_{inf})^n = \frac{CPI_{ref+n}}{CPI_{ref}}. \quad (9.2)$$

Suppose $\$_{2010}\, 1.00$ has been invested at an interest rate $r_{int} = 5\%$, the *nominal* or *market* rate, as usually quoted by financial institutions. Then after one year this dollar has grown to $\$_{2011}\, 1.05$, but it is worth only $\$_{2010}\, 1.05/1.025 = \$_{2010}\, 1.024$. To show the increase in the real value, it is convenient to define the real interest rate r_{int0} by the relation,

$$1 + r_{int0} = \frac{1 + r_{int}}{1 + r_{inf}}, \quad (9.3)$$

---

* For simplicity we write the equations as though all growth rates were constant. Otherwise the factor $(1 + r)^n$ would have to be replaced by the product of factors for each year $(1 + r_1)(1 + r_2) \ldots (1 + r_n)$. Such a generalization is straightforward but tedious, and of dubious value in practice as it is chancy enough to predict averages trends without trying to guess a detailed scenario.

or,

$$r_{int0} = \frac{r_{int} - r_{inf}}{1 + r_{inf}}.$$

The simplest way of dealing with inflation is to eliminate it from the analysis right at the start by using so-called *constant currency*, and expressing all growth rates (interest, energy price escalation, etc.) as real rates, relative to constant currency. After all, one is concerned about the real value of cost flows, not about their nominal values in a currency eroded by inflation. Constant currency is obtained by expressing the *current* or *inflating* currency of each year (i.e. the nominal value of the currency) in terms of equivalent currency of an, arbitrarily chosen, reference year *ref*. Thus the current dollar of year *ref+n* has a constant dollar value of,

$$\$_{ref} = \frac{\$_{ref+n}}{(1 + r_{inf})^n}. \tag{9.4}$$

A *real rate* of change r_0 is related to the *nominal rate* r in a way analogous to Eq. (9.3),

$$r_0 = \frac{r - r_{inf}}{1 + r_{inf}}. \tag{9.5}$$

For low inflation rates one can use the approximation,

$$r_0 \approx r - r_{inf} \text{ if } r_{inf} \text{ small}. \tag{9.6}$$

For social cost analysis one uses constant currency and real rates, as we do throughout (more on that in Section 9.1.7).

9.1.2   The discount rate

Since a future cash amount F is not equal to its **present value** P; it must be discounted. The relation between P and its future value F_n in year n from now is determined by the **discount rate** r_{dis}, defined such that,

$$P = F_n/(1 + r_{dis})^n. \tag{9.7}$$

The higher the discount rate, the lower the present value of future transactions. The ratio P/F_n is called present worth factor.

To determine the appropriate value of the discount rate, one has to ask, at what value of r_{dis} is one indifferent about the amount P today and an amount $F_1 = P/(1 + r_{dis})^1$ a year from now? The answer depends on who is posing the question and under what circumstances: an individual, a company or society. Roughly speaking, and neglecting the complication of taxes, for an individual who would put his money into a savings account, the discount rate is the interest rate, and for a company, the discount rate is the rate of return on its investments. The rate to be used for evaluating choices by society is called the **social discount rate**. The appropriate value is difficult to determine and uncertain; it is somewhat controversial, because it affects the ranking of different choices. For a good review of the issues and results we cite, see Harrison (2010).

There are two approaches that are generally used for the social discount rate. The first one sets the rate equal to the pre-tax rate of return on risk-free investments. The second is based on a formula due to Ramsey. That formula follows from comparing the utility that a consumer can gain from using a certain amount of money at different times. The result is,

$$r = r_{prtp} + \eta r_{growth}, \tag{9.8}$$

where r_{prtp} is the "pure rate of time preference" (which arises from the impatience to consume now rather than in the future and from uncertainty about the future), r_{growth} is the rate at which consumption is expected to grow and η is the elasticity of the marginal utility of consumption. With this approach, r is determined by estimating the three parameters on the right-hand side of Eq. (9.8).

Some governments choose the first approach, some the second, some take a weighted average of the two. There is no agreement among economists, either on the method to use or on the parameters for the Ramsey formula. Table 9.1 shows the social discount rates used by various governments. There are two general patterns: the rates are higher in developing countries, and in industrialized countries there has been a tendency in recent years to move to lower rates.

9.1.3 The intergenerational discount rate

The long-term or intergenerational discount rate is a crucial, and controversial, parameter for the assessment of the damage costs of climate change. Tol (2005) carried out a review of damage cost estimates and found that the most important determinants of the results are the discount rate and whether to apply equity weighting (see Section 10.3.3). For example, the Stern report (Stern *et al.*, 2006) came up with very high damage costs ($85/$t_{CO2}$ compared with mainstream estimates in the range $15 to $30/$t_{CO2}$), but was widely criticized for using a discount

Table 9.1 *Social discount rates in practice. Adapted from Harrison (2010).*

Country	Agency	Discount rate (%)
Canada	Treasury Board	8^c
		From 1976–2007 used 10 (and test 8–12)^{ab}
China (People's Republic)		8^a
France	Commissariat General du Plan	4 From 1985–2005 used 8^{ab}
European Union	European Commission	5 From 2001–2006 used 6^a
Germany	Federal Finance Ministry	3 From 1999–2004 used 4^{ab}
India		12^a
Italy	Central Guidance to Regional Authorities	5^a
The Netherlands	Ministry of Finance	4^e
Norway		3.5 From 1978–98 used 7^{ab}
Philippines		15^a
Pakistan		12^a
South Africa		8 (and test 3 and 12)^d
United Kingdom	HM Treasury	3.5 (declining to 1 after 300 years)
		From 1969–78 used 10^a
United States	Environmental Protection Agency	2–3 (and test 7)^a
United States	Office of Management and Budget	7 (and test 3) Used 10 until 1992^a
International Multi-lateral Development Banks	World Bank	$10–12^a$
	Asia Development Bank	$10–12^a$
	Inter-American Development Bank	12^a
	European Bank for Reconstruction and Development	10^a
	African Development Bank	$10–12^a$

[a] Zhuang *et al.* (2007, table 4, pp. 17–18, 20).
[b] Spackman (2006, table A.1, p. 31).
[c] Treasury Board of Canada (2007, p. 37, 1998, p. 45).
[d] South African Department of Environmental Affairs and Tourism (2004, p. 8).
[e] van Ewijk and Tang (2003, p. 1).

rate much below conventional values of the social discount rate.[1] There have been endless debates about the choice of the discount rate for climate change, especially about the rate of pure time preference.

[1] See for example the section on discounting in http://en.wikipedia.org/wiki/Stern_Review

Table 9.2 *Schedule of standard discount rate of the Green Book of the UK (2008).*

Time period, years after present	0 – 30	31 – 75	76 – 125	126 – 200	201 – 300	>300
Discount rate	3.5%	3.0%	2.5%	2.0%	1.5%	1.0%

Weitzman (2001) surveyed professional Ph.D. level economists about their "professionally considered gut feeling" for the discount rate that should be used to evaluate the costs and benefits of climate change mitigation. The 2160 responses he received ranged from -3% to 27%, with a mean of about 4%, a standard deviation of about 3%, a median of 3% and a mode of 2%. In recent years, more and more economists have come to the conclusion that the long-term social discount rate should be significantly lower than the conventional short-term rate (see e.g. Weitzman, 1998, Gollier and Weitzman, 2010, Cline, 2008). In particular the UK Treasury (2008) recommends declining discount rates in the Green Book. This discount rate schedule is reproduced here in Table 9.2.

The purpose of evaluating the costs of climate change is to compare the abatement costs (for reducing greenhouse gas emissions) with the benefits of avoided damage. In this context, the discount rate is not a fundamental quantity of optimal growth models, but merely a parameter for the cost–benefit analysis (CBA), needed to allow a correct weighting of costs and benefits that occur at different times.

A simple accounting argument shows that there will be a contradiction if one chooses a discount rate r for climate change that is greater than the growth rate r_{gro} of global world product. Consider an abatement option that imposes a cost C at time t = 0 and brings a benefit B at some future time t_f, the discount rate being r; the latter being chosen as the pre-tax rate of return on risk-free investments. A CBA evaluated at time t = 0 says that the decision is beneficial if,

$$\text{Beneficial if } C < B \exp(-r t_f). \qquad (9.9)$$

But it could equally well be evaluated at time t_f,

$$\text{Beneficial if } C \exp(r t_f) < B. \qquad (9.10)$$

This can be interpreted to say that if this latter criterion is not satisfied, it would be more profitable for society to invest the amount C elsewhere, because it would grow at rate r and in year t_f it would be worth more than B. By that logic the investment would bring an annual return of r C exp(r t) in any year t for the indefinite future. Obviously there is a contradiction if r > r_{gro}, because eventually such a return would exceed the total GDP of the world. As an example, take a typical value of r = 4% for the conventional social discount rate and r_{gro} = 1% for the growth rate of world GDP, at constant population. If the investment C is a 0.1% of world GDP, it will exceed the total GDP after about 184 years, well within the time span during which the analysis is supposed to be valid. The explanation of this contradiction is that the difference between the conventional rate r and r_{gro} (due to time preference) does not create wealth; it only redistributes it. Since climate change is a global problem, the CBA has to be approached from a global perspective. The world cannot borrow from itself and get rich on the interest. Since an average investment brings long-term benefits at a rate r_{gro}, the intergenerational rate should be equal to r_{gro}.

Had we considered the approach where the discount rate is based on the Ramsey formula, the argument would involve the utility of the amounts C and B. The same contradiction would arise, because eventually the utility of the annual return on any investment would exceed the utility of the world GDP. Clearly the justification for discount rates greater than r_{gro} breaks down sooner or later, as already pointed out by one of us (Rabl, 1996), recommending a transition to the growth rate of GDP per capita at the end of the present generation. The contradiction also arises for individual countries, although not quite as soon because they may be able to earn interest from other countries at a rate greater than r_{gro}.

Since there is no agreement among economists and a rate higher than r_{gro} leads to a contradiction, let us ask whether there is a universal ethical principle that all could accept for the weighting of costs and benefits that affect different people at different times. Rawls, the philosopher of social justice, proposed the veil of ignorance as a guiding principle for the formulation of laws: if an individual does not know how s/he will end up in her/his own conceived society, s/he is likely to develop a scheme of justice that treats everybody fairly, rather than privileging any one class of people. The idea that justice should treat everybody fairly is implicit in the Declaration of Human Rights and in the constitutions of democratic countries.

Application of the veil of ignorance to CBA is straightforward. One simply has to ask how one should weight the costs and benefits if one did not know which, among the individuals affected by a decision, one might end up being. If a decision is made today, with long-term effects, would you find an effect (e.g. a probability p of getting malaria, all else being equal) any different if it struck you ten years from now or a hundred years from now (all else being equal, including the severity of that malaria)?

Once one accepts the principle that all effects should be weighted equally, regardless of when they occur, the remaining question is how to weight them in monetary terms. Following the general rule of monetary valuation, the weighting should be based on the willingness to pay (WTP) for the effect in question. Thus a reasonable principle is to weight them equally in terms of fraction of disposable income, adjusted for a possible change due to a change in income if the income elasticity of WTP is different from unity. For an average consumption basket, the income elasticity can be taken as unity. If the value of an average consumption basket remains constant while disposable income grows at a rate r_{gro}, the weighting factor decreases and the decrease is equivalent to a discount rate of r_{gro}. That is therefore the appropriate intergenerational discount rate. As an approximation of that rate the growth rate r_{gro} of GDP per capita seems to be the best choice.

Rabl (1996) reviewed the data for average growth rates that have been observed in countries where long-term data are available, and found that long-term average growth rates of GDP per capita in industrialized countries have been in the range 1 to 2%. One might wonder if giving equal weight to all future generations could overwhelm the decision process in cases where an action might have effects for a very long time. However, that is unlikely because the long-term average of r_{gro} is positive, thus sufficiently reducing the contribution of distant times in any CBA.

How about the transition between the conventional short-term social discount rate and the intergenerational rate? Pure time preference is a factor that almost everyone applies in their everyday financial decisions, impatient to consume now rather than in the future and fully willing to live with the consequences. Thus it seems fair to use the conventional rate, with its pure time preference component, for costs and benefits that affect those who could participate in the decision, either directly or by expressing their preferences in elections. But if a decision today has an effect, say, 30 years from now, part of the affected population may not have participated in the decision and it may not be fair to weight their effects according to a rate that includes pure time preference. Thus the social discount rate r_{dis} should decrease with time, towards r_{gro}, by incorporating the fraction f(t) of the total population that was above voting age at the time of the decision and is still alive t years later. These considerations lead to the recommendation that the social discount rate $r_{dis}(t)$ should be calculated via

$$\text{Exp}[-t r_{dis}(t)] = f(t)\,\text{Exp}[-t(r_{prtp} + \eta\, r_{gro})] + [1 - f(t)\,\text{Exp}[-t r_{gro}]],$$

$$(9.11)$$

where the term $(r_{prtp} + \eta\, r_{gro})$ is of course the conventional short-term rate according to Eq. (9.8). Note that it is not the discount rates that should be weighted by $f(t)$ but the corresponding exponentials, because it is the exponentials that are the weighting factors for future costs and benefits, as Weitzman (1998) had pointed out. In any case, both the accounting argument and the Rawlsian veil of ignorance imply the same result for the intergenerational discount rate.

9.1.4   Evolution of costs with time

Even if there were agreement on the appropriate discount rate, another rate is just as important: the rate r_{dam} at which the damage costs change over time. Before discounting a future cost, one must predict what that cost will be. Only the difference,

$$r_{eff} = r_{dis} - r_{dam} \qquad (9.12)$$

matters. Even though this point is obvious and has been noted by some economists, almost all of the environmental literature has been preoccupied with the discount rate, while neglecting r_{dam}. The evolution can be very different for different impacts. Both physical impacts and their monetary valuation can change, in different ways.

Forecasting the evolution necessarily involves subjective judgments about the progress of science, medicine and society. For example, if the progress of medicine renders cancers as harmless as the common cold at some time in the future, it would be absurd to estimate their damage after that time according to today's assumptions. A look at the history of cancer therapy is instructive. A century ago almost all cancers were fatal, but today, already more than half can be cured (with the customary convention of considering a cancer cured when there is no relapse within five years).

One may expect the environment to be a superior good, i.e. one whose income elasticity is greater than unity. In that case, the valuation will grow faster than income. However, for the valuation of mortality, the observed income elasticity has been found to be less than unity, generally in the range 0.4 to 0.7. On the other hand, the demand for unspoiled nature is likely to increase faster than income as more and more land becomes built up.

9.1.5   Net present value and equivalent annual values

In most projects a variety of different costs and benefits occur at different times. In order to compare different projects one needs to express such costs on a common basis, namely as equivalent present values; they are calculated by means of Eq. (9.7). The net present value of a project is the sum of the present values of all the costs and benefits that will occur during the life cycle of the project.

Often it is useful to express the present value P as an equivalent series of equal annual payments A during the duration of the project, N years, each payment being made at the end of the respective year. The present value of the first payment is $A/(1 + r_{dis})$, that of the second payment is $A/(1 + r_{dis})^2$, etc. Adding the present values from year 1 to year N, one finds the total present value,

$$P = A/(1 + r_{dis}) + A/(1 + r_{dis})^2 + \ldots A/(1 + r_{dis})^N. \qquad (9.13)$$

This is a simple geometric series, and the result is readily summed to

$$P = A \frac{1 - (1 + r_{dis})^{-N}}{r_{dis}} \text{ for } r_{dis} \neq 0. \qquad (9.14)$$

For a zero discount rate, the right-hand side is indeterminate, but its limit $r_{dis} \rightarrow 0$ is A N, reflecting the fact that the N present values all become equal to A in that case. The ratio of A and P is a very useful quantity, often called the capital recovery factor, and a convenient notation for it is $A/P = (A/P, r_{dis}, N)$,

$$A/P = (A/P, r_{dis}, N) = \begin{cases} \dfrac{r_{dis}}{1 - (1 + r_{dis})^{-N}} & \text{for } r_{dis} \neq 0 \\ \dfrac{1}{N} & \text{for } r_{dis} = 0 \end{cases}. \qquad (9.15)$$

Its inverse is known as the series present worth factor, since P is the present value of a series of equal payments A.

The reason for the name, capital recovery factor, lies in its use for the calculation of loan payments. In principle, a loan could be repaid according to any arbitrary schedule, but in practice the most common arrangement is based on constant payments in regular intervals. The portion of A due to interest varies from year to year, but to find the relation between A and the loan amount L this need not worry us. Let us first consider a loan of amount L_n that is to be repaid with a single payment F_n at the end of n years. With n years of interest, at loan interest rate r_1, the payment must be,

$$F_n = L_n (1 + r_1)^n. \qquad (9.16)$$

This means that the loan amount is the present value of the future payment F_n, discounted at the loan interest rate. A loan that is to be repaid in N equal installments can be considered as the sum of N loans, the nth loan to be repaid in a single installment A at the end of the nth year. Discounting each of these payments at the loan interest rate and adding them, we find the total present value; it is equal to the total loan amount,

$$L = P = A/(1 + r_l) + A/(1 + r_l)^2 + \ldots A/(1 + r_l)^N. \qquad (9.17)$$

This is just the series of the capital recovery factor. Hence the relation between annual loan payment A and loan amount L is,

$$A = L(A/P, r_1, N). \qquad (9.18)$$

Example 9.1   A home buyer obtains a mortgage of 100,000 € at an interest rate of 8% over 20 years. What are the annual payments?

Solution: The capital recovery factor is (A/P, 0.08, 20) = 0.1019/yr, and the annual payments are 10,190 €/yr, approximately one-tenth of the loan amount.

   With the help of present worth factor and capital recovery factor, any single expense C_n that occurs in year n, for instance, a major repair, can be expressed as an equivalent annual expense A that is constant during each of the N years of the life of the system. The present value of C_n is $(1 + r_{dis})^{-n}$ C_n, and the corresponding annual cost is,

$$A = (A/P, r_{dis}, N)(1 + r_{dis})^{-n} C_n. \qquad (9.19)$$

Example 9.2   A system has a salvage value of 1000 € at the end of its useful life of N = 20 years. What is the equivalent annual value if the discount rate is 8%?

Solution: Insert into Eq. (9.19),
   $A = (A/P, 0.08, 20)(1 + 0.08)^{-20} 1000 € = 21.86 €$ per year.

Example 9.3   Consider a large coal fired power plant with an annual output of 5×10^9 kWh. If it emits 0.5 g/kWh of SO_2 and the damage cost is 6 €/kg of SO_2, what is the present value of the total SO_2 damage cost during 30 yr? Assume a social discount rate of 5%.

Solution: The annual damage cost is 15 million €. To find the present value, divide the annual cost by (A/P, 5%, 30 yr) = 0.0651. The result is 231 million €.

In some situations, recurring annual costs or benefits may accrue for very long times, for instance the costs of maintaining a storage site for high level nuclear waste. Because of discounting, the present value of such a series is finite, even if the series continues to infinity. Summing the right-hand side of Eq. (9.13) in the limit $N \to \infty$, one readily obtains,

$$P = A/r_{dis}. \tag{9.20}$$

For example, if $r_{dis} = 4\%$, the present value P is 25 times the annual amount. However, one has to be careful when it comes to valuing a permanent loss. Consider an annual loss A that is incurred during each of 50 years: at $r_{dis} = 4\%$, the present value of the total loss is 21.5 times A according to Eq. (9.15). If the loss becomes permanent, the present value increases to only 25 times A. In reality most permanent losses would be valued much higher; there is a premium for irreversibility.

9.1.6 The rule of 70 for doubling times

Most of us do not have a good intuition for exponential growth. As a helpful tool, therefore, we present the rule of 70 for doubling times. The doubling time T_2 is related to the continuous growth rate r_{cont} by,

$$2 = \exp(T_2 r_{cont}). \tag{9.21}$$

Solving the exponential relation for T_2, we obtain,

$$T_2 = \ln(2)/r_{cont} = 0.693\ldots/r_{cont}. \tag{9.22}$$

The product of doubling time and growth rate in units of percent is very close to 70 years:

$$100\,T_2\,r_{cont} = 69.3\ldots \text{yr} \approx 70\,\text{yr}. \tag{9.23}$$

In terms of annual rates r_{ann} the relation would be,

$$T_2 = \ln(2)/\ln(1 + r_{ann}), \tag{9.24}$$

numerically very close to Eq. (9.22) for small rates, but less convenient.

Example 9.4 Population growth rates average at around two percent for the world as a whole and reach four percent in certain countries. What are the corresponding doubling times?

Solution: 70 yr/2 = 35 yr for the world and 70 yr/4 = 17.5 yr for countries with four percent growth.

9.1.7   Comments on cost–benefit analysis (CBA)

A cost–benefit analysis of environmental policy choices should be done from the perspective of society. Thus the economic analysis is quite different from that of a private investor or consumer. In particular, the treatment of discount rate, inflation and taxes is all very different, and the equations are much simpler.

Taxes should not be included because the net cost to society is zero: resources are simply being transferred rather than consumed. Society neither pays taxes[2] nor receives subsidies. As for inflation, the simplest way of dealing with it is to eliminate it from the analysis by using real rates (for discount rate and growth rates of specific costs). After all, what matters is the real value of cost flows, not their nominal values in a currency eroded by inflation. The reason why inflation should be included in the analysis of a private investor lies in tax deductions for items such as interest payments and depreciation, since tax deductions are based on inflating currency. There are no such items in social CBA.

The discount rate should be the social discount rate, much lower than typical discount rates for private enterprises. The latter are higher because of corporate taxes and because of greater uncertainties about the future performance or profitability of an investment. From a social perspective, the uncertainties are lower because the risks are pooled over all investments. In a social cost analysis a single discount rate for all investments is appropriate to account for the cost of money, whereas uncertainties about the future performance or profitability of a particular investment should be explicitly treated by considering the probability distributions of all relevant parameters, such as the performance and durability of equipment for pollution abatement. All costs should be based on the cost to society rather than on prices paid by private consumers. Thus they should include external costs wherever they are significant.

Most CBAs assume changes to be small enough to allow linear approximations throughout. In particular, they assume that the external costs are proportional to changes in the corresponding environmental burdens and that relative prices in the economy do not change. This is called a partial

---

[2] However, in some cases taxes might be an acceptable proxy for real costs. For example, part of certain payments associated with vehicles (gasoline tax, title and registration) is used for road construction and repair; of course, a direct estimation of such costs is better than the use of a proxy.

equilibrium approach. When the changes under consideration are large, for example a doubling of atmospheric CO_2, a general equilibrium analysis may be needed to account for changes in relative prices and behavior of economic agents.

9.2   Valuation methods

9.2.1   Market and non-market costs

The goal of the monetary valuation of impacts is to assess the total economic value, including the value of marketed and non-marketed goods and services. The total economic value of health impact is expressed in monetary units (or costs). It consists of resource costs (such as medical treatment), opportunity costs (lost income) and dis-utility costs (such as pain and suffering, reduced quality of life). For example, the value of avoiding an asthma attack includes not only the cost of the medical treatment but also the willingness to pay (WTP) to avoid the residual suffering. If the WTP for a non-market good has been determined correctly, it is like a price, consistent with prices paid for market goods.

Economists have developed a variety of tools for measuring non-market costs; they can be grouped into two categories: revealed preferences and stated preferences. Revealed preference methods determine the WTP by analyzing consumer choices. Stated preference methods ask individuals directly to state their preferences.

9.2.2   Revealed preference methods

In this category, one approach is the hedonic price method,[3] which analyzes the relation between the price and a non-market attribute of a good. For example, the price of houses is higher if the ambient air is less polluted, all else being equal. Thus the WTP for clean air could be determined, at least in principle, by regressing house prices against air pollution. In practice, the difficulty lies in the "all else being equal" assumption. One never finds sets of houses that are equal in all aspects other than air pollution. Careful and complicated statistical tools have to be used to disentangle the many factors that may affect the price, and usually the results are very uncertain. And then there is the problem that people cannot calculate the most important impacts of air pollution, namely the impact on their health. Therefore one

---

[3] For a more detailed discussion of hedonic methods, we recommend the book by P. Graves (2013).

Table 9.3 *Values of cost of noise, recommended for France by Boiteux (2001).*

Range of noise level, decibels	55–60	60–65	65–70	70–75	>75
% decrease of rent per decibel	0.4%/dB	0.8%/dB	0.9%/dB	1%/dB	1.1%/dB

cannot determine a meaningful cost of air pollution by such an approach, even if one had perfect "all else being equal" data.

9.2.2.1 The cost of noise

One good for which the hedonic price method works quite well is traffic noise, because people understand what matters to them, namely the annoyance. For the relation between noise and rent or house prices, suitable data are relatively easy to find and the resulting numbers are fairly reliable. As an example, in Table 9.3 we cite values recommended for France by Boiteux (2001); they are based on a review of the international literature and expressed in terms of % reduction in rent as a function of the ambient noise level. Road noise is an externality imposed by drivers on the adjacent homeowners. The following example illustrates how such information can be used.

Example 9.5   Estimate the value of a noise barrier for a suburban highway where the sound level for adjacent houses would be 75 dB in the absence of a barrier. A sound barrier can reduce the level by about 5 dB if its height reaches the line of sight between the source and the ear; higher barriers can achieve up to about 10 dB. Suppose that each house along the highway occupies a site that is 25m wide and their average value is 400,000 € without a noise barrier.

Solution: Assume that Table 9.3 is equally applicable to the owners of houses. On each side of the highway the value of the houses per km is 40 × 400,000 € = 16 M€, and a 5 dB reduction of the noise is worth 5% or 0.8 M€ per km.

Comments:

(1) Cost data indicate that a sound barrier could be built for about 150 €/m^2, hence for a height of 5 m the area would be 5000 m^2 and the cost 0.75 M€. With these numbers the barrier would be worth building, but just barely. However, this calculation omits an important additional benefit of noise reduction, namely a reduction in blood pressure and cardiovascular disease (probably due to noise induced stress), amply demonstrated by epidemiological studies. Since renters and homeowners are not aware of this benefit, it is not reflected in the results of hedonic price studies.

(2) It is interesting to express the cost of noise as external cost per vehicle km. Suppose this highway has two lanes in each direction and the traffic is equivalent to a constant stream at 80 km/h and a distance between vehicles of 80 m.

The time between vehicles is $80 \text{ m}/(80 \text{ km/hr}) = 10^{-3}$ hr, hence there are $365.25 \times 24 \text{ hr}/(10^{-3} \text{ hr}) = 8.766 \times 10^{6}$ vehicles per lane per yr.

With 4 lanes and a length of 1 km, the total number of vehicle km (vkm) is, $4 \text{ km} \times 8.766 \times 10^{6}$ vehicles/yr $= 35 \times 10^{6}$ vkm per yr.

Since the cost of 2×0.8 M€ (for both sides of the highway) is a lifetime cost, we have to express it as an equivalent annual cost. The lifetime cost is the sum of the stream of annual costs, discounted at the discount rate r_{dis}. Houses last a long time, long enough that with a social discount rate $r_{dis} = 4\%$ we can approximate its value by an infinite series of annual values. Thus, according to Eq. (9.8), the equivalent annual cost is r_{dis} 1.6 M€ = 0.064 M€/yr. The cost per vkm is 0.064 M€/$(35 \times 10^{6}$ vkm$) = 0.002$ €/vkm for travel on this highway. Inside cities the population density is much higher and so is the cost of noise. Compared to "quiet" (< 55 dB) the numbers for the other intervals in Table 9.3 have to be added, yielding a total of 0.006 €/vkm.

Some hedonic price studies have tried to evaluate the external costs imposed by an incinerator or landfill on the local population (see e.g. Walton et al., 2006). A clear decrease in house prices has been found in the vicinity of the installation, often up to a distance of about five km. This reflects several so-called amenity costs, in particular odor, truck traffic, visual intrusion, and, for poorly managed sites, flying waste. It may also reflect fear of health impacts (however poorly understood). Whereas the resulting values are indicative of the sites that have been studied, one has to be very careful when trying to apply them to new installations, because there have been major improvements. Also, the values are the total of the amenity impacts, without any possibility of disaggregating the separate contributions of odors, visual intrusion, etc.

9.2.2.2 Value of prevented fatality (VPF)

Revealed preference methods have been used extensively to determine VPF. One approach is based on the relation between wage and risk of dying in different professions. Workers in jobs with higher risk demand higher wages, at least in theory and all else being equal. In spite of the practical difficulties of finding good data that allow the separation of risk from all the other aspects of a job, this approach has been quite successful, and results are used, in particular by governmental agencies in the USA. For a good review and meta-analysis of wage–risk studies, we refer to

Mrozek and Taylor (2002). Another approach is based on consumer purchases of goods that reduce the risk of accidents, for example the purchase of smoke detectors for houses. Here the difficulty lies in the fact that most consumers have no good information on the risk reduction provided by the good. The following hypothetical example illustrates the basic idea as well as some of the practical difficulties.

Example 9.6 Consider the purchase of air bags for cars in the USA before the date when air bags became required equipment. For the purpose of this example, let us suppose that an air bag (driver side only) at the time cost $1000, and that the buyers had this safety information: air bags reduce the risk of dying in a car accident by 10% and the accident rate is 40,000 deaths/yr in a fleet of 200 million passenger vehicles. What can you say about the VPF of car buyers who choose this option?

Solution: The information is not exactly what is needed, because it does not indicate the number of deaths per car. Let us assume that each fatal accident involves only the driver of the car and that the car is not driven again. Then for a car owner who keeps the car for 12 years (average life of a car) the risk of a fatal crash is $12 \times 40,000/200,000,000 = 2.4 \times 10^{-3}$. The air bag avoids therefore 2.4×10^{-3} deaths for $1000, implying a VPF of at least $\$1000/(2.4 \times 10^{-3}$ deaths$) = \$0.42$ million/death for a buyer of such an air bag.

Comments: Revealed preference methods suppose that consumers are rational and well-informed about the attributes of the goods they purchase. In this case that is a dubious assumption, since pertinent information on the benefits of safety products is not usually easy to obtain or even available at all. Furthermore, most people have a very poor understanding of small probabilities. Even if the appropriate safety information were available, the analyst does not know how it is interpreted by the consumer. An additional problem lies in the fact that for a single good with a single price (or a narrow range of prices) this method yields only the percentage of consumers whose VPF is above or below the implied value of the good; to estimate the VPF of the population, distributional assumptions are needed. In practice there are few, if any, consumer products that could be used for a determination of VPF, and almost none of the VPF studies are based on consumer purchases.

As an interesting example of revealed preference methods for VPF, we cite a study by Ashenfelter and Greenstone (2004), who infer VPF from decisions by state governments in the USA about speed limits. It is based on the fact that in 1987 the US Federal Government allowed the states to raise the speed limit on rural interstate roads from 55 mph to 65 mph (88 km/hr to 105 km/hr). Since information on the increase in accidents with speed was widely available, the states that adopted the higher speed limit valued the time saved more than the resulting fatalities. Ashenfelter and Greenstone find that average speeds increased by approximately 4%, or 2.5 mph, and fatality rates by roughly 35% in the states that adopted the higher speed limit. These numbers imply that about 125,000 hours were saved per lost life. Valuing the time saved at the average hourly wage, the estimates imply that these states were willing to accept risks that resulted in a saving of $\$_{1997}1.54$ million per fatality. Since the decision concerned only a single value of the speed increase, the result is only an upper bound for VPF. Another serious limitation of this study is that it is based on data after the introduction of the new speed limit. Really the results reflect the information available to state legislators before the decision. If the legislators thought that the average speed would increase by 10 mph, the implied lower bound of VPF would be four times larger.

9.2.2.3 Travel cost method

Another type of revealed preference method is the travel cost method, suitable for determining the value of a tourist attraction, such as a national park. The basic idea is that the total expenditure to visit the site is a reflection of the WTP. Unfortunately this method is very limited in the type of good that it can evaluate. It can be used for estimating the external costs of amenity losses or gains of reservoirs for hydropower.

9.2.3 Stated preference methods

9.2.3.1 The principle

For many if not most non-market goods, revealed preference methods are either not suitable at all or not sufficiently reliable. Here the alternative is to use so-called stated preference methods, where individuals are asked to state their preferences directly. A classic tool is contingent valuation (CV) (Mitchell and Carson, 1989); over the past two decades, it has enjoyed increasing popularity. The basic idea of a CV is to ask people to state their willingness to pay (WTP) for a certain good if they could buy it, i.e. the maximum amount they would be willing to pay for the good in question.

Example 9.7 As a simple example of a CV, consider a questionnaire that asks, "For your next vacation you have the choice between a plane ticket for $1000 from a regular company with negligible accident risk and a ticket from a charter company where you have a 1/5000 probability of dying in a plane crash. At what price of the charter ticket are you neutral between the two?" Suppose the mean of the answers is $400, what is the implied VPF?

Solution: The WTP is $1000–400 = $600 to avoid 1/5000 death, hence VPF = $600 × 5000 = $3 million per death.

Comment: this is a hypothetical example, for the purpose of illustration only. Real plane accident rates are orders of magnitude lower.

CV questionnaires involve much more than just such a simple question. To be credible, a CV study must take great care to ensure that the respondents fully understand the proposed scenario and find it realistic. The wording is crucial. Careful testing in so-called focus groups is necessary to find out how the respondents interpret the questionnaire. Often their interpretation is quite different from what the investigator intended.

A WTP is not meaningful if the individuals do not understand the nature of the good in question. In particular for health impacts of pollution, some economists have misunderstood what should be measured. Asking people how much they would pay to avoid dioxins makes no sense because the typical respondent cannot calculate the relation between dioxin exposure and health. Instead one should try to determine the value of items people can understand, for instance a fatal cancer, noise, bad odor or poor visibility. After all, the objective of CBA is to evaluate the real costs and benefits, not the fantasies of uninformed people. Once a decision has been made, one reaps the real costs and benefits, not what was imagined.

As a variant of CV, we mention contingent ranking studies, where respondents are asked to choose between different options. For example, in order to investigate people's time preference one could ask at what number x they are neutral between saving 1000 lives today and x lives ten years from now.

In the early nineties, to resolve controversies after the valuation of the costs of the Exxon Valdez oil spill, the US government convened a panel of experts to assess the reliability of CV (Arrow *et al.*, 1993). This panel specified several requirements that a CV study should satisfy and it concluded that the results of well conducted stated preference studies are considered sufficiently reliable for policy applications.

9.2.3.2 Difficulties

The practical implementation of CV is fraught with many difficulties and great care is required to ensure that the results are meaningful. We briefly discuss some of these difficulties. To begin with, there is the hypothetical bias due to the hypothetical nature of the CV. Since the responses are not exposed to the rigor of the market, how can one make sure that the respondents would really spend the stated amounts? In many CVs some respondents do indeed state unrealistically high values. Numerous studies have examined this problem, by giving the respondents the opportunity to actually purchase the good in question (for example a donation for the protection of the environment) at a later time. The conclusion is that this bias can be kept within acceptable limits by (i) obliging the respondents to think carefully about their budget constraints (for instance by asking what expenditures they would reduce to pay for their WTP amount), and (ii) by removing outliers from the analysis. Of course, removing outliers is problematic if one has no clear criteria for doing so; usually the outliers can be identified by looking at the distribution of the WTP responses and by comparing WTP with income. Many researchers use the median rather than the mean, since the median is far less affected by high values of outliers.

In most CV surveys, the mean WTP is much higher than the median WTP, often by a factor of about 1.5 to 2. The median is in effect a voting system where the WTP of each individual is counted only as being above or below a reference value, i.e. the median; this is analogous to typical yes/no choices in democratic elections. By contrast, the mean takes the strength of the vote into account: an individual A whose WTP is twice that of individual B carries twice as much weight in calculating the mean WTP. We argue that the strength of the vote should be taken into account for issues that clearly involve a matter of degree, a frequent situation for environmental policy choices, and thus it seems more appropriate to use the mean, provided the effect of unrealistic outliers has been eliminated.

In many CV studies a significant fraction of the respondents refuse the scenario. For example, when asked for the WTP to reduce the pollution emitted by waste incinerators, quite a few state zero, even though they value clean air; rather they believe that someone else ought to pay (government or owner of incinerator). Such protest zeroes can be distinguished from answers with true WTP = 0 by asking additional questions; for example in the case of pollution emitted by waste incinerators, one would ask whether the WTP = 0 respondents believe that the pollution should be reduced and who they think should pay.

A major problem with stated preference methods is the lack of proportionality between the quantity of the good and the stated WTP. For example, in a CV study of traffic safety (Desaigues and Rabl, 1995), the respondents were told that the number of traffic deaths in France was 10,000 per year (at the time of the study, 1994) and then they were asked how much they were willing to pay for programs that would save 50, 100, 500, 1000, 2000 and 5000 lives per year, respectively. Figure 9.1 displays four typical patterns of responses as a log–log plot of WTP versus number of lives saved. There are two extremes. One is the straight line labeled "constant WTP/life"; these respondents calculate their WTP response by multiplying a constant value per life by the number of lives saved. The other extreme is the horizontal line labeled "lump sum"; these respondents give a fixed value, regardless of the number of lives saved.

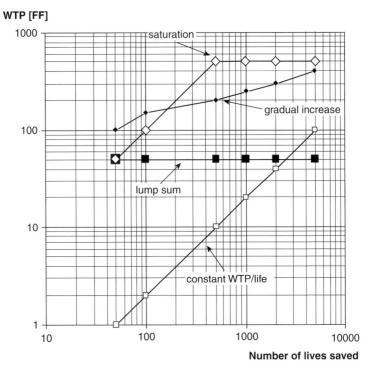

WTP [FF]

Number of lives saved

Fig. 9.1 Variation of individual WTP values as a function of number of lives saved: the four typical patterns. 1 FF = 0.15 €.

In addition to these two extreme patterns, the curve labeled "saturation" represents a very common pattern. These individuals begin with a fairly high WTP relative to their budget when faced with the first question about 50 lives saved, but as the number of lives saved increases from question to question, a point is reached where these respondents seem to say "I can't pay any more." Finally, the curve labeled "gradual increase" shows a pattern intermediate between the "lump sum" and "constant WTP/life"; these respondents appear to increase their WTP in some intuitive manner without performing a calculation, and less than the "constant WTP/life" pattern.

The number of responses for each of the patterns is shown in Table 9.4. 55% of the nonzero responses ("lump sum" and "saturation") are totally inconsistent with the economic model: these individuals in effect assign zero value to an improvement in safety (from 2000 to 5000 lives saved). The average WTP for all respondents increases far less than in proportion to the number of lives saved. It is obviously problematic to extract a single VPF from a data set with such a mixture of different response types.

Table 9.4 *The number of responses for each of the patterns in Fig. 9.1.*

Pattern	Number of responses	% of nonzero responses
"constant WTP/life"	39	4%
"gradual increase"	368	41%
"saturation"	315	36%
"lump sum"	165	19%

The lack of a reasonable relation between stated WTP amounts and scope of the good in question is especially striking in the valuation of ecosystem impacts. When asked to value the preservation of a bird species in a particular nature area, people declare more or less the same amount as for the preservation of all birds or even the preservation of all species in the area. The valuation of ecosystem impacts by CV is problematic.

One of the reasons for the lack of proportionality may lie in the possibility that people perceive the magnitude of their WTPs on a logarithmic rather than a linear scale. Logarithmic scales are well known in the study of perception, as Fechner's law, and an established fact for vision and hearing;[4] they are a necessity if sufficient sensitivity to changes is to be achieved over a very large range of stimuli. It seems plausible that the perception of monetary amounts, or other quantities, may also be more logarithmic than linear. For example, a change from one million to two million does not appear much larger than a change from one hundred thousand to two hundred thousand, if such changes are perceived more in relative than in absolute terms. Perception in relative terms may be especially likely when the good in question is not at all familiar, the typical situation in CV studies.

9.3   Valuation of mortality

There are four main types of mortality for which different monetary values are needed:
(1) accidents
(2) cancers
(3) small reductions in life expectancy
(4) infant mortality
Cancers are discussed below in Section 9.4.6.

---

[4] e.g. http://en.wikipedia.org/wiki/Dynamic range

In the past, VPF values were often based on the human capital approach, which considers a human being in terms of productive value, more or less by setting VPF equal to the total remaining expected lifetime income. The human capital approach considers, in effect, the contribution of an individual to society and represents a collective perspective, in contrast with the individual preferences approach now used in the individualistic societies of North America and Europe. While we follow the individual preferences approach, we note that there is an intrinsic conflict with the preferences of future generations. A society that chooses the individual preferences approach favors investments for the protection of current lives, whereas the human capital approach favors investments in education. The former is an item of consumption by the present generation, the latter an investment for the future. Numerically, a human capital VPF tends to be smaller than one based on individual preferences. The difference between the two approaches is dramatic for the value of a life year lost due to air pollution (VOLY), because much of this impact occurs after retirement.

9.3.1   Accidents

Studies of the value of prevented fatality (VPF) involve accidental deaths that, on average, cause the loss of about half a life span, about 35 to 40 years. The main methods are wage–risk studies and CV. Countless studies have been devoted to measuring this quantity. Already in 1993 a review by Ives et al. (1993) covered 78 VPF studies published between 1973 and 1990. As more recent reviews, we cite Mrozek and Taylor (2002) and Viscusi (2004). Even though the spread of values is very large, there is a general consensus that in Europe and North America, VPF is in the range 1 to 5 M€. In the USA, the EPA uses $6 million, whereas the Department of Transport chose $3 million in 2002; however, more recently the latter was updated to $5.8 million. In Europe, the DG Environment of the European Commission (EC, 2000) recommended values in the range 1 to 1.5 M€, depending on the nature of the death. Since 2000, ExternE has been using 1.5 M€.

However, that mortality valuation by ExternE (2008) appears too low by a factor of about two, in view of a recent systematic review and meta-analysis of all available mortality valuation studies, carried out by Lindhjem et al. (2011) under the auspices of the OECD. Specifically the official report by OECD (2012) recommends a VPF of US$ 3.6 million for the EU27, with a range US$ 1.8–5.4 million.

Even though OECD (2012) did not update the value of a life year (VOLY) because there are too few studies, we believe that the VOLY of ExternE should likewise be doubled, for the following reasons. CV for small changes in life expectancy is very uncertain because of the difficult nature of the valuation question. The results should not be taken too literally. Since the ultimate objective of monetary valuation is to render decision-making more consistent, the VPF/VOLY ratio should remain reasonably consistent. Equation (9.26) below implies a ratio in the range about 20 to 30, and for ExternE (2008), the ratio is 25 to 37.5 with the VOLY of 40,000 €, obtained with the CV described in the following section. We find it appropriate to maintain the VPF/VOLY ratio, and in Chapters 12 to 15 we therefore indicate how the damage costs increase with a doubling of the mortality valuation.

9.3.2 VOLY for small changes in life expectancy

Most public health measures, in particular reductions in air pollution, bring about relatively small increases in life expectancy (LE), on the order of months or at most a few years. For air pollution mortality, VPF is not directly relevant, for two unrelated reasons. On the side of epidemiology, the total number of air pollution deaths cannot be determined, as shown in Section 3.4 of Chapter 3. On the side of valuation, VPF numbers are derived from accidental deaths, very different from deaths due to pollution. Since air pollution deaths tend to involve far fewer years of life lost (YOLL) per death than accidents, consideration of the loss of life expectancy (LE) is appropriate. Thus one needs to know the value of a life year (VOLY). Cancer deaths are also very different from accidents, in terms of suffering and LE loss.

- ExternE had already recognized in 1998 that mortality due to the classical air pollutants had to be evaluated in terms of LE loss rather than number of premature deaths. ExternE (1998 and 2000) calculated VOLY on theoretical grounds by considering VPF as the net present value of a series of discounted annual values. If an accident causes the loss of N years and the discount rate is r_{dis}, VPF and VOLY are thus related by,

$$\text{VPF} = \text{VOLY} \sum_{i=1}^{N} (1 + r_{dis})^{-i}. \tag{9.25}$$

Since different individuals have different survival probabilities, a more realistic equation is,

$$\text{VPF} = \text{VOLY} \sum_{i=1}^{N} p_i (1 + r_{dis})^{-i}, \tag{9.26}$$

where p_i is the conditional probability of surviving another year, conditional upon having survived until year i. The ratio of VPF and the value of a YOLL thus obtained depends on the discount rate; it is typically in the range 20 to 30. However, doubts have been raised as to whether such an accounting approach correctly represents the WTP for an extension of LE. Therefore it would be desirable to have studies that measure VOLY directly.

But in contrast with the vast literature on VPF, there have been few VOLY studies until recently. The NewExt phase (2001–2003) of ExternE used the questionnaire of Krupnick *et al.* (2002) for a contingent valuation study in France, Italy and the UK to determine VOLY, with the result of 50,000 € for a year of life lost due to air pollution. In the UK another CV for VOLY was carried out by DEFRA (2004).

In the UK, Mason *et al.* (2008) provide estimates of VOLY (from £20,000 to £58,000) and QALY (£24,000 to £71,000), based on the VPF (£$_{2005}$ 1.42 million) that has been adopted by several departments of the UK government; the ranges correspond to different choices of models and discount rates, and the values of VOLY and QALY are somewhat different because of different accounting for quality of life.

The NEEDS phase (2004–2008) of ExternE carried out a new CV study to measure VOLY, this time with a questionnaire specifically designed to elicit the WTP, to avoid the loss of life expectancy due to air pollution (Desaigues *et al.*, 2011). This questionnaire was applied in nine countries: France, Spain, UK, Denmark, Germany, Switzerland, Czech Republic, Hungary and Poland, with a total sample size of 1463.

One of the challenges for developing such a questionnaire is to explain the nature of the LE gain: it is not a few months of misery at the end of life, but the avoidance of accelerated aging. There is a gain at all ages, although it is more pronounced for older people. The accelerated aging due to pollution is illustrated schematically in Fig. 9.2. "Ability to survive" expresses the state of youth and health of an individual; it is used only as a qualitative concept, without any need for a quantitative definition. The figure shows the ability to survive as a function of age, for two different levels of air pollution. If the level of air pollution decreases (increases), the survival curve will expand (shrink) in the horizontal direction. LE gains arise as small increases in the probability of survival throughout life due to a slowing of the aging process.

In the light of this graph, it is interesting to reflect on the question of how much an air pollution death shortens life. Discussions on this question have usually focused on the issue of frailty: air pollution deaths occur among individuals who are sufficiently frail because of illness or old age to succumb to an air pollution peak. But in Fig. 9.2, the issue of frailty is irrelevant. We will all die and most of us will be very frail near the end. Air

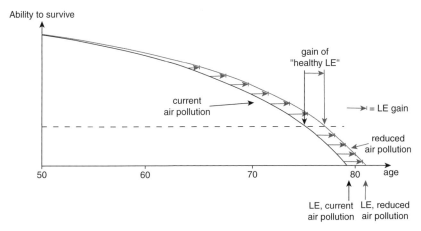

Fig. 9.2 Gain of life expectancy (LE) when air pollution is reduced (example of someone who is aged 50 now). To improve readability the gain is shown larger than is typical of the EU or North America.

pollution does not change that frailty, it merely advances it. What we are valuing really is the quantity of life itself, all else being equal.

Since an expansion of the "ability to survive" curve implies, at any given age, an improvement in health and thus of quality of life, the concept of VOLY also involves a change in the quality of life before death. This is intrinsic in any meaningful definition of VOLY. Whatever the exact shape of the curves in Fig. 9.2, an LE gain implies an upward shift of the "ability to survive" curve, at least near its end. The only exception would be an unrealistic scenario where the curve is extended just along the horizontal axis (and the entire LE gain would be lived hovering at the threshold of death). All the evidence on the effects of air pollution (just like for smoking) indicates that accelerated aging is a good way of describing the effects, and thus Fig. 9.2 is appropriate. For the VOLY questionnaire, we consider a pure stretching of the curve along the horizontal (a pure stretching along the vertical, by contrast, would be a pure change of quality of life).

Here we briefly summarize the key features of the questionnaire. The respondents were given information on average life expectancy in the various countries and Fig. 9.2 (personalized for the age and gender of each respondent) was shown to explain how air pollution affects life expectancy. Attention was also drawn to their own life expectancy, given how old they were at the time of the interview, and factors that affect individual life expectancy such as genetic, behavioral and environmental conditions.

In order to elicit WTP responses, two potential policies to reduce air pollution and hence generate gains in life expectancy were described:

"Policy I (II), will impose a 1.5% (3%) reduction per year in the emission of air pollutants for 20 years, leading to a total reduction of 30% (60%) by 2025. Afterwards the emission of air pollutants will be maintained at this lower level whatever the economic growth. The benefit in terms of life expectancy would be an average increase of 3 (6) months." The interviewer explained that the cost of any measures to reduce air pollution would increase prices and hence their cost of living. Then the respondents were asked how much they would be willing to pay for these policies, as a constant monthly payment for the rest of their life.

To calculate VOLY from the WTP one has to consider how the respondents perceived the payments. Did they see a future payment as having the same nominal value as one made today, or did they implicitly discount the future? During the interview, a special effort was made to explain that the payments must be made for the entire remaining lifetime. Thus, Desaigues *et al.* (2011) argue that the reported WTP are already implicitly discounted and that no further discounting should be applied.

To illustrate the calculation of VOLY, let us take a shortcut and use a single average value for the remaining LE of the respondents, 35.18 yr, and the mean WTP for the entire sample, 24.7 €/month for the 3 month gain. Then we obtain,

VOLY = 4 × 24.7 €/month × 12 months/yr × 35.18 yr = 41,709 €;

(the factor 4 is needed so that WTP and VOLY are for an entire life year).

Desaigues *et al.* obtained several different results for VOLY, depending on the choice of the LE gain (3 or 6 months) and whether the mean or the median WTP was used; in addition, two different formulas were used for relating VOLY to WTP. One of these equations was,

$$VOLY_k = \frac{1}{n_k} \sum_{i=1}^{n_k} (4 \times WTP_{k,i} \times 12) \times \Delta LE_{k,i} \qquad (9.27)$$

for the three-month LE gain, where

n_k = number of respondents in country k,

$WTP_{k,i}$ = WTP of respondent i in country k, and

$\Delta LE_{k,i}$ = remaining life expectancy of respondent i (calculated by means of the life tables of each country k according to the gender and age of respondent i).

As result of the detailed calculations, Desaigues *et al.* obtained a VOLY of 40,000 €$_{2006}$ and this has been applied in the NEEDS and CASES projects of ExternE.

The above mentioned lack of proportionality between WTP and the quantity of the good in question, here the magnitude of the LE gain, is especially striking in a CV to determine VOLY. The ratio of the WTP for six months and the WTP for three months in the study of Desaigues *et al.* was 1.3, not 2 as it should be according to economic theory. In this case, the lack of proportionality is due in part to a budget constraint. Just think about how many years of life most people could afford to buy at 40,000 € per life year. In this study we chose three and six months as a compromise between two conflicting goals: having an LE gain large enough for people to take seriously, yet small enough so the budget constraint is not felt too much.

Some analysts interpret this lack of proportionality as a reflection of diminishing returns, arguing that additional LE gains do indeed have less value. Such an interpretation is patently wrong because WTPs are essentially the same, regardless of the current level of life expectancy. In fact, in most countries, and particularly in the EU, LE has been increasing by about two to three years per decade, yet people do not cease their efforts to increase LE further. Witness for example the heroic efforts most cancer victims make to prolong their lives at any cost.

Furthermore, a literal interpretation of the declining WTP per LE change between three and six months (and similar results in other CV studies) would imply absurd conclusions for CBA. Consider a policy A that increases LE by three months, followed by another policy B that brings about a further increase by three months. Each of these policies would be evaluated according to the WTP for a three-month gain and the combined gain of A followed by B would be valued much higher than a policy C that achieves a six-month gain in one step and is valued according to the WTP for a six-month gain. With such a literal interpretation A followed by B could turn out to look beneficial, whereas C would not, even though it is equivalent.

9.3.3 *Infant mortality*

Infant mortality is one of the impacts of air pollution and therefore monetary values are needed for this end point. The first question is whether the value should be based on VPF or VOLY times YOLL. If, for example, an infant would have died at age five in the absence of pollution, a premature death at age one due to pollution implies that the loss attributable to pollution is only four years. A valuation with a fixed VPF cannot account for this. Thus a VOLY valuation is in principle the correct approach. However, there is a total lack of the necessary information: the loss of life expectancy is extremely difficult, if not impossible, to determine. There are individuals who barely survived an extremely frail infancy and yet end up living to a ripe old age. An air pollution peak might have killed them as an infant, causing a loss of eighty years. Another problem for a VOLY approach is the lack of any VOLY estimates for children. Therefore VPF is a more solid basis.

If the LE loss is a full life, some 80 years, it may indeed be appropriate to take 2 × VPF, since the latter corresponds to accidental deaths with a loss of 35 to 40 years. Valuation at 2 × VPF is the choice made by ExternE (2008). It seems reasonable in view of the central importance of children to families and to society, and it seems to be the majority view among health economists. Parents tend to be much more concerned about their children's safety than about their own, and as a society we tend to provide better protection for children than for adults. In any case, the valuation of infant mortality is more uncertain than that of adults. For further information on the valuation of children's health, we refer to the *Children's Health Valuation Handbook* (EPA, 2003).

9.4   Valuation of morbidity

9.4.1   Cost components

For morbidity costs there are three components to consider:
  (i) Direct costs incurred by the health service plus any other personal out-of-pocket expenditures by an individual (or family).
  (ii) The cost of lost productivity (work loss days or work at reduced capacity).
  (iii) Loss of utility due to discomfort or inconvenience (pain or suffering), restrictions on or reduced enjoyment of desired leisure activities, anxiety about the future, and concern and inconvenience to family members and others.

The first two components can be approximated by market prices for these items. The loss of utility can be measured by CV studies that ask for the WTP or WTA to avoid such a loss. There may be some overlap between components since, for example, an individual may include both financial and non-financial concerns in his/her WTP. Direct health costs are often shared through health insurance and public health care. Since most studies of utility loss do not identify financial concerns separately, it is difficult to avoid some overlap between the cost components. In any case the data, especially the results of CV studies, are quite uncertain, and in view of the uncertainties one may not be able to do better than simply adding the three cost components.

9.4.2   Costs of work loss days

ExternE (2005) estimated the costs of absenteeism using data in a report by the Confederation of British Industry (CBI, 1998). This report is based on a survey conducted by the CBI, covering a wide range of organizations:

45% from manufacturing, 34% from services, 19% from the public sector and 2% from other types. Respondents to the survey were asked to quantify the direct cost of absence, due to the salary costs of absent individuals, replacement costs (i.e. the employment of temporary staff or additional overtime) and lost service or production time. The mean direct cost to business per employee-day absence was found to be 114 €. However, the mean is too high because a small number of employers have very high costs. Considering the structure of the survey, the authors recommended the median, 85 €, as a better indicator of average costs (CBI, 1998, p. 13).

The survey also provides an estimate of the indirect costs of absence, due to lower customer satisfaction and poorer quality of products or services. The CBI report estimates indirect cost at an average of 168 €/ day, although with considerable uncertainty, because of a relatively low response rate for this question in the survey. ExternE (2005) does not include this item, considering it too uncertain.

For a crude alternative estimate of absentee costs, ExternE (2005) has looked at EUROSTAT data on mean annual gross earnings paid to EU employees and divided this by data on the size of the labor force. The result, 56 €, is a lower bound on absentee costs because it includes none of the indirect and not even all of the direct costs.

To derive country-specific estimates, ExternE (2005) scaled the UK numbers of the CBI report in proportion to the country data from EUROSTAT on mean annual gross earnings per EU employee. Where such data were not available, the country purchasing power parity relative to the UK was used to derive appropriate scaling factors. The resulting mean across the EU is 82 €$_{2000}$ for a work loss day. More recently ExternE (2008) has increased this to 295 € per work loss day.

9.4.3   Hospitalization

Data from ExternE (2005) for direct costs of hospital-based health care are presented in Table 9.5; they are derived from Netten and Curtis (2000) and Ready et al. (2004). The hospital costs are generic values, averaged over a wide variety of specialist treatments, for use when information about the nature of the individual's hospital stay is not known. For example, the inpatient unit cost for cardiology is 1.92 times higher than the generic unit cost. The outpatient cost in the UK is significantly higher than in the other countries in this table, perhaps due to a different cost definition. Medical costs have been increasing faster than general inflation and the numbers in Table 9.5 are much too low; we cite them only because they have been used by ExternE (2008) for hospital admissions, see Table 12.3 in Chapter 12.

Table 9.5 *Generic unit costs of hospital health care, in €$_{2000}$. From ExternE (2005).*

Country	Emergency room or outpatient: cost/visit	Hospitalization: cost/inpatient day
Belgium	19	241
France	29	375
Germany	24	321
Italy	20	256
Netherlands	30	390
Spain	27	345
UK	96	330
Mean	**35**	**323**

Using a CV survey, Ready *et al.* (2004) estimated the WTP to avoid a respiratory hospital admission (HA). The questionnaire presented a scenario where the patient stays in the hospital for treatment for three days, followed by five days in bed at home. The mean WTP is 468 € per occurrence. In addition there is a productivity loss of 8 days × 88 €$_{2003}$/day = 704 €$_{2003}$ and the hospital cost of 3 days × 323 €$_{2003}$/day = 969 €$_{2003}$. The total amounts to 2,141 €$_{2003}$ per respiratory hospital admission, or 2,000 €$_{2000}$. Since this estimate is very similar to that derived by Otterström *et al.* (1998) for a general HA (respiratory, cardiac or cerebrovascular), ExternE uses this common value for any HA.

Ready *et al.* (2004) also determine the WTP to avoid a respiratory emergency room visit (ERV). The scenario in the questionnaire describes a visit to a hospital emergency room, with oxygen and medication to assist breathing, followed by five days in bed at home. The mean WTP for the five countries in the study is 242 €. Adding the productivity loss, 5 days × 88 €$_{2003}$/day = 440 €$_{2003}$, and the medical costs of an ERV, 35 €$_{2003}$, one obtains the total cost of an ERV as 717 €$_{2003}$ or 670 €$_{2000}$.

9.4.4   Chronic bronchitis

The valuation of chronic bronchitis (CB) is particularly important because this end point is one of the largest contributors to the damage costs of ExternE. For the WTP to avoid CB there only seem to be two studies, Viscusi *et al.* (1991) and Krupnick and Cropper (1992). To interpret their results, one needs to take into account the severity of different CB cases. Some cases are mild and clear up after a few years, others are truly debilitating and long lasting or even permanent. The questionnaire of Viscusi *et al.* described severe cases and was addressed to the general population. Krupnick and Cropper used a modified version of the questionnaire of Viscusi *et al.*, but addressed only to individuals who knew someone with CB. Thus the results of Krupnick and Cropper reflect a typical distribution of different severity levels. ExternE relies more on the study of Krupnick and Cropper because direct experience of knowing someone with CB gives a much better understanding than a mere description of symptoms in the questionnaire of Viscusi *et al.*

In applying the results, one should make sure that the assumptions about severity levels are consistent between ERFs, background rates and monetary valuation. The ERF used by ExternE is based on Abbey *et al.* (1995) who considered mild cases of CB. Nonetheless it is most plausible to assume that the relative risk (RR) of Abbey *et al.* is approximately the same for all severity levels, because quite generally the change in RR per $\mu g/m^3$ is not very different, even between totally different end points. Data for incidence rates are based on an average of severity levels. Therefore, the assumptions about severity levels are consistent to the extent that the distribution of severity levels in the study of Krupnick and Cropper is typical.

Krupnick and Cropper provide several estimates and extraction of an appropriate WTP to avoid CB is not straightforward. ExternE uses the estimate of $0.4 million which is based on a trade-off between CB and the risk of dying, combined with a VPF of $2 million (used by Krupnick and Cropper to convert the risk–risk trade-off to monetary values). Multiplying $0.4 million by the ratio of 1.0 million €$_{2000}$ (the VPF of the 2004 version of ExternE) and $ 2 million, ExternE (2005) obtains a WTP of 0.2 million €$_{2000}$ for avoiding a case of CB. Incidentally, the externality study of the US EPA (Abt 2000 and 2004) assumes a WTP of $0.33 million for CB, also based on Krupnick and Cropper, albeit with different assumptions.

In view of the uncertainty of such estimates, it would be desirable to have an independent check. It is therefore interesting to look at DALY and QALY scores for CB to obtain complementary information for the valuation of CB. The "Catalog of Preference Scores" at the Cost Effectiveness Analysis (CEA) Registry of Tufts-New England Medical Center (www. tufts-nemc.org/cearegistry/index.html) shows QALY scores for various conditions of CB, ranging from 0.68 to 0.79. Assuming that the monetary value of a QALY is identical to a VOLY, the forgone value of a year lived with CB, compared with a life year in normal health, is in the range 0.21 to 0.32 VOLY. Of course, the uncertainties of QALY scores are very large, see for instance the review by Arnesen and Trommald (2004). Mathers *et al.* (2003) show a DALY score of 0.266 for chronic obstructive pulmonary disease (COPD). COPD comprises several severe respiratory conditions, but mostly CB. A DALY is roughly equivalent to 1 – QALY, although their precise definitions involve differences such as discounting (for DALY but not QALY). In view of these numbers we assume that, on average, a year lived with CB corresponds to the loss of 0.26 VOLY.

To obtain the total cost of a case of CB, the annual cost is to be multiplied by the discounted duration. The onset of CB is typically at around the age of fifty, and the condition is difficult or impossible to cure, especially if it is advanced (see e.g. Priez and Jeanrenaud, 1999). Symptoms can be alleviated but the prognosis is poor. We have not been able to find sufficient data for a firm estimate, to say nothing about the large differences in severity between different cases. However, it seems reasonable to assume a discounted duration of about 20 years. This implies a monetary value of $20 \times 0.26 = 5.2$ VOLY for CB. With a VOLY = 40,000 €, the monetary value of chronic bronchitis would therefore be 208,000 €. We conclude that the current ExternE (2008) value of 200,000 € for a case of CB is plausible.

9.4.5   Other end points

Visit to a doctor for lower respiratory symptoms or asthma
For the medical cost of a visit to a general practitioner in the UK, Netten and Curtis (2000) indicate values between 25 € (for a consultation of 9.4 minutes) and 42 € (for a consultation of 12.6 minutes). Considering the latter more realistic, ExternE (2008) recommends a value of 42 €, to be added to the WTP.

For lower respiratory symptoms, the CV scenario of Ready et al. (2004) is described as a persistent phlegm cough occurring every half-hour or so, and lasting one day. The corresponding WTP was found to be 38 €. Together with the medical cost of a doctor visit of 42 €, the total for lower respiratory symptoms is 80 €$_{2003}$ or 75 €$_{2000}$.

Ready et al. (2004) also report that the WTP to avoid a day of asthma is 67 €, 139 € and 295 € for adult non-asthmatics, adult asthmatics and children, respectively. But when asked to value one additional day of asthma in addition to a 14-day asthma episode, the WTP was only 14 €, 15 € and 42 € respectively. As a central unit value 139 € for an asthma episode and 15 €/day for each additional day are recommended.

New cases of asthma
Note that the latter end point involves asthma related visits to a doctor, rather than new cases of asthma. For new cases, ExternE (2005) cites work in the UK (www.hse.gov.uk/ria/chemical/asthma.htm) that estimated the cost of a new case of asthma at between £42,000 and

£45,000, including loss of income through absence from work or having to change jobs; medical treatment; and pain and suffering. For the EU, ExternE (2008) recommends a value of 60,000 € per new case of asthma.

Restricted activity days (RAD)

For RADs, Ready *et al.* (2004) use a scenario of three days in bed, with shortness of breath on slight exertion. They find a WTP of 148 €$_{2003}$. Since this is for three days, the WTP to avoid a RAD is 49 €$_{2003}$. Adding a productivity loss of 88 €$_{2003}$/day, the total cost of a RAD is 137 €$_{2003}$ or 130 €$_{2000}$. For minor restricted activity days (MRAD), ExternE (2008) uses 38 €.

Respiratory medication use by children and adults

Respiratory medication for asthma involves the use of bronchodilators. The costs of bronchodilator drugs range from 0.5 to 1 € per day, according to whether Terbutaline or Albuterol is used. ExternE (2005) recommends using 1 € per day.

Symptom day and minor restricted activity day

Ready *et al.* (2004) also report that the WTP to avoid one cough day is 41 €. Converting to year 2000 prices, this is 38 €$_{2000}$. ExternE also uses this value for symptom days and for minor restricted activity days.

9.4.6 *Cancers*

There are many different cancers and their severity, in terms of suffering, cost and life expectancy reduction, can vary enormously from case to case. For example, most cases of skin cancer, if detected early enough, are treated successfully during a doctor's visit. Lung cancer, in contrast, cannot yet be detected early enough and is fatal in about 90% of patients (with the customary definition of classifying a cancer as non-fatal if the patient survives at least five years after detection). Of course, valuation for application to environmental policy is based on typical or average cases.

For non-fatal cancers. the monetary value is the sum of the cost of illness (COI), the loss of income during the illness, and the WTP to avoid the suffering. However, the available information is not clear. ExternE (1998) used a value of 450,000 € for the cost of a non-fatal cancer, based on a survey of American data. This includes COI and lost earnings and an adjustment for the WTP to avoid the suffering (by assuming a ratio WTP/COI of 1.5 for non-fatal cancers). European data indicate much lower COI than in the USA. For example, Borella *et al.* (2002) state that the total cost of cancers in France is 10 billion €/yr for treatment and 15 billion

€/yr including lost productivity. Since the incidence of cancers is 240,000 new cases/year in France, this implies a cost per case of approximately 42,000 €/cancer for treatment, and 63,000 €/cancer including lost productivity.

For the WTP to avoid fatal cancers there are two different approaches. One is a lump sum, essentially VPF, possibly increased by a premium beyond an accident-based VPF, because cancers are feared as an especially dreadful form of death. Typical estimates of the cancer premium are on the order of 50% of VPF. However, there is much uncertainty because many CV studies have not detected a significant cancer premium. ExternE (2004) assumed a total cost of 2 M€/cancer death.

The other approach is to multiply the YOLL due to the cancer by VOLY (including a correction for discounting. The loss of life per cancer death is in the range of about 6 to 15 years (depending on the type of cancer), intermediate between accidents and air pollution; for more detailed data, see Table 9.6, based on Bachmann (2006). However even without discounting, the resulting value for a fatal cancer would be very low, 500,000 € for a loss of 12.5 years with VOLY = 40,000 €, whereas VPF is at least twice as large. Such an approach is problematic because it applies to cancer deaths, a VOLY that has been obtained by means of a questionnaire based on the very short loss of life due to air pollution (where losses of 3 and 6 months were proposed), and it does not account for the loss of quality of life during the illness. For these reasons, we find a valuation based on VPF plus cancer premium more appropriate than one based on YOLL and VOLY. For fatal cancers, ExternE (2008) assumes an average LE loss of 15.95 YOLL, valued at 40,000 € per YOLL. Together with a COI of 481,050 € the result is 15.95 × 40,000€ + 481,050 € = 1.12 million €.

Whatever the valuation of a cancer, one should account for discounting because of the lag between emission and exposure and the lag between exposure and development of cancer.

Table 9.6 *Years of life lost (YOLL) per world-average cancer death, for selected cancers. From Till Bachmann, personal communication.*

Type of cancer	YOLL due to mortality	YOLL-equivalents due to morbidity
Skin cancer	6.09	0.19
Lung cancer	15.95	0.26
Average cancer	12.5	0.3

9.4.7 *End points without monetary valuation*

Many morbidity end points, for instance kidney disease, are not fatal but impair the quality of life. For most morbidity end points there are few direct estimates of monetary values, or none at all. CV studies are expensive and cannot evaluate more than a few end points at a time. An even greater difficulty lies in the ground rule for valuation studies, namely that they should be based on the preferences of individuals familiar with the good in question. Few have sufficient knowledge and experience to offer a meaningful valuation of the countless possible health end points, each with a wide range of possible severity levels. People do not have a mental catalog of values for the innumerable goods among which they might have to choose; rather, values tend to be formed while thinking about the goods. The challenge for CV studies is to provide sufficient information to give the respondents an adequate appreciation of the impact in question. But how can a verbal description (in a paragraph, short enough to be read during the interview) ever convey the suffering of a terminal cancer to someone who has never been close to a patient with such a condition?

As an alternative, two indicators have been developed for measuring both the quality and the quantity of life lived, as a means of quantifying the benefit of medical or public health interventions. One is the QALY (quality adjusted life years), the other the DALY (disability adjusted life years). For QALY, each year is assigned a value between 1 for perfect health and 0 for death, whereas for DALY, death corresponds to 1 and perfect health to 0. The scores are for an entire year lived with the respective condition. To apply the scores, one also needs information on the duration of the condition. DALYs for a disease are the sum of the equivalent years lived with disability (YLD) and, in the case of premature mortality, of the years of life lost due to premature mortality (YOLL).

The DALY, developed by the World Health Organization (www.who.int) (see also, Murray and Acharya (1997)), takes into account two aspects that affect the contribution of an individual to the economy: age and discounting. A loss in the future is discounted relative to one now, and the loss of a young or old person counts less than a loss at midlife. QALY, by contrast, does not discount and treats all ages equally. Probably the most complete listing of QALY weights is the "Catalog of Preference Scores," formerly at the Center for Risk Analysis of Harvard University, now at the Cost Effectiveness Analysis (CEA) Registry of Tufts-New England Medical Center (CEA 2006). DALY scores have been published by Mathers *et al.* (2003), and a recent very thorough update by Salomon *et al.* (2012).

QALY and DALY scores have been developed by health care professionals for a wide variety of end points. Even though that does not satisfy the requirement that the valuation be made by the concerned individuals themselves, it seems to be the only viable option for most end points. Health care professionals do have deep knowledge and experience with a wide range of impacts, and most also have the necessary empathy to make a valid assessment. Furthermore, some of their valuations are based on questionnaires filled out by the patients themselves. For the sake of illustration we have selected a small subset of QALY and DALY scores, in Table 9.7. The scores are evolving and for any serious work the reader should consult the latest results.

If explicit data for specific end points cannot be found, Pennington *et al.* (2002) suggest the categories in Table 9.8 as a first rough approximation (actually, the YOLL-equivalents in this table have been corrected by Bachmann (2006) and are 1.9 times larger than the original values of Pennington *et al.*).

If there were a generally accepted monetary value of a DALY or QALY, one could quantify the costs of most morbidity end points. Health professionals have tended to avoid economic considerations in decisions about treatment choices; only in recent years has there been a recognition of the need for guidelines about the value of a DALY or QALY, but there is still no official consensus of an appropriate monetary value and many health experts remain opposed to monetary valuation (e.g. the Institute of Medicine's Committee to Evaluate Measures of Health Benefits for Environmental, Health, and Safety Regulation (Miller *et al.*, 2006)). Furthermore, the underlying assumptions of DALYs and QALYs are not entirely consistent with those of a VOLY. There are also unresolved questions about variation with age, number of years of life at stake, etc. Nonetheless, as an interim solution it seems reasonable to set the value of a DALY or QALY equal to a VOLY of about 40,000 $€_{2006}$. In Section 9.4.4 we have already illustrated the use of DALY and QALY scores for the valuation of chronic bronchitis.

9.5   Neurotoxic impacts

There is a wide variety of possible neurotoxic impacts, for instance loss of memory or reduced sensitivity of limbs or senses, but for most of these no monetary values are available. However, loss of IQ points is a good general indicator of neurotoxic damage, and fairly reliable estimates for the associated costs can be found in the literature, mostly based on lost earnings and, in severe cases, the cost of remedial education. The basic approach was developed by Schwartz (1994), who estimated that an incremental one IQ point in cognitive ability would raise annual earnings by approximately 1.8%. Subsequently, Salkever (1995) calculated that a 1-point IQ difference is associated with a roughly 2.4% difference in earnings, and

Table 9.7 *A small subset of QALY and DALY scores.*

(a) *selected DALY scores for region Euro A of WHO (Mathers* et al. *2003)*

Cause	Sequelae[a]	DALY
Lower respiratory infections	Episodes	0.278
Upper respiratory infections	Episodes	0.241
	Pharyngitis	0.070
Hypertensive heart disease	Cases	0.201
Ischaemic heart disease	Acute myocardial infarction	0.405
	Angina pectoris	0.108
	Congestive heart failure	0.186
Chronic obstructive pulmonary disease	Symptomatic cases	0.266
Asthma	Cases	0.036
Nephritis and nephrosis	Acute glomerulonephritis	0.099
	End-stage renal disease	0.098

[a] sequela = a condition which is the consequence of a previous disease or injury.
Euro A: Andorra, Austria, Belgium, Croatia, Czech Republic, Cyprus, Denmark, Finland, France, Germany, Greece, Iceland, Ireland, Israel, Italy, Luxembourg, Malta, Monaco, Netherlands, Norway, Portugal, San Marino, Slovenia, Spain, Sweden, Switzerland, United Kingdom

(b) *selected DALY scores for cancers (Mathers* et al. *2003)*

Site	Diagnosis/therapy	Waiting	Metastasis	Terminal
Mouth and oropharynx	0.09	0.09	0.75	0.81
Oesophagus	0.20	0.20	0.75	0.81
Stomach	0.20	0.20	0.75	0.81
Colon and rectum	0.20	0.20	0.75	0.81
Liver	0.20	0.20	0.75	0.81
Trachea bronchus and lung	0.15	0.15	0.75	0.81
Melanoma and other skin	0.05	0.05	0.75	0.81
Breast	0.09	0.09	0.75	0.81
Prostate	0.13	0.13	0.75	0.81
Leukaemia	0.09	0.09	0.75	0.81

(c) *QALY scores (CEA 2006, accessed July 2006).*

Health State	QALY score
Lung cancer	0.52 to 0.65
Colon cancer	0.63 to 0.74
Chronic Bronchitis	0.68 to 0.79
Hypertension	0.8 to 1
Myocardial infarction	0.7
Congestive Heart Failure	0.5 to 0.71

Table 9.8 *International Life Sciences Institute classification scheme for human health impact categories and YOLL-equivalents per end point (adapted from Pennington et al., 2002 and the literature cited therein, with corrected values for YOLL-equivalents.)*

Criteria	Category 1 Irreversible / Life-shortening effects	Category 2 Probably irreversible / Life-shortening effects	Category 3 Reversible / Non life-shortening effects
Examples	Cancer Reproductive effects Teratogenic effects (birth defects) Acute fatal or acute severe and irreversible effects (e.g. fatal poisoning) Mutagenicity	Immunotoxicity Neurotoxicity Nephrotoxicity (kidney damage) Hepatotoxicity (liver damage) Pulmonary toxicity (lung damage) Cardiotoxicity (heart damage)	Irritation (eye, skin, mucosal; that is, transient) Sensitization (allergy) Reversible acute organ or system effects (gastrointestinal inflammation)
Weight	1	0.1	0.01
YOLL-equivalents	12.8	1.28	0.128

this estimate has been used in regulatory analyses in the USA. There are of course uncertainties in these percentages, and the resulting cost of an IQ point depends on additional assumptions, in particular the discount rate and the future growth of earnings. Furthermore, people like to be smart and value a high IQ not only for the earnings potential; however, such additional benefits are difficult to quantify. In practice, most studies use loss of IQ points and earnings potential and/or remedial education as proxy for the real cost of neurotoxic impacts, even though it is far from ideal.

For numerical results we cite these estimates:

- Muir and Zegarac (2001) $\$_{1999}14,700$/IQpoint,
- Grosse *et al.* (2002) $\$_{2000}14,500$/IQ point,
- Rice and Hammitt (2005) $\$_{2000}16,500$/IQpoint,
- Trasande *et al.* (2005) $\$_{2000}22,200$/IQpoint, and
- Griffiths *et al.* (2007) $\$_{2000}11,245$/IQpoint.

Adjusting these figures to $\$_{2005}$ by means of the CPI (consumer price index), we obtain a mean of about $\$_{2005}18,000$/IQpoint. To apply this cost in different countries one can modify the cost in proportion to the GDP_{PPP}/capita, the per capita GDP adjusted for purchase power parity (www.cia.gov/cia/publications/factbook/). Thus Spadaro and Rabl (2008b) find a worldwide average cost of $3,890/IQpoint.

In the following example, we calculate the value of an IQpoint, using as a basis the GDP per capita, rather than an average salary, because we believe that GDP per capita is a better indicator of the general productivity. Salary does not take into account items such as unemployment, profit, and investment activities. Also, many non-market activities contribute to well-being and wealth and can benefit from higher intelligence, for instance reading about healthy lifestyles or smart purchase decisions, the instruction that parents can give their children at home, etc.

Example 9.8　Estimate the value of an IQpoint in the EU15, given that the average per capita GDP in 2010 was 28,400 €, and assuming that a difference of 1 is associated with a 2.4% difference in earnings.

Solution:　The present value P of the lifetime earnings are calculated as a discounted series of annual values A = 28,400 €, over a period of activity for which we assume N = 40 years. Let us assume a growth rate of r_{gro} = 1%/yr for GDP per capita and a discount rate r_{dis} = 5%/yr. Then P is given by,

$$P = A(1 + r_{gro})/(1 + r_{dis}) + A(1 + r_{gro})^2/(1 + r_{dis})^2$$
$$+ \ldots A(1 + r_{gro})^N/(1 + r_{dis})^N. \tag{9.28}$$

If we approximate $(1 + r_{gro})/(1 + r_{dis})$ by $1/(1 + r_{eff})$ with $r_{eff} = r_{dis} - r_{gro}$, this series has the same form as Eq. (9.13) which we used for the derivation of the capital recovery factor of $(A/P,r,N)$ of Eq. (9.15). Thus we obtain the present value P of the lifetime earnings,

$$P = 28,400 \text{\euro}/(A/P, r_{eff}, N). \tag{9.29}$$

The result for P is 562,115 € and the loss per IQ point is 13,491 €. Since the parameters are very uncertain, let us vary them one at a time. We find,

18,645 € if r_{dis} = 3% instead of 5%,
15,755 € if r_{gro} = 2% instead of 1%,
8,432 € if earnings difference = 1.5% instead of 2.5% per IQpoint.

9.6   Transfer of values

9.6.1   Methods for transfer of values

Sometimes values are needed in a different country from the one in which they have been determined. Obviously they need to be adjusted for inflation and converted to the local currency (see Section 9.6.2). To convert the currency, purchase power parity (PPP) is a better indicator of the real value than the actual exchange rate. PPP rates also have the advantage of being more stable from year to year.

But there is also the question of whether the underlying preferences are the same or whether additional adjustments are called for. Economists have carried out much work to develop and test procedures for the transfer of monetary values from one country to another. Needless to say, transferring the results of WTP studies to other countries involves large uncertainties, especially if the standard of living is very different.

There are three main approaches to transfer (Navrud, 2004).
1. Unit value transfer, with or without adjustments
2. Function transfer of the benefit function.
3. Meta analysis.
Unit transfer of e.g. an estimate of mean WTP/household/year, is the simplest approach. It can be used when one can assume that the WTP is the same at the study site (where the primary valuation study was conducted) as at the policy site (which is the site we want to transfer the value to). Value estimates are then transferred directly, *without* adjustments. This is often termed "naïve (unit) value transfer." Between countries with different income levels and costs of living one usually uses unit transfer

with adjustments. This is done by using purchase power parities (PPP),[5] i.e. rates of currency conversion that eliminate the differences in price levels between countries.

When there are large differences in the income levels between the study and policy sites (e.g. from EU15 to the new member states of EU27), the adjusted benefit estimate Bp' at the policy site can be calculated as,

$$B_p' = B_s (Y_p/Y_s)^\beta, \qquad (9.30)$$

where B_s is the original primary benefit estimate from the study site, Y_s and Y_p are the income levels at the study and policy site, respectively, and β is the income elasticity of demand for the environmental good in question. There is, however, little empirical evidence on how the demand for different environmental goods and health impacts varies with income. There is also no clear connection between income elasticity of WTP and income elasticity of demand. However, in most CV studies one observes WTP at the study site, and income elasticity of WTP at the study site, rather than income elasticity of demand. This information can be used to estimate WTP at the policy site using Eq. (9.30) (by substituting in Eq. (9.30), WTP for B and the income elasticity of WTP for β).

The default assumption in adjusting WTP values in proportion with some measure of income is that the income elasticity of WTP is 1.0. However, Krupnick *et al.* (1996) note in their benefit transfer exercise for impacts of air pollution in Central and Eastern Europe (CEE), that there is no reason to think that WTP for environmental quality varies proportionally with income. They note that empirical evidence from the USA shows that the premature mortality risk (from e.g. air pollution) is an inferior good (defined as a good with an income elasticity < 1.0). The default approach will then understate the WTP of lower income countries (as shown by the following example). Thus, Krupnick *et al.* use an income elasticity of 0.35 (with 1.0 as a sensitivity analysis) when transferring mortality values from the USA to CEE. Alberini and Krupnick (2003) conclude from their value transfer comparison, that assuming an income elasticity of WTP of 1.0, or even making other adjustments, does not appear to be reliable for valuing morbidity and mortality risks in developing countries. As part of the CV for mortality valuation, carried out by Desaigues *et al.* (2011) in nine countries of the EU25, the income elasticity of WTP was found to be less than 1 in all models and countries, and typically in the range 0.4–0.7.

---

[5] See e.g. http://www.oecd.org/dataoecd/32/34/2078177.pdf

Example 9.9   To evaluate the damage costs of air pollution in China one needs the value of a life year (VOLY) in China. Lacking any direct information, one can transfer the EU VOLY of 40,000 € per life year.

Solution: According to World Bank data, as cited by http://en.wikipedia.org/wiki/List_of_countries_by_GDP_(PPP)_per_capita (accessed 7 July 2011), the GDP per capita of China is about $7,500 in terms of PPP equivalent, whereas that of France is $33,800. We take the latter as proxy for the EU because this database has no entry for the EU. Using Eq. (9.30), the Chinese VOLY is

40,000 €$_{2000}$ ($7500/$33,800)$^\beta$
The result is 23,616 €$_{2000}$ for $\beta = 0.35$
and 13,943 €$_{2000}$ for $\beta = 0.7$.
and 8,876 €$_{2000}$ for $\beta = 1.0$.

Lacking any additional information, we might choose 14,000 €$_{2000}$ as the central value, with an uncertainty range from 8900 to 23,600 €$_{2000}$.

Function transfer might seem more appealing than unit transfer, because more information can be taken into account in the transfer. For a CV study, the benefit function can be written as:

$$WTP_{ij} = b_0 + b_1 G_j + b_2 H_{ij} + e, \qquad (9.31)$$

where WTP_{ij} = the willingness to pay of household i at site j, G_j = the set of characteristics of the environmental good at site j, and H_{ij} = the set of characteristics of household i at site j, and b_0, b_1 and b_2 are sets of parameters and e is the random error.

To implement this approach, the analyst would have to find a primary study with estimates of the constant b_0 and the sets of parameters b_1 and b_2. Then the analyst would have to collect data on the two groups of independent variables G and H, at the policy site, insert their mean values in Eq. (9.31), and calculate households' WTP at the policy site.

The main problem with the benefit function approach is due to the exclusion of relevant variables in the WTP function estimated in a single study. When the estimation is based on observations from a single study of one or a small number of recreational sites, or a particular change in environmental quality, a lack of variation in some of the independent variables usually prohibits inclusion of these variables.

In meta-analyses, several original studies are analyzed as a group. Results from each study are used as single observations for a regression analysis. The resulting regression function, together with collected data on the independent variables in the model that describes the policy site, is used to construct an adjusted unit value (Navrud, 2004). The transfer approaches are summarized in Table 9.9.

Table 9.9 *Transfer approaches and their applicability*

Transfer approach	When to apply	Comments
1. Unit value transfer, with or without income adjustment (from a primary valuation study of a similar type site).	Can be used if the study and policy sites are considered to be very close in all respects.	Recommended as the simplest and most transparent way of transfer both within and between countries. Just as reliable as function transfer and meta analysis.
2. Function transfer (from a primary valuation study of a similar type site).	Can be used if the value functions have high explanatory power, and contain variables for which data are readily available at the policy site.	Low explanatory power of WTP functions of stated preference studies can result in larger rather than smaller transfer errors compared with unit value transfer.
3. Meta analysis (i.e. function transfer based on a meta-regression function of many primary valuation studies).	Can be used when many valuation studies exist for an impact. Should be limited in scope, i.e. including valuation studies using as similar a valuation methodology as possible (as many meta analyses seem to be dominated by methodological choices of the primary studies they consider).	Methodological choice, rather than the characteristics of the site, substitute sites and the affected population, has a large explanatory power. Errors can be reduced by limiting the scope of the meta analyses.

In the EU, values for mortality have been determined only in certain member countries and the question arises of what value to use in other member countries. For ExternE, a decision was taken right at the beginning to use a single value for mortality for all countries of the EU. While this choice could well be criticized on grounds of strict economic considerations, it was wise for political reasons. As we have been disseminating the results of ExternE, nobody has questioned that choice – but one can well imagine the angry reaction of Greeks or Portuguese if they saw that their valuation was lower. Such controversies did in fact arise when IPCC tried to calculate mortality costs for global warming during the mid 1990s, using lower values in developing countries. Since then IPCC has shied away from any monetary valuation of mortality.

9.6.2   Conversion of currencies

In transferring values from other countries one needs to convert the currency, and usually one also needs to adjust for inflation. There are several possible ways of doing this. For currency conversion one can use ordinary exchange rates or purchasing power parity (PPP) exchange rates. The latter are more appropriate because they reflect what a good in one country is worth in terms of the purchases a consumer would normally make. When adjusting for inflation, there are also two choices: adjust for inflation in the country of origin and then convert, or else first convert and then adjust for inflation. A priori the choice is not clear. However, when making several conversions, for different time periods or different countries, it is preferable for the sake of consistency to first adjust for inflation and then make all the currency conversions in the same year.

It is interesting to take as an example the conversion of 1 $\$_{2000}$ to €$_{2010}$. The required data are shown in part (a) of Table 9.10. Even though for OECD member countries such data can be found at the OECD website, we also had to consult other sources because of gaps or errors in the OECD data (no data for 2000 and erroneous data for inflation). For inflation in the Euro zone, Eurostat uses an index called HICP (harmonized index of consumer prices) and lists several choices, depending on the countries taken into account. Here we have chosen HICP for 16 countries that use the Euro. In the USA, inflation is measured by the CPI (consumer price index); it is normalized to have the value 100 in the period 1982–1984. Part (b) of Table 9.10 shows the results of the four methods. With ordinary exchange rates the results are very different, 0.96 €$_{2010}$/$\$_{2000}$ for one order and 1.33 €$_{2010}$/$\$_{2000}$ for the other. With PPP exchange rates the results are much closer,

Table 9.10 *Conversion of 1 $\$_{2000}$ to $€_{2010}$. The required data are in part (a), the results in part (b).*

(a) *Data*

	US CPI[a]	€ zone (16 countries) HICP[b]	Exchange rate € per $\$^c$	PPP exchange rate € per $\$^c$
2000	172.2	89.6	1.08[d]	0.87[e]
2001	177.1	91.7	1.12	0.87
2002	179.9	93.8	1.06	0.87
2003	184.0	95.8	0.89	0.87
2004	188.9	97.9	0.81	0.87
2005	195.3	100.0	0.81	0.86
2006	201.6	102.2	0.80	0.83
2007	207.3	104.4	0.73	0.82
2008	215.3	107.8	0.68	0.81
2009	214.5	108.1	0.72	0.80
2010	218.1	109.9	0.76	0.81
2011	224.9	112.9	0.72	0.80
2012	229.6	115.7	0.78	0.81

[a] ftp://ftp.bls.gov/pub/special.requests/cpi/cpiai.txt
[b] http://epp.eurostat.ec.europa.eu/tgm/table.do?tab=table&init=1&language=en&pcode=tec00027&plugin=1
[c] http://stats.oecd.org/Index.aspx?datasetcode=SNA_TABLE4
[d] www.federalreserve.gov/releases/H10/hist/dat00_eu.htm
[e] our guess since the rate is essentially constant 2001–2004

(b) *Results of four methods (in bold face).*

Convert $\$\to€$ in 2000, then HICP	Convert $\$\to€$ PPP in 2000, then HICP
1.08 $€_{2000}/\$_{2000}$	0.87 $€_{2000}/\$_{2000}$
1.23 $€_{2010}/€_{2000}$	1.23 $€_{2010}/€_{2000}$
1.33 $€_{2010}/\$_{2000}$	**1.07 $€_{2010}/\$_{2000}$**

CPI, then convert $\$\to€$ in 2010	CPI, then convert $\$\to€$ PPP in 2010
1.27 $\$_{2010}/\$_{2000}$	1.27 $\$_{2010}/\$_{2000}$
0.76 $€_{2010}/\$_{2010}$	0.81 $€_{2010}/\$_{2010}$
0.96 $€_{2010}/\$_{2000}$	**1.02 $€_{2010}/\$_{2000}$**

1.02 $€_{2010}/\$_{2000}$ for one order and 1.07 $€_{2010}/\$_{2000}$ for the other; the latter is preferred because it uses one consistent exchange rate when converting several values. At the present time (2012 and beginning of 2013) the $\$/€$ exchange rate is about $ 1.28/€.

402   Monetary valuation

References

Abbey, D. E., Lebowitz, M. D., Mills, P. K. *et al.* (1995) Long-term ambient concentrations of particulates and oxidants and development of chronic disease in a cohort of non-smoking Californian residents. *Inhalation Toxicology* 7: 19–34.

Abt 2000. *The Particulate-Related Health Benefits of Reducing Power Plant Emissions.* October 2000. Prepared for EPA by Abt Associates Inc., 4800 Montgomery Lane, Bethesda, MD 20814-5341.

Abt 2004. *Power Plant Emissions: Particulate Matter-Related Health Damages and the Benefits of Alternative Emission Reduction Scenarios.* Prepared for EPA by Abt Associates Inc. 4800 Montgomery Lane. Bethesda, MD 20814-5341.

Alberini, A. and Krupnick, A. 2003. Valuing the health effects of pollution. In, Tietenberg, T. and Folmer, H., *The International Yearbook of Environmental and Resource Economics 2003/2004. A Survey of Current Issues.* Cheltenham, UK. Edward Edgar, 233–277.

Arnesen, T. and Trommald, M. 2004. Roughly right or precisely wrong? Systematic review of quality-of-life weights elicited with the time trade-off method. *Journal of Health Services Research & Polic,* 9(1): 43–50.

Arrow, K., Solow, R., Leamer, E. *et al.* 1993. Report of the NOAA Panel on Contingent Valuation, Federal Register, 58, n°10.

Ashenfelter, O. and Greenstone, M. 2004. Using mandated speed limits to measure the value of a statistical life, *Journal of Political Economy* 112 (S1): S226-S267.

Bachmann, T. M. 2006. *Hazardous substances and human health: exposure, impact and external cost assessment at the European scale.* Elsevier, Amsterdam.

Boiteux, M. 2001. *Transports: choix des investissements et coût des nuisances.* Commissariat General du Plan, Paris.

Borella, L., Finkel, S., Crapeau, N. *et al.* 2002. Volume et coût de la prise en charge hospitalière du cancer en France en 1999. *Bull. Cancer* 89 (9): 809–821.

CBI 1998. *Missing Out: 1998 Absence and Labour Turnover Survey,* London: Confederation of British Industry (CBI)

CEA 2006. Catalog of Preference Scores. Cost Effectiveness Analysis (CEA) Registry of Tufts-New England Medical Center. Downloaded 2 July 2006 from www.tufts-nemc.org/cearegistry/index.html

Cline, W. (5 January 2008). Comments on the Stern Review. Peter G. Peterson Institute for International Economics. Retrieved 20 May 2009.

DEFRA 2004. Chilton, S., Covey, J., Jones-Lee, M., Loomes, G. and Metcalf, H. *Valuation of health benefits associated with reductions in air pollution.* Department for Environment, Food and Rural Affairs. London.

Desaigues, B, Ami, D., Bartczak, A. *et al.* 2011. Economic Valuation of Air Pollution Mortality: A 9-country contingent valuation survey of value of a life year (VOLY). *Ecological Indicators* 11 (3): 902–910. For more detail see also, by the same authors, *Final report on the monetary valuation of mortality and morbidity risks from air pollution,* Framework VI Research Programme (Project no: 502687 'New Energy Externalities Developments for Sustainability' [NEEDS]).

Desaigues, B. and Rabl, A. 1995. Reference values for human life: an econometric analysis of a contingent valuation in France. In Nathalie Schwab & Nils Soguel, editors (1995) *Contingent Valuation, Transport Safety and Value of Life*, Kluwer, Boston. This paper can also be downloaded from www.arirabl.org

EC 2000. *Recommended Interim Values for the Value of Preventing a Fatality in DG Environment Cost Benefit Analysis*. Recommendations by DG Environment, based on a workshop for experts held in Brussels on November 13th 2000.

EPA 2003. *Children's Health Valuation Handbook*. United States Office of Children's Health Protection EPA 100-R-03-003 Environmental Protection Office of Policy, Economics, and Innovation Agency National Center for Environmental Economics Health Valuation. October 2003.

ExternE 1998. ExternE: Externalities of Energy. Vol.7: Methodology 1998 Update (EUR 19083); Vol.8: Global Warming (EUR 18836); Vol.9: Fuel Cycles for Emerging and End-Use Technologies, Transport and Waste (EUR 18887); Vol.10: National Implementation (EUR 18528). Published by European Commission, Directorate-General XII, Science Research and Development. Office for Official Publications of the European Communities, L-2920 Luxembourg.

ExternE 2000. External Costs of Energy Conversion - Improvement of the ExternE Methodology and Assessment of Energy-Related Transport Externalities. Final Report for Contract JOS3-CT97-0015, published as *Environmental External Costs of Transport*. R. Friedrich & P. Bickel, editors. Springer Verlag Heidelberg 2001.

ExternE 2004. Final report for Project NewExt *New Elements for the Assessment of External Costs from Energy Technologies*, European Commission DG Research, Contract No. ENG1-CT2000-00129, coordinated by R. Friedrich, IER, University of Stuttgart; and final report for project ExternE-Pol *Externalities of Energy: Extension of accounting framework and Policy Applications*, European Commission DG Research, Contract No. N° ENG1-CT2002-00609. EC DG Research, coordinated by A. Rabl. Available at www.externe.info and at www.arirabl.org

ExternE 2005. *ExternE: Externalities of Energy, Methodology 2005 Update*. Edited by P. Bickel and R. Friedrich. Published by the European Commission, Directorate-General for Research, Sustainable Energy Systems. Luxembourg: Office for Official Publications of the European Communities. ISBN 92-79-00429-9.

ExternE 2008. With this reference we cite the methodology and results of the NEEDS (2004–2008) and CASES (2006–2008) phases of ExternE. For the damage costs per kg of pollutant and per kWh of electricity, we cite the numbers in the data CD that is included in the book edited by Markandya, A., Bigano, A. and Porchia, R. in 2010: *The Social Cost of Electricity: Scenarios and Policy Implications*. Edward Elgar Publishing Ltd, Cheltenham, UK. They can also be downloaded from www.feem-project.net/cases/ (although in the latter, some numbers have changed since the data CD in the book).

Gollier, C. and Weitzman, M. L. 2010. How should the distant future be discounted when discount rates are uncertain?, *Economics Letters*, Elsevier, **107**(3): 350–353, June.

Graves, P. 2013. *Environmental Economics: An Integrated Approach*. CRC Press/ Taylor & Francis, Boca Raton, FL 33487.

Griffiths, C., McGartland, A. and Miller, M. 2007. A comparison of the monetized impact of IQ decrements from mercury emissions. *Environ Health Perspect.* 115(6): 841–847.

Grosse, S. D., Matte, T. D., Schwartz, J. and Jackson, R. 2002. Economic gains resulting from the reduction in children's exposure to lead in the United States. *Environmental Health Perspectives* 110(6): 563–569.

Harrison, M. 2010. Valuing the Future: the social discount rate in cost-benefit analysis. Visiting Researcher Paper, Productivity Commission, Canberra. Media and Publications, Productivity Commission, Locked Bag 2 Collins Street East, Melbourne VIC 8003. Email: maps@pc.gov.au. www.pc.gov.au/research/visiting-researcher/cost-benefit-discount

Ives, D. P., Kemp, R. V. and Thieme, M. 1993. *The Statistical Value of Life and Safety Investment Research.* Environmental Risk Assessment Unit, University of East Anglia, Norwich, Report n°13 February 1993.

Krupnick, A. and Cropper, M. 1992. The effect of information on health risk valuation. *Journal of Risk and Uncertainty* 5: 29–48.

Krupnick, A., Alberini, A., Cropper, M. *et al.* 2002. Age, health, and the willingness to pay for mortality risk reductions: A contingent valuation survey of Ontario residents. *J Risk and Uncertainty* 24(2): 161–186.

Krupnick, A., Harrison, K., Nickell E. and Toman, M. 1996. Value of health benefits from ambient air quality improvements in Central and Eastern Europe, an exercise in benefits transfer, *Environmental and Resource Economics* 7: 307–332.

Lindjhen, H., Navrud, S., Braathen, N. A. *et al.* 2011. Valuing mortality risk reductions from environmental, transport and health policies: A global meta-analysis of stated preference studies. *Risk Analysis,* 31(9): 1381–1407.

Mason, H., Jones-Lee, M. and Donaldson, C. 2008. Modelling the monetary value of a QALY: A New Approach Based on UK Data. *Health Econ* 18 (8): 933–950.

Mathers, C. D., Bernard, C., Iburg, K. *et al.* 2003. The Global Burden of Disease in 2002: data sources, methods and results. Geneva, World Health Organization (GPE Discussion Paper No. 54). Downloaded 16 July 2006 from www.who.int/healthinfo/boddalysmphreferences/en/index.html.

Miller, W., Robinson, L. A., Lawrence, R. S., 2006. Valuing health for regulatory cost-effectiveness analysis. Institute of Medicine's Committee to Evaluate Measures of Health Benefits for Environmental, Health, and Safety Regulation. p. 382. Online publication at: www.nap.edu/catalog.php?record_id=11534#toc.

Mitchell, R. C., and Carson, R.T. 1989. Using Surveys to Value Public Goods: the Contingent Valuation Method. Resources for the Future. Washington, DC.

Mrozek, J. R., and Taylor, L. O. 2002. What determines the value of life? A meta-analysis. *Journal of Policy Analysis and Management* 21(2): 253–270.

Muir, T. and Zegarac, M. 2001. Societal costs of exposure to toxic substances: economic and health costs of four case studies that are candidates for environmental causation. *Environmental Health Perspectives* 109, Supplement 6 (December): 885–903.

Murray, C. J. L., and Acharya, A. K. 1997. Understanding DALYs. *Journal of Health Economics* 16(6): 703–730.

Navrud, S. 2004. Value Transfer and Environmental Policy. I: T. and H. Folmer (red.) *The International Yearbook of Environmental and Resource Economics 2004/*

2005. A Survey of Current Issues. New Horizons in Environmental Economics Series. Edward Elgar Publishing, Cheltenham, UK and Northampton, MA, US, 189–217.

Netten, A. and Curtis, L. 2000. *Unit Costs of Health and Social Care 2000*. Personal Social Services Research Unit (PSSRU). www.pssru.ac.uk/index.htm.

OECD 2012. *Mortality Risk Valuation in Environment, Health and Transport Policies*, OECD Publishing. http://dx.doi.org/10.1787/9789264130807-en

Otterstrom, T., Gynther, L. and Vesa, P. 1998. *The willingness to pay for better air quality*, Ekono Energy Ltd.

Pennington, D., Crettaz, P., Tauxe, A. *et al*. 2002. Assessing human health response in life cycle assessment using ED10s and DALYs: part 2 – noncancer effects. *Risk Analysis* 22 (5): 947–963.

Priez, F. and Jeanrenaud, C. 1999. Human costs of chronic bronchitis in Switzerland. *Swiss J. Economics and Statistics* 135(3): 287–301.

Rabl, A. 1996. Discounting of long term costs: what would future generations prefer us to do? *Ecological Economics* 17: 137–145.

Ready, R., Navrud S., Day B. *et al*. 2004. Benefit transfer in Europe: How reliable are transfers across countries?, *Environmental & Resource Economics* 29: 67–82.

Rice, G. and Hammitt, J. K. 2005. *Economic Valuation of Human Health Benefits of Controlling Mercury Emissions from US Coal-Fired Power Plants*. Northeast States for Coordinated Air Use Management (NESCAUM). Boston, MA. February 2005.

Salkever, D. S. 1995. Updated estimates of earnings benefits from reduced exposure of children to environmental lead. *Environmental Research* 70: 1–6.

Salomon, J. A., Vos, T., Hogan, D. R., Gagnon, M. *et al*. 2012. Common values in assessing health outcomes from disease and injury: disability weights measurement study for the Global Burden of Disease Study 2010. *Lancet* 380: 2129–2143.

Schwartz, J. 1994. Societal benefits of reducing lead exposure. *Environmental Research* 66: 105–124.

South African Department of Environmental Affairs and Tourism. 2004. *Cost Benefit Analysis*. Integrated Environmental Management Information Series 8, Pretoria, http://www.deat.gov.za

Spackman, M. 2006. *Social Discount Rates for the European Union: An Overview*. Università degli Studi di Milano, Dipartimento di Scienze Economiche, Aziendali e Statistiche, Working Paper no. 2006–33, October.

Spadaro, J. V. and Rabl, A. 2008b. Global health impacts and costs due to mercury emissions. *Risk Analysis* 28 (3): 603–613.

Stern, N. *et al*. 2006. *The Economics of Climate Change: The Stern Review*, Cambridge University Press, Cambridge, UK. Available at www.hm-treasury. gov.uk/stern_review_report.htm

Tol, R. S. J. 2005. The marginal damage costs of carbon dioxide emissions: an assessment of the uncertainties. *Energy Policy* 33: 2064–2074.

Trasande, L., Landrigan, P. J. and Schechter, C. 2005. Public health and economic consequences of methyl mercury toxicity to the developing brain. *Environmental Health Perspectives* 113(5): 590–596.

Treasury Board of Canada 2007. Canadian Cost-Benefit Analysis Guide Regulatory Proposals, www.tbs-sct.gc.ca.

UK 2008. *Intergenerational wealth transfers and social discounting: Supplementary Green Book guidance.* HM Treasury, 1 Horse Guards Road, London SW1A 2HQ. Available at http://hm-treasury.gov.uk

Van Ewijk, C. and Tang, P. 2003. How to price the risk of public investment?, *De Economist* **151**, no. 3: 317–328.

Viscusi, W. K., Magat, W. A. and Huber, J. 1991. Pricing environmental health risks: Survey assessments of risk-risk and risk-dollar tradeoffs for chronic bronchitis. *Journal of Environmental Economics and Management* **21**(1): 32–51.

Viscusi, W. K. 2004. The value of life: Estimates with risks by occupation and industry. *Economic Inquiry* **42**.1: 29–48. http://www.law.harvard.edu/faculty/viscusi/pubs/245_2004_EI-42-1.pdf.

Walton, H., Boyd, R., Taylor, T. and Markandya, A. 2006. *Explaining Variation in Amenity Costs of Landfill: Meta-Analysis and Benefit Transfer.* Presented at the Third World Congress of Environmental and Resource Economists, Kyoto, July 3rd-7th 2006.

Weitzman, M. L. 1998. Why the far distant future should be discounted at its lowest possible rate, *Journal of Environmental Economics and Management* **36**: 201–208.

Weitzman, M. L. 2001. Gamma Discounting, *American Economic Review* **91**, no. 1, March: 260–271.

Zhuang, J., Liang, Z., Lin, T. and De Guzman, F. 2007. *Theory and Practice in the Choice of Social Discount Rate for Cost–Benefit Analysis: A Survey*, ERD Working Paper No. 94, Asia Development Bank, May.

10   The costs of climate change

Summary

After a brief explanation of the greenhouse effect, we present some data from the 2007 assessment by the IPCC (2007a), the principal international body that is working on climate change. These data show the main anthropogenic contributions to climate change, as well as the increases in global average temperature and sea level that have been occurring since the industrial revolution. Since the impacts depend on cumulative emissions and involve long time constants, one needs to define emission scenarios before one can estimate the corresponding impacts, a topic addressed in Section 10.2. We then describe, in Section 10.3, the impacts that can be expected and discuss some of the difficulties in estimating the corresponding damage costs. In Section 10.4 we review damage cost estimates in the literature. It is also of interest to look at abatement costs, see Section 10.5. Finally, we discuss some of the implications of a CO_2 tax in the light of emission reductions required to stabilize the climate at acceptable levels.

10.1   Greenhouse gases (GHG) and their effects: some data

Climate change is a vast subject and we cannot do it justice with a single chapter. Here we merely give an introduction to the problem of estimating the damage costs of GHG.

That anthropogenic emissions of CO_2 would increase global temperatures had been recognized at the end of the nineteenth century, when the great chemist Arrhenius attempted a first estimate of the temperature increase that could be expected if the atmospheric CO_2 concentration doubles relative to the pre-industrial level: he found that the average temperature at the surface of the earth would increase by about 5 to 6 K (Weart, 2008), not very far from current estimates, generally around 2.5 K.

Indeed, the basic mechanism is simple and clear: atmospheric CO_2 and other greenhouse gases absorb some of the infrared radiation emitted

407

from the earth's surface, radiation that would otherwise escape to outer space. Thus the radiative heat loss from the earth's surface is reduced and the earth gets warmer.[1] The effect is very much like putting an extra blanket on one's bed. For a given source of heat (the person in the bed), the temperature in the bed increases when the heat loss is reduced. However, the climate system is far more complex and a change in CO_2 induces countless feedback loops, some positive, some negative. In particular, water in the atmosphere plays a crucial role. With increasing temperature there is more evaporation, which has two opposite effects. On the one hand, water vapor is a powerful greenhouse gas, on the other hand, when it condenses as clouds it reduces the input of solar radiation. Whereas the net effect of adding greenhouse gases to the atmosphere is necessarily an increase in temperature, the magnitude of the increase is difficult to calculate precisely, and even with the most sophisticated climate models currently available, significant uncertainties remain. More detailed effects, such as the geographic and seasonal distributions of changes in temperature and precipitation, are even more uncertain. However, enough is known to be able to say that not only will the continued release of GHGs (greenhouse gases) at current rates lead to a warming of the planet, but some of the impacts will be severe. In view of the inability to deal with a number of crises in recent decades (e.g. the Ethiopian famines), it seems that mankind is not well equipped to deal with a system undergoing dramatic change.

We begin with some graphs from the 2007 Assessment by the IPCC (Intergovernmental Panel on Climate Change). Figure 10.1 shows the contribution of the principal greenhouse gases to global warming, all expressed as radiative forcing in W/m^2. Among the long-lived gases, CO_2 is the largest anthropogenic contribution, but CH_4, N_2O and halocarbons are also important. Whereas the halocarbons (F-gases in Fig. 10.2) are mostly a legacy of the past, N_2O from nitrogen fertilizer is still increasing, as can be seen from Fig. 10.2.

It is often convenient to combine the contributions of all the different GHGs by using CO_2 equivalents. The CO_2-equivalent emission of a long-lived GHG or mixture of GHGs is the amount of CO_2 emission that would cause the same time-integrated radiative forcing, over a given time horizon. The equivalent CO_2 emission of a GHG is obtained by multiplying its emission by its global warming potential (GWP) for the given time horizon. GWP values are shown in Table 10.1. IPCC uses the 100 yr time horizon. Likewise the CO_2-equivalent concentration of a GHG is the concentration of CO_2 that would cause the same amount of

---

[1] It is called greenhouse effect because the CO_2 acts like the cover of a greenhouse.

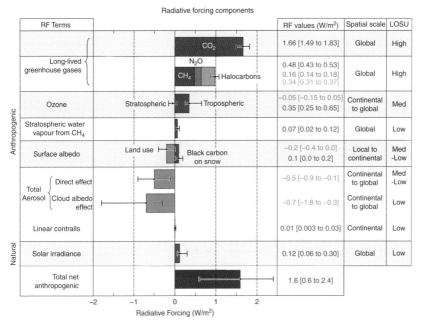

Fig. 10.1 Global average radiative forcing (RF) in 2005 (best estimates and 5 to 95% uncertainty ranges) with respect to 1750 for CO_2, CH_4, N_2O and other important agents and mechanisms, together with the typical geographical extent (spatial scale) of the forcing and the assessed level of scientific understanding (LOSU). Aerosols from explosive volcanic eruptions contribute an additional episodic cooling term for a few years following an eruption. The range for linear contrails does not include other possible effects of aviation on cloudiness. From IPCC (2007b Figure SPM.2).

radiative forcing (attention: sometimes CO_2-equivalent concentrations include only GHGs, sometimes also aerosols). CO_2 equivalent is a standard and useful metric for comparing emissions and concentrations of different GHGs, but it can be misleading because the different time constants imply different climate impacts. For example, the discounting involved in damage cost calculations is not included in GWP and therefore it is not quite correct to estimate the damage costs of other GHGs by multiplying that of CO_2 by the respective GWP.

An especially troubling aspect of climate change lies in the long time constants of the various processes that are involved. The most important greenhouse gases, CO_2, CH_4 and N_2O, remain in the atmosphere for many years; the time constants for their removal are 12 yr for CH_4, 114 yr for N_2O,

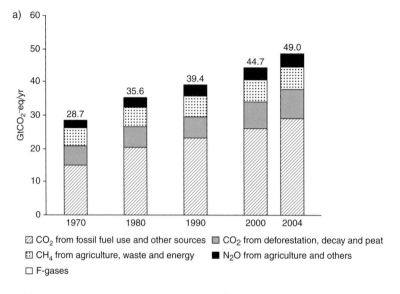

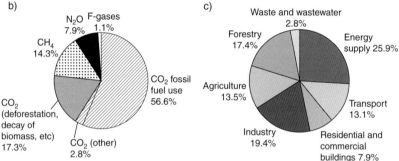

Fig. 10.2 Global anthropogenic GHG emissions, from IPCC (2007a, Figure 2.1).
(a) Global annual emissions of anthropogenic GHGs from 1970 to 2004 (contribution of F-gases is too small to be visible).
(b) Share of different anthropogenic GHGs in total emissions in 2004 in terms of carbon dioxide equivalents ($CO_{2\text{-eq}}$).
(c) Share of different sectors in total anthropogenic GHG emissions in 2004 in terms of $CO_{2\text{-eq}}$. (Forestry includes deforestation.)

and for CO_2 there are several different removal processes with time constants from years to centuries. Halocarbons also have long time constants. Furthermore, the full warming impact of the CO_2 emitted today will not be felt until several centuries from now, because of the large thermal inertia of the oceans. Of course, a rational policy for greenhouse gas emissions should be based on a consideration of costs and benefits. But the long time

Table 10.1 *GWP values and lifetimes. From 2007 IPCC (2007a).*

	Lifetime (years)	Time horizon		
		20 years	100 years	500 years
Methane	12	72	25	7.6
Nitrous oxide	114	289	298	153
HFC-23 (hydrofluorocarbon)	270	12,000	14,800	12,200
HFC-134a (hydrofluorocarbon)	14	3,830	1,430	435
Sulfur hexafluoride	3200	16,300	22,800	32,600

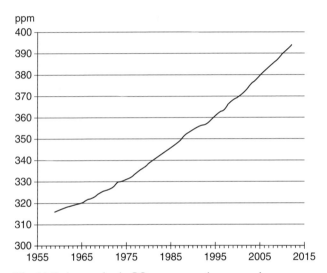

Fig. 10.3 Atmospheric CO_2 concentration, annual averages measured at Mauna Loa Observatory, Hawaii. Downloaded 1 June 2013 from ftp:// ftp.cmdl.noaa.gov/ccg/co2/trends/co2_annmean_mlo.txt

constants, coupled with the difficulties and uncertainties of estimating the costs of climate change, render the challenge of choosing an appropriate policy for greenhouse gas emissions extremely difficult and controversial. All the other pollutants are trivial by comparison.

The increase in atmospheric CO_2 concentration is not easy to calculate because of the role of the oceans in absorbing part of the emission. But the data, monitored since the late 1950s and shown here in Fig. 10.3, show a dramatic trend. The present concentration is close to 400 ppm, compared with the pre-industrial level of 280 ppm.

At the same time, the global average temperature and the sea level have been rising, as shown in Fig. 10.4. The temperature rise has been

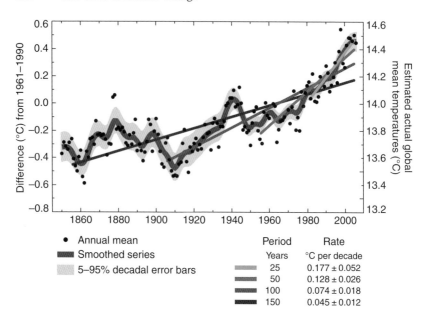

Period	Rate
Years	°C per decade
25	0.177 ± 0.052
50	0.128 ± 0.026
100	0.074 ± 0.018
150	0.045 ± 0.012

- Annual mean
- Smoothed series
- 5–95% decadal error bars

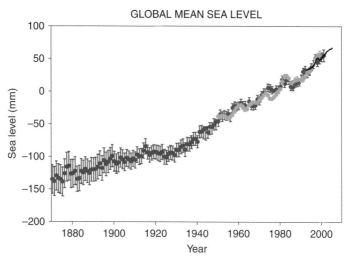

Fig. 10.4 Observed changes in
(a) global average surface temperature;
(b) global average sea level.
All differences are relative to corresponding averages for the period 1961–1990.
From: www.ipcc.ch/pdf/assessment-report/ar4/wg1/ar4-wg1-ts.pdf
part (a) Figure TS.6, part (b) Figure TS.18.

accelerating. The sea level rises because of two effects, the thermal expansion of water and the melting of glaciers, principally in Greenland and Antarctica. Climate models explain the temperature rise as a consequence of greenhouse gas emissions, as can be seen from Fig. 10.5, which compares model predictions with and without anthropogenic greenhouse gas emissions. Of course there are doubters, especially among industries

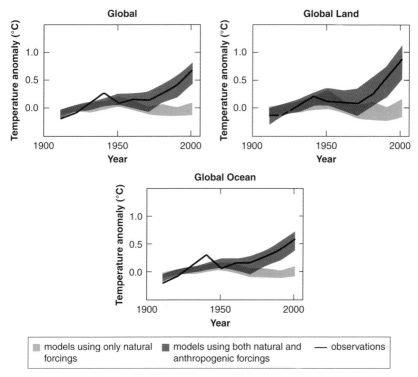

Fig. 10.5 Comparison of observed continental- and global-scale changes in surface temperature with results simulated by climate models using either natural or both natural and anthropogenic forcings. The term "temperature anomaly" refers to the difference to the long run average. Decadal averages of observations are shown for the period 1906–2005 (black line) plotted against the center of the decade and relative to the corresponding average for 1901–1950. Lines are dashed where spatial coverage is less than 50%. Blue (lighter) shaded bands show the 5 to 95% range for 19 simulations from five climate models using only the natural forcings due to solar activity and volcanoes. Pink (darker) shaded bands show the 5 to 95% range for 58 simulations from 14 climate models using both natural and anthropogenic forcings.
From IPCC (2007a, syn Figure 2.5).

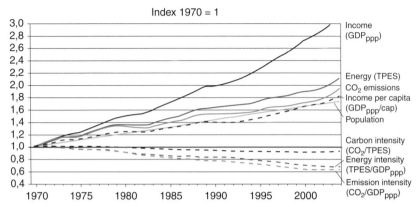

Fig. 10.6 The evolution of several indicators between 1970 and 2008, in relative units normalized to unity in 1970: World gross domestic product measured in PPP (GDP$_{ppp}$), Total Primary Energy Supply (TPES), CO_2 emissions (from fossil fuel burning, gas flaring and cement manufacturing) and Population. The dotted lines show Income per capita, Energy Intensity (TPES/GDP$_{ppp}$), Carbon Intensity of energy supply (CO_2/TPES), and Emission Intensity of the economic production process (CO_2/GDP$_{ppp}$). (From Fig.1.5 of WG3 IPCC, 2007b)

that do not want to be constrained by greenhouse gas regulations. But the general impression of a causal link is difficult to avoid given observed data and our knowledge of the mechanisms involved.

It is instructive to look at some of the driving forces behind the emissions. Figure 10.6 plots the evolution of several interesting indicators between 1970 and 2008, in relative units normalized to unity in 1970. The solid lines show data for emissions, income, energy use and population, the dashed lines show some ratios. During this period the emissions increase at more or less the same rate as the population: the emission per capita increases only about ten percent. That could change dramatically as developing countries that have had mostly stagnating economies during that period move towards the living standard of the rich nations. To avoid a dangerous build-up of greenhouse gases, a serious worldwide effort is required to reduce the emission intensity per unit of GDP, far more than the trend of the last 30 years.

10.2   Scenarios

To predict future emissions and impacts one needs to make assumptions about the driving forces, especially the evolution of the population, the

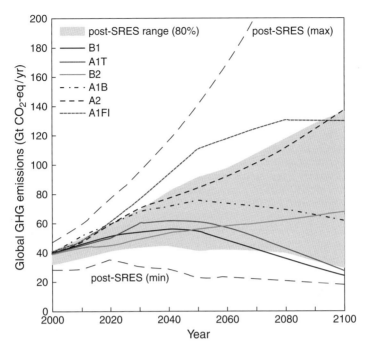

Fig. 10.7 SRES (2000) scenarios for greenhouse emissions in the absence of new climate policies.
From IPCC (2007a, Figure 3.1).

growth of the economy, and the technologies that will be used. Obviously, one cannot predict, but one can formulate scenarios and evaluate the consequences. For that purpose IPCC, in a report called Special Report on Emissions Scenarios (SRES, 2000), has developed a set of standard scenarios that span the range of plausible futures. They are plotted in Fig. 10.7, together with an indication of the range of additional scenarios that have been published since then.

The SRES scenarios do not include additional climate policies beyond those in effect in 2000. They are grouped into four scenario families (A1, A2, B1 and B2) that cover a wide range of demographic, economic and technological driving forces and resulting GHG emissions:

The A1 scenarios assume a world where the global population peaks in mid-century and economic growth is very strong, with rapid introduction of new and more efficient technologies. There is convergence among regions, with increased cultural and social interactions and substantial reduction of regional differences in per capita income. A1 is divided into three groups with alternative development of energy technologies:

A1FI   = fossil intensive,
A1T   = non-fossil energy resources,
A1B   = a balance across all sources.

The B1 scenarios assume a convergent world, with the same global population as A1, but more rapid changes in economic structures toward a service and information economy; the emphasis is on global solutions to economic, social and environmental sustainability.

B2 assumes a world with intermediate economic growth, emphasizing local solutions to economic, social, and environmental sustainability; population continues to grow, although more slowly than in A2.

A2 assumes a very heterogeneous world with high population growth, preservation of local identities, regional fragmentation, slow economic development and slow technological change.

The post-SRES scenarios tend to imply somewhat lower emissions.

The implications of several scenarios for temperature rise are shown in Fig. 10.8. Even if the concentrations could be held constant at the level of 2000, the temperature would keep increasing because of the long time constants of the climate processes.

In the long run the temperature must not be allowed to increase indefinitely. For the design of appropriate policies it is therefore helpful to look at scenarios where the atmosphere eventually reaches stable conditions. Many such scenarios have been developed and a summary is presented in

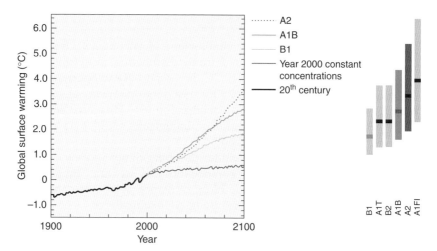

Fig. 10.8 Global temperature rise above the average level of 1980–1999 for several SRES (2000) scenarios. The bars at the right indicate the best estimate and the likely range.
From IPCC (2007, syn Figure 3.2).

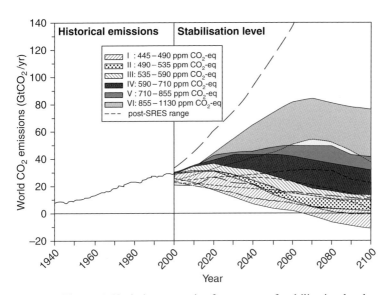

Fig. 10.9 Emission scenarios for a range of stabilization levels.
From IPCC (2007b, Synthesis rep Figure 5.1a).

Fig. 10.9. The scenarios are grouped in bands corresponding with the greenhouse gas concentrations that will ultimately be reached. For example, if the concentrations are not to exceed 590 ppm CO_{2eq}, the emissions must be reduced below the levels of 2000 by 2050 at the latest.

The increase in global average temperature above pre-industrial level that will ultimately result from these scenarios is indicated in Fig. 10.10. The central line is the best estimate, the other lines indicate an uncertainty range. For example, if the temperature increase according to the best estimate is not to exceed 3°C, the stabilization level must not be above 590 ppm; it needs to be below 445 ppm if the true temperature increase corresponds to the upper limit in Fig. 10.10.

More detailed characteristics of these stabilization scenarios are presented in Table 5.1 of IPCC (2007a), reproduced here in Table 10.2. In 2005 the CO_2 concentration was 379 ppm; the CO_{2eq} concentration was 445 ppm if all long-lived greenhouse gases are included. The ranges correspond to the 15th and 85th percentiles of the scenario distribution. The best estimate of the climate sensitivity, defined as the increase for a doubling of atmospheric CO_2, is 3°C. For the majority of scenarios, the GHG concentrations reach stable levels between 2100 and 2150, but because of inertia of the climate system, the global average

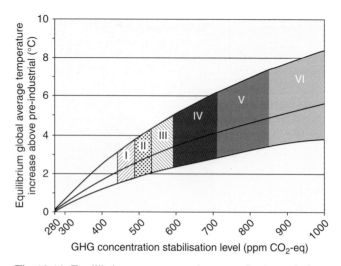

Fig. 10.10 Equilibrium temperature increases for the emission scenarios of Fig. 9.9.
From IPCC (2007b Synthesis rep Figure 5.1b).

temperature at equilibrium is different from expected global average temperature at the time of stabilization of the concentrations. The entries for sea level include only thermal expansion, not the melting of ice. Note that because of the long time constants the equilibrium sea level will only be reached after many centuries.

10.3   The challenge of estimating the damage costs of climate change

10.3.1 Predicting the impacts

There is a wide variety of different impacts, many of which are indicated in Fig. 10.11 as a function of the temperature rise where they become significant. Not mentioned in this figure is the change in energy consumption for heating and cooling of buildings; of all the impact categories the one whose costs involve the least uncertainties. One should note that not all impacts are harmful. For example, heating loads will decrease and agricultural production in cold zones is likely to improve. However, the overall balance of impacts is forecast to be very negative.

In Table 10.3, we indicate the key difficulties encountered in estimation of the damage costs of the various impact categories. In some areas,

Table 10.2 *Characteristics of the stabilization scenarios of Fig. 10.10 from IPCC (2007a).*

Category	CO_2 concentration at stabilization	CO_2-equivalent concentration at stabilization[a]	Peaking year for CO_2 emissions	Change in global CO_2 emissions in 2050 (% of 2000 emissions)	Global average temperature increase above pre-industrial at equilibrium	Sea level rise above pre-industrial at equilibrium[b]	Number of assessed scenarios
	ppm	ppm	Year	%	°C	meters	
I	350–400	445–490	2000–2015	−85 to −50	2.0–2.4	0.4–1.4	6
II	400–440	490–535	2000–2020	−60 to −30	2.4–2.8	0.5–1.7	18
III	440–485	535–590	2010–2030	−30 to +5	2.8–3.2	0.6–1.9	21
IV	485–570	590–710	2020–2060	+10 to +60	3.2–4.0	0.6–2.4	118
V	570–660	710–855	2050–2080	+25 to +85	4.0–4.9	0.8–2.9	9
VI	660–790	855–1130	2060–2090	+90 to +140	4.9–6.1	1.0–3.7	5

[a] including GHGs and aerosols
[b] from thermal expansion only

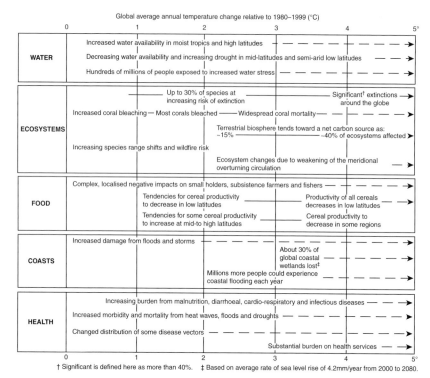

Fig. 10.11 Examples of impacts associated with global average temperature change. Illustrative examples of global impacts projected for climate changes (and sea level and atmospheric CO_2 where relevant) associated with different amounts of increase in global average surface temperature in the twenty-first century. The black lines link impacts; broken-line arrows indicate impacts continuing with increasing temperature. Entries are placed so that the left-hand end of the text indicates the approximate level of warming that is associated with the onset of a given impact. Quantitative entries for water scarcity and flooding represent the additional impacts of climate change relative to the conditions projected across the range of SRES scenarios A1FI, A2, B1 and B2. Adaptation to climate change is not included in these estimations. From IPCC (2007 synthesis Figure 3.6).

scientific understanding is not yet sufficiently advanced and unpleasant surprises may turn up. For example, the absorption of CO_2 by the oceans increases their acidity, with disastrous consequences for aquatic ecosystems, and the severity of this problem is only beginning to be recognized (Hardt and Safina, 2010, Fabry *et al.*, 2008). The table does not mention several fundamental difficulties that affect all categories: uncertainties

Table 10.3 *Impact categories of climate change and key difficulties encountered in the estimation of their costs. The last column shows our personal assessment of the uncertainties. Some of the impacts depend not only on the absolute level of ΔT but also on the rate of change.*

Category	Key difficulties	Uncertainty
Heating and cooling		low
Agriculture	Predicting the success of possible adaptive measures	medium
Forests	Predicting the change of forest ecosystems	medium
Migration due to sea level rise	Predicting the amount of sea level rise	medium
Coastal damage due to sea level rise	Predicting the amount of sea level rise and the value of coastal properties	medium
Temperature effects on mortality	Depends on distribution of population and on implementation of adaptive measures (temperature inside buildings)	medium
Water supplies	Depends not only on the geographical and temporal distribution of precipitation but also on the evolution of the demand for water	high
Extreme weather events	The current climate models are not sufficiently detailed and reliable	high
Tropical diseases	Depends on sanitary conditions of tropical populations and on success of vaccines and of treatments. Monetary valuation of mortality in poor countries is controversial	very high
Ecosystems	The world's ecosystem is so complex that its evolution under climate change is extremely difficult to foresee	extreme
Changes in ocean circulation	The current models for climate and ocean circulation are not sufficiently detailed and reliable	extreme
Social unrest, migrations (other than due to sea level) and wars	There is no solid basis for estimating such effects	extreme

about the appropriate scenario and uncertainties of modeling impacts in the distant future. The monetary valuation encounters further problems that we address in the following section.

10.3.2 *Problems in monetary valuation*

Even if the physical impacts could be predicted reliably, their monetary valuation is difficult and controversial. Determining monetary values for today's impacts is challenging enough; for climate change one needs values for a future that is many decades, even centuries ahead, and we don't even know what our society will look like at that time.

The long time horizon has also led to fierce debates about the choice of an appropriate discount rate. We have discussed long-term discount rates in Section 9.1.3. With conventional values of the social discount rate the long-term costs shrink to insignificance. Even though a consensus seems to have emerged that lower rates should be used for the long term, the detail is still controversial. This matters greatly because for such long time horizons the difference between a rate of, say, 1% and 3%, is enormous.

For several of the categories a further difficulty lies in the monetary valuation of mortality in poor countries for which no data on VPF (value of a prevented fatality) are available. This issue is important because many, if not most, of the mortality impacts are expected to occur in poor countries. However, the valuation is a hot potato politically, as highlighted by the controversy that arose when the IPCC included mortality valuation in its assessment in 1995; something they have not dared to address again (even the latest assessment, of 2007, does not attempt any new estimate of damage costs and merely offers a brief review of the literature). At one level the economically rational approach may appear to be to estimate values for such countries using benefit transfer, as discussed in Section 9.6. Typically this means taking the VPF of North America or the EU and adjusting it according to Eq. (9.30) with the appropriate value for the elasticity of demand. This would provide values that reflect the WTP of those impacted. However, such an approach pays little regard to the fact that emissions in North America and the EU are much higher, both in absolute terms and per capita, than in most of the poor countries. As such, establishment of willingness to accept risk would appear more appropriate than measures based on willingness to pay.

Yet another difficulty lies in the fact that many impacts depend not only on the level of CO_2 but also on its rate of change. It is much easier to adapt to changes that are slow than those that are sudden. That is especially true for ecosystems. For instance many plants may be able to move their habitat if the change is sufficiently slow, but disappear if it is too fast. Therefore a single indicator of cost per tonne of CO_2 cannot correctly represent all impacts.

10.3.3 Equity weighting

Many of the impacts of climate change are expected to occur in the poorer countries of the world. If the corresponding damage costs are calculated according to the standard procedure, there is a serious problem with equity, because the same cost will affect a poor person much more than someone who is rich with more disposable income. Give a thousand Euros to a billionaire and he/she will hardly notice any difference, whereas the same amount would be a fortune for someone who has to live on one Euro per day. Economists use the term "utility" to express how much a given amount of money really matters to an individual.

Equity weighting is a procedure to modify the damage costs incurred by different countries to reflect the loss of utility rather than the loss of a monetary amount. Whether or not to use equity weighting is a matter of ethics. Equity weighting appears logical if one accepts a principle of justice (called "veil of ignorance") developed by Rawls, a great philosopher of social justice: if an individual does not know how he will end up in his own conceived society, he is likely to develop a scheme of justice that treats all fairly, rather than privileging any one class of people.

To derive equity weights, one needs to examine the relation between utility u and wealth, income or consumption. It is often appropriate to consider u as a function of consumption c, because to a large extent the utility derives from the goods one consumes. Frequently one assumes that the elasticity η of the marginal utility of income, $\eta = -\frac{u''(c)c}{u'(c)}$, is constant. In that case straightforward integration yields the iso-elastic utility function,

$$u(c) = a\,c^{1-\eta} + b \quad \text{for } \eta \neq 1 \tag{10.1a}$$

and

$$u(c) = a\,\ln(c) + b \quad \text{for } \eta = 1, \tag{10.1b}$$

where a and b are constants. The form Eq. 10.1b appears quite natural, because the additional utility Δu due to an additional consumption Δc is inversely proportional to what one has already, in other words, it is the relative change that matters.[2]

Now consider two individuals or countries, one rich (R), and one poor (P). Suppose the inhabitants of R have a consumption level c_R 10 times higher than c_P, that of P. The marginal utility Δu of an extra consumption Δc takes the form,

$$\Delta u(c) = \text{constant } c^{-\eta}\,\Delta c. \tag{10.2}$$

---

[2] This functional form was first proposed by one of the Bernoulli brothers in 1738 (Bernoulli, 1738).

Table 10.4 *Ratio of marginal utilities of the same monetary value for the example of a person or country R that is 10 times as rich as person or country P.*

η	0	0.5	0.8	1.0	1.2	1.5	2.0
Δu for R as a fraction of Δu for P	1.0	0.31	0.16	0.1	0.06	0.03	0.01

For $\eta = 1$, the marginal utility of the same monetary amount is 10 times higher in P than in R. The higher η, the more rapidly marginal utility falls with additional wealth. If η is high, there is little additional utility gained from additional consumption by people who are already rich. Table 10.4 illustrates this example with a range of values of η.

Equity weighting simply involves the weighting of monetary values in different countries in such a way that they have the same utility. In other words, one multiplies the monetary value by the ratio of the marginal utility per marginal consumption, $\Delta u(c)/\Delta c$, between a rich person (or country) R and a poor person (or country) P,

$$\text{equity weight} = \frac{[du/dc]_P}{[du/dc]_R} = (c_R/c_P)^{\eta}. \tag{10.3}$$

On average and apart from changes in savings or investments, consumption equals GDP per capita. As a measure of the wealth of a country it is appropriate to use the purchasing power parity (PPP) adjusted GDP/capita Y. Using for reference (often called *numeraire*) a world average income level Y_N, the equity weight for a country X with income Y_X is thus,

$$\text{equity weight} = (Y_N/Y_X)^{\eta}. \tag{10.4}$$

For $\eta = 0$ the utility is exactly proportional to consumption and there is no equity weighting.

Example What is the equity weight for a country whose GDP/capita is 1/5 the world average, if one chooses $\eta = 1$?

Solution: The equity weight is 5.

With equity weighting one can correct the main effect of income inequality. There is some uncertainty about the appropriate choice of η, but a value of unity seems reasonable. An additional effect of inequality is not addressed by equity weighting: the rich can cope with a given loss much better than the poor; but that is probably secondary and much more difficult to quantify.

10.3.4 A simple alternative that is useless

In view of all these difficulties some might wonder if it would not be simpler to just do a contingent valuation (CV), asking people for their willingness to pay (WTP) to avoid the impacts of climate change. Indeed, this would be simpler. But totally beside the point, because CV is meaningful only to the extent that the interviewees understand the good in question. Climate change is so complex that even the experts have trouble understanding all the links and all the consequences. Asking for the WTP to avoid an environmental burden makes no sense if people do not understand the relation between that burden and its impacts. We already have that sort of simple-minded CV in our political system in one sense: the carbon taxes that the various countries have or have not implemented are a reflection of the respective WTPs, as perceived by the politicians.

10.4   Damage cost estimates

10.4.1 Review of the literature

A number of integrated assessment models have been developed to assess climate change damage costs and their evolution over time. Some are based directly on the impact of ΔT on gross domestic product (GDP) or gross world product, by assuming a relation between ΔT and a change in GDP, the latter being estimated by an analysis of the various sectors of the economy. Many models use a quadratic term in ΔT in addition to a linear one, because at higher temperatures the damages increase more rapidly. One of the pioneering models, the DICE (dynamic integrated climate and economy) model of Nordhaus and Boyer (2000) is of this type. Another widely used climate policy assessment model is MERGE, a multi-region Ramsey–Solow optimal growth model including greenhouse gas emissions and a global climate module (Manne and Richels, 2004). This model assumes that for a ΔT of 2.5 °C an economic loss of 2% of GDP is incurred in high-income countries, while in low-income countries that percentage is only 1% in the range where per capita annual income is between $5000 and $50,000, going up to 2% when per capita income rises above $50,000.

Other models, in particular FUND (climate framework for uncertainty, negotiation and distribution) (Tol, 1995) and PAGE (policy analysis of the greenhouse effect) (Plambeck and Hope, 1996), attempt to simulate climatic impacts in more detail, according to the categories or economic sectors to which they apply. FUND and PAGE are often used in this field. Both are integrated assessment models designed to jointly simulate economic growth, CO_2 emissions, and climate change impacts. However, these models do not cover all climate change damage costs; they exclude, for example, major climatic events, socially contingent effects and many of the non-market impacts. In particular, they do not consider the possibility of catastrophic climate change impacts, something for which other types of analysis may be more appropriate (see e.g. Weitzman, 2009 and Yohe, 1996).

Two studies (DEFRA, 2004 and 2005) have produced an important overview of the climate change damage literature and made a comparison of two modeling exercises determining the marginal damage cost of CO_2. In Fig. 10.12 we show a main result, damage cost estimates for six points in time until 2050 (for this figure we have transformed the original data into units of €/t_{CO2}, using an exchange rate of 1.5 €/£). The thick solid line in the middle shows the central estimates; they represent the average of the time-dependent means as calculated by FUND and PAGE.

Damage cost estimates strongly depend on the discounting method, and on assumptions about equity weighting and about risk aversion or willingness to pay to avoid damages. The central curve in Fig. 10.12 is based on relatively low and declining discount rates, and includes moderate global assumptions on equity weighting and risk aversion.

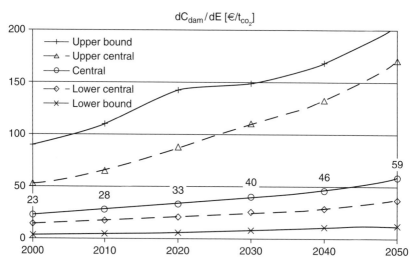

Fig. 10.12 The marginal damage cost of CO_2, in €/t_{CO2}, versus year of emission.
Data based on DEFRA (2004 and 2005).

The other four curves indicate the uncertainty range of the central curve. The dashed curves directly below and above the middle curve represent lower and upper estimates for the central curve. For the lower central curve a value of 14 €/t$_{CO2}$ is taken in 2000, as proposed by DEFRA (2005). For its evolution over time it is assumed that the ratio (central/lower central) remains fixed. The lower central estimate of 14 €/t$_{CO2}$ is considered reasonable for a global decision context committed to reduce the threat of dangerous climate change under relatively low time discounting and equity weighting and a modest level of risk aversion to extreme damages. The thin outer bound curves reflect an even larger cost range uncertainty. The lower bound is the 5% confidence level (CL) estimate of PAGE, and the upper bound the average between the 95% CL simulations of FUND and PAGE. Even though the range thus obtained is enormous, it still does not fully capture all published estimates. Negative dC_{dam}/dE values have been reported (implying net climate change benefits rather than costs), as well as values several times higher than the top dC_{dam}/dE value of 200 €/t$_{CO2}$ in 2050.

DEFRA recommends using the curves of Fig. 10.12 in a multi-level approach. The central range is intended for most day-to-day policy appraisal (e.g. of relatively minor short-term infrastructure construction projects or local environmental impact assessments) and the full range is for major long-term purposes (such as the national planning of an energy security strategy) or uncertainty analysis in a global cost–benefit framework. However, guidance on interpretation and hence use of ranges in a policy forum is lacking.

Tol (2005) also reviews a large number of climate change impact studies, and combines over 100 estimates for the marginal damage cost of CO$_2$ to form an overall probability density function. The distribution is strongly right-skewed, with a median of $3.8/t$_{CO2}$, a mean of $25.4/t$_{CO2}$, and a 95% CL of $95/t$_{CO2}$. According to Tol, under standard assumptions of time discounting, equity weighting, and risk aversion, the marginal damage cost is unlikely to exceed $14/t$_{CO2}$, and is probably even smaller. This value is significantly lower than the $85/t$_{CO2}$ reported by the widely publicized Stern review (Stern et al., 2006). The discrepancy can to a large extent be explained by the very low value of the social discount rate employed by Stern et al., in comparison with the discounting conventionally used (Dasgupta, 2006; Nordhaus, 2006). Also, there are inevitable biases in any such assessment, not least that many of the studies included will not be fully independent of each other.

The highly skewed distribution of damage cost estimates in the literature is fairly consistent with a lognormal distribution, even though a few studies claim negative damages. Indeed, global climate change probably produces both winners and losers, at least at moderate temperature increases, but we do not believe that the net worldwide damage cost could be negative for any increase of the atmospheric CO_2 concentration. As a representation of the estimates found in the literature we therefore take a lognormal distribution, and choose for its parameters a median $\mu_g = \$3.8/t_{CO2}$ and upper limit $\mu_g \, \sigma_g^2 = \$95/t_{CO2}$ according to the review by Tol (2005), which implies $\sigma_g = 5$. These numbers are reasonably consistent with those of DEFRA. In view of the limitations of currently available studies – notably the fact that especially some of the most troubling potential impacts, such as a change in the thermohaline circulation, rapid non-linear ice-sheet disintegration, or methane release from permafrost melting, have not yet been adequately, or even hardly at all, taken into account – we realize that the uncertainty range may well be larger than $\sigma_g = 5$.

In the ExternE project series, a first assessment of CO_2 damage costs was published by ExternE (1998); it indicated a range of values from 3.8 to 139 €/t_{CO2}, with a geometric mean of 29 €/t_{CO2}. Two years later, a new assessment by the ExternE team came up with 2.4 €/t_{CO2}, but that value was only applied in the transport sector. In the following years ExternE used a value of 19 €/t_{CO2}, based on abatement costs implied by the Kyoto Protocol for the EU and, in view of the uncertainties, fairly consistent with the estimates in Fig. 10.12 (the logic for using abatement cost estimates in certain situations has been explained in Section 2.2.4.3 of Chapter 2). More recent evaluations by ExternE are more detailed, with different numbers for different assumptions (time period, discount rate, equity weighting etc.). ExternE (2008) has chosen a value of 21 €/t_{CO2eq} for emissions in 2010.

Hanemann (2008) offers an instructive critique of the results of Nordhaus and Boyer, pointing out several features that the latter, like most analysts of climate change costs, do not take into account. In particular, they use the average global temperature rise instead of consideration of regional and seasonal variations of climate change. In effect, basing the analysis on global averages is a bit like saying "what's your problem?" to someone who has one hand in ice, the other in boiling water. With a more realistic analysis he estimates that the damage cost imposed on the USA is four times higher than what Nordhaus and Boyer find for the USA.

10.4.2   Damage cost as a function of temperature change

Many integrated assessment studies on energy, climate change, and the economy, use a damage cost function with the shape:

Table 10.5 *Parameter values for ρ of Eq.*
(10.5) as assumed in several widely employed
integrated assessment models of climate
change.

Source	ρ
Cline (1992)	0.0023
Fankhauser (1995)	0.0028
Manne and Richels (2004)	0.0032
Nordhaus (1991, 1994)	0.0015
Plambeck and Hope (1996)	0.0028
Titus (1992)	0.0021
Tol (1995)	0.0032

N.B. Most of these authors report damages relative to
GDP in the USA for one temperature increase level
only, typically as associated with a doubling of the
atmospheric CO_2 concentration.

$$C_{dam} = \rho(\Delta T_{stab})^{\theta}, \qquad (10.5)$$

in which C_{dam} is the damage cost expressed as fractional loss of world
GDP, ΔT_{stab} the global average temperature change with respect to the
pre-industrial atmospheric temperature in stabilized conditions, i.e. when
equilibrium is reached in the climate system, and with ρ and θ coefficients
characterizing the shape of the damage function. Most studies take $\theta = 2$.
Roughgarden and Schneider (1999) investigate values of θ other than 2
(both $1 < \theta < 2$ and $\theta > 2$) on the basis of a set of expert views. They
conclude, however, that $\theta = 2$ is the most plausible choice. World GDP is
a commonly employed measure for the size of the global economy, but by
definition it only covers tangible markets. It is therefore incomplete, as not
all market activity always enters national accounts, especially in developing economies, and, of course, non-market impacts are not included at all
in Eq. (10.5). Note that ΔT_{stab} is the stabilized global average temperature
change obtained after a specified emission profile leads to new equilibrium values of the atmospheric CO_2 concentration and the corresponding increase in atmospheric temperature. There are large time lags both
between the CO_2 emissions level and the stabilized CO_2 concentration,
and between this new atmospheric CO_2 concentration and ΔT_{stab}, typically in each case of at least several decades up to a century. Parameter
values for ρ are shown in Table 10.5.

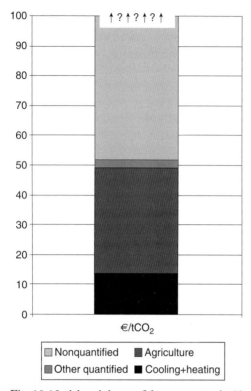

Fig. 10.13 A breakdown of damage costs, in €/t$_{CO2}$, by impact category, estimated by FUND, for pure rate of time preference 0.1% and no equity weighting.
Quantified costs based on Anthoff *et al.* (2010).

10.4.3 Damage cost by impact category

Detailed models such as PAGE and FUND calculate the damage cost for a wide variety of impact categories or sectors of the economy. As an example we show in Fig. 10.13 a breakdown of FUND results for a particular set of underlying assumptions, namely the numbers in Figure 6 of Anthoff *et al.* (2010), for the case of pure time preference = 0.1%; on top of their costs we have added non-quantified costs.

What we find the most striking in this graph is the relative magnitude of cooling and agriculture compared with all the other sectors that have been quantified (for pure rate of time preference = 1%, the relative contribution of cooling in Figure 6 of Anthoff *et al.* would be even larger). Would climate change be such a worrisome problem if it were merely a matter of an increase in the cost of cooling and food production? Such costs would

Uncertainty in Valuation				
		Market	**Non Market**	**(Socially Contingent)**
Uncertainty in Predicting Climate Change	**Projection** (e.g. sea level rise)	Coastal protection Loss of dryland Energy (heating/cooling)	Heat stress Loss of wetland	Regional costs Investment
	Bounded Risks (e.g. droughts, floods, storms)	Agriculture Water Variability (drought, floods, storms)	Ecosystem change Biodiversity Loss of life Secondary social effects	Comparative advantage & market structures
	System change & surprises (e.g. major events)	Above, plus Significant loss of land and resources Non-marginal effects	Higher order social effects Regional collapse Irreversible losses	Regional collapse

Fig. 10.14 Uncertainties of climate change damage costs.
From Watkiss and Downing (2008).

not be a serious burden for our GDP; we would pay and go on living without agonizing about hurricanes, changing ocean currents and destruction of ecosystems. Rather, we interpret results such as Fig. 10.13 as a warning that current damage cost models are not able to account for the most troubling impacts.

Watkiss and Downing (2008) provide an interesting perspective on the current state of damage cost assessments by means of a matrix where impact types are arranged in rows (in order of increasing uncertainty of the valuation) and columns (in order of increasing uncertainty in predicting climate change). This is reproduced in Fig. 10.14.

The first row covers impacts that can be predicted with relative confidence and where the confidence in the direction of effect is certain (e.g. average temperature).

The second row covers impacts where prediction is more uncertain, and where models often give different levels of impacts, or even predictions of a different sign (positive/negative), as with for example regional estimates of levels of precipitation, or frequency or magnitude of extreme events.

The third row covers impacts where prediction is highly uncertain, notably around the major "tipping points" commonly identified (major climate discontinuities or irreversibilities, such as the West Antarctic ice sheet, methane hydrates, etc.).

The current damage cost estimates are more or less limited to the three top left elements of this matrix (Coastal protection ..., Heat stress ..., Agriculture ...). None have succeeded in assessing the remaining six elements. But those are likely to have by far the largest costs.

10.5   Abatement costs

Compared with damage costs, the cost of abatement[3] options is straightforward, at least in principle: just prepare a list of the options and estimate their costs. But in practice, the cost and performance of most options are unknown or extremely uncertain. For example, carbon capture and sequestration is a new technology that has not yet been tested on full scale power plants. Even if the performance can be estimated quite well, the cost cannot because it involves economies of scale, difficult to foresee if one does not know the implementation rate. The implementation rate depends not only on technical aspects such as capital cost, operating cost and reliability, but also on the policy context. How large will the carbon tax be, if any? How difficult will it be to obtain a permit for carbon sequestration? How many sites for CO_2 storage will be considered safe? The further into the future one tries to estimate abatement costs and potentials the greater the difficulties and uncertainties. Another problem is that estimated costs rarely account for the co-benefits and tradeoffs associated with climate mitigation, for example through reducing emissions of local and regional air pollutants, or impacting on landscapes through extensive development of wind farms.

There are two approaches for estimating abatement costs, bottom-up and top-down:
- Bottom-up studies are based on detailed engineering analysis of each option, taking into account technical characteristics as well as any regulations that may be applicable. Each sector of the economy is analyzed by itself, assuming that the macro-economy is unchanged.
- Top-down studies use macro-economic models with aggregated information about abatement options and take into account macro-economic and market feedbacks.

Bottom-up studies can be used for the assessment of specific policy options in each sector, e.g. transportation; whereas top-down studies are convenient for assessing economy-wide climate change policies, e.g. tradable permits. Ideally there should be only one approach, combining both

[3] IPCC uses the term "mitigation" instead of "abatement."

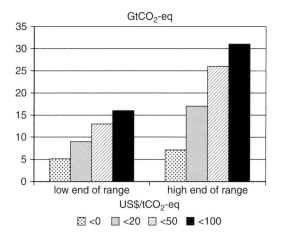

(a) Bottom-up studies

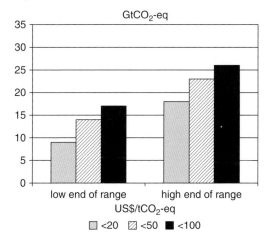

(b) Top-down studies

Fig. 10.15 Estimates of global abatement potential in 2030, in Gt_{CO2eq}/yr. From IPCC (2007b, Figure SPM.5A).

aspects of the problem. Some recent studies have moved in that direction, for instance by incorporating macro-economic feedbacks in a bottom-up approach.

Several different formats have been used for the presentation of the results. We have already shown examples of abatement cost curves in Figs. 1.1 and 1.3 of Chapter 1. Here we show, in Fig. 10.15, estimates of abatement potential published by IPCC (2007b): in part (a) for bottom-up and in part (b) for top-down studies. Both scope and format are very different from the curve in Fig. 1.1, which covers only the EU, for 2020, and plots marginal abatement cost versus abatement potential with fairly fine resolution. By contrast, Fig. 10.15 covers the world in 2030 and shows abatement potential versus marginal abatement cost with rather crude resolution, indicating upper and lower bounds for each abatement cost range.

Like most bottom-up studies, Fig. 10.15a indicates significant potential at negative cost. Such options are found especially in the building sector where there are many opportunities for savings from energy conservation. In the words of energy guru Amory Lovins, it's not only a free lunch but a lunch you are paid to eat. So, why haven't these options been implemented already if their costs are really negative? The explanation is that such results come from pure engineering estimates that do not take into account transaction costs or particular circumstances that so often prevent the implementation of these options. Implementation decisions involve many considerations besides direct costs, in particular comfort, convenience and esthetics, as well as the cost of information and perceived or real uncertainties about the savings that would actually be obtained. The very fact that most people have not implemented all of these options proves that such considerations and transaction costs are crucial in practice. For these reasons, the engineering estimates of $/t_{CO2}$ in these studies can be considered optimistic, and it is difficult to estimate the real reduction potential.

For some abatement options the implementation can be greatly increased by means of government regulations. For example, in the EU the replacement of incandescent light bulbs by fluorescent or other high efficiency lamps is being brought about by simply outlawing the sale of most types of incandescent bulb.

10.6   Emission reductions and implications of a CO_2 tax

To appreciate the magnitude of the challenge facing mankind, let us look at the GHG emissions per capita in the context of the reductions needed for the stabilization scenarios in Table 10.2. Figure 10.16 shows the greenhouse gas emissions per capita in major countries/regions of the world in 2004, in t_{CO2eq}/yr/capita. The total GHG emissions are much higher than the CO_2 emissions from fuel use, because they cover all GHG covered by the Kyoto agreement, including those from land use change: in 2004 they were about 49 Gt_{CO2eq}/yr, of which only about 30 Gt_{CO2}/yr was CO_2 from fuel use, as was shown in Fig. 10.2. The world average is about 7.5 t_{CO2eq}/yr per capita. The USA and Canada have the highest emissions,

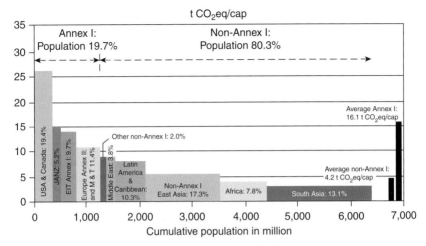

Fig. 10.16 The GHG emissions per capita in major countries/regions of the world in 2004, in t_{CO2eq}/yr/capita (all GHG, including those from land-use change). The percentages in the bars indicate a region's share in global GHG emissions. The countries in the groupings are shown in Table 10.6. From IPCC (2007a, Figure SPM.3a).

about 26 t_{CO2}/yr per capita. In Europe, the group Annex II and M&T (essentially the old member states of the EU) have an average of about 11 t_{CO2eq}/yr per capita, while the former Eastern Bloc countries are around 14 t_{CO2eq}/yr per capita. In 2008, the EU27 average was 10.7 t_{CO2eq}/yr/capita, Germany had 11.7 t_{CO2eq}/yr/capita and France 8.2 t_{CO2eq}/yr/capita.[4] France is much lower thanks to nuclear power which provides about 80% of her electricity, whereas about half of the electricity in Germany comes from coal.

Table 10.2, p. 419, indicates the emission reductions needed by 2050 for the stabilization categories I to VI of Fig. 10.10. Let us focus on categories I to IV, as being more desirable, and express the reductions in terms of per capita emissions, as shown in Table 10.7. The reductions are indicated as a range because each category comprises several different scenarios with different emission paths. In line 4 of Table 10.7 we show the emissions in 2050 as a fraction of the emissions in 2000, for the bounds of the respective ranges for each category. The corresponding total global emissions are in line 5. Dividing by the world population in 2050, estimated at about 9.2 billion,[5] yields the emissions per capita in line 6. For example, if the

[4] www.eea.europa.eu/data-and-maps/figures/greenhouse-gas-emissions-as-tonnes
[5] http://esa.un.org/wpp/other-information/faq.htm

Table 10.6 *The country groupings in Fig. 10.16.*

Annex I	Australia, Austria, Belarus, Belgium, Bulgaria, Canada, Croatia, Cyprus, Czech Republic, Denmark, Estonia, European Union, Finland, France, Germany, Greece, Hungary, Iceland, Ireland, Italy, Japan, Latvia, Liechtenstein, Lithuania, Luxembourg, Malta, Monaco, Netherlands, New Zealand, Norway, Poland, Portugal, Romania, Russian Federation, Slovakia, Slovenia, Spain, Sweden, Switzerland, Turkey, Ukraine, United Kingdom of Great Britain and Northern Ireland, United States of America.
EIT Annex I	Belarus, Bulgaria, Croatia, Czech Republic, Estonia, Hungary, Latvia, Lithuania, Poland, Romania, Russian Federation, Slovakia, Slovenia, Ukraine.
Europe Annex II & M&T	Austria, Belgium, Denmark, Finland, France, Germany, Greece, Iceland, Ireland, Italy, Liechtenstein, Luxembourg, Netherlands, Norway, Portugal, Spain, Sweden, Switzerland, United Kingdom; Monaco and Turkey.
JANZ	Japan, Australia, New Zealand.
Non-Annex I East Asia	Cambodia, China, Korea (DPR), Laos (PDR), Mongolia, Republic of Korea, Viet Nam.
South Asia:	Afghanistan, Bangladesh, Bhutan, Comoros, Cook Islands, Fiji, India, Indonesia, Kiribati, Malaysia, Maldives, Marshall Islands, Micronesia (Federated States of), Myanmar, Nauru, Niue, Nepal, Pakistan, Palau, Papua New Guinea, Philippine, Samoa, Singapore, Solomon Islands, Sri Lanka, Thailand, Timor-L'Este, Tonga, Tuvalu, Vanuatu.

Table 10.7 *GHG emission reductions required by the year 2050, for the stabilization scenarios I to IV of Table 10.2 and Fig. 10.10.*

Scenario category	I		II		III		IV	
Global average ΔT, °K	2.0– 2.4		2.4 – 2.8		2.8 – 3.2		3.2 – 4.0	
Emission change from 2000	−85 to −50%		−60 to −30%		−30 to +5%		+10 to +60%	
Emission reduction factor	0.15	0.5	0.4	0.7	0.7	1.05	1.1	1.6
Gt_{CO2eq}/yr in 2050	6.7	22.4	17.9	31.3	31.3	46.9	49.2	71.5
t_{CO2eq}/yr/capita in 2050	0.7	2.4	1.9	3.4	4.3	5.1	5.3	7.8

temperature rise ΔT is to be kept in the range 2.4 to 2.8 K (category II), the per capita emissions have to be brought down to the range 1.9 to 3.4 t_{CO2eq}/yr/capita. By comparison, the world average was about 7.5 t_{CO2eq}/yr/capita in 2004 and has been increasing. Much of the increase is occurring in developing economies such as China and India, an evolution that can be expected to continue as the poorer countries are trying to reach the standard of living of the EU, America and Japan. Since that evolution reduces the inequality of emissions between countries, even the emissions of the rich countries will have to approach the range 1.9 to 3.4 t_{CO2eq}/yr/capita if we are to stay within stabilization category II.

One wonders how such a low level can be reached, given the current lack of concern about climate change. Even if we had a carbon tax of, say, 30 €/t_{CO2}, would it significantly alter our emissions? In the EU27 the CO_2 emissions are about 11 t_{CO2}/yr/capita and the average cost of such a tax would be 330 €/yr per person, roughly 1.3% of per capita GDP.

Consider electricity production, the largest single source of GHG. The effect on the price of electricity would be an increase of 0.9 kg_{CO2}/kWh$_e$ × 30 €/t_{CO2} = 2.7 €cents/kWh$_e$ for coal (steam) and 0.4 kg_{CO2}/kWh$_e$ × 30 €/t_{CO2} = 1.2 €cents/kWh$_e$ for natural gas (combined cycle), compared with an average retail price in the EU 27 of about 13 €cents/kWh$_e$ for residential customers in 2012. The price elasticity of demand for electricity, even in the long run where it is much higher than in the short run, has been estimated to be in the range −0.7 to −1.0 (Fell *et al.*, 2012). For the sake of illustration, let us assume an average price increase of 1.3 €cents/kWh$_e$ or 10%, and the greatest elasticity, i.e. −1.0, in which case the electricity demand would decrease by 10%. That would hardly make a dent in the CO_2 emissions. It seems that far stronger actions are required than a carbon tax or permit prices at the levels currently in place or under consideration. Policy measures such as portfolio standards for electricity or vehicles (i.e. requiring that a certain percentage of the sales must come from low emission technologies) can be very effective because they can force changes that would not occur when the price elasticity of demand is too low.

References

Anthoff, D., Rose, S., Smith, J. B., Tol, R. S. J. and Waldhoff, S. 2010. *Regional and Sectoral Estimates of the Social Cost of Carbon: An Application of FUND.* Economic and Social Research Institute, Dublin, Ireland. 15 Feb. 2010.

Bernoulli, D. 1738. Reprinted as exposition of a new theory on the measurement of risk. *Econometrica* **22**, No. 1. (Jan., 1954): 23–36.

Cline, W. R. 1992. *The economics of global warming.* Institute for International Economics, Washington D.C.

Dasgupta, P. 2006. Comments on the Stern Review's Economics of Climate Change, mimeo, see www.econ.cam.ac.uk/faculty/dasgupta.

DEFRA 2004. *The Social Costs of Carbon (SCC) Review – Methodological approaches for using SCC estimates in policy assessment*, Department for Environment, Food and Rural Affairs, London.

DEFRA 2005. *Social Cost of Carbon: A closer look at uncertainty*, Department for Environment, Food and Rural Affairs, London.

ExternE 1998. ExternE: Externalities of Energy, European Commission, DG-XII, Science Research and Development, http://ExternE.jrc.es/publica.html.

ExternE 2008. With this reference we cite the methodology and results of the NEEDS (2004–2008) and CASES (2006–2008) phases of ExternE. For the damage costs per kg of pollutant and per kWh of electricity we cite the numbers of the data CD that is included in the book edited by Markandya, A., Bigano, A. and Porchia, R. in 2010: *The Social Cost of Electricity: Scenarios and Policy Implications*. Edward Elgar Publishing Ltd, Cheltenham, UK. They can also be downloaded from www.feem-project.net/cases/ (although in the latter some numbers have changed since the data CD in the book).

Fabry, V. J., Seibel, B. A., Feely, R. A. and Orr, J. C. 2008. Impacts of ocean acidification on marine fauna and ecosystem processes. *ICES Journal of Marine Science* **65**: 414–432.

Fankhauser, S. 1995. *Valuing Climate Change: The economics of the greenhouse*. Earthscan, London.

Fell, H., Li, S. J. and Paul, A. 2012. *A New Look at Residential Electricity Demand Using Household Expenditure Data*. Working Paper 2012–04, Colorado School of Mines, Division of Economics and Business. http://econbus.mines.edu/working-papers/wp201204.pdf

Hanemann, W. M. 2008. *What is the Economic Cost of Climate Change?* eScholarship. University of California – Berkeley, URL: http://repositories.cdlib.org/are_ucb/1071/.

Hardt, M. J. and Safina, C. 2010. Threatening ocean life from the inside out. *Scientific American*, August: 52–59.

IPCC 2007a. Climate Change 2007: Synthesis Report. Intergovernmental Panel on Climate Change. Downloaded 1 Oct. 2010 from www.ipcc.ch/publications_and_data/ar4/syr/en/contents.html

IPCC. 2007b. Summary for Policymakers. In: Climate Change 2007: Mitigation. Contribution of Working Group III to the Fourth Assessment. *Report of the Intergovernmental Panel on Climate Change*, B. Metz, O. R. Davidson, P. R. Bosch, R. Dave, L. A. Meyer (eds), Cambridge University Press, Cambridge, United Kingdom and New York, NY, USA.

Manne, A. S. and Richels, R. G. 2004. MERGE: an integrated assessment model for global climate change. www.stanford.edu/group/MERGE.

Nordhaus, W. D. 1991. To slow or not to slow: the economics of the greenhouse effect, *The Economics Journal* **101**: 920–937.

Nordhaus, W. D. 1994. *Managing the Global Commons*. MIT Press, Cambridge, Massachusetts.

Nordhaus, W. and Boyer, J. 2000. *Warming the World: Economic Models of Global Warming*. Cambridge, MA: MIT Press.

Nordhaus, W. D. 2006. The Stern Review on the Economics of Climate Change, mimeo, see http://nordhaus.econ.yale.edu.

Plambeck, E. L. and Hope, C. W. 1996. PAGE-95: an updated valuation of the impacts of global warming. *Energy Policy* **24**, 9: 783–794.

Roughgarden, T. and Schneider, S. H. 1999. Climate change policy: quantifying uncertainties for damages and optimal carbon taxes, *Energy Policy* **27**: 415–429.

SRES 2000. *IPCC Special Report on Emissions Scenarios*. Intergovernmental Panel on Climate Change.

Stern, N. *et al.* 2006. *The Economics of Climate Change: The Stern Review*, Cambridge University Press, Cambridge, UK. Available at www.hm-treasury. gov.uk/stern_review_report.htm

Titus, J. G. 1992. The cost of climate change to the United States, in: S. K. Majumdar *et al.* (Eds.), *Global Climate Change: implications, challenges, and mitigation measures*, Pennsylvania Academy of Science, Easton, PA.

Tol, R. S. J. 1995. The damage costs of climate change: towards more comprehensive calculations. *Environmental and Resource Economics* **5**: 353–374.

Tol, R. S. J. 2005. The marginal damage costs of carbon dioxide emissions: an assessment of the uncertainties, *Energy Policy* **33**, 16: 2064–2074.

Watkiss, P. and Downing, T. E. 2008. The social cost of carbon: Valuation estimates and their use in UK policy. *IAJ The Integrated Assessment Journal: Bridging Sciences & Policy* **8** (1): 85–105.

Weart, S. 2008. *The Discovery of Global Warming*, 2nd edition. www.aip.org/ history/climate/index.htm

Weitzman, M. L. 2009. On modeling and interpreting the economics of catastrophic climate change. *Review of Economics and Statistics* **91**: 1–19.

Yohe, G. 1996. Exercises in hedging against extreme consequences of global change and the expected value of information. *Global Environmental Change* **6**, 2: 87–100.

11   Uncertainty of damage costs

Summary

This chapter presents an analysis of the uncertainties of damage costs, all the more important because their uncertainties are large. Two methods for the analysis of uncertainties are presented. One is the customary Monte Carlo approach; it is general and powerful, but opaque because it produces only numbers. As an alternative we present an analytical approach that is suitable for multiplicative models, in particular the "uniform world model" (UWM) for damage costs; it has the advantage of being transparent and easy to modify if one wants to test different assumptions about the various sources of uncertainty. We show results, based on a literature review of the various sources of uncertainty in the steps of the damage cost calculation. We find that the uncertainty of damage costs can be characterized, with a sufficiently good approximation, by a lognormal probability distribution with multiplicative confidence intervals around the median estimate μ_g (a random variable has a lognormal distribution if the distribution of the logarithm of the variable is normal). The width of the confidence intervals is given by the geometric standard deviation σ_g, such that the 68% confidence interval ranges from μ_g/σ_g to $\mu_g\,\sigma_g$. For the classical air pollutants (PM, NO_x, SO_2, VOC) we find that σ_g is approximately 3; for toxic metals we estimate that it is about 4 and for dioxins and greenhouse gases about 5. We also present a simple method for the uncertainty of the sum of damage costs due to different pollutants, for instance the damage cost of a kWh of electricity.

We also address, in Section 11.5, the problem of cases where costs and benefits cannot be estimated with sufficient reliability. We examine three examples: nitrates in drinking water, NH_3 emissions to the atmosphere, and greenhouse gases, and we discuss possible approaches for dealing with such cases.

Finally, we ask what effect the uncertainties have on the decisions that are based on ExternE. The key question is, "how large is the cost penalty if one makes the wrong choice because of errors or uncertainties in the cost or benefit estimates?" The answer turns out to be very encouraging: the cost penalties are surprisingly small despite the large uncertainties.

11.1   Introduction

11.1.1   General remarks

The uncertainties of environmental damages are far too large for the usual error analysis of physics and engineering (using only the first term in a Taylor expansion). Rigorous systematic assessment of the uncertainties is difficult and few studies have attempted this. Most have merely indicated an upper and a lower value, but based on the range of just one input parameter or by simply combining the upper and lower bounds of several inputs, without taking into account the combination of uncertainties (e.g. of atmospheric dispersion, exposure–response function and monetary valuation). Many damage assessments involve so many different inputs that an analytical solution is usually not considered, and of the uncertainty analyses that have been done, almost all have used Monte Carlo techniques and numerical calculations (see e.g. Morgan and Henrion, 1990). The Monte Carlo method is powerful, and in principle capable of treating any problem, but the result is "black box": it is difficult to see how important each of the component uncertainties are or how the result would change if a component uncertainty were to change, unless one does an extensive sensitivity analysis.

Another drawback of the Monte Carlo method is that it is difficult to implement in the complex models of environmental dispersion and chemistry. The number of parameters and input data whose uncertainty would have to be estimated and whose contribution to the uncertainty of the result would have to be evaluated is far too large. In practice people only carry out a small number of sensitivity studies.

As a simple and transparent alternative, we therefore present an analytic method based on simple environmental models (the UWM of Section 7.4, or a more detailed model described in Section 11.3.1.2) that represent the key features of the phenomena. Their uncertainty can be characterized in terms of geometric standard deviations, σ_g, which have an immediate interpretation as multiplicative confidence intervals. Their justification lies in the observation that the calculation of the damage cost of a pollutant involves essentially a product of uncorrelated factors x_i and that σ_g of the product can readily be obtained via the square root of the sum of squared logarithms of the $\sigma_{g,i}$ of the factors x_i (Eq. (11.22) in Section 11.2.3.2). Thus it suffices to specify the geometric standard deviations $\sigma_{g,i}$ for each of the factors. We show that they can be estimated very simply from the respective confidence intervals (Eq. (11.31) in Section 11.2.4). Furthermore, the uncertainty distribution of the product is approximately lognormal for most damage cost calculations. Compared with a Monte Carlo analysis, this method is approximate and yields typical answers that are easy to apply and communicate. The calculation is simple enough to allow the reader to modify the assumptions and see the consequences. The contribution of the uncertainties of each of the factors is immediately apparent. Furthermore, the analytic method can be combined with Monte Carlo results for certain parts of the calculation, thus benefiting from the best features of each approach.

Whatever the method, an assessment of the uncertainties of damage costs must begin with a detailed examination of the uncertainties of each of the inputs to the impact pathway analysis, to estimate standard deviation and the shape of the probability distribution of their uncertainties. These component uncertainties are then combined to obtain the total uncertainty of the damage cost. Unfortunately, in most cases one does not have enough information and has to substitute assumptions that are unavoidably subjective. Such subjectivity is not unscientific, despite what some people may think. On the contrary, it would be unscientific to pretend that all is objective. Of course, it is the duty of the analyst to indicate when subjective choices are made.

11.1.2   Validation of models

Ideally one would like to evaluate the uncertainty of a model by comparing its results with observations. That is the standard approach in the physical sciences, and it has been used for atmospheric models to the extent that it is feasible. But the feasibility is limited by the complexity of the phenomena and the large number of parameters and variables that are involved. The models are complex and their validation is usually limited to some components of the model or to some special situations.

One can emit a tracer gas and measure the resulting concentrations at many different points. Unfortunately such an approach is not feasible for the problem of damage costs because they require long-term averages, at both the local and the regional scale. The quantity of tracer gas would be excessive. An interesting validation has been carried out following the Chernobyl nuclear accident, where the emitted quantities were sufficient for validation at the regional and global scale, but the results are relevant for episodes, not for long-term averages; besides, that validation was limited by uncertainties in emitted quantities and the detailed worldwide meteorological conditions.

To validate regional models for long-term averages the simplest way is to use regional emission inventories for one of the classical air pollutants as a model input and compare with regional ambient concentration data. Such a comparison, for sulfate concentrations calculated by the Windrose Trajectory Model (WTM) of EcoSense, has been reported in Section 4.4 of ExternE (2000). Measured SO_2 concentrations for the year 1990 at German monitoring sites were compared with calculated concentrations in the respective 50 x 50 km grid cells. For most grid cells the deviation between the calculated and observed concentration was less than a factor of 2. Some exceptions in the eastern part of Germany were blamed on the monitoring sites being too close to actual point sources and therefore not representative of the average concentration of the grid cells. Additionally, a similar comparison was also made for sulfate concentrations across Europe. There was a tendency towards overestimation by WTM, but 85% of the data points fell within a factor of 2 of the data.

Such a validation for epidemiology is not feasible. Intervention studies are the closest one can get, namely situations where a large and abrupt change in emissions has occurred and the resulting health impacts measured. There have only been three such cases (closure of a steel mill in Utah for an entire year in 1986/87; interdiction of the burning of coal in Dublin starting in July 1990; interdiction of high sulfur oil in Hong Kong starting in July 1990). These provided clear evidence for the reality of the health impacts of air pollution, but the information is not sufficient to validate specific ERFs.

For monetary values of non-market goods validation is not possible. The best one can do is to carry out new and improved studies.

11.2 Methodologies for estimation of uncertainty

11.2.1 General formulation and Monte Carlo calculation

The general problem with uncertainty analysis can be stated as follows. The quantity y to be estimated is a function $y = f(x_1, x_2, \ldots x_n)$ of the input parameters x_i. The x_i are uncertain, i.e. they are random variables with probability distributions $p_i(x_i)$ and with joint probability distribution $p(x_1, x_2, \ldots x_n)$. What is the resulting probability density distribution $p(y)$ of y?

The solution is straightforward in principle: one integrates the distribution $p(x_1, x_2, \ldots x_n)$ over $n-1$ of the variables (it does not matter which) subject to the constraint $y = f(x_1, x_2, \ldots x_n)$,

$$p(y) = \int dx_1 \int dx_2 \ldots \int dx_{n-1} \, p(x_1, x_2, \ldots x_n) \Big|_{f(x_1, x_2, \ldots, x_n) = y}. \qquad (11.1)$$

If the x_i are uncorrelated, the typical case for external cost calculations, the analysis becomes much simpler because the joint distribution is simply the product of the individual $p_i(x_i)$,

$$p\,(x_1, x_2, \ldots x_n) = p_1\,(x_1)\,p_2\,(x_2) \ldots p_n\,(x_n)$$
$$\text{if the } x_i \text{ uncorrelated.} \qquad (11.2)$$

Once one knows $p(y)$ one can readily calculate confidence intervals, standard deviation (also called standard error or simply error) or any other indicator of the uncertainty of the estimated value.

In practice an analytic solution is usually not feasible because the number of input parameters is too large and/or the p_i cannot be integrated in closed form. Here the Monte Carlo method offers a numerical solution. One chooses values $\{x_1, x_2, \ldots x_n\}$ at random, according to the respective probability distributions, calculates the corresponding y, and repeats this process many times. If the samples $\{x_1, x_2, \ldots x_n\}$ are chosen according to the respective probability distributions, the distribution of the resulting y approximates the integral of Eq. (11.1). The larger the number of samples, the better the approximation. To choose the samples, in the typical situation where one assumes that the variables are uncorrelated, one needs the inverse $P_i^{-1}(x_i)$ of each of the cumulative probability distributions $P_i(x_i)$,

$$P_i(x_i) = \int_{-\infty}^{x_i} p_i(x_i')\,dx_i'. \qquad (11.3)$$

One obtains random values x_i with the required probability distribution by setting $P_i(x_i)$ equal to a random number RAND between 0 and 1 and solving for the corresponding,

$$x_i = P_i^{-1}(\text{RAND}). \qquad (11.4)$$

That is the principle. However, simply choosing each of the x_i independently at random is not the most efficient method of obtaining a representative sample set, especially when the number of variables is large. Latin hypercube sampling and orthogonal sampling yield sample sets that, for a given size, are more representative of the full probability distributions. For Latin hypercube sampling, the sample space for each variable is divided into M intervals of equal probability. In two dimensions, a sample grid is a Latin square if (and only if) there is only one sample in each row and each column. A Latin hypercube is the generalization of this concept to an arbitrary number of dimensions, whereby each sample is the only one in each axis-aligned hyperplane that contains it.

Let us illustrate the Monte Carlo method with a very simple example. If the benefit B of a new regulation that limits the emission of SO_2 has been estimated to be B = 4 M€, while the required abatement cost is C = 2 M€, the regulation seems to be clearly justified. But if both of these numbers are quite uncertain, there is a risk that the real cost exceeds the benefit. Suppose that the distributions $p_B(B)$ of B and $p_C(C)$ of C are lognormal with geometric means μ_{gB} = 4 M€ and μ_{gC} = 2 M€ and geometric standard deviations σ_{gB} = 3 and σ_{gC} = 1.5. What is the probability that the regulation is indeed justified economically? A geometric standard deviation of 3 for B is fairly realistic in that the distribution of SO_2 damage costs is approximately lognormal with $\sigma_g \approx 3$, as we will show in Section 11.3 (the lognormal distribution is, loosely speaking, normal at a logarithmic scale, see Section 11.2.4). Abatement costs for new regulations are also quite uncertain because they tend to involve changes in technologies or their use. Thus we are faced with a cost-benefit analysis (CBA) where both C and B have probability distributions and we want to know the probability distribution of either the difference B – C or the ratio B/C.

Applying Eq. (11.1) to the difference B – C one obtains the distribution p(y) of y = B−C as,

$$p(y) = \int dB \ p_B(B)p_C(B - y). \tag{11.5}$$

This integral could be evaluated by standard techniques of numerical integration. But one can also obtain the distribution by the above Monte Carlo calculation. For this example that is fairly easy to do in Excel (versions for Mac 2007 to 2011) if one uses the function LOGNORM.INV, with the random number generator RAND() as one of its arguments. LOGNORM.INV (p, LN(μ_g), LN(σ_g)) is the inverse of the lognormal cumulative probability distribution with geometric mean μ_g and geometric standard deviation σ_g, evaluated at probability p. The function RAND() returns a random number between 0 and 1 with equal probability. Thus according to Eq. (11.4),

$$\text{LOGNORM.INV(RAND(), LN}(\mu_{gB}), \text{LN}(\sigma_{gB}))$$

returns random values of B with the lognormal distribution with parameters μ_{gB} and σ_{gB}. The probability distribution of B can be obtained by evaluating this quantity a large number of times and grouping the results in bins by means of the function COUNTIF(,). With the latter function one can plot histograms, as the Analysis Toolpack is no longer available in the Mac version of Excel. Since there are only two variables, B and C, we do not bother with Latin hypercube sampling.

Figure 11.1 shows the distributions we have obtained with 10,000 sets of random values {B, C}, using bins of width 0.25. The distributions have long tails on the right. The integral of the B − C curve from −∞ to 0 is the probability that the benefit B is smaller than the cost C; the numeric

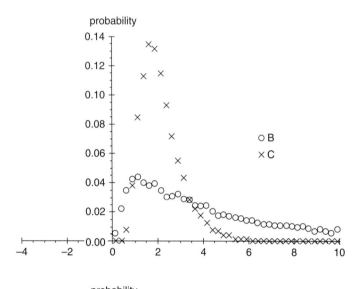

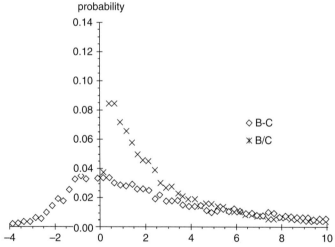

Fig. 11.1 Probability distributions of B, C, B − C and B/C for a simple CBA. B is lognormal with μ_{gB}= 4 and σ_{gB} = 2; C is lognormal with μ_{gC} = 2 and σ_{gC} = 1.5.
   (a) Probability distributions of B and C
   (b) Probability distributions of B − C and B/C

result is 0.27. There is a significant risk of the benefit being smaller than the cost even though the geometric mean of B is twice as large as that of C.

Had we chosen a normal instead of a lognormal distribution for C, some of the values would be negative and so would the corresponding B/C. In such a case the usual criterion that B/C should be greater than unity is modified by noting that the project is beneficial if B/C is outside the interval [0,1]. If both B and C can be negative, the B/C criterion is not relevant.

In practice one frequently assumes triangular probability distributions, a choice which can readily exclude unrealistic outliers or negative values; see Section 11.2.5. In Excel, the inverse of the cumulative triangular distribution is not available as a built-in function but it is not difficult to program. There are also Excel based software packages, e.g. @RISK® and Crystal Ball®, that can readily carry out Monte Carlo calculations with the most common probability distributions.

11.2.2 The calculation of damage costs

For damage costs of pollution there is also an analytic alternative to a Monte Carlo calculation. To explain why, we recall the equations used for calculating the damage costs. The impact rate $\dot{I}$ due to inhalation of an air pollutant can be written in the form,

$$\dot{I}(\dot{m}) = \int dx \int dy \; \rho(\mathbf{x}) \; S_{ERF}(\mathbf{x}) \; c(\mathbf{x}, \dot{m}), \qquad (11.6)$$

where,

$\dot{I}$ ($\dot{m}$) = impact rate (cases/yr),

$\dot{m}$ = emission rate of pollutant (kg/yr),

$c(\mathbf{x}, \dot{m})$ = increase in concentration ($\mu g/m^3$) at a point $\mathbf{x} = (x,y)$ due to the emission $\dot{m}$,

$\rho(\mathbf{x})$ = density of receptors (population, buildings, crops, ...) (receptors/m^2) at $\mathbf{x}$, and

$S_{ERF}(\mathbf{x})$ = slope of the exposure–response function (ERF), at $\mathbf{x}$ ((cases/yr)/(receptor·($\mu g/m^3$))).

For some persistent pollutants, in particular As, Hg, Pb, and dioxins, some or most of the damage is due to ingestion, and the ingestion dose can be one to two orders of magnitude larger than the inhalation dose (Spadaro and Rabl, 2004). The calculation of impacts due to ingestion has the same form as Eq. (11.6), with $c(\mathbf{x}, \dot{m})$ being replaced by the concentration in the food or drinking water. If the ERF is based on the ingested dose, an additional factor is needed to convert from concentration to dose.

The marginal unit impact I_u (in physical units) is the ratio of $\dot{I}$ ($\dot{m}$) and $\dot{m}$ in the limit of small $\dot{m}$,

$$I_u = \dot{I}(\dot{m})/\dot{m}, \tag{11.7}$$

and multiplication by the unit cost P (€/case) yields the unit damage cost in €/kg for the impact in question. The total unit damage cost D_u of the pollutant is obtained by summing the individual $I_i P_i$ over all impacts i caused by this pollutant (for health the various impacts are called end points),

$$D_u = \sum_i I_{u,i} P_i. \tag{11.8}$$

Assessments by EPA (Abt, 2000 and 2004) and by the ExternE project series have found that more than 95% of the total quantified damage cost (for each air pollutant with the exception of O_3 and greenhouse gases) is due to health impacts. Since the ERF slopes S_{ERF} for health impacts are assumed to be independent of $\mathbf{x}$, S_{ERF} can be taken outside the integral. Let us designate the remaining integral, divided by the emission rate, as exposure E per emitted quantity of pollutant,

$$E = \int dx \int dy \; \rho(\mathbf{x}) \; c_{air}(\mathbf{x}, \dot{m})/\dot{m}. \tag{11.9}$$

This is the contribution of inhalation to the intake fraction as defined by Bennett et al. (2002). Thus for health impacts (and any other impact whose S_{ERF} is independent of $\mathbf{x}$) the unit damage cost due to inhalation can be written as,

$$D_u = E \sum_i S_{ERFi} \; P_i. \tag{11.10}$$

Since E involves the integration of a complicated function, its uncertainty is more difficult to evaluate, although one can simplify by using approximations, in particular replacing the integral by a sum over finite areas. Furthermore, as shown in Section 7.4 of Chapter 7, there is a very simple approximation, the "uniform world model" (UWM) of Eq. (7.51) that yields results for typical situations. With this model, the uncertainty can be estimated with an explicit formula. A more detailed Monte Carlo calculation is described in Section 11.3.1.2.

For pollutants that are harmful by ingestion or dermal contact, the calculations are more involved but the basic structure of the equations (combination of sums and products) is similar if transfer factors are used, as shown in Section 8.4.

11.2.3 Uncertainty of sums and products

11.2.3.1 Sums

The UWM for the damage cost of a single impact (or end point) involves a simple product. For many pollutants a single impact, mortality, contributes more than two-thirds of the total damage cost (ORNL/RFF, 1994; Rowe *et al.*, 1995; Abt, 2000 and 2004; ExternE, 1998, 2000 and 2008), and the uncertainty of the mortality cost can be taken as a first estimate for the sum of the impacts. If several impacts make a significant contribution, one also has to sum over such products for the total damage cost. For sums and products an analytical solution is possible.

To begin, consider the sum,

$$y = x_1 + x_2 + \ldots + x_n \qquad (11.11)$$

of n uncorrelated random variables x_i. For example, the total cost due to health impacts, Eq. (11.10), is a sum over end points i where the costs P_i and the ERF slopes $S_{ERF,i}$ are uncorrelated between different end points i. Suppose that each of these x_i has a probability density distribution $p_i(x_i)$ with mean μ_{xi} and standard deviation σ_{xi}, μ_{xi} and σ_{xi} having been estimated from the available data. The mean of y can be calculated as,

$$\mu_y = \int dx_1 \int dx_2 \ldots \int dx_n \, p_1(x_1) \, p_2(x_2) \ldots p_n(x_n)(x_1 + x_2 \ldots + x_n), \quad (11.12)$$

with the result,

$$\mu_y = \mu_{x1} + \mu_{x2} + \ldots + \mu_{xn}. \qquad (11.13)$$

The square of the standard deviation σ_x of a variable x with distribution $p(x)$ is the variance,

$$\sigma_x^2 = VAR(x) = \int dx(x - \mu_x)^2 p(x) . \qquad (11.14)$$

Because the x_i are uncorrelated, the standard deviation σ_y of the sum y is given by the usual quadratic combination,

$$\sigma_y^2 = \sigma_{x1}^2 + \sigma_{x2}^2 + \ldots + \sigma_{xn}^2 \quad \text{for sum of uncorrelated } x_i \qquad (11.15)$$

of the standard deviations σ_{xi} of the x_i.

Even though these relations are exact, regardless of the size of the standard deviations, they do not yield an interpretation of σ_y in terms of confidence intervals (CI). For that one also needs the probability distribution of y. Fortunately in many cases of practical interest the distributions are approximately Gaussian. In particular, in the limit where the number of terms in the sum becomes large, the central limit theorem of statistics implies that the distribution of y approaches a Gaussian regardless of the individual distributions of the terms in the sum. In practice, the distribution of y is close to a Gaussian unless one or several of the terms have distributions that have large standard deviations and are very different from Gaussian. When the distribution of y is nearly Gaussian one can say that,

$$[\mu_y - \sigma_y, \mu_y + \sigma_y] \text{ is approximately the 68\% CI}$$

and

$$[\mu_y - 2\sigma_y, \mu_y + 2\sigma_y] \text{ is approximately the 95\% CI.}$$

It is instructive to illustrate with some examples why a simple method can be sufficiently accurate for estimating the uncertainty of the sum. Table 11.1 shows what happens for the sum w of two uncorrelated random variables v_1 and v_2, with means μ_i and standard deviation σ_i. Both terms have the same relative error σ_i/μ_i, here taken as 300% to illustrate a case of large uncertainties. In part (a) the uncertainty σ of the smaller term

Table 11.1 *Examples of a combination of errors in a sum* $w = v_1 + v_2$ *of two uncorrelated random variables* v_1 *and* v_2, *each with mean* μ_i *and relative error* $\sigma_i/\mu_i = 300\%$.

(a) *First term is 80% of total.*

Variable	μ	σ	σ^2	Relative error
v_1	0.8	2.4	5.76	300%
v_2	0.2	0.6	0.36	300%
$w = v_1 + v_2$	1.0	2.47	6.12	247%

(b) *First term is 65% of total.*

Variable	μ	σ	σ^2	Relative error
v_1	0.65	1.95	3.80	300%
v_2	0.35	1.05	1.10	300%
$w = v_1 + v_2$	1.0	2.21	4.91	221%

increases the standard error of the sum only to 2.47 compared with 2.4 for the larger term alone. In part (b) the magnitude of the means μ_1 and μ_2 is chosen to correspond roughly with the relative contributions of mortality and the other impacts on the total damage cost of PM; even here the second term increases the error only from 1.95 to 2.21.

In these examples the contribution of the error of the smaller term to the total error of the sum is quite small, as shown by the column for the variance $\sigma^2 = \sigma_1^2 + \sigma_2^2$. Thanks to the quadratic combination of errors, terms with small errors make negligible contributions to the total error. If a single term dominates, as is the case with mortality for the damage cost of the classical air pollutants, the standard deviation of the sum is not much larger than that of the largest summand.

As for *relative* errors, even in part (b) the difference between the relative errors of the larger term (300%) and of the sum (221%) is not very significant in view of the subjectivity of any uncertainty estimate in this domain. Therefore one obtains a first approximation and upper limit for the relative errors if one considers only the uncertainty of the largest term (mortality) and takes its σ_g as an appropriate estimate of the uncertainty of the total damage cost of these pollutants; it is an upper limit not much above the exact result.

If the sum in Eq. (11.11) contains terms that are partially correlated, Eq. (11.15) is replaced by,

$$\sigma_y^2 = \sum_i^n \sum_j^n COV(x_i, x_j), \qquad (11.16)$$

with the covariance matrix,

$$COV(x_i, x_j) = \int dx_i \int dx_j \, p(x_i, x_j)(x_i - \mu_i)(x_j - \mu_j). \qquad (11.17)$$

For $i \neq j$ one has,

$$COV(x_i, x_j) = 0 \quad \text{for uncorrelated variables,}$$

and

$$COV(x_i, x_j) = \sigma_i \sigma_j \quad \text{for perfect correlation.}$$

For example, if v_1 and v_2 are perfectly correlated, the standard deviation of their sum is $\sigma = \sigma_1 + \sigma_2$, larger than the square root of $\sigma_1^2 + \sigma_2^2$ for the uncorrelated case. In practice, we find that many cases are perfectly *uncorrelated*, e.g. the sum of the costs of CO_2 and of the classical air pollutants. By contrast, for the sum of the costs of PM, NO_x and SO_2 we assume perfect correlation, since the same exposure–response functions and monetary values are used, and the dispersion part of the analysis is also correlated.

11.2.3.2 Products

These considerations apply also to the product z of *uncorrelated* variables x_i if one looks at the logarithm,

$$z = x_1 x_2 x_3 \ldots x_n \quad \text{or} \quad \ln(z) = \ln(x_1) + \ln(x_2) + \ldots + \ln(x_n). \quad (11.18)$$

For example, the factors of the UWM are uncorrelated with each other. The mean of the logarithm of a random variable is the logarithm of the geometric mean μ_g; specifically, if $p(z)$ is the probability distribution of z, the geometric mean is given by,

$$\ln(\mu_{g,z}) = \int_0^\infty p(z)\ln(z)dz. \quad (11.19)$$

Since the mean of $\ln(\mu_{g,z})$ is the sum of the logarithms of the geometric means $\mu_{g,xi}$ of the x_i, $\mu_{g,z}$ is given by the product,

$$\mu_{g,z} = \mu_{g,x1}\, \mu_{g,x2} \cdots \mu_{g,xn}. \quad (11.20)$$

Let us now define the geometric standard deviation $\sigma_{g,z}$ as,

$$\left[\ln(\sigma_{g,z})\right]^2 = \int_0^\infty p(z)\left[\ln(z) - \ln(\mu_{g,z})\right]^2 dz \quad (11.21)$$

and analogously for the x_i. The right-hand side is ≥ 0 and therefore σ_g is always ≥ 1; a value of 1 indicates zero uncertainty. Applying Eq. (11.15) to the sum of the logarithms one finds that the geometric standard deviation $\sigma_{g,z}$ of the product z is given by,

$$\left[\ln(\sigma_{g,z})\right]^2 = \left[\ln(\sigma_{g,x1})\right]^2 + \left[\ln(\sigma_{g,x2})\right]^2 + \ldots + \left[\ln(\sigma_{g,xn})\right]^2$$
$$\text{for product of uncorrelated } x_i. \quad (11.22)$$

In addition to Eq. (11.20) for the geometric means, an analogous relation holds for the ordinary means if the variables are uncorrelated,

$$\mu_z = \int dx_1 \int dx_2 \ldots \int dx_{n-1}\; p_1(x_1)p_2(x_2)\ldots p_n(x_n)(x_1 x_2 \ldots x_n)$$
$$= \mu_{x1}\, \mu_{x2} \cdots \mu_{xn} . \quad (11.23)$$

Since the UWM approximates the damage costs as a product of factors, one can characterize its uncertainty by its geometric standard deviation, calculated by means of Eq. (11.22). All one needs to do is to estimate the geometric standard deviation of each of the factors in the UWM. This will be described in Section 11.3.

11.2.4 The lognormal distribution

In practice the lognormal distribution is far more common than most people realize, a situation highlighted in an interesting review by Limpert et al. (2001). Many distributions in nature are asymmetrical, especially when means are small, variances large and values cannot be negative. Often such distributions are approximately lognormal. Examples are species abundance, latency periods of infectious diseases, distributions of mineral resources in the earth's crust, and concentrations of air pollutants.

The lognormal distribution of a variable x is obtained by assuming that the logarithm of x has a normal distribution. Consider a normal distribution with mean ξ and standard deviation ϕ,

$$g(u) = \frac{1}{\phi\sqrt{2\pi}} \exp\left(-\frac{(u - \xi)^2}{2\phi^2}\right), \tag{11.24}$$

and replace the variable u by u = ln(x). Since g(u) is normalized to unity when u is integrated from $-\infty$ to $+\infty$, the normalization integral becomes,

$$\int_0^\infty \frac{g\left(\ln(x)\right)}{x} \, dx = 1, \tag{11.25}$$

which allows one to interpret the function,

$$p(x) = \frac{g\left(\ln(x)\right)}{x} = \frac{1}{\phi x\sqrt{2\pi}} \exp\left[-\frac{\left(\ln(x) - \xi\right)^2}{2\phi^2}\right] \tag{11.26}$$

as the probability density of a new distribution between 0 and $+\infty$. This is the lognormal probability density distribution. Its geometric mean μ_g and geometric standard deviation σ_g are related to ξ and ϕ by,

$$\mu_g = \exp(\xi) \quad \text{and} \quad \sigma_g = \exp(\phi). \tag{11.27}$$

Invoking the central limit theorem for the product z, one sees that the lognormal distribution is the "natural" distribution for multiplicative processes, the same way that the Gaussian distribution is natural for additive processes. Although the lognormal distribution becomes rigorous only in the limit of infinitely many factors, in practice it can be a good approximation even for a few factors, provided the distributions with the largest spread are not too far from lognormal. For many environmental impacts the lognormal model for the result is quite relevant, to the extent that the impact is a product of factors and the distributions of the individual factors are not too far from lognormality. All one has to do is estimate the geometric standard deviations of the individual factors and combine them according to Eq. (11.22).

An example is shown in Fig. 11.2. For the lognormal distribution, the geometric mean μ_g is equal to the median. If a quantity with a lognormal distribution has a geometric mean μ_g and a geometric standard deviation σ_g, the probability is approximately 68% for its value to be in the interval $[\mu_g/\sigma_g, \mu_g \sigma_g]$ and 95% for it to be in the interval $[\mu_g/\sigma_g^2, \mu_g \sigma_g^2]$, in other words the confidence intervals (CI) are:

$$68\% \text{ CI is } [\mu_g/\sigma_g, \ \mu_g \sigma_g] \quad \text{and} \quad 95\% \text{ CI is } [\mu_g/\sigma_g^2, \ \mu_g \sigma_g^2]. \quad (11.28)$$

One can show that the ordinary mean μ and standard deviation σ of the lognormal variable are given by

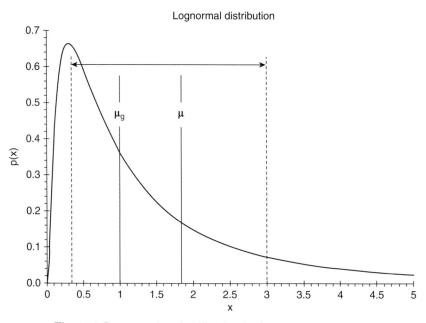

Fig. 11.2 Lognormal probability density function ($\mu_g = 1$, $\sigma_g = 3$, mean $\mu = 1.83$). The arrows indicate the 68% confidence interval, [1/3, 3].

$$\mu = \exp(\xi + \phi^2/2) = \mu_g \exp\left(\frac{[\ln(\sigma_g)]^2}{2}\right) \qquad (11.29)$$

and

$$\sigma = \sqrt{[\exp(\phi^2) - 1]\exp(2\xi + \phi^2)} = \mu \sqrt{(\mu/\mu_g)^2 - 1}. \qquad (11.30)$$

Given any one of the pairs $\{\xi, \phi\}$, $\{\mu, \sigma\}$ or $\{\mu_g, \sigma_g\}$, the others can be determined by means of Eqs. (11.27), (11.29) and (11.30).

If a distribution of a variable x is approximately lognormal and one knows the lower and upper limits of the 68% CI $[x_{68\%l}, x_{68\%u}]$ (or of the 95% CI $[x_{95\%l}, x_{95\%u}]$), one can estimate μ_g and σ_g as,

$$\mu_g = \sqrt{x_{68\%l}\, x_{68\%u}} \quad \text{or} \quad \mu_g = \sqrt{x_{95\%l}\, x_{95\%u}} \qquad (11.31)$$

and

$$\sigma_g = \sqrt{x_{68\%u}/x_{68\%l}} \quad \text{or} \quad \sigma_g = \sqrt[4]{x_{95\%u}/x_{95\%l}} \qquad (11.32)$$

11.2.5 The triangular distribution

In practice, one rarely has sufficient information for a rigorous determination of the geometric standard deviation σ_g. To begin with, for most data one does not know enough about the underlying probability distribution. Many distributions are neither normal nor lognormal but somewhere in between. For the normal distribution the geometric standard deviation is complex, because part of the variable range is negative. In reality for many variables negative values are unphysical. In practice, one frequently assumes triangular probability distributions. The triangular distribution has the advantage of avoiding very large or small values, in particular the negative contributions of the Gaussian distribution. Also, in contrast with the box distribution (equal probability within specified range), it expresses the idea that the probability is highest for the central value. It can be symmetric or asymmetric.

Let us look at the symmetric triangular distribution, shown in Fig. 11.3, with the equation,

$$p_{tri}(x) = \begin{cases} 0 & \text{for x} < 0 \text{ and x} > 1 \\ 4x & \text{for } 0 < x < 0.5 \\ 4(1-x) & \text{for } 0.5 < x < 1 \end{cases} \qquad (11.33)$$

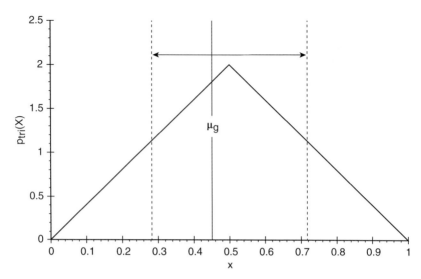

Fig. 11.3 Triangular distribution of Eq. (11.33). The arrows indicate the 68% confidence interval.

Using a program such as Mathematica it is easy to evaluate μ_g and σ_g as well as the confidence intervals for 68% and 95%. Inserting $p_{tri}(x)$ into Eq. (11.19) we obtain,

$$\mu_g = 0.44626 \tag{11.34}$$

and inserting this with $p_{tri}(x)$ into Eq. (11.21) we obtain,

$$\sigma_g = 1.71202. \tag{11.35}$$

The confidence intervals can be found by solving,

$$\int_{x_{68\%l}}^{1-x_{68\%l}} p_{tri}(x)dx = 0.68 \quad \text{and} \quad \int_{x_{95\%l}}^{1-x_{95\%l}} p_{tri}(x)dx = 0.95 \tag{11.36}$$

with the result,

$$[x_{68\%l}, x_{68\%u}] = [0.2828, 0.7172] \text{ and } [x_{95\%l}, x_{95\%u}] = [0.1118, 0.8882]. \tag{11.37}$$

If instead of Eq. (11.21) (which is exact) we had used Eq. (11.32) to estimate σ_g we would have found,

$$\sqrt{x_{68\%u}/x_{68\%l}} = 1.592 \quad \text{or} \quad \sqrt[4]{x_{95\%u}/x_{95\%l}} = 1.679\,, \qquad (11.38)$$

sufficiently close to the exact answer in Eq. (11.35). We have repeated these calculations for a triangular distribution that is shifted to the right by one unit, i.e. where in the intervals of Eq. (11.33) the 0 is replaced by 1, the 0.5 by 1.5 and the 1 by 2. In this case the confidence intervals are simply shifted to the right by one unit and σ_g becomes 1.14872. Now Eq. (11.32) yields the σ_g estimates 1.15696 and 1.14158, respectively, for the 68% and 95% confidence intervals, again sufficiently close to the exact value. Since all these estimates of σ_g are sufficiently close to the exact answer, even for a distribution very different from lognormal, we recommend Eq. (11.32) as a general rule for estimating the geometric standard deviation even when the distribution is not known.

11.2.6 Application to CBA

At this point an example may be helpful, illustrating the application of these ideas to a simple CBA. Let us consider a proposed policy that would reduce the emission of PM_{10} from a cement kiln that burns waste (e.g. old tires and waste oil) as fuel, as discussed later in Section 11.7. The available information on the costs and benefits is shown in Fig. 11.12 of that section. Reading from that graph the benefit B, i.e. the avoided damage of PM_{10} emissions, is about 0.2 €/$t_{clinker}$. For the abatement cost C the low and high estimates are 0.7 and 2.3 €/$t_{clinker}$, respectively. Clearly not a very attractive policy option. But what is the probability that the abatement option might be justified after all?

The lack of sufficient information is typical. For the uncertainty of the benefit B we anticipate a result from Section 11.3.4, namely that the uncertainty of PM damage costs can be characterized by a distribution that is approximately lognormal with geometric standard deviation of about 3, and so we take $\sigma_{g,B} = 3$. Not knowing the probabilities of the low and high estimates of C, let us consider them as limits of either the 68% or the 95% confidence intervals (CI). Thus Eq. (11.32) yields,

$$\sigma_{g,C} = \sqrt{2.3/0.7} = 1.81 \quad \text{if taken as } 68\% \text{ CI, or}$$
$$\sigma_{g,C} = \sqrt[4]{2.3/0.7} = 1.35 \quad \text{if taken as } 95\% \text{ CI.}$$

Inserting $\sigma_{g,B}$ and $\sigma_{g,C}$ into Eq. (11.22),

$$[\ln(\sigma_{g,B/C})]^2 = [\ln(\sigma_{g,B})]^2 + [\ln(\sigma_{g,C})]^2,$$

we obtain $\sigma_{g,B/C}$ of the B/C ratio as,

$$\sigma_{g,B/C} = 3.5 \quad \text{if} \quad \sigma_{g,C} = 1.81 \quad \text{and} \quad \sigma_{g,B/C} = 3.1 \quad \text{if} \quad \sigma_{g,C} = 1.35,$$

as shown in this little table:

	$\sigma_{g,i}$	$\ln(\sigma_{g,i})^2$	$\sigma_{g,i}$	$\ln(\sigma_{g,i})^2$
$\sigma_{g,B}$	3	1.2069	3	1.2069
$\sigma_{g,C}$	1.81	0.3538	1.35	0.0884
$\sigma_{g,B/C}$	3.49	1.5607	3.12	1.2954

By far the largest contribution to the total uncertainty, in terms of $\ln(\sigma_{g,i})^2$, comes from B. Since we do not know how to interpret the cost limits for C, let us simply take the average of these two estimates and say that $\sigma_{g,B/C} = 3.3$.

To proceed we also need to know the distribution of B/C. Since by far the largest uncertainty comes from B for which the distribution is approximately lognormal, the distribution of B/C is also not far from lognormal. Let us interpret the benefit B in Fig. (11.12) as geometric mean,

$$\mu_{g,B} = 0.2 \,\text{€} / t_{clinker},$$

and let us take the costs C to have a geometric mean of,

$$\mu_{g,C} = \sqrt{0.7 \times 2.3}\,\text{€}/t_{clinker} = 1.269\,\text{€}/t_{clinker}.$$

Then the geometric mean of B/C is,

$$\mu_{g,B/C} = 0.2/1.269 = 0.158.$$

So we estimate the CI of B/C as,

the 68% CI from $0.158/3.3 = 0.048$ to $0.158 \times 3.3 = 0.52\ \text{€}/t_{clinker}$

and

the 95% CI from $0.158/3.3^2 = 0.014$ to $0.158 \times 3.3^2 = 1.72\ \text{€}/t_{clinker}.$

This allows us to estimate the probability that the abatement option might be justified after all. For that purpose one can use the function, LOGNORM. DIST(x,mean,standard_dev,) of Excel, which calculates the cumulative lognormal distribution of x, here = B/C, where ln (x) is normally distributed with parameters mean = mean of ln(x), here $= \ln(\mu_{g,B/C}) = \ln(0.158)$, and standard_dev of $\ln(x)$, here $= \ln(\sigma_{g,B/C}) = \ln(3.3)$. The result,

$$\text{LOGNORMDIST } (1, \ln (0.158), \ln (3.3), 1) = 0.939,$$

says that the probability of the abatement option being justified is $1 - 0.939$, about 6%.

This example illustrates how a rather simple analysis can yield estimates of confidence intervals and probabilities that are just about as reliable as one could obtain from the scant information that is given. We had to make somewhat arbitrary assumptions, for instance for the choice of the geometric means of B and C (we will say more about the choice between ordinary means and geometric means in Section 11.3.5.7). The calculations are admittedly rough, but what else could one do with the available information? Estimating uncertainties is an uncertain business.

11.3 Component uncertainties and results for air pollution

11.3.1 *Models*

11.3.1.1 Atmospheric models

For air pollution, by far the most complex part of the analysis concerns the dispersion and chemistry of the pollutant in the atmosphere. Many models are available, ranging from simple Gaussian plume models to very detailed and computationally intensive Eulerian models. In this section we review those aspects that affect the uncertainty. For the uncertainty of atmospheric models a geometric standard deviation in the range from two to five is sometimes cited, but without making a distinction between episodic values and averages over space or time. In fact, atmospheric models are far more accurate for averages than for episodic values. This is an important consideration, since for policy applications one needs long-term average values rather than episodic values. For example, the European tracer experiment (ETEX) (van Dop *et al.*, 1998) has provided validation for a variety of dispersion models, but on an episodic basis; the relatively large discrepancies between measured and calculated values are therefore no indication of the accuracy that can be expected for long-term averages. Generally models for non-reactive species have greater accuracy than models that include chemical reactions; the reactions of nitrogen compounds are especially difficult to model.

To compare long-term average concentrations between models and measured data one needs either models at the continental scale, with a complete emissions inventory, or else special situations where a single source dominates the concentrations in a local zone (another possibility is to emit a tracer gas, but the quantity tends to be unacceptably large for long-term tests). Comparisons for long-term averages at the regional scale can be found in the reports of the EMEP Program (see for instance Barrett, 1992). They generally indicate agreement within a factor of two or better. At the local scale, we found an interesting confirmation of the Gaussian plume model ISC (Brode and Wang, 1992) with SO_2 data for a refinery in Donges, near Nantes, France. The refinery is by far the dominant source of SO_2 in the region. Our ISC calculations, with detailed emissions data for the year 2000, including stack heights of all the sources at the refinery, indicate that the refinery contributes about 3.5 μg/m^3 on average at the measuring station some 2 km from the source. The annual averages measured in 2000 were 8.8 μg/m^3 at this station, and 3.6 μg/m^3 and 4.7 μg/m^3, respectively, in the nearby cities, St.-Nazaire (about 15 km upwind) and in Nantes (about 25 km downwind); the latter two values can be taken as an approximation of the background. The difference between the measured value 8.8 μg/m^3 near the refinery and the background of approximately 4.1 μg/m^3, i.e. 4.7 μg/m^3, agrees reasonably well with the calculated contribution of 3.5 μg/m^3 from the refinery, considering the uncertainties of such calculations.

Among the many parameters and input data from an atmospheric model, most have only a relatively minor effect on the calculation of long-term average concentrations. To see which parameters are the most important, it is instructive to look at the UWM of Section 7.4, Eq. (7.51), because it yields the damage costs for typical conditions. The UWM for the unit damage cost D_u of a particular impact due to the inhalation of a primary pollutant is,

$$D_u = P \, S_{ERF} \, \rho_{eff} / k_{dep}, \qquad (11.39)$$

where,

> P = cost per case ("price") (€/case),
> S_{ERF} = ERF slope ((cases/yr)/(pers·(μg/m^3))),
> ρ_{eff} = average receptor density (pers/km^2) within 1000 km of source, and
> k_{dep} = depletion velocity of pollutant (dry + wet) (m/s).

D_u is the damage cost due to the emission of a specified quantity of the pollutant and has units of €/kg (the units indicated here are customary for the respective quantities, but in the equation they must of course be converted to a consistent set). For secondary pollutants the equation has the same form, but k_{dep} is interpreted as depletion velocity, a quantity that includes the transformation rate of the primary into the secondary pollutant, as well as the deposition velocities of the primary and secondary pollutants.

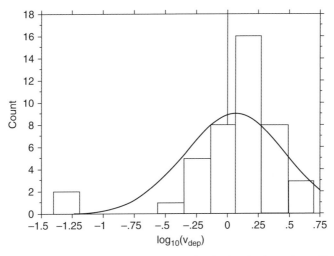

Fig. 11.4 Distribution and lognormal fit of values in the review of Sehmel, (1980), for dry deposition velocity (in cm/s) of SO_2 over different surfaces. From Rabl and Spadaro (1999), with permission of Elsevier.

For the UWM, the key parameters of a dispersion model are those that affect the deposition and/or depletion velocity k_{dep}. For an indication of the kind of distribution that can be expected, we show in Fig. 11.4 a histogram of dry deposition velocity data for SO_2, based on a review by Sehmel (1980). Visibly, a logarithmic scale is much more appropriate for these data than a linear scale. The geometric standard deviation is approximately 2.5 for this sample. The variability of this sample is due to different surface materials, atmospheric conditions and variation with time of day and year. More recent data may have a smaller standard deviation, but we have not been able to find a survey as comprehensive as that of Sehmel. Even though variability does not imply uncertainties if the model accounts correctly for all of its causes, in practice, most models cannot treat all the necessary detail and so the variability increases the uncertainties.

For particles, we refer to Fig. 16.20 of Seinfeld (1986), which likewise suggests a lognormal distribution. Dry deposition of small particles has been reviewed by Nicholson (1988), who points out the large variability of measured deposition velocities with the nature of the surface and the conditions of the observation. The spread of values seems to be comparable with Fig. 11.4 for SO_2. Deposition velocities for reactive nitrogen compounds have been reviewed by Hanson and Lindberg (1991); here the variability with the conditions of the absorbing surface is further enhanced by the high chemical reactivity of nitrogen compounds.

The possibility of low values of dry deposition velocities could imply very large damages under dry conditions. However, for the wet climates typical of Europe, long distance dispersion will be limited by wet deposition. We have verified this for particulate matter by varying the dry deposition velocity in the EcoSense model, the model used by ExternE. Thus the uncertainty of total deposition in Europe appears to be significantly smaller than suggested by dry deposition data.

11.3.1.2 A Monte Carlo analysis of exposure

In contrast with typical results for the UWM, for site-specific calculations, the exposure involves integration over the entire region of the product of receptor density and concentrations, the latter calculated using detailed dispersion models. We have estimated the uncertainty of exposure by means of a Monte Carlo calculation, taking into account the uncertainties of the numerous input data (Spadaro and Rabl, 2008a). Since the probability distributions for some of the distributions are not well known, we have considered several possible cases. Only dispersion is taken into account, without chemical reactions.

The analysis for primary air pollutants starts from the mass balance for the average pollutant concentration in a column of air that moves with the wind from source to receptors. For an initial analysis we have made several assumptions to calculate the ground level concentrations; they are generally made by models that calculate collective exposure and are believed to be acceptable for that purpose:

- A1: the pollutant moves along straight trajectories away from the source;
- A2: the wind speed does not vary with height (although for a given stack height we take this wind speed to be the value calculated at the stack height via a power-law relationship);
- A3: wind speed, mixing layer height and atmospheric stability class are constant for a puff moving along its trajectory;
- A4: there is no exchange with the upper atmosphere above the mixing layer height;
- A5: the distributions of the parameters are statistically independent;
- A6: for the ratio of the ground level and the column-average concentrations one can take the ratio of a Gaussian plume dispersion model, multiplied by a random number with a lognormal distribution (this is to take into account the fact that the real concentrations are not vertically uniform).

We then carried out a sensitivity analysis to show that the results do not change significantly when these assumptions are relaxed. This approach provides a model-independent assessment of the uncertainty of any dispersion model that satisfies the above assumptions, including the model used by the ECOSENSE (www.externe.info) software of ExternE.

We obtained results for the dispersion of PM_{10} with stack height 75 m and typical plume rise at three locations: a very large population center (Paris), an intermediate site (Lauffen near Stuttgart), and a rural site (Albi in the southwest of France). The uncertainty of the collective exposure, expressed as geometric standard deviation σ_g, ranges from about 1.2 for Paris and 1.5 for Stuttgart to about 1.9 for Albi. The uncertainty is larger for rural sites, because for a rural site the regional impacts dominate and the regional impacts are very sensitive to assumptions about the deposition velocity, whereas deposition is almost negligible in the local zone.

Based on these results, we assume a σ_g of 1.5 for the dispersion modeling of primary non-reactive air pollutants, since pollution sources tend to be located more in or around cities than in rural areas. For the dispersion of secondary pollutants we take a larger σ_g of 1.7, because their formation takes place over large distances and the impacts are almost entirely regional, a situation more akin to Albi than Stuttgart. These numbers are consistent with estimates by McKone and Ryan (1989). For secondary pollutants there is additional uncertainty due to chemistry, especially in the case of ozone, but the uncertainty due to inaccuracies in the spatial distribution of concentration values relative to receptors is much smaller, because secondary pollutants form only gradually at distances removed from the source. Since the chemical reactions depend on the background concentrations which are not sufficiently well known, we also introduce a σ_g for the effect of background emissions.

These considerations lead us to assume:

$\sigma_g = 1.5$ for the dispersion of non-reactive primary pollutants,

$\sigma_g = 1.7$ for the dispersion of SO_2 and sulfates,

$\sigma_g = 1.2$ for the formation of sulfates from SO_2,

$\sigma_g = 1.05$ for the effect of background emissions on the formation of sulfates from SO_2,

$\sigma_g = 1.7$ for the dispersion of NO_x and nitrates,

$\sigma_g = 1.4$ for the formation of nitrates from NO_x,

$\sigma_g = 1.15$ for the effect of background emissions on the formation of nitrates from NO_x.

The key parameters for these processes enter in approximately multiplicative fashion (in the UWM, their combination is exactly multiplicative).

11.3.2 Exposure–response functions

The uncertainty of exposure–response functions varies widely from case to case. The most established ones are those for health impacts from radionuclides, the ERFs for certain health impacts from the classical pollutants (PM_{10}, SO_2, NO_2 and O_3), and the ERFs for impacts of SO_2, NO_2 and O_3 on certain crops whose economic importance has prompted laboratory studies.

The ERFs for health impacts are based on RR (relative risk) data in epidemiological studies. Their confidence intervals are reported for 95% probability, and they are approximately symmetric (of the form $\mu \pm 2\sigma$) around the mean, μ. The underlying probability distributions (implicit in the regression software used in the respective studies) are usually not lognormal, hence it is necessary to estimate equivalent geometric standard deviations, σ_g.

If one knows the probability distribution of the residuals in the respective studies, one could calculate the geometric standard deviation exactly from its definition in Eq. (11.21). If one does not, but the reported confidence intervals are symmetric, it is reasonable to assume a Gaussian distribution. Strictly speaking, the resulting σ_g is complex because the Gaussian is non-zero at negative values. However, negative values are not plausible on physical grounds (for health impacts of air pollutants a beneficial effect is not plausible), and the distribution should be cut off at zero. Furthermore, if one uses only ERFs that are statistically significant at the 95% level, the contribution of the negative values represents at most 2.5% of the normalization integral of the Gaussian, and the effect on the resulting σ_g would be negligible.

A much simpler, albeit approximate, alternative is to use Eq. (11.32). Specifically, suppose that $\mu \pm \sigma$ corresponds to a 68% confidence interval, as for a Gaussian distribution. Then Eq. (11.32) yields σ_g as,

$$\sigma_g = \sqrt{\frac{\mu + \sigma}{\mu - \sigma}}. \qquad (11.40)$$

We have evaluated Eq. (11.40) for all of the ERFs of ExternE for NO_x, SO_2, PM and O_3; typically, σ_g is in the range 1.2 to 1.8. Specifically for chronic mortality, Table 2 of Pope et al. (2002) indicates, for the average exposure during the observation period, a relative risk RR given by $RR - 1 = 0.06$ with 95% confidence interval [0.02, 0.11]. With $\mu = 0.06$ and $\sigma = 0.023$, we find $\sigma_g = 1.48$. In the following we will take 1.5 as a typical value.

For chronic mortality, one also needs to determine the relation between the YOLL (years of life lost) and the change in the age-specific mortality rate that has been reported by studies of chronic mortality. Leksell and Rabl (2001) have examined the uncertainties of this calculation; their results suggest a σ_g of 1.3 for the calculation of the YOLL, given the relative risk.

There is, however, another type of uncertainty due to the difference between the PM in ambient air on which epidemiology is based and the primary and secondary PM in the damage calculations. Ambient PM is a mix of primary PM from combustion and secondary PM, especially nitrates (due to NO_x emissions) and sulfates (due to SO_2 emissions). For the damage calculations one needs assumptions about the relative toxicity of the different components of ambient PM. The uncertainty of these assumptions is difficult to estimate, because of the lack of data. To deal with this issue we introduce a factor for the respective toxicities of primary particles, nitrates and sulfates, relative to ambient PM, and we assume a σ_g for these toxicities (our choices are a subjective judgment based on extensive discussions with epidemiologists and toxicologists). There could also be important direct impacts of NO_2 and SO_2, but currently the dominant thinking among epidemiologists is that they are negligible compared with those of PM and O_3 (see e.g. WHO, 2003).

To sum up, we assume:

$\sigma_g = 1.5$ (range 1.2 to 1.8) for the morbidity ERFs due to ambient PM,

$\sigma_g = 1.5$ for the mortality risk (RR-1) due to ambient PM,

$\sigma_g = 1.5$ for the toxicity of primary particles relative to ambient PM,

$\sigma_g = 2$ for the toxicity of nitrates and sulfates relative to ambient PM,

$\sigma_g = 1.3$ for the calculation of the YOLL for a given mortality risk.

These elements enter the calculation in multiplicative fashion.

11.3.3 Monetary valuation

Some physical impacts can be easily valued by their price on the market, e.g. the price of crops. There is little uncertainty in these prices as quoted at any particular place and time; uncertainty comes mainly from their future evolution (since the interesting policy applications concern the future) and from possible errors in collecting the information. Geometric standard deviations around 1.1 to 1.3 seem reasonable.

Non-market goods are difficult to value economically. This is especially true for mortality, i.e the value of a prevented fatality (VPF) or the value of a life year (VOLY). It is a difficult good to monetize, and the uncertainty is large. The distribution of VPF results from various studies of individual preferences tends to be lognormal, as illustrated for example by Fig. 11.5 which is based on the Ives et al. (1993) survey of 78 VPF studies published between 1973 and 1990. Figure 11.5 gives equal weight to all studies, regardless of quality or age. The resulting large spread of values could

Table 11.2 *Uncertainty of VPF: distributional characteristics from the survey by Ives* et al. *(1993).*

mean	Million £ 2.76 (£$_{1990}$ 1 = $ 1.78)
standard deviation	Million £ 3.00
median	Million £ 1.59
geometric mean μ_g	Million £ 1.49
geometric standard deviation σ_g	**3.4**

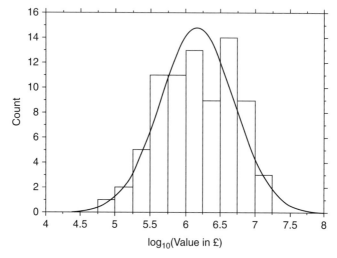

Fig. 11.5 Example of lognormal distribution for economic valuation: value of a prevented fatality (VPF), in £$_{1990}$, as determined by 78 studies reviewed by Ives *et al.* (1993), histogram and lognormal fit plotted on log scale. From Rabl and Spadaro (1999), with permission of Elsevier.

probably be reduced by applying reasonable selection criteria. The spread reflects both uncertainty (due to methodology) and variability (due to study population and type of mortality risk).

The distribution is asymmetrical and has a large tail of high outliers, a situation typical of such valuation studies. As a consequence, if the spread is large, the mean is much larger than the median. This is illustrated in Table 11.2, which summarizes the distributional characteristics of Ives *et al.* The spread is so large that an interval of plus/minus one ordinary standard deviation extends to negative values. Clearly it does not make much sense to use ordinary mean and standard deviation in such cases. The median is far less affected by outliers, and in Table 11.2 it is fairly close to the geometric mean; the spread is best expressed by the geometric standard deviation.

Example 11.1   If the distribution of VPF values were exactly lognormal with $\mu_g = £\ 1.49$ million and $\sigma_g = 3.4$, what would be the mean? What is the $1\ \sigma_g$ confidence interval?

Solution: From Eq. (11.29), we have,

$$\mu = \mu_g \exp\left(\frac{[\ln(\sigma_g)]^2}{2}\right) = £\ 1.49\,\text{million} \times 2.1144 = £\ 3.15\,\text{million},$$

somewhat different from the £ 2.76 million in Table 11.2, because the distribution is not exactly lognormal. The $1\ \sigma_g$ confidence interval is £ [0.438, 5.07] million.

Mortality valuation is a good example with which to highlight the subjective judgment involved in estimating uncertainties. The task would be straightforward only if all studies looked at the same population and applied the same methodology. But in reality there is such a variation in methodologies and underlying data that one has no clear criterion for selecting the studies to include in a meta-analysis. By applying strict selection criteria, a meta-analysis can reduce the range of plausible values, as was done for example by Mrozek and Taylor (2002). But the choice of such criteria does not find universal agreement. For example, in the USA, the Department of Transportation had been using $ 3 million for many years while the Environmental Protection Agency was using $ 6 million; only recently did the Department of Transportation adopt the same higher value. In the EU, lower values have been in use (EC, 2000a), in particular, ExternE (2008) takes VPF to be 1.5 million €. However, they are likely to be updated soon in view of the meta-analysis by Lindhjem *et al.* (2011), on the basis of which OECD (2012) recommends $3.6 million (2.8 million €) for the EU27, almost twice as high. We find it difficult to estimate σ_g for mortality or morbidity. The uncertainty has been reduced since the review by Ives *et al.*, but in view of the fact that even during recent times, recommended values have changed by a factor of about two, we believe that large uncertainties remain, and we assume $\sigma_g = 2$ for mortality. We also assume $\sigma_g = 2$ for the cost of suffering due to morbidity.

To sum up, we assume,

  $\sigma_g = 1.1$ to 1.3 for market prices (cost of medical treatment, crop losses, repair cost for materials, etc.),

  $\sigma_g = 2$ for the WTP to avoid suffering or death, in particular for chronic bronchitis and for mortality.

Table 11.3 *Uncertainty of damage cost estimates per kg of pollutant for mortality. Sample calculations of geometric standard deviation* σ_g*, inserting the component uncertainties* $\sigma_{g,i}$ *into Eq. (11.22). The relative contributions of the* $\sigma_{g,i}$ *to total can be seen under* $\ln(\sigma_{g,i})^2$*.*

	lognormal?	$\sigma_{g,i}$ PM	$\ln(\sigma_{g,i})^2$	$\sigma_{g,i}$ SO$_2$ via sulfates	$\ln(\sigma_{g,i})^2$	$\sigma_{g,i}$ NO$_x$ via nitrates	$\ln(\sigma_{g,i})^2$
Exposure calculation							
Dispersion	yes	1.5	*0.164*	1.7	*0.282*	1.7	*0.282*
Chemical transformation	yes	1	*0.000*	1.2	*0.033*	1.4	*0.113*
Background emissions	no	1	*0.000*	1.05	*0.002*	1.15	*0.020*
Total σ_g *for exposure*		1.50	*0.164*	1.76	*0.32*	1.90	*0.415*
ERF							
Relative risk	no	1.5	*0.164*	1.5	*0.164*	1.5	*0.164*
Toxicity of PM components	?	1.5	*0.164*	2	*0.480*	2	*0.480*
YOLL, given relative risk	?	1.3	*0.069*	1.3	*0.069*	1.3	*0.069*
Total σ_g *for ERF*		1.88	*0.40*	2.33	*0.71*	2.33	*0.71*
Monetary valuation							
Value of YOLL (VOLY)	yes	2	*0.480*	2	*0.480*	2	*0.480*
Total (Eq. (11.22))		**2.78**	***1.04***	**3.42**	***1.51***	**3.55**	***1.61***

11.3.4 Total σ_g for the classical air pollutants

Based on the preceding Sections 11.3.1 to 11.3.3, we show in Table 11.3 our choices for $\sigma_{g,i}$ of the components of the damage cost calculation for mortality. The bottom line shows the resulting σ_g of the damage cost. The contributions of the individual $\sigma_{g,i}$ to the total σ_g can be seen in the column with $[\ln(\sigma_{g,i})]^2$.

For sulfates and nitrates, ExternE assumes the same ERFs as for PM (apart from an overall scale factor), therefore we take the contributions to the uncertainty to be the same for each of these pollutants, with the exception of

(i) atmospheric dispersion and chemistry (we assume different geometric standard deviations for PM, NO$_x$ and SO$_2$),

(ii) the toxicities of primary PM, sulfates and nitrates relative to ambient PM$_{10}$, as discussed at the end of Section 11.3.2.

The resulting geometric standard deviations are 2.78 for primary PM, 3.42 for SO$_2$ and 3.55 for NO$_x$. The distribution of a product is exactly lognormal if each of the factors is lognormal. In practice, it is sufficient for the factors with the largest widths to be approximately lognormal. In the present case, lognormality for the distribution of the result is plausible for the damage costs.

We show three significant figures in Table 11.3, just to bring out the differences between these pollutants and the larger uncertainties of the secondary pollutants. But in view of the subjective and rather uncertain assumptions we had to make about the component uncertainties, we believe that it is best to simply sum up the results by saying that the geometric standard deviation of the damage costs of PM, NO_x and SO_2 is approximately 3.

Because of the quadratic combination, small σ_g make a negligible contribution. Thus one has to focus on the major uncertainties without worrying too much about the less uncertain items. Equation (11.22) is simple and transparent, and it is easy to change the $\sigma_{g,i}$ and see the result.

We have taken the mortality cost as an indicator for the uncertainty of the total damage cost, because this end point makes the dominant contribution and the $\sigma_{g,i}$ for the other end points are not very different. It would be more rigorous, but also more work, to first calculate the uncertainty of the sum over all end points according to Eq. (11.15), before using Eq. (11.22).

11.3.5 Total σ_g for dioxins and toxic metals

11.3.5.1 Total σ_g for Hg

A detailed site-specific assessment of the impacts of Hg emissions is extremely difficult and uncertain, because of the very complex pathways for the dispersion of this pollutant in the environment. An estimate of the global average damage cost, on the other hand, is relatively simple, as shown in Section 8.3, and it is relevant because much if not most of the emitted Hg disperses globally because of the long residence time of Hg in the atmosphere.

The starting point of the analysis is the observation that the dominant pathway is the ingestion of fish and seafood, and the transfer factor from emission to ingestion can be estimated by comparing the global average dose and emissions, both of which have been estimated in a number of studies. Such a comprehensive transfer factor is much easier to determine and less uncertain than a detailed calculation of all the individual transfer factors, in particular the fraction transformed to methyl-Hg (MeHg) and the bioconcentration factor in seafood. The most troubling and currently best understood health impacts of Hg are neurotoxic (see Axelrad *et al.* (2007) and references therein), and we evaluate them using as proxy recent dose–response function data for IQ decrement due to MeHg ingestion.

Table 11.4 *Our estimates of the uncertainties of the marginal damage cost of Hg for the case with threshold (Section 8.3). The geometric standard deviations $\sigma_{g,i}$ of the individual factors are estimated as the 4th root of the ratios high/low. Some numbers for the results have been rounded in the text.*

Factor	Unit	Median	Low	High	$\sigma_{g,i}$	$ln^2(\sigma_{g,i})$
Emission rate	t/yr	6000	3000	9000	1.32	0.08
Average dose, d_{av}	µg/day	2.40	0.60	4.00	1.61	0.22
Fraction of dose above threshold $d_{av}(d_{th})/d_{av}$	–	0.44	0.20	1.20	1.57	0.20
S_{ERF} based on Hg in hair	IQpt/ppm	0.18	0.009	0.38	2.55	0.87
Hair conc/cord blood conc	ppm/(µg/L_{cord})	0.20	0.10	0.30	1.32	0.08
Ratio of cord to maternal blood	(µg/L_{cord})/(µg/L_{mat})	1.65	0.41	3.14	1.66	0.26
Ratio of maternal blood to dose intake	(µg/L_{mat})/(µg/day)	0.61	0.40	1.00	1.26	0.05
Cost of IQ pt in USA	$/IQpt	18,000	10,000	25,000	1.26	0.05
f_{world} [a]	1/yr	0.00315	0.0014	0.0071	1.50	0.16
Discount factor, f_{disc}	–	0.64	0.30	0.82	1.29	0.07
World population, p_{world}	persons	6.43E+09			1.00	0.00
Results for damage cost D	**$/kg**	**1,487**			**4.2**	**2.04**

[a] f_{world} = fraction of the world population p_{world} that is affected per year by Hg, weighted by GDP/capita.

Our estimates of the uncertainties are listed in Table 11.4 for the case with threshold; without threshold the uncertainty would be lower. The table lists our assumptions for the uncertainty of each step of the analysis. We have taken the geometric standard deviations $\sigma_{g,i}$ of the individual factors as 4th root of the ratios high value/low value, roughly interpreting the low to high range as width of the 95% confidence interval. For the estimation of the average dose above threshold, we have considered a wide range of possible dose distributions and their effect on the ratio $d_{av}(d_{th})/d_{av}$; we believe that this accounts for the uncertainty. The uncertainty of f_{world} includes the GDP/capita adjustment of the cost of an IQpoint in different countries. The resulting σ_g is 4.2.

Of course, our choice of the $\sigma_{g,i}$ is somewhat arbitrary, since the probability distributions of most of the parameters may be quite different from lognormal and our high and low values may not be the correct 95% confidence intervals, but even with different and equally plausible choices the σ_g of the result would not be very different as the reader can readily verify. Note that because of the quadratic combination of the individual terms (see last column of the table), terms with relatively small uncertainty make negligible contributions to the total. In view of the great uncertainty of the estimation of uncertainties, we round the numbers to summarize by saying that σ_g is about 4.

11.3.5.2 Total σ_g for As, Cd, Cr-VI and Ni

The metals As, Cd, Cr-VI and Ni are considered carcinogenic via inhalation. For these metals only cancers have been included in the damage cost estimates of ExternE (2008). The cancers due to Cd, Cr-VI and Ni are mostly lung cancers and thus essentially fatal (even if about 10% of the victims survive more than five years, the conventional threshold for considering a cancer cured). Thus the cost per cancer death is appropriate.

For impacts due to inhalation, the damage cost can be estimated by means of the "uniform world model." The most uncertain factors of that calculation for cancers due to inhalation are the depletion velocity of atmospheric dispersion, the slope S_{ERF} of the exposure–response function and the cost of a fatal cancer. The latency and discount rate r_{dis} introduce some additional but much smaller uncertainty. Our assumptions for the uncertainties of these factors are listed in Table 11.5, stated in terms of their $\sigma_{g,i}$. Our estimate of 3 for σ_g of S_{ERF} is a rough typical value for this type of carcinogen, obtained by a simple assessment of the information readily available on the IRIS website of EPA, rather than a detailed analysis of each of the respective studies on which the S_{ERF} is based. The resulting geometric standard deviation of the damage cost is 3.9, to be rounded to 4, because the estimation of uncertainties is at best an approximate matter with unavoidable subjective judgment.

11.3.5.3 Total σ_g for dioxins

Dioxins have a variety of impacts on human health, but cancers are the only impacts for which both dose–response functions and monetary values are

Table 11.5 *Assumptions for the uncertainties of the key input parameter for the calculation of the damage cost of cancers due to inhalation of As, Cd, Cr and Ni.*

	lognormal?	$\sigma_{g,i}$	$[\ln(\sigma_{g,i})]^2$
Dispersion	yes	1.5	0.16
S_{ERF}	probably yes	3	1.21
Cost of cancer	yes	2	0.48
r_{dis} and latency	probably yes	1.2	0.03
Total σ_g		**3.9**	**1.88**

sufficiently well established for quantification of damage costs. Dioxins (2,3,7,8-TCDD and HxCDD) were said by EPA to be "the most potent carcinogen(s) evaluated by the EPA's Carcinogen Assessment Group." In calculating the dose for dioxins, one needs to take non-inhalation pathways into account, because dioxins are persistent and bioaccumulate, becoming concentrated in milk, meat and fish. Figure II-5, p. 37 of the report "Estimating exposure to dioxin-like compounds" vol. I Executive Summary (EPA, 1994) indicates a typical value of,

$$\frac{\text{total dose}}{\text{inhalation dose}} = \frac{119}{2.2} = 54 \tag{11.41}$$

for the ratio of total dose to inhalation dose for dioxins. The precise value can vary strongly with diet and local conditions, but for calculation of typical values, the use of Eq. (11.41) is more appropriate than a site-specific calculation.

In any case the non-inhalation dose involves dispersion over large distances (e.g. by transport of cattle feed, followed by transport of meat). This has the effect of making the receptor distribution more uniform. Therefore we estimate the unit damage cost D_u of dioxin by multiplying the "uniform world model" for inhalation, Eq. (11.39) of Section 11.3.1, by the ratio of Eq. (11.41),

$$D_u = 54 \, P \, S_{ERF} \, \rho_{eff}/k_{dep}, \tag{11.42}$$

where,

P = unit cost of cancer ("price"),
S_{ERF} = slope of ERF,
ρ_{eff} = average population density of the region,
k_{dep} = mean depletion velocity.

Of course, several of these factors are extremely uncertain and the result should be rounded to one or at most two significant figures. Our assumptions for the uncertainties are listed in Table 11.6. The total geometric standard deviation is 4.7, to be rounded to 5.

11.3.5.4 Total σ_g for Pb

Pb has many harmful effects on human health. For one of the most important end points, loss of IQ points, the damage cost has been calculated by Spadaro and Rabl (2004), with a recent update for the NEEDS project. The ERF is relatively well established, but there are major uncertainties in the modeling of the ingestion dose. The resulting geometric standard deviation is 4.1, as shown in Table 11.7; this can be rounded to 4.

Table 11.6 *Assumptions for the uncertainties of cancers due to dioxins.*

	lognormal?	$\sigma_{g,i}$	$[\ln(\sigma_{g,i})]^2$
Atmospheric dispersion	Yes	1.5	0.16
Total dose/inhalation dose	probably yes	2	0.48
S_{ERF}	probably yes	3	1.21
Cost of cancer	Yes	2	0.48
r_{disc} and latency	probably yes	1.2	0.03
Total σ_g		**4.7**	**2.36**

Table 11.7 *Assumptions for the uncertainties of IQ loss due to Pb.*

	lognormal?	$\sigma_{g,i}$	$[\ln(\sigma_{g,i})]^2$
Atmospheric dispersion	Yes	1.5	0.16
Total dose/inhalation dose	probably yes	3	1.21
S_{ERF}	probably yes	1.5	0.16
Cost of IQ point	Yes	2	0.48
Total σ_g		**4.1**	**2.02**

11.3.5.5 Greenhouse gases

For greenhouse gases we have used an entirely different approach, because the pathways and models are far too complex for a detailed uncertainty analysis by anyone not intimately familiar with all the details of the models. Instead we have estimated the uncertainty by considering the spread of values found in the review by Tol (2005), who carried out a critical review of all publications that have tried to estimate the damage cost of greenhouse gases. The probability distribution of the values can be approximated by a lognormal with geometric standard deviation $\sigma_g = 4.5$ (estimated as the average of the square roots of the ratios of 95% confidence interval and median in his Table 3), if one omits publications with negative damage costs – cases we consider to be unrealistic, because the total worldwide damage cost will certainly be positive, even if some countries may obtain a benefit. More recently, Tol (2008) has updated his review, with similar results; for this more recent review, we find that σ_g of a lognormal approximation is 4.4. Since few if any of the publications in these two reviews succeed in assessing problematic impacts (socially contingent or potentially catastrophic ones), we round up and assume $\sigma_g \approx 5$ as an estimate of the uncertainty of global warming costs.

Table 11.8 *Summary of geometric standard deviations σ_g for the damage costs.*

Pollutant	σ_g
NO$_x$, NMVOC, PM and SO$_2$	3
As, Cd, Cr-VI, Hg, Ni, Pb	4
Dioxins	5
Greenhouse gases	5

11.3.5.6 Summary

Since the estimation of uncertainties is in itself uncertain, we believe that it would be appropriate to cite just a single set of results for air pollutants such as PM, SO$_2$ and NO$_x$ that act via inhalation. Likewise we round off the σ_g for the other pollutants. In any case, for policy applications typical uncertainties are more instructive than detailed values of geometric standard deviation for each source and each impact. Thus we summarize results of this section with the numbers in Table 11.8.

11.3.5.7 Placement of confidence intervals

We should comment on the placement of the confidence intervals (CI) relative to the damage cost estimates. That is a problem whenever there are asymmetric distributions because their means are different from their geometric means. Therefore one needs to decide whether the key parameters of the calculations have been estimated as means, medians or something else, for instance modes (= the point where the probability distribution of possible parameter values has its maximum). Since our CIs for the damage costs are symmetric around the geometric mean on a logarithmic scale, their placement relative to the quoted damage costs would have to be modified if the results have not been calculated as geometric means. For large σ_g there is a sizeable difference between mean and geometric mean. Recalling Eq. (11.29) for the ratio of mean μ and median μ_g of a lognormal distribution,

$$\mu/\mu_g = \exp\left(\frac{[\ln(\sigma_g)]^2}{2}\right), \tag{11.29}$$

we have plotted in Fig. 11.6, the ratios median/mean, upper/mean and lower/mean as a function of the geometric standard deviation σ_g. For example, with $\sigma_g = 3$ the median/mean ratio $\mu_{g/\mu}$ is 0.55.

The difficulty with applying Eq. (11.29) or Fig. 11.6 lies in the general lack of knowledge about the distributional assumptions behind the various

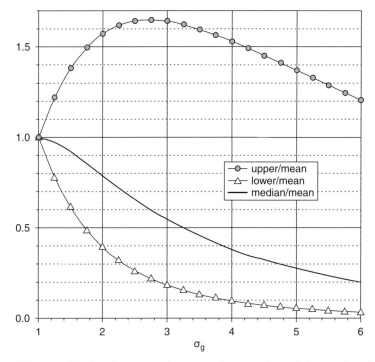

Fig. 11.6 Median (= geometric mean μ_g), upper bound ($\mu_g \times \sigma_g$) and lower bound (μ_g/σ_g), all divided by the mean μ, as function of the geometric standard deviation σ_g for a lognormal distribution. From Spadaro and Rabl (2008), with permission of Elsevier.

parameters of the calculation. There seems to be a general tendency for scientists to quote means when distributions are narrow, in which case the ratio mean/geometric mean is close to unity anyway. In distributions with relatively large widths (e.g. contingent valuations), outliers tend to be disregarded or their weight is reduced by taking the median rather than the mean; in that case the quoted values are closer to geometric means than to ordinary means. In the absence of more information about distributional assumptions, we assume therefore that the parameter values of the UWM are geometric means. Then the results of the damage cost calculation are also geometric means, and on a log scale the confidence intervals are symmetrical about the calculated damage costs. But if instead one assumes that the damage costs are ordinary means, one has to calculate the corresponding geometric means via Eq. (11.29) before placing the CIs. In that case the CIs are shifted downwards, implying a higher probability for the calculated value to be an underestimate rather than an overestimate.

11.4   Sum of lognormal variables

11.4.1   Combination of classical air pollutants and greenhouse gases

In practice, the largest environmental externalities are due to classical air pollutants and greenhouse gases. When they are added to obtain the total external cost per product, e.g. per kWh of electricity, the uncertainty of the result depends on the pollutant mix that is emitted. Let us see what happens when two uncorrelated costs with different uncertainties are added, as is the case for the sum of classical air pollutants and greenhouse gases, for which we take geometric standard deviations $\sigma_{g,1} = 3$ and $\sigma_{g,2} = 5$ respectively. Figure 11.7 shows a Monte Carlo calculation of the resulting distribution, together with a lognormal fit, for the sum of two equal costs (geometric means $\mu_{g,1} = \mu_{g,2} = 1$). The lognormal fit has $\sigma_g = 2.6$, smaller than for each of the summands alone.

We have found that in external cost calculations, lognormal fits are usually a sufficiently good approximation for sums of lognormal variables, and we have developed a simple procedure for calculating the resulting σ_g; it

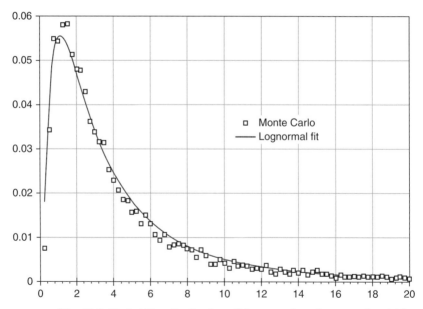

Fig. 11.7 Probability distribution of the sum of two lognormal variables with $\{\mu_{g,1}=1, \sigma_{g,1}=3\}$ and $\{\mu_{g,2}=1, \sigma_{g,2}=5\}$: data points (Monte Carlo calculation) and fit by a single lognormal distribution with $\{\mu_g=2.9, \sigma_g=2.6\}$.

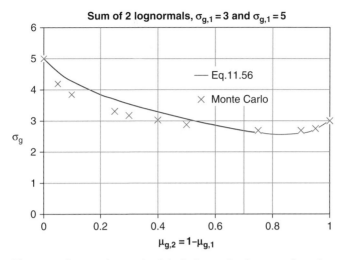

Fig. 11.8 Geometric standard deviation σ_g for the sum of two lognormal variables, one with $\sigma_{g,1} = 3$, the other with $\sigma_{g,2} = 5$, for a range of different contributions of each, expressed by their respective geometric means $\mu_{g,1}$ and $\mu_{g,2} = 1 - \mu_{g,1}$. Lognormal fit to Monte Carlo calculation and results of Eq. (11.56).

is presented in the following Section 11.4.2. Here we anticipate some results of this procedure by applying it to the sum of two lognormals, one with $\sigma_{g,1} = 3$ (for classical air pollutants) the other with $\sigma_{g,2} = 5$ (for greenhouse gases). In Fig. 11.8, we plot the results of this procedure for a range of different contributions of each, expressed by their respective geometric means $\mu_{g,1}$ and $\mu_{g,2} = 1 - \mu_{g,1}$. The fits to detailed Monte Carlo calculations are shown by crosses, the results of the procedure of Section 11.4.2 are labeled Eq. (11.56). σ_g is significantly smaller than for greenhouse gases alone, and the procedure of Section 11.4.2 yields results that are quite acceptable in view of the overall uncertainties.

11.4.2 *The general case*

More generally, we have developed the following approximation for the sum of several lognormal variables. To begin with we have verified with Monte Carlo calculations for a large number of realistic cases that the sum of lognormally distributed damage costs is, to a fair approximation, also lognormal, as illustrated in Fig. 11.7 and Fig. 11.9. We have then derived two analytical estimates for the geometric standard deviation of the sum, one over- and one underestimation. Finally, we have found that the average of these two estimates turns out to be remarkably close to the correct answer.

The first estimate is based on the relation between $\{\mu, \sigma\}$ and $\{\mu_g, \sigma_g\}$ of the lognormal distribution; its mean is (recalling Eq. (11.29)),

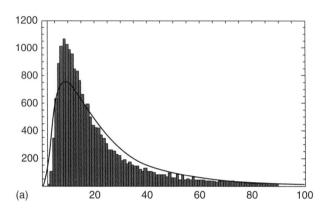

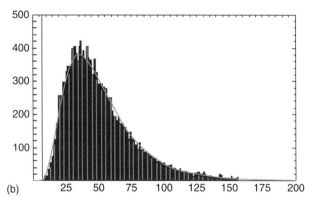

Fig. 11.9 Three examples of frequency distribution of the sum of lognormal variables, as well as fits by a lognormal distribution. Geometric mean and geometric standard deviation of the summands are indicated in the form $\{\mu_g, \sigma_g\}$.

(a) $\{2, 1.2\}$, $\{4, 2\}$ and $\{8, 4\}$. Lognormal fit has $\{\mu_g, \sigma_g\} = \{17.9, 2.30\}$.

(b) $\{2, 2\}$, $\{10, 2\}$ and $\{30, 2\}$. Lognormal fit has $\{\mu_g, \sigma_g\} = \{46.57, 1.67\}$.

(c) $\{0.56, 5\}$, $\{0.66, 3.42\}$, $\{1.1, 3.55\}$ and $\{0.22, 2.78\}$. Lognormal fit has $\{\mu_g, \sigma_g\} = \{4.24, 2.32\}$.

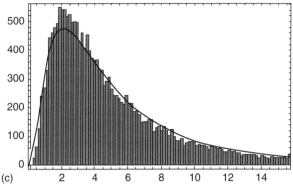

(c)

Fig. 11.9 (cont.)

$$\mu = \mu_g \exp\left(\frac{[\ln(\sigma_g)]^2}{2}\right) \tag{11.43}$$

and its standard deviation is (recalling Eq. (11.30)),

$$\sigma = \mu \sqrt{\left(\frac{\mu}{\mu_g}\right)^2 - 1} . \tag{11.44}$$

Using Eq. (11.43), we calculate the ordinary mean μ_j for each of the lognormal distributions j, and then we add them according to the usual rule to obtain the mean of the sum,

$$
\begin{aligned}
\mu_{\text{est1}} &= \sum_j \mu_j \\
&= \sum_j \mu_{g,j} \exp\left(\frac{[\ln(\sigma_{g,j})]^2}{2}\right).
\end{aligned}
\tag{11.45}
$$

Then, using Eq. (11.44) for the standard deviation of each of the lognormal distributions j, together with the usual rule for combining ordinary standard deviations for a sum, one obtains the standard deviation of the sum,

$$\sigma_{est1} = \sqrt{\sum_j \sigma_j^2} = \sqrt{\sum_j \mu_j^2 \left[\left(\frac{\mu_j}{\mu_{g,j}} \right)^2 - 1 \right]}. \qquad (11.46)$$

Now the first estimate of the geometric mean $\mu_{g,est1}$ and geometric standard deviation $\sigma_{g,est1}$ of the sum can be found by inverting Eqs. (11.43) and (11.44),

$$\mu_{g,est1} = \frac{\mu_{est1}^2}{\sqrt{\mu_{est1}^2 + \sigma_{est1}^2}} \qquad (11.47)$$

and

$$\sigma_{g,est1} = \exp\left[\sqrt{2 \ln\left(\mu_{est1}/\mu_{g,est1}\right)}\right]. \qquad (11.48)$$

For the second estimate, we calculate the quantities $\mu_{eq,j}$ and $\sigma_{eq,j}$ for the distribution of each term j in the sum according to the equations,

$$\mu_{eq,j} = 0.5 \times [\mu_{g,j} \times \sigma_{g,j} + \mu_{g,j}/\sigma_{g,j}] \qquad (11.49)$$

and

$$\sigma_{eq,j} = 0.5 \times [\mu_{g,j} \times \sigma_{g,j} - \mu_{g,j}/\sigma_{g,j}], \qquad (11.50)$$

i.e. we calculate an equivalent mean $\mu_{eq,j}$ as mid-point of the confidence interval of the lognormal distribution and an equivalent standard deviation $\sigma_{eq,j}$ as half width of this confidence interval. For $\mu_{eq,sum}$ and $\sigma_{eq,sum}$ of the sum we follow the usual rules, as we did for Eqs. (11.45) and (11.46),

$$\mu_{eq,sum} = \sum_j \mu_{eq,j} \qquad (11.51)$$

and

$$\sigma_{eq,sum} = \sqrt{\sum_j \sigma_{eq,j}^2}. \qquad (11.52)$$

We then apply Eqs. (11.49) and (11.50) to the sum (taking them with the subscript sum instead of j) and solve for $\mu_{g,est}$ and $\sigma_{g,est}$ of the sum to obtain the equations,

$$\mu_{g,est2} = \mu_{eq,sum} \sqrt{1 - \left(\frac{\sigma_{eq,sum}}{\mu_{eq,sum}}\right)^2} \qquad (11.53)$$

and

$$\sigma_{g,est2} = \sqrt{\frac{1 + \left(\frac{\sigma_{eq,sum}}{\mu_{eq,sum}}\right)}{1 - \left(\frac{\sigma_{eq,sum}}{\mu_{eq,sum}}\right)}} \qquad (11.54)$$

as an estimate of its geometric mean and geometric standard deviation.

We have compared these two approximations with the results of Monte Carlo calculations for 28 cases covering the ranges of the individual $\sigma_{g,j}$ that one is likely to encounter for air pollution damages, and we find that the first approximation overestimates σ_g of the sum by about 20%, whereas the second underestimates it by a comparable amount. Taking the simple averages,

$$\mu_{g,est} = (\mu_{g,est1} + \mu_{g,est2})/2 \qquad (11.55)$$

and

$$\sigma_{g,est} = (\sigma_{g,est1} + \sigma_{g,est2})/2, \qquad (11.56)$$

one comes sufficiently close to the exact result, as shown by the examples in Fig. 11.10. For someone without the tools and experience to do Monte Carlo calculations, the equations in this section thus offer a convenient alternative.

11.5   Problematic uncertainties

The uncertainty analysis presented above assumes that the component uncertainties of the impact pathway analysis can be characterized in terms of probability distributions. Even though that is a reasonable assumption for most damages of most pollutants, there are some problematic cases. In Section 11.5.1, we explain what the problems are and in Sections 11.5.2 and 11.5.3 we suggest methods of dealing with such cases.

Here it is appropriate to distinguish between different types of uncertainty. One type is the uncertainty of data and of model parameters; this can usually be treated in terms of probability distributions, as we have done above. Another type (sometimes called model uncertainty, although the definition of uncertainty types is not without ambiguity) occurs where an impact is suspected or even fairly likely, but the evidence is not sufficiently clear and one has to invoke expert judgment (really a euphemism for subjectivity). In such a situation exposure–response functions may be available in the literature and make it possible to calculate an

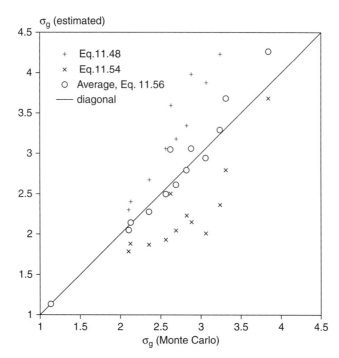

Fig. 11.10 Comparison of the geometric standard deviations estimated by Eq. (11.48), by Eq. (11.54), and by their average, Eq. (11.56), with a Monte Carlo calculation, for nine examples of sums of three lognormal distributions and four examples of sums of two lognormal distributions.

impact – but is it real? In Section 11.5.1, we illustrate the problem with three examples of problematic uncertainties, and in the following sections we discuss possible approaches to dealing with them.

11.5.1 Examples: CO_2, NH_3, and nitrates in drinking water

11.5.1.1 Nitrates in drinking water

As a first example let us consider the risk of cancer due to consumption of drinking water with nitrates, a case that has been examined in the EXIOPOL project by Hansen and Andersen (2009), who calculate external costs for nitrates in drinking water with the hypothesis that the effect is indeed real. With that hypothesis the calculation is straightforward and the uncertainty comparable to other pollution costs. But the mechanisms of carcinogenesis by nitrates, if any, are complex and difficult to establish, see e.g. van Grinsven et al. (2010), and at the present time the reality of a causal link is controversial, see e.g. L'hirondel *et al.* (2006). In such a case, the analyst ought to add a clear warning that the results have to be interpreted and used with great caution. The analyst should summarize the evidence for or against to give the reader an idea of the weight of the evidence or the lack of it.

11.5.1.2 Atmospheric NH_3 emissions

A somewhat similar question arises on the impact of NH_3 emissions to the atmosphere. Whereas NH_3 at typical atmospheric concentrations is totally harmless, it is a contributor to the formation of secondary pollutants, namely nitrates and sulfates, that are most probably harmful. Although the toxicity of $PM_{2.5}$ has been established beyond doubt, the relative toxicity of different components of the particle mass is difficult to determine. At the present time most health impact assessments (EPA in the USA (Abt, 2004), the World Health Organization (WHO, 2003), and the GAINS model of IIASA (www.iiasa.ac.at)) assume that the impacts are simply proportional to the mass of the suspended primary and secondary particles. ExternE (2005) has tried to differentiate between different components by assuming higher toxicity for primary PM from combustion, but in more recent phases (NEEDS and CASES), ExternE makes no distinction on the basis of chemical composition.

Whereas several studies, in particular the very important cohort study of mortality by Pope *et al.* (2002), have reported positive associations between health impacts and sulfates, the evidence for nitrate particles is far less convincing. For example, the review by Reiss *et al.* (2007) states, *For nitrate-containing PM, virtually no epidemiological data exist. Limited toxicological evidence does not support a causal association between particulate nitrate compounds and excess health risks.* The authors mention the possibility of indirect effects, via activation of the toxicity of other PM components, something that seems quite plausible. Even if nitrates are relatively harmless, one could argue that the current ExternE assumption of the same toxicity per mass for nitrates as for primary combustion PM is reasonable, in view of the numerous epidemiological studies that find similar associations for NO_2 as for PM_{10}.

But the issue of NH_3 is more complicated. Concentration data from sulfate particles used by epidemiological studies include, without distinction, droplets of H_2SO_4 (because it is captured by the detectors of PM monitoring stations), as well as NH_4HSO_4 and $(NH_4)_2SO_4$ which are solid. NH_3 emissions decrease the acidity by neutralizing the H_2SO_4. They also convert gaseous HNO_3, due to NO_x emissions, to solid NH_4NO_3. Acidity and, especially oxidizing potential, have sometimes been mentioned as factors that enhance the toxicity of PM. Since NH_3 is the only major air pollutant that is both a base and a reducing agent, one may wonder whether it can render ambient PM more toxic. NH_3 emissions increase the mass of nitrate aerosols, but are NH_4NO_3 particles really more toxic than gaseous HNO_3?

If NH_3 emissions do not change the toxicity of aerosols, their health impacts would be negligible (except for a totally different effect: since the deposition velocity of NH_4NO_3 is different from HNO_3, the impact on the total population is different). In view of the uncertainties about the health impacts of nitrates, it would be advisable, as with the case of nitrates in drinking water, to add a warning to the external cost estimates of NH_3 to the effect that the real numbers might well be much lower.

11.5.1.3 Global warming

The damage costs of greenhouse gases are problematic because some of the most troubling and potentially most important impacts are extremely difficult to assess; in particular, ecosystem impacts and impacts due to societal stress. Quite apart from any monetary valuation these impacts are extremely complex and uncertain. Ecosystem impacts involve not just the extinction of a few species but the very functioning of an environment on whose services we depend. To cite just a minor example, if bees were to disappear so would most of the fruit that we eat and all the associated products (fruit juices, desserts with fruit, ...). The consequences of societal stress (due to migrations, shortage of agricultural land in some regions, general poverty, ...) are even more difficult to evaluate. What if the Gulf Stream stops maintaining a comfortable climate in Europe? What if methane clathrates in Siberia release methane at a rapidly increasing rate? What if the ecosystem of the world undergoes an abrupt and irreversible shift with catastrophic consequences; a possibility highlighted by Barnosky *et al.* (2012)? Even if the likelihood of truly catastrophic impacts is very small, they must not be overlooked. As Weitzman (2007) points out, a conventional cost–benefit analysis (CBA) has difficulties with extreme events of low and unknown probability.

To think about the uncertainties of global warming costs, it is helpful to recall the matrix in Fig. 10.14 (in Chapter 10) where the different impact categories are arranged according to uncertainty in predicting the effect and uncertainty in monetary valuation. Until now most studies have been limited to the top left portion of the matrix. Hardly anyone has attempted to quantify the costs due to socially *contingent effects* or due to *system change and surprises,* and no plausible estimates are available. Even in the *bounded risks* and *non-market* fields the quantification of many costs has remained incomplete and problematic, in particular ecosystem changes and loss of biodiversity. But the non-quantified costs are likely to dominate. Therefore we are not sure whether our estimate of $\sigma_g = 5$ captures the full uncertainty.

11.5.2   Graphical presentation of impacts with very different uncertainties

If a CBA involves items with very different uncertainties, a good approach is to present it in graphical format as a stacked bar chart, the different items being stacked on top of each other in the order of increasing uncertainty, a format proposed by Holland and King (1998). In many cases the most certain items already suffice, by themselves, to justify a policy choice, and the answer is clear. For the example of climate change, the stacked bar chart in Fig. 10.13 (of Chapter 10) suggests an approach for the near future: for the time being, set the level of a carbon tax equal to the sum of the items that are relatively certain, namely *cooling+heating* and *agriculture*, and adjust the level as further research leads to better assessment of the other categories. Unfortunately even that approach is difficult to implement as long as there is no agreement on the appropriate choice of the discount rate (more precisely the pure rate of time preference) and on the use of equity weighting.

For the example of nitrates in drinking water, it could be that the costs of eutrophication are sufficiently large and certain to justify major limitations of nitrate pollution, even if any additional costs due to cancers are less certain. Unfortunately so far we do not have sufficient information to proceed even on this basis. But impact assessment is an evolving art, and as new results emerge, we can re-examine possible recommendations for environmental policies.

11.5.3   Sensitivity studies

For another perspective on uncertainties one can carry out sensitivity studies, a classical approach for dealing with uncertain input parameters or assumptions. One varies the input and sees how the result changes. Here we illustrate the technique with the example of climate change costs, because a sensitivity study is particularly appropriate for global warming. We use a graphical format that can show the effect of a large number of input parameters in a simple and concise manner.

Since it is not so much the external cost itself that matters, but the consequence for decisions, we have examined the effect of uncertainties on the choice of the optimal global emission ceiling for CO_2. To make the problem simple and transparent enough for a sensitivity study of all the important parameters, Rabl and van der Zwaan (2009) used a very simple dynamic model of global warming. Only a small number of parameters are important for the uncertainties. They are listed in Table 11.9, together with the equations for damage and abatement costs. The goal is to determine the

Table 11.9 *Central values and ranges of the parameters p for the damage and abatement costs in the optimization problem of Rabl and van der Zwaan (2009). The uncertainties of ΔT are included in the uncertainty range of ρ. N = time horizon for the calculation of costs, r_{dis} = discount rate, $G(t)$ = gross world product in year t, r_{gro} = growth rate of G(t), $E(t)$ = global emissions with emission reductions, in year t, $E_s(t)$ = global emissions without emission reductions, in year t.*

Parameters p	p_{min}	$p_{central}$	p_{max}
	Global parameters		
N (yr)	100	150	200
r_{dis}	0.01	0.02	0.03
r_{gro}	0.00	0.01	0.02
	Damage cost $C_{dam}(t) = G(t)\rho\left(\frac{\Delta T(t)}{2.5^\circ}\right)^2$		
ρ	0.01	0.02	0.03
	Abatement cost $C_{ab}(t) = \frac{\alpha E_s(t)}{\gamma+1}\left[\left(\frac{E_s(t)-\beta}{E_s(t)}\right)^{\gamma+1} - \left(\frac{E(t)-\beta}{E_s(t)}\right)^{\gamma+1}\right]$		
α [€/tCO$_2$]	5.0	7.5	10
β [GtCO$_2$/yr]	2	3	4
γ	−1.5	−1.3	−1.1

optimal emissions scenario $E_o(t)$ as a uniform reduction of the reference scenario $E_s(t)$ in the absence of reductions.

To present the results of such a sensitivity study, the normalized format of Fig. 11.11 is particularly instructive. Such a graph can display a large number of parameters by representing all parameters w that are varied (from w_{min} to w_{max}) in non-dimensional form as,

$$x = (2w - w_{max} - w_{min})/(w_{max} - w_{min}).$$

In Fig. 11.11, the optimal emission level is about a third of the current level E_s, and even very large changes in the parameters of the model do not drastically change this result.

Even though the range of parameters in Fig. 11.11 is too limited to account for catastrophic impacts, the line for the parameter ρ (which represents the magnitude of the damage cost) becomes quite flat as ρ increases. Even if the damage cost turns out larger than currently expected, not much can be done to reduce the emissions even further, because the abatement costs would become prohibitive, at least with the abatement cost assumptions in this paper. However, such emission levels would be so

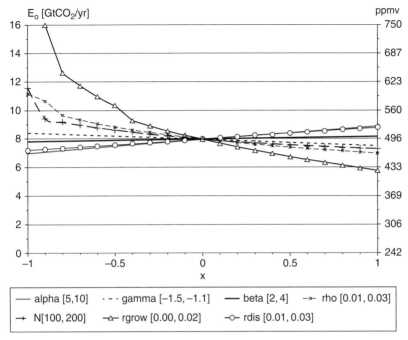

Fig. 11.11 Dependence of the optimal emissions level E_o (in GtCO$_2$/yr) (left-hand scale) on the parameters α, β, γ, ρ, r_{gro} and r_{dis}. The right-hand scale indicates the corresponding stabilized CO$_2$ concentrations. The uncertainties of ΔT are included in the uncertainty range of ρ. Each curve shows the effect of varying the parameter under consideration, while keeping the others fixed at their central value. The x-axis shows the variation of each parameter w in non-dimensional form as,

$$x = (2w - w_{max} - w_{min})/(w_{max} - w_{min}).$$

From Rabl and van der Zwaan (2009), with permission.

far below current levels that the corresponding abatement costs cannot be sufficiently well estimated at the present time for a meaningful CBA. Rather, the message of this analysis is that one needs already to aim for large emission reductions now; the detailed evolution of the emissions should then be determined by continually updating such a CBA in the light of new information.

Of course, a sensitivity analysis is no better than the model on which it is based. The equations of Table 11.9 do not include the possibility of a threshold beyond which the damage becomes catastrophic. The risk of such a threshold for global ecosystem impacts due to greenhouse gas emissions has been highlighted in a recent paper by Barnosky *et al.*

(2012). Cost–benefit analysis is not well suited for catastrophic impacts and it is appropriate to turn to the precautionary principle.

11.5.4   The precautionary principle and external costs

The precautionary principle is sometimes invoked in situations where costs and benefits cannot be estimated with sufficient reliability. Unfortunately it is not an operational principle because it gives no specific guidance, other than "think before you act," i.e. try to estimate cost, benefits and uncertainties as well as can reasonably be achieved. There is no easy answer, since all depends on the specifics of a situation. For example, how should one weigh the risks of nuclear power against the risks of global warming? One person's precaution is another person's folly. That very observation leads us to a bit of advice for dealing with uncertainty: see how others, with whom you don't agree, assess the problem in question. Try to understand why, because they may have a point that you have overlooked.

In spite of its vagueness the precautionary principle can offer useful advice in some cases, although, again, without indicating definite answers.

 (i) If a very rough estimate of costs and benefits suggests that costs could be large but benefits are definitely small (or vice versa), the prudent choice is fairly clear.
 (ii) If there is a small risk of extremely high damage, it is advisable to seek ways to avoid such risk.
 (iii) If damage is irreversible or will last a very long time, it is much more serious than short-term damage and should be weighted much more heavily in a CBA.
 (iv) If a decision can be postponed or implemented gradually, wait or proceed with caution (e.g. reintroduction of nuclear). Unfortunately that may not be a good option for CO_2, even though it is advocated by many: because of the long time constants major emission reductions may already be needed now before it is too late.
 (v) Diversify to reduce the total risk – a classic strategy, well known by investors. For example, develop several technologies of solar PV (crystalline, amorphous, thin film, and systems with high optical concentration), rather than favoring a single type. Build a diversified portfolio of power plants: solar, wind, coal with carbon capture, and nuclear.

11.5.5   Integrating the uncertainty into the decision process

If cost and benefits cannot be estimated well enough, rather than making (probably futile) attempts to improve the estimates, it may be better to integrate the uncertainty into the decision process. For example, sometimes the most uncertain items in a CBA are not the costs but the probabilities of various scenarios. In such cases it may be more instructive to look at the critical probability (probability for which cost = benefit) and ask whether the estimated probability is higher or lower. This is an illustration of the value of looking at risk and uncertainty from different perspectives. The best understanding of risk and uncertainty comes from having as many different presentations and perspectives as possible.

MCA (multicriteria analysis) (Belton and Stewart, 2002) is another good way to integrate the uncertainty into the decision process. The essence of MCA lies in stakeholder consultation to measure preferences among a series of environmental outcomes. Stakeholder outputs are then processed to determine which option provides the best balance among those preferences. The method is particularly useful where it is not possible, for whatever reason, to quantify the damage associated with a given option. Unfortunately in practice it is often difficult to find enough qualified participants who are willing to spend the necessary time to familiarize themselves with the issues of concern, in which case the results may not be sufficiently representative. But with sufficient suitable participants MCA can be an excellent approach for dealing with uncertainty, by rendering the decision process and the accounting for uncertainty more explicit. As an interesting use of MCA for external costs, we mention the SusTools project of EC DG Research (Rabl et al., 2004) where stakeholders were convened in a workshop to choose among environmental options. Even though MCA is unlikely to reduce the uncertainties, it can be very effective in promoting a consensus about the most desirable outcome.

11.6   Presentation of uncertainty

Communicating the uncertainties of external costs is very important, to ensure that users understand the limitations. Since 1998, ExternE has made a concerted effort to show the uncertainties. Unfortunately it is much more difficult to deal with both a number and its uncertainty than just with the number. Several formats have been tried, for example, giving a high and a low estimate in addition to the central estimate. In some cases the sensitivity to certain assumptions was also shown, for instance the use of VSL or of VOLY for mortality valuation in the 1998 reports. However, doing so complicates the presentation of the results, with an awkward exponential proliferation of numbers if more than a few sensitivity studies are shown. Often users focus on just the central estimate, without paying attention to the uncertainties no matter how clearly they are displayed. To prevent readers from doing so, ExternE (1998) showed the global warming costs (whose uncertainty is notoriously large) as a range rather than a single number. That does not seem to be a good approach either, because many users extracted a single number by taking an average – not appropriate since the probability distribution is lognormal and the average is much larger than the geometric mean which is more correct.

11.7   Consequences of the uncertainties for decisions

Since many people have questioned the usefulness of the externality studies because of their large uncertainty, we emphasize that the uncertainties should not be looked at by themselves; rather one should ask what effect the uncertainties have on the choice of policy options. The key question to be asked is "how large is the cost penalty if one makes the wrong choice because of errors or uncertainties in the cost or benefit estimates?". Rabl *et al.* (2005) have looked at the uncertainties from this perspective and their findings are very encouraging: the risk of cost penalties is surprisingly small, even with the very large uncertainties of ExternE.

It is instructive to distinguish between policy decisions that are binary (e.g. choice between nuclear or coal fired power plant) and those that are continuous (e.g. what limit to set for the SO_2 emissions from a power plant). For binary decisions, the situation is sometimes quite simple because the uncertainty, even if very large, has no effect if it does not change the ranking. For example, in France the market cost of nuclear is lower than that of coal for baseload electricity, and that ranking does not change if external costs are included (DGEC, 2008); furthermore, the external costs of nuclear are so much lower than those of coal, that the ranking is not affected by the uncertainties. Even if one were to take the upper limit of the confidence interval for nuclear and the lower limit of the confidence interval for coal, the ranking would still remain the same.

As another example of a binary choice, we show in Fig. 11.12 a graph from a cost–benefit analysis of a proposed new regulation for the emissions of particles from cement kilns that use waste as fuel (Rabl, 2000), a study carried out at a time when the details of the EC Directive (EC, 2000b) for the incineration of waste were still under discussion. In particular, Fig. 11.12 compares costs and benefits of a proposed reduction in the emission limit for PM_{10} in the exhaust gas from 50 to 15 mg/Nm^3 (which for a typical ratio of average and peak emissions of PM_{10} from cement kilns, implies a reduction of the annual average from 20 to 5 mg/Nm^3). Since both the benefits and the abatement costs are uncertain, the graph shows high and low estimates. Even with the upper limit of the error bar, the benefit is less than the lower estimate of the abatement cost: in all probability such an emission reduction would not have been

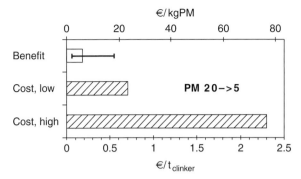

Figure 11.12 Comparison costs and benefits of a reduction in the PM_{10} emission limit for cement kilns that use waste as fuel. The reduction of annual average emissions is from 20 to 5 mg/m^3 (reduction of peak emission from 50 to 15 mg/m^3). Benefit = avoided damage cost, shown as €/kg$_{PM10}$ on upper scale and as €/t$_{clinker}$ (product of cement kilns) on lower scale.

justified.[1] And indeed, as we have already shown in Section 11.2.6, the probability that it would be justified after all is only 6%.

For continuous choices, the effect of uncertainty turns out to be surprisingly small, because near an optimum, the total social cost varies only slowly as individual cost components are varied. Specifically, using abatement cost curves for NO_x, SO_2, dioxins and CO_2, Rabl *et al.* (2005) have evaluated the cost penalty for errors in the following choices: national emission ceilings for NO_x and SO_2 in each of 12 countries of Europe, an emission ceiling for dioxins in the UK, and limits for the emission of CO_2 in Europe. As a generic example, Fig. 11.13 shows the extra social cost (as % of the optimum with perfect information) for NO_x abatement as a function of the error in the estimate of the damage cost. The error, expressed as the ratio C_{true}/C_{est} of true and estimated damage cost, and the extra social cost, which we also call the cost penalty, is stated as a percentage of the total cost (abatement + damage) at the optimum with perfect information. The vertical dashed lines indicate the 1 σ_g confidence interval of C_{est}. The cost penalty is less than 15% within the confidence interval, but gets very large for larger errors. If C_{est} is too small ($C_{true}/C_{est} > 1$), one would pay too little for abatement and too much for damage, and vice versa if $C_{true}/C_{est} < 1$. Even if the damage cost is overestimated by a factor of $\sigma_g = 3$, the cost

---

[1] We were told that the Commission did not impose this limit as a result of this study.

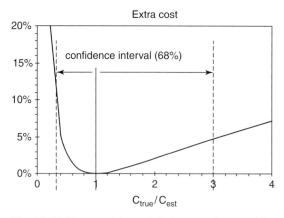

Fig. 11.13 Extra social cost (relative to optimum with perfect information) for NO_x abatement as a function of the error in the estimate of the damage cost.

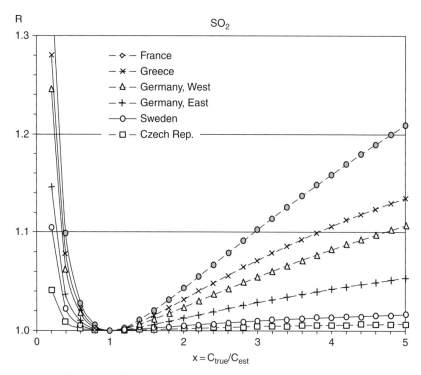

Fig. 11.14 The cost penalty ratio R versus the error $x = C_{true}/C_{est}$ in the damage cost estimate for several countries, selected to show extremes as well as intermediate curves. The labels are placed in the same order as the curves. Dashed lines correspond to extrapolated regions of the cost curves. From Rabl *et al.* (2005) with permission.

penalty is less than 15%. But the cost penalty increases dramatically at small C_{true}/C_{est}, and without an analysis of the ExternE type, one could easily end up paying a very large extra cost. In other words, thanks to such analysis one can avoid costly mistakes. Knowing the damage cost within a factor of 3 is incomparably better than the range 0 to ∞ in the absence of any analysis.

As a specific and more detailed example, Fig. 11.14 shows the cost penalty ratio R for national emission ceilings for SO_2 in different countries, if a wrong level is chosen because of an error in the damage cost estimate. The format is essentially the same as in the preceding figure, except that here the y-axis is shown as the cost penalty ratio R, i.e. the total social cost (abatement + damage) at C_{est} divided by the total social cost at the true optimum, rather than $R - 1$ as previously. The curves are quite different for different countries: damage costs are different because of population densities, and abatement costs are different because different technological options are cost-effective in different countries.

References

Abt 2000. The Particulate-Related Health Benefits of Reducing Power Plant Emissions. October 2000. Prepared for EPA by Abt Associates Inc., 4800 Montgomery Lane, Bethesda, MD 20814–5341.

Abt 2004. Power Plant Emissions: Particulate Matter-Related Health Damages and the Benefits of Alternative Emission Reduction Scenarios. Prepared for EPA by Abt Associates Inc. 4800 Montgomery Lane. Bethesda, MD 20814–5341.

Axelrad, D. A., Bellinger, D. C., Ryan, L. M. and Woodruff, T. J. 2007. Dose-response relationship of prenatal mercury exposure and IQ: an integrative analysis of epidemiologic data. *Environ Health Perspect* **115**(4): 609–615.

Barnosky, A. D., Hadly, E. A., Bascompte, J. *et al.* 2012. Approaching a state shift in Earth's biosphere. *Nature* **486** (7 June 2012): 52–58.

Barrett, K. 1992. Dispersion of nitrogen and sulfur across Europe from individual grid elements: marine and terrestrial deposition. EMEP/MSC-W Note 3/92. August 1992. Norwegian Meteorological Institute, P.O.Box 43, Blindern, N-0313 Oslo 3.

Belton, V. and Stewart, T. J. 2002. *Multiple Criteria Decision Analysis: An integrated approach.* Kluwer Academic Publishers, Dordrecht.

Bennett, D. H., McKone, T. E., Evans, J. S. *et al.* 2002. Defining intake fraction. *Environmental Science and Technology* **36**: 206 A–211 A.

Brode, R. W. and Wang, J. 1992. *User's Guide for the Industrial Source Complex (ISC2) Dispersion Model.* Vols. 1–3, EPA 450/4-92-008a, EPA 450/4-92-008b, and EPA 450/4-92-008c. US Environmental Protection Agency, Research Triangle Park, NC 27711.

DGEC 2008. *Synthèse publique de l'étude des coûts de référence de la production électrique* (Public summary of the study of the reference costs of electricity production). Direction Générale de l'Energie et du Climat (DGEC), Ministère de l'Ecologie, du Développement Durable et de l'énergie, France. (www.developpement-dura ble.gouv.fr/IMG/pdf/cout-ref-synthese2008.pdf)

EC 2000a. *Recommended Interim Values for the Value of Preventing a Fatality in DG Environment Cost Benefit Analysis*. Recommendations by DG Environment, based on a workshop for experts held in Brussels on November 13th 2000.

EC 2000b. Directive 2000/76/EC of the European Parliament and of the Council of 4 December 2000 on the incineration of waste.

EPA 1994. *Estimating exposure to dioxin-like compounds* Report EPA/600/6-88/005Ca, b and c. June 1994. United States Environmental Protection Agency. Washington, DC 20460.

ExternE 1998. ExternE: Externalities of Energy. Vol.7: Methodology 1998 Update (EUR 19083); Vol.8: Global Warming (EUR 18836); Vol.9: Fuel Cycles for Emerging and End-Use Technologies, Transport and Waste (EUR 18887); Vol.10: National Implementation (EUR 18528). Published by European Commission, Directorate-General XII, Science Research and Development. Office for Official Publications of the European Communities, L-2920 Luxembourg.

ExternE 2000. *External Costs of Energy Conversion – Improvement of the ExternE Methodology and Assessment of Energy-Related Transport Externalities*. Final Report for Contract JOS3-CT97-0015, published as Environmental External Costs of Transport. R. Friedrich & P. Bickel, editors. Springer Verlag Heidelberg 2001.

ExternE 2004. New results of ExternE, reported by Rabl, A., Spadaro, J., Bickel, P., *et al. Externalities of Energy: Extension of accounting framework and Policy Applications*. Final Report ExternE-Pol project, contract N° ENG1-CT2002-00609. EC DG Research. See also, www.externe.info.

ExternE 2005. *ExternE: Externalities of Energy, Methodology 2005 Update*. Edited by P. Bickel and R. Friedrich. Published by the European Commission, Directorate-General for Research, Sustainable Energy Systems. Luxembourg: Office for Official Publications of the European Communities. ISBN 92-79-00429-9.

ExternE 2008. With this reference we cite the methodology and results of the NEEDS (2004–2008) and CASES (2006–2008) phases of ExternE. For the damage costs per kg of pollutant and per kWh of electricity we cite the numbers of the data CD that is included in the book edited by Markandya, A., Bigano, A. and Porchia, R. in 2010: *The Social Cost of Electricity: Scenarios and Policy Implications*. Edward Elgar Publishing Ltd, Cheltenham, UK. They can also be downloaded from http://www.feem-project.net/cases/ (although in the latter some numbers have changed since the data CD in the book).

Hansen, M. S. and Andersen, M. S. 2009. External costs of nutrients (N and P) – first estimates. Deliverable DII.2b.-1 of EXIOPOL project of the EC DG Research.

Hanson, P. J. and Lindberg, S. E. 1991. Dry deposition of reactive nitrogen compounds: a review of leaf, canopy and non-foliar measurements. *Atmospheric Environment* **25A**: 1615–1634.

Holland, M. R. and King, K. 1998. *Economic evaluation of air quality targets for tropospheric ozone*. Part C: Economic benefit assessment. Contract report for European Commission DG XI. http://ec.europa.eu/environment/enveco/air/pdf/tropozone-c.pdf

Ives, D. P., Kemp, R. V. and Thieme, M. 1993. *The Statistical Value of Life and Safety Investment Research*. Environmental Risk Assessment Unit, University of East Anglia, Norwich, Report n°13 February 1993.

L'hirondel, J. L., Alexander, A. A. and Addiscott, T. 2006. Dietary nitrate: where is the risk? *Environ Health Perspect* **114**: 458–459.

Leksell, L. and Rabl, A 2001. Air pollution and mortality: Quantification and valuation of years of life lost. *Risk Analysis* **21**(5): 843–857.

Limpert, E., Stahel, W. A. and Abbt, M. 2001. Lognormal distributions across the sciences: Keys and Clues. *BioScience* **51**(5), 341–352.

Lindhjem, H., Navrud, S., Braathen, N. A. and Biausque, V. 2011. Valuing mortality risk reductions from environmental, transport, and health policies: A global meta-analysis of stated preference studies. *Risk Analysis* **31** (9): 1381–1407.

McKone, T. E. and Ryan, P. B. 1989. Human exposures to chemicals through food chains: an uncertainty analysis. *Environmental Science and Technology* **23**: 1154–1163.

Morgan, M. G. and Henrion, M. 1990. *Uncertainty: A Guide to Dealing with Uncertainty in Quantitative Risk and Policy Analysis*. Cambridge University Press. Cambridge, UK.

Mrozek, J. R. and Taylor, L. O. 2002. What determines the value of life? A meta-analysis. *Journal of Policy Analysis and Management* **21**, No. 2: 253–270.

Nicholson, K. W. 1988. The dry deposition of small particles: a review of experimental measurements. *Atmospheric Environment* **22**: 2653–2666.

OECD 2012. Mortality Risk Valuation in Environment, Health and Transport Policies. OECD Publishing. http://dx.doi.org/10.1787/9789264130807-en

ORNL/RFF 1994. *External Costs and Benefits of Fuel Cycles*. Prepared by Oak Ridge National Laboratory and Resources for the Future. Edited by Russell Lee, Oak Ridge National Laboratory, Oak Ridge, TN 37831.

Pope, C. A., Burnett, R. T., Thun, M. J. *et al.* 2002. Lung cancer, cardiopulmonary mortality, and long term exposure to fine particulate air pollution. *J. Amer. Med. Assoc.* **287**(9): 1132–1141.

Rabl, A. 2000. *Criteria for limits on the emission of dust from cement kilns that burn waste as fuel*. ARMINES/Ecole des Mines de Paris, Paris. March 2000. Available at www.arirabl.org.

Rabl, A. and van der Zwaan, B. 2009. Cost–benefit analysis of climate change dynamics: uncertainties and the value of information. *Climatic Change* **96**, No. 3, October, 2009.

Rabl, A., Spadaro, J. V. and van der Zwaan, B. 2005. Uncertainty of pollution damage cost estimates: to what extent does it matter?. *Environmental Science & Technology* **39**(2): 399–408.

Rabl, A., Zoughaib, A., von Blottnitz, H. *et al.* 2004. *Tools for sustainability: Development and application of an integrated framework*. Final Technical Report for project SusTools, contract N° EVG3-CT-2002–80010. EC DG Research. Available at www.arirabl.com/sustools.htm

Reiss, R., Anderson, E. L., Cross, C. E. *et al.* 2007. Evidence for health impacts of sulfate and nitrate containing particles in ambient air. *Inhalation Toxicology* **19**: 419–449.

Rowe, R. D., Lang, C. M., Chestnut, L. G. *et al.* 1995. *The New York Electricity Externality Study*. Oceana Publications, Dobbs Ferry, New York.

Sehmel, G. 1980. Particle and gas dry depostion: a review. *Atmospheric Environment* **14**: 983.

Seinfeld, J. H. 1986. *Atmospheric Chemistry and Physics of Air Pollution*. John Wiley and Sons, Somerset, NJ.

Spadaro, J. V. and Rabl, A., 2004. Pathway analysis for population-total health impacts of toxic metal emissions. *Risk Analysis* **24**(5): 1121–1141.

Spadaro, J. V. and Rabl, A. 2008a. Estimating the uncertainty of damage costs of pollution: a simple transparent method and typical results. *Environmental Impact Assessment Review* **28** (2): 166–183.

Spadaro, J. V. and Rabl, A. 2008b. Global health impacts and costs due to mercury emissions. *Risk Analysis* **28** (3): 603–613.

Tol, R. S. J. 2005. The marginal damage costs of carbon dioxide emissions: an assessment of the uncertainties. *Energy Policy* **33**, 16: 2064–2074.

Tol, R. S. J. 2008. The Social Cost of Carbon: Trends, Outliers and Catastrophes. *Economics* **2**: 2008–2025.

van Dop, H., Addis, R., Fraser G. *et al.* 1998. ETEX, a European tracer experiment: observations, dispersion modelling and emergency response. *Atmospheric Environment* **32**(24): 4089–4094.

van Grinsven, H. J. M., Rabl, A., de Kok, T. M. and Grizzetti, B. 2010. Assessing social cost of cancers due to nitrate in drinking water in the EU for the case of colon cancer. *Environmental Health* **9**: 58 (12 p).

Weitzman, M. L. 2007. The role of uncertainty in economics of catastrophic climate change. Working paper, Harvard University.

WHO. 2003. *Health aspects of air pollution with particulate matter, ozone and nitrogen dioxide*, Report on a WHO Working Group, Bonn, Germany, 13–15 January 2003. World Health Organisation report EUR/03/5042688: Available at www. euro.who.int/document/e79097.pdf; accessed November 2004.

12 Key assumptions and results for cost per kg of pollutant

Summary

In Chapters 12 to 15 we present results. We focus on results obtained within the ExternE series, because we are most familiar with that and have access to the complete documentation. But we also cite results from studies in the USA. The underlying assumptions, especially the ERFs and monetary values, have been evolving in the light of new research, and a variety of different damage cost numbers can be found in the literature. The most recent European results for electricity production and for LCA are those published by ExternE (2008), and we present them in Chapters 12 to 14. We also show how to adjust them for increases in the valuation of mortality and greenhouse gases that we believe are appropriate.

A complication lies in the site dependence, in particular the variation with emission site and stack height. In these chapters we try to present results for typical applications, with indications of how they might differ for different situations. Variation with emission site is especially strong for transport emissions, discussed in Chapter 15.

In the present chapter, we present a summary of the key assumptions and the resulting damage costs per kg of pollutant for typical European conditions. We also show results of assessments in the USA. The implications for electricity production, waste treatment and vehicles are presented in Chapters 13 to 15.

12.1 Assumptions and models of ExternE

12.1.1 Overview

For the calculation of results, EcoSense has been the central software of ExternE. It contains all the required databases, in particular for meteorological data, receptor distributions, ERFs and monetary values. It also contains models for atmospheric dispersion and chemistry. The key assumptions for the calculations of ExternE are summarized in Table 12.1.

Table 12.1 *Key assumptions for the calculations of ExternE.*

Atmospheric dispersion models	
Local range:	Gaussian plume models ISC (for elevated sources) and ROADPOL (for ground level sources).
Regional range (Europe):	Source-receptor (SR) matrices calculated by EMEP were used for ExternE (2008), without any local modeling. SR matrices were calculated for primary and secondary pollutants, including O_3.
	(Before 2006 the results for PM, NO_x and SO_2 were calculated by Harwell Trajectory Model as implemented in EcoSense software of ExternE)
Impacts on health	
Form of ERFs	Straight line for all health impacts, without threshold for PM (primary and secondary), with threshold for O_3.
Chronic mortality	The ERF for life expectancy loss is derived from increase in age-specific mortality due to $PM_{2.5}$ (Pope *et al.*, 2002), by integrating over age distribution.
Acute mortality	For ozone, assuming 0.75 YOLL per death.
Nitrate and sulfate aerosols	The ERFs for $PM_{2.5}$ and PM_{10} are applied to nitrates and sulfates on the basis of mass and size range, without distinction of composition or other characteristics.
	Sulfates are assumed to be entirely $PM_{2.5}$ whereas for nitrates the $PM_{2.5}$ and PM_{10} fractions are calculated by EMEP.
Radionuclides	Linear ERFs without threshold:
	0.05 fatal cancers/personSv,
	0.12 non-fatal cancers/personSv,
	0.01 severe effects hereditary/personSv.
Micropollutants	Cancers due to As, Cd, Cr, Ni, dioxins, benzene, butadiene, formaldehyde. Neurotoxic impacts of Hg and Pb.
Impacts on plants	ERFs for crop loss due to SO_2 and ozone.
Impacts on buildings and materials	Corrosion and erosion due to SO_2 and soiling due to particles.
Impacts not quantified but potentially significant	Reduced visibility due to air pollution; disposal of residues from fossil fuels.
Monetary valuation	
Accidents	Value of prevented fatality VPF = 1.5 M€
Loss of life expectancy	Proportional to loss of life expectancy, with Value of life year
	VOLY = 40,000 € for chronic and 60,000 € for acute mortality
Cancers	0.48 M€ per non-fatal cancer, 1.12 M€ per fatal cancer.
Discount rate	3% until 2030, 2% thereafter

12.1.2 *Atmospheric modeling*

For dispersion modeling in the local zone, Gaussian plume models have been used, either ISC for elevated sources (see Brode and Wang (1992)) or ROADPOL for ground level sources (see Vossiniotis *et al.* (1996)). Chemical reactions are not considered in ISC because for nitrate and sulfate aerosols they are significant only beyond the local zone; however, the user has the option of including pollutant removal. ROADPOL can take nitrogen reactions into account because some of them are fast enough to be significant even in the local zone.

For regional modeling, i.e. beyond the local zone, ExternE used until 2006 the Windrose trajectory model of Krewitt *et al.* (1995), which is derived from the Harwell trajectory model (Derwent and Nodop, 1986). It includes the chemistry for the creation of sulfate and nitrate aerosols, but not ozone. The regional grid cells have a resolution of about 50 km × 50 km (the precise values varying with latitude and longitude).

The local zone modeling is done in a 100 km × 100 km cell placed symmetrically around the source, with 10 km × 10 km resolution. The local impacts are counted fully. To avoid double counting in the cells of the regional grid that overlap the local cell, the impacts of the regional cells are multiplied by the fraction of their areas outside the local cell.

In order to obtain a better and more complete modeling of the chemical reactions, ExternE (2008) has used source–receptor (SR) matrices that have been provided specifically for ExternE by EMEP, the official modeling program for transboundary pollution in Europe.[1] The grid cell size is approximately 50 km × 50 km for the model of the European region, i.e. the continent plus adjacent areas. The resolution has been improved since then. Concentrations or depositions are calculated for the pollutants in Table 12.2. In addition to the European region, results from the Northern hemispherical model of EMEP are also included.

The procedure used by EMEP to calculate these SR matrices for ExternE is as follows. First the actual concentrations are calculated for the actual emissions, using the emissions inventories for each of the countries in the model. Then the calculations are repeated for each pollutant by reducing the emission of that pollutant by 15%. For the actual emissions EMEP used the emissions that had been predicted for 2010. The meteorological data for the years 1996, 1997, 1998 and 2000 were used to obtain results representative of the long-term average. As an example, we have shown in Table 7.17 in Section 7.4.2 an SR matrix for $PM_{2.5}$. An element (x, y) of this matrix indicates how much an emission in

---

[1] www.emep.int/

Table 12.2 *Primary and secondary air pollutants for which EMEP provides concentrations or depositions.*

Short name	Explanation	Unit
aNH$_4$	ammonium particles (ammonium nitrate and sulphate)	μgN/m^3
aNO$_3$	nitrate particles with diameter below 2.5 μm	μgN/m^3
DDEP_OXN	total dry deposition of oxidized nitrogen	mg/m^2
DDEP_RDN	total dry deposition of reduced nitrogen	mg/m^2
DDEP_SOX	total dry deposition of sulphur	mg/m^2
NO$_x$	NO$_x$ = NO$_2$ + NO	μgN/m^3
pNO$_3$	nitrate particles with diameter above 2.5 a. below 10 μm	μgN/m^3
SIA	secondary inorganic aerosols < 10 μm	μg/m^3
SO$_4$	sulphate, also includes ammonium sulphate	μgS/m^3
SOMO35	sum of means over 35 ppb	ppb day
tNO$_3$	total coarse and fine nitrate aerosols	μgN/m^3
WDEP_OXN	wet deposition of oxidized nitrogen	mg/m^2
WDEP_RDN	wet deposition of reduced nitrogen	mg/m^2
WDEP_SO$_x$	wet deposition of sulfur	mg/m^2
PPM$_{25}$	primary particles with diameter below 2.5 μm	μg/m^3
PPM$_{co}$	primary particles with diameter between 2.5 and 10 μm	μg/m^3

cell x changes the concentrations in cell y. The magnitudes in Table 7.17 correspond to a 15% change in all emission cells.

The emissions are provided according to the SNAP (selected nomenclature for reporting of air pollutants) categories of the various sectors of the economy. Since one of these sectors, combustion in energy and transformation industry, involves tall stacks, the emissions from this sector were used to obtain separate SR matrices for tall sources. This allowed the calculation of separate sets of SR matrices and damage costs, one for "high height of release", the other for "low height of release." An additional set of damage costs for "unknown height of release" was also calculated, corresponding to the emissions from all SNAP sectors.

For PM, the emissions are provided in terms of PM$_{2.5}$ and PM$_{co}$ (= PM fraction between 2.5 and 10 μm) rather than PM$_{10}$, and so are the SR matrices and the damage costs. Damage costs for PM$_{10}$ can be calculated if one knows the respective fractions are $f_{2.5}$ and f_{co} of PM$_{2.5}$ and PM$_{co}$ in the PM$_{10}$:

$$\text{€/kg}_{PM10} = f_{2.5}\,\text{€/kg}_{PM2.5} + f_{co}\,\text{€/kg}_{PMco}. \quad (12.1)$$

Users also have the possibility of performing their own calculations with the software available at the ExternE website.[2] As of May 2013, the

[2] www.externe.info

EcoSenseWeb tool now allows calculations with the short-term version of the industrial source complex model (ISC) for local air quality modeling, coupled with the source–receptor matrices of EMEP for regional modeling; thus the effect of stack height can be analyzed. However, the ExternE results presented in Chapters 12 to 14 do not include any local modeling.

12.1.3 ERFs and monetary values

The ERF slopes of ExternE (2008) are listed in Table 12.3 (Preiss *et al.*, 2008). The entries in the ERF column include the fraction of the total population that is affected by the respective pollutant and end point, thus they are applied directly to the general population. Their derivation has been explained in Sections 3.4 and 3.5 of Chapter 3.

Since EMEP provides source–receptor matrices for $PM_{2.5}$ and PM_{co}, but not for PM_{10}, one has to decide how to apply the ERFs for the set of effects that are expressed against PM_{10} concentration. For that one would need to know how much of the impact of the PM_{10} ERFs is due to $PM_{2.5}$ and how much due to PM_{co}. Such information is not available, and so the following somewhat literal interpretation has been used by ExternE (2008) for applying the ERFs for PM_{10}: both $PM_{2.5}$ and PM_{co} being by definition also PM_{10}, the PM_{10} exposure costs are applied to both $PM_{2.5}$ and PM_{co}. To calculate the damage cost of $PM_{2.5}$, the concentration of $PM_{2.5}$ is multiplied by each of the exposure costs for $PM_{2.5}$ and for PM_{10}, and the resulting terms are summed; thus the $PM_{2.5}$ concentrations are multiplied by $(32.79 + 5.963)$ ($€/yr$) per $(person.\mu g/m^3)$. For the damage cost of PM_{co}, the concentration of PM_{co} is multiplied by the exposure costs for PM_{10}; thus the PM_{co} concentrations are multiplied by 5.963 ($€/yr$) per $(person.\mu g/m^3)$. This procedure, documented in Preiss *et al.* (2008), can be summarized as,

$$\text{cost per } PM_{2.5} \text{ exposure} = (32.79 + 5.963)(€/yr)/(person.\mu g/m^3)$$
$$= 38.753(€/yr)/(person.\mu g/m^3)$$

(12.2a)

and

$$\text{cost per } PM_{co} \text{exposure} = 5.963(€/yr)/(person.\mu g/m^3)$$

(12.2b)

and with $f_{2.5}$ = fraction of PM_{10} that is $PM_{2.5}$

$$\text{cost per } PM_{10} \text{ exposure} = (f_{2.5}\ 32.79 + 5.963)\ (€/yr)/(person.\mu g/m^3)$$
$$= 25.637\ (€/yr)/(person.\mu g/m^3)\ for\ f_{2.5} = 0.60.$$

(12.2c)

Table 12.3 *Exposure–response function slopes and unit costs, assumed by ExternE (2008).*

Pollutant	End point	ERF slope	Unit cost	Exposure cost	% of total
		(cases/yr) per (person.μg/m³)	€₂₀₀₀/case	(€/yr) per (person.μg/m³)	
PM₂.₅					
	Chronic mortality, YOLL	6.51E-04	40,000[a]	26.04	79.4%
	Net restricted activity days (netRAD)	9.59E-03	130	1.25	3.8%
	Work loss days (WLD)	1.39E-02	295	4.10	12.5%
	Minor restricted activity days (MRAD)	3.69E-02	38	1.40	4.3%
	Total			**32.79**	**100%**
PM₁₀					
	Infant mortality, deaths	6.84E-08	3,000,000	0.205	3.4%
	New cases of chronic bronchitis	1.86E-05	200,000	3.710	62.2%
	Respiratory hospital admissions	7.03E-06	2,000	0.014	0.2%
	Cardiac hospital admissions	4.34E-06	2,000	0.009	0.1%
	Medication/bronchodilator use, children	4.03E-04	1	0.000	0.0%
	Medication/bronchodilator use, adults	3.27E-03	1	0.003	0.1%
	LRS, children, days	2.08E-02	38	0.792	13.3%
	LRS, adults, days	3.24E-02	38	1.230	20.6%
	Total			**5.963**	**100%**
O₃, SOMO35					
	Acute mortality, YOLL	2.23E-06	60,000[b]	0.134	15.1%
	Respiratory hospital admissions	1.98E-06	2,000	0.004	0.4%
	Minor restricted activity days (MRAD)	7.36E-03	38	0.280	31.7%
	Medication use/bronchodilator use	2.62E-03	1	0.003	0.3%
	LRS excluding cough, days	1.79E-03	38	0.068	7.7%
	Cough days	1.04E-02	38	0.396	44.8%
	Total			**0.885**	**100%**

YOLL = years of life lost
LRS = lower respiratory symptoms
SOMO35 = sum of ozone means over 35 ppb (Eq. (12.3)).
[a] value of life year lost due to chronic exposure (discounted because of delay between exposure and death).
[b] value of life year lost due to acute exposure (not discounted).

For another look at the interpretation of the PM_{10} ERFs, it is instructive to ask how they would have been reported if the respective studies had used $PM_{2.5}$. There are two extreme possibilities. The first is that $PM_{2.5}$ and PM_{co} are equally toxic per mass of PM, the second that only the $PM_{2.5}$ component is toxic. Let us assume a concentration ratio $c_{PM2.5}/c_{PM10} = 0.6$, typical at the time of these studies. If $PM_{2.5}$ and PM_{co} are equally toxic per mass of PM, only 0.6 of the cases would have been attributed to $PM_{2.5}$ and the corresponding ERF slopes would have been reported as $S_{ERF2.5} = 0.6$ cases/$c_{PM2.5} = 0.6$ cases/$(0.6\ c_{PM10}) = S_{ERF10}$. The remainder would have been attributed to PM_{co}, with ERF slopes $S_{ERFco} = 0.4$ cases/$(0.4\ c_{PM10}) = S_{ERF10}$. That is in effect the interpretation chosen by ExternE (2008).

But if only the $PM_{2.5}$ component is toxic, the ERF slopes would have been reported as $S_{ERF2.5} = $ cases/$c_{PM2.5}$ instead of $S_{ERF10} = $ cases/c_{PM10}, hence $S_{ERF2.5} = S_{ERF10}/0.6$ instead of S_{ERF10}. Then the exposure cost of $PM_{2.5}$ would be $32.79 + 5.97/0.6 = 42.74$ (€/yr)/(person.μg/m^3), and that of PM_{co} would be zero. This second possibility may be closer to the truth.

For O_3, ExternE (2008) assumes a hockey stick ERF with zero impact below a threshold of 35 ppb and linear increase above. This is implemented by stating the ERF slope in terms of SOMO35, the sum of ozone means over 35 ppb. More explicitly, SOMO35 is defined as the sum of the differences between maximum daily 8-hour moving average concentrations greater than 35 ppb (= 70 μg/m^3) and 35 ppb:

$$SOMO35 = \sum_{i=1}^{365} Max[0, c_i - 35ppb], \qquad (12.3)$$

where c_i is the maximum daily 8-hour running mean ozone concentration in ppb and the summation is over all days per calendar year. SOMO35 has a dimension of ppb·days.

Multiplying the ERF slope by the unit cost, one obtains the exposure cost, i.e. the incremental cost per $1/(\mu g/m^3)$ per person per yr, to be multiplied by the population in the grid cells of EcoSense and by the concentration increase per kg of emitted pollutant to calculate the cost per kg of pollutant. In the UWM, the incremental cost per $1/(\mu g/m^3)$ per person per yr is multiplied by the regional population density and divided by the depletion velocity k_{dep}; see the following example.

Example 12.1 Use the UWM of Section 7.4 to calculate the contribution of chronic mortality to the cost of $PM_{2.5}$ if the regional population density $\rho = 100$ person/km^2 and the depletion velocity $k_{dep} = 0.005$ m/s.

Solution: The UWM for the health impact rate (cases/yr) due to inhalation can be written in the form of Eq. (8.31) as

$$\dot{I}_{inhal}(\dot{m}) = S_{ERF} \; \rho_{eff} \; \dot{m}/k_{dep}, \qquad (12.4)$$

where,

S_{ERF} = slope of ERF ((cases/yr)/(person.$\mu g/m^3$)),
$\dot{m}$ = rate at which pollutant is emitted to the air (kg/yr),
ρ_{eff} = regional average population density (pers/km^2), and
k_{dep} = depletion velocity of pollutant (m/s).

Using the entries for chronic mortality in Table 12.3 and an emission rate of $\dot{m} = 1$ kg/yr $= 1$ kg/(365.25 × 24 × 3600 s) we find an impact rate of

$$\dot{I}_{inhal}(\dot{m}) = S_{ERF}* \; \dot{m} * \rho/k_{dep}$$

$$= \frac{6.51E-04 \; \frac{YOLL}{yr}}{person \; \left(\frac{\mu g}{m^3}\right)} \times \frac{10^{-4} persons}{m^2} \times \frac{1 \; kg}{365.25 \times 24 \times 3600 \; s} \times \frac{10^9 \; \frac{\mu g}{kg}}{0.005 \; \frac{m}{s}}$$

$$= 4.13E-04 \; YOLL/yr.$$

Since both emission rate and impact rate are stated per yr, the impact per kg is 4.13E-04 YOLL/kg. Multiplying by 40,000 €/YOLL we obtain a damage cost of 16.52 €/kg$_{PM2.5}$.

12.1.4 *Variation with site and stack height*

For primary pollutants, the damage cost varies strongly with source location and stack height, as can be seen with the examples in Fig. 7.16 of Section 7.4. Local dispersion models are needed to properly account for that. Since ExternE (2008) did not include any local modeling, the results cannot be used for ground-level sources, in particular transport. Rather, for power plants the results for *high height of release* are appropriate. For LCA applications, the choice between the height classes depends on the type of process under consideration. As for sites, typical source locations in the respective emitter countries or grid cells are assumed.

12.2 Cost per kg of pollutant

12.2.1 *Results of ExternE (2008)*

Figure 12.1 shows the damage costs calculated by ExternE (2008) for average emissions in the EU27. The error bars indicate the uncertainties,

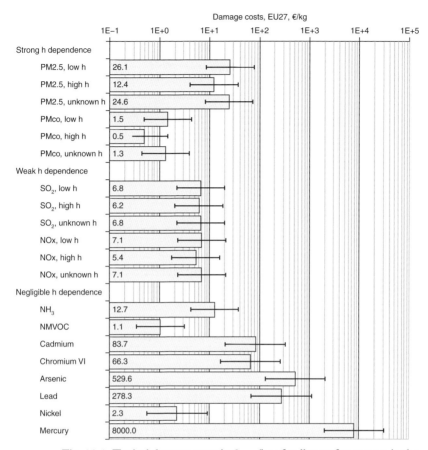

Fig. 12.1 Typical damage costs in €$_{2000}$/kg of pollutant for sources in the EU27, according to ExternE (2008). The error bars show 68% confidence intervals.

as 68% confidence intervals (corresponding to 1 geometric standard deviation), estimated in Section 11.3.4. The graph does not include the damage cost of dioxin because it is very much larger: it is 3.7E07 €/kg (toxic equivalent 2,3,7,8-TCDD). Entries for "high h" are for fossil power plants. Entries for "low h" are for the lower stack heights of most industrial sources, but not for transport emissions in cities. Results for the latter can be found in Table 15.2.

Not shown in Fig. 12.1 are the greenhouse gases. The value used by ExternE (2008), 21 €/t$_{CO2eq}$ for 2010, can be found in the data CD distributed with the CASES book or on the corresponding website of the CASES project.[3] In our book, we use the numbers of the data CD because they are a fixed reference, unlike the website where some numbers have been changed in the meantime.

Example What is the damage cost of PM_{10} emissions from low stacks in the EU27 if they have a composition 60% $PM_{2.5}$ and 40% PM_{co}?

Solution: Inserting the respective numbers from Fig. 12.1 into Eq. (12.1) we obtain,

$$0.6 \times 26.1 \, €/kg_{PM2.5} + 0.4 \times 1.5 \, €/kg_{PMco} = 16.3 \, €/kg_{PM10}.$$

12.2.2   Comparison with previous phases of ExternE

Damage cost estimates change with time because of scientific progress. During the past decades there has been intense worldwide research in air pollution epidemiology, so changes in ERFs should not come as a surprise. In addition, there have been major changes in the monetary valuation of air pollution mortality. Many estimates have been published in recent decades, but we do not consider any external cost publications before 1998 to be relevant (except some of the assessments of nuclear power), because the methodology was not yet sufficiently developed and stable.

The evolution of damage cost estimates by ExternE since 1998 is shown in Fig. 12.2. The numbers are not perfectly comparable, for instance those for 2008 are for EU27, whereas the others are for EU15. Some of the increase for SO_2 and NO_x in 2008 is due to the inclusion of ecosystem impacts, not evaluated in preceding phases. Since PM_{10} was not calculated in 2008, we have calculated a value, assuming a composition of 60% $PM_{2.5}$ and 40% PM_{co}. In Fig. 12.2, we show what we think comes closest to a meaningful comparison for the major publication dates of ExternE. In any case, the main message is that all of these changes are well within a factor of three, i.e. within the confidence intervals that we have estimated.

12.2.3   Adjustment for higher mortality and GHG costs

As already mentioned before, in Section 9.3.1, the recent meta-analysis of mortality valuation studies by Lindhjem et al. (2011) led OECD (2012) to recommend that VPF for the EU should be around 3 M€, twice as high as what had been assumed by the most recent phases of ExternE. To

---

[3] www.feem-project.net/cases/project.php

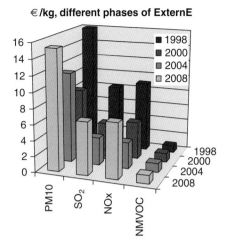

Fig. 12.2 Evolution of damage cost estimates by ExternE.

maintain the same ratio of VPF and VOLY, the latter should be doubled as well. For the damage costs of $PM_{2.5}$ the adjustment is straightforward because the costs in Fig. 12.1 include only health impacts. Looking at Table 12.3 and doubling the chronic mortality cost, one readily finds that the exposure cost of 32.79 (€/yr)/(person.µg/m^3) for $PM_{2.5}$ becomes 58.83 (€/yr)/(person.µg/m^3), an increase by a factor of 1.8. Doubling likewise the infant mortality cost, one finds that the exposure cost of PM_{10} increases from 5.963 to 6.168 (€/yr)/(person.µg/m^3). In the logic of Eq. (12.2a), the damage cost of $PM_{2.5}$ increases by a factor,

$$PM_{2.5} \text{ damage cost increase for 2} \times \text{mortality} = \frac{58.83 + 6.168}{32.79 + 5.963} = 1.68.$$

$$(12.5)$$

For the damage costs of PM_{co}, only the change in infant mortality contributes,

$$PM_{co} \text{ damage cost increase for 2} \times \text{mortality} = \frac{6.168}{5.963} = 1.03. \quad (12.6)$$

For O_3, the exposure cost increases from 0.885 to 1.019 and the damage cost by a factor,

$$O_3 \text{ damage cost increase for 2} \times \text{mortality} = \frac{1.019}{0.885} = 1.15. \quad (12.7)$$

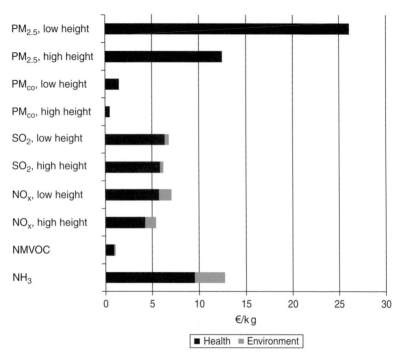

Fig. 12.3 The cost contributions of health and environment for the classical air pollutants.

Thus, for example, the damage cost of 12.4 €/kg in Fig. 12.1 for $PM_{2.5}$ emitted from tall stacks becomes 1.68×12.4 €/kg $= 20.8$ €/kg.

In the following chapters we will indicate how the damage costs of electricity, waste and transport will increase for a doubling of mortality. Since for that purpose one needs to know what fraction of the damage cost of the respective pollutants is due to mortality, we plot in Fig. 12.3 the contributions of health and of environment for the classical air pollutants. This graph highlights once again the importance of health in the total damage costs.

The PM costs are entirely due to health. For SO_2 all the health impacts are assumed to be due to sulfates, considered to be $PM_{2.5}$. The health contribution is 95%. Thus, if the mortality cost is doubled, the damage cost of SO_2 increases by a factor of

$$SO_2 \text{ damage cost increase for } 2 \times \text{mortality}$$
$$= (1 - 0.95) + 0.95 \times 1.68 = 1.65. \tag{12.8}$$

For NO_x, health contributes 80% of the total, but no information is provided for the relative contribution of O_3 and of the fine and coarse fractions of the nitrates to the health impacts. The smallness of the NMVOC cost (which is entirely due to O_3) suggests that the contribution of O_3 is relatively small. Assuming that about half of the nitrates are coarse and neglecting the O_3, this suggests that a doubling of the mortality costs would increase the damage cost of NO_x by a factor of,

$$NO_x \text{ damage cost increase for } 2 \times \text{mortality}$$
$$= (1 - 0.8) + 0.8 \times (1.68 + 1.03)/2 = 1.28. \qquad (12.9)$$

The damage cost of 21 €/t_{CO2eq} for GHG used by the CASES project of ExternE (2008) was obtained with the conventional discount rate. As we have argued in Section 9.1.3, we believe that the intergenerational discount rate should not include the pure rate of time preference. Therefore we find the damage cost estimate, \$85/tonne$_{CO2}$ (65 €/tonne$_{CO2}$), of the Stern review (Stern *et al.*, 2006) far more appropriate, because at 0.1% its choice for the pure rate of time preference is practically zero. Thus the cost of GHG emissions increases by a factor,

$$\text{damage cost increase of GHG if Stern instead of ExternE} = \frac{65}{21} = 3.1.$$
$$(12.10)$$

12.3   Assessments in the USA

The beginnings of ExternE were conducted in a joint program with the USA from 1991 to 1994, and a report on the American part was published by Oak Ridge National Laboratory and Resources for the Future (ORNL/ RFF, 1994). Analogous work was also commissioned by the State of New York (Rowe *et al.*, 1995). But after these pioneering studies, work on this subject fell victim to the general wave of deregulation in the electric power industry: with increasing privatization, power companies were no longer required to internalize environmental damages, and funding for new studies dried up. Nevertheless EPA continued such assessments to evaluate costs and benefits of new environmental regulations for power plants (Abt, 2000 and 2004). A similar study was carried out by Levy *et al.* (1999). The latter studies did not consider greenhouse gases. More recently the subject of external costs was picked up again as the Government commissioned the National Research Council (NRC, 2010) to evaluate the external costs of transport, heat and power; this study includes greenhouse gases. The

Table 12.4 *Epidemiology studies employed in NRC (2010).*

Health end point	Pollutant	Study author	$/case
COPD hospital admissions	NO_2	Moolgavkar (2000)	11,276
IHD hospital admissions	NO_2	Burnett *et al.* (1999)	18,210
ER-Visits asthma	O_3	Stieb *et al.* (1996)	240
Respiratory hospital admissions	O_3	Schwartz (1995)	8300
Chronic asthma	O_3	McDonnell *et al.* (1999)	30,800
Acute exposure mortality	O_3	Bell *et al.* (2004)	?
Asthma hospital admissions	SO_2	Sheppard (1999)	6700
Cardiac hospital admissions	SO_2	Burnett *et al.* (1999)	17,526
Chronic bronchitis	PM_{10}	Abbey *et al.* (1993)	320,000
Infant chronic exposure mortality	$PM_{2.5}$	Woodruff *et al.* (2006)	?
All-cause adult chronic exposure mortality	$PM_{2.5}$	Pope *et al.* (2002)	5,910,000

APEEP model used for the calculations has been described by Muller and Mendelsohn (2007), together with some results.

Generally the methodology and key assumptions of these studies have been fairly similar to those of ExternE at the corresponding time. The atmospheric modeling is similar to that of EMEP and accounts for the creation of O_3 and nitrate and sulfate aerosols. The same epidemiological studies were reviewed and the same studies were used for chronic bronchitis and for chronic mortality (the single most important impact). As with ExternE, the toxicity of these aerosols is assumed to be the same as that of ambient PM. The choice of specific ERFs for health impacts is somewhat different, for example, some direct impacts of SO_2 and of NO_2 are evaluated, see Table 12.4.

A major difference between ExternE and the assessments in the USA lies in the valuation of air pollution mortality. Whereas ExternE evaluates air pollution mortality in terms of life expectancy loss and multiplies by the value of a life year (in 2008 at 40,000 €/VOLY), the studies in the USA count deaths and multiply by VPF (with $ 5.91 million per premature death in NRC (2010)). These two approaches have been compared by the CAFE program of the EU DG Environment (CAFE, 2005). Since the key authors of CAFE were or had been members of the ExternE team, the results are similar to those of ExternE, with one difference: economists from the USA on the review committee requested an evaluation of air pollution mortality in terms of number of deaths, and the CAFE team complied by presenting two sets of results, one for each of the two valuation methods (Hurley *et al.*, 2005). A VPF calculation with VPF = 1 M€ yields damage costs that are only about 50% higher than a VOLY calculation with VOLY = 52,000 €.

Table 12.5 *Distribution of air pollution damages per ton of emissions from coal fired and gas fired plants, in $\$_{2007}$/ton of pollutant. The range indicates the variability due to site dependence, from the 5th percentile to the 95th percentile of plants; the range does not take into account the uncertainties. Data from NRC (2010).*

	Mean, coal	Range, coal	Mean, gas	Range, gas
SO_2	5800	1800 to 11,000	13,000	1800 to 44,000
NO_x	1600	680 to 2800	2200	460 to 4900
$PM_{2.5}$	9500	2600 to 26,000	32,000	2600 to 160,000
PM_{co}[a]	460	140 to 1300	1700	170 to 7800

[a] We had some trouble interpreting the PM data in NRC (2010) because different contributors used different definitions for PM_{10}: whereas the emissions data seem to be specified with the usual definition of PM_{10}, the impacts seem to be reported according to the awkwardly unconventional definition given in the footnote on p. 313 of their Appendix C, where PM_{10} is defined as PM between 2.5 and 10 μm, a quantity designated PM_{co} in ExternE and in this book.

Impacts on visibility, crop yields, timber yields, building materials and recreation have also been evaluated in the USA. The NRC chapter on uncertainty addresses only variability due to the different sites (and surrounding population densities) and different emission rates, but uncertainties due to atmospheric modeling, ERFs and monetary valuation are not discussed. A few sensitivity studies have been reported by Muller and Mendelsohn (2007).

The EPA assessments (Abt, 2000 and 2004) do not explicitly show damage costs per unit emission of pollutant, and thus it is not straightforward to make a direct comparison with the results of ExternE. However, Muller and Mendelsohn (2007) and the NRC report do show the damage costs per unit emission for the classical air pollutants, called "criteria air pollutants".

In Table 12.5 we show some data from Tables 2–8 (coal plants) and 2–17 (gas plants) of the NRC report for the damage costs of coal and gas fired power plants, per ton of pollutant.[4] These damages include only emissions from the power plants, not the entire fuel chain. For the means and the ranges all plants are weighted equally, rather than by the fraction of electricity they produce. The range indicates the variability due to site dependence, from the 5th percentile to the 95th percentile of plants; it does not account for uncertainties. The difference between gas and coal

---

[4] Presumably these are short tons which equal 907 kg, because the APEEP model uses units of short tons (Nick Muller, personal communication 21 May 2013), but the NRC report is not entirely clear; in some places it also uses metric tons.

Table 12.6 *Results for the total damage attributable to the criteria air pollutants emitted in 2002 in the USA, in $billion/year. From Muller and Mendelsohn (2007), with permission from Elsevier.*

Pollutant	Mortality	Morbidity	Agriculture	Timber	Visibility	Materials	Recreation	Total
PM$_{2.5}$	14.4	2.6	0	0	0.4	0	0	17.4
PM$_{10}$	0	7.8	0	0	1.3	0	0	9.1
NO$_x$	4.4	0.8	0.7	0.05	0.2	0	0.03	6.2
NH$_3$	8.3	1.5	0	0	0.2	0	0	10
SO$_2$	16.1	2.9	0	0	0.4	0.1	0	19.5
VOC	9.6	1.8	0.5	0.03	0.2	0	0	12.1
Total	**52.8**	**17.4**	**1.2**	**0.08**	**2.7**	**0.1**	**0.03**	**74.3**
%	**71.1%**	**23.4%**	**1.6%**	**0.1%**	**3.6%**	**0.1%**	**0.0%**	**100%**

Table 12.7 *Comparison of the APEEP results with those of USEPA (1999), for the benefit of the 1990 amendments of the Clean Air Act in the year 2000. Data from Muller and Mendelsohn (2007).*

	APEEP		USEPA	
	$billion/yr	%	$billion/yr	%
Mortality	42.8	90.1%	63	89.2%
Morbidity	4.1	8.6%	5	7.1%
Agriculture	0.2	0.4%	0.5	0.7%
Timber	0.03	0.1%	0.08	0.1%
Visibility	0.3	0.6%	2	2.8%
Materials	0.04	0.1%	0	0.0%
Recreation	0.01	0.0%	0.03	0.0%
Total	47.5	100.0%	70.6	100.0%

plants reflects the fact that most coal fired power plants are located farther away from population centers than are natural gas plants.

We also cite, in Table 12.6, interesting results of Muller and Mendelsohn (2007) for the total damage attributable to the criteria air pollutants in the USA. They are somewhat older than what was calculated for the NRC report, but based on the same model. It is interesting that loss of visibility contributes about 3.6% of the total damage, an impact category that unfortunately has not been evaluated in Europe.

The assessments of USEPA use different models. Muller and Mendelsohn provide a comparison of the APEEP results with those of USEPA, shown here in Table 12.7. Note that these numbers show an evaluation of the benefit of the 1990 amendments of the Clean Air Act in the year 2000. They are therefore very different from Table 12.6, which shows the damage from the entire air pollution in 2002 rather than the change in 2000 attributable to the 1990 amendments. In terms of relative contributions from different impact categories, the largest difference arises from visibility for which the USEPA benefit is seven times larger than the benefit calculated by APEEP.

12.4   Comments on the use of the results

12.4.1   *Assessment of process chains*

In many situations one wants to compare different policy options, for example, the choice between nuclear and coal for the production of electric power. One has to define carefully what one wants to compare and on what basis: this determines the boundaries of the analysis. Many policy choices affect the entire life cycle of a process (for which we prefer the term "process chain," because in most cases there is nothing cyclical). For example, an assessment of recycling of glass should take into account the entire chain of producing glass. Thus the first step of the assessment has to be the specification of the boundaries of the analysis: which activities should be included? In the next three chapters, we present examples for the production of electricity, for the treatment of waste, and for transportation.

Once the boundaries have been fixed, one has to obtain an inventory of all the relevant environmental burdens. Even though this step is simple in theory, thanks to available LCA databases, in practice it can be very difficult to determine which technologies/activities are involved in the upstream and downstream stages of a process or product. Coal for a power plant in Germany might come from underground mines in Poland, from open pit mines in the USA etc, in combinations that change over time in response to market conditions. The burdens and external costs are totally different for the different sources. This kind of situation is a classic problem for any LCA – and usually swept under the rug when analysts use LCA databases. Whereas CO_2 emissions are relatively well determined, the emissions of other pollutants are very uncertain; few LCA studies mention the uncertainties of their inventories.

One of the largest and most detailed LCA databases, EcoInvent,[5] takes a big step towards solving this problem by disaggregating the inventory according to major regions of the world. But the data are still retrospective, based on current or past practice, rather than the future. The

---

[5] www.ecoinvent.ch

emission of most pollutants (with the important exception of CO_2) by most processes has been reduced dramatically thanks to more stringent regulations, often by a factor of 2 to 10 per decade. Really one needs prospective databases because almost all applications of LCA concern choices for the future. Unfortunately that is extremely difficult, as demonstrated by an attempt by Heck et al. (2009). The authors conclude with a cautionary remark: *"The generalization of the experience curve approach to environmental burdens opens an interesting possibility for the modeling of future scenarios. Nevertheless, the approach has to be handled with care because the influence of political decisions on emission limits may have a more erratic impact on the development of environmental burdens than the permanent pressure of the market has on reductions of costs. It can also be concluded that external costs, contrary to internal costs, cannot be expected to follow a simple experience curve approach in general although it might be the case for some technologies and certain time periods."*

The quality of emissions data varies enormously from one pollutant and from one source to another. Data for CO_2 from combustion are usually very accurate because the emission rate follows from fuel input and mass balance. For major pollutants such as particulate matter (PM), NO_x, SO_x, and CO, the emission rates are controlled by government regulations; barring accidents or lapses in enforcement, the regulatory values are upper bounds, and the real performance may be better. Unregulated pollutants, on the other hand, are rarely measured and thus quite uncertain. In particular the heavy metal content of coal and oil is highly variable from one source to another and difficult to estimate.

12.4.2   How to account for CO_2 emissions from biomass

In a part of the LCA community a special convention has been established according to which CO_2 emissions should not be counted if emitted by the combustion of biomass. For example, many studies on waste incineration do not take into account CO_2 from biomass within the incinerated waste, arguing that the creation of the biomass has removed as much CO_2 as is emitted during its combustion. Of course, the latter statement is correct, as is the recommendation of the IPCC guidelines for national CO_2 inventories to the effect that CO_2 emissions from biomass combustion be reported as zero in the energy sector. But one has to be very careful with the interpretation of that recommendation in other contexts, in particular LCA, CBA and external costs.

Naïve interpretation would imply absurd conclusions, for instance that burning a tropical forest would have the same climate impact as preserving it, because the CO_2 is from biomass and therefore not to be counted.

Likewise, the benefit of adding carbon capture and sequestration (CCS) to a biomass fueled power plant would not be considered because that CO_2 is totally omitted from the analysis.

To avoid such conclusions, we recommend that emission and removal of CO_2 be counted explicitly at each stage of the life cycle of a process or product (Rabl *et al.*, 2007), with careful attention to the choice of the boundaries for the analysis. For example, in a study of a biomass fuel chain (where biomass is grown as fuel to be burned in a power plant), the removal of CO_2 should be counted explicitly for the biomass plantation, and the emission of CO_2 explicitly for the power plant. The net effect is of course zero, or almost zero in this case: the biomass has been produced only to provide fuel for the power plant, and the time difference between carbon fixing and re-emission is negligible. If the power plant includes CCS, the net emissions are negative.

For an LCA or CBA of waste treatment, the appropriate system boundary is at the point where the waste has been produced, since it has been produced regardless of the chosen treatment method. Thus the CO_2 emitted during waste incineration has to be counted fully. To determine the appropriate carbon tax for an incinerator, the boundary is at the incinerator. A tax provides the correct price signal if it is based on the total CO_2 emission. Of course, at the present time CCS for such applications would be prohibitively expensive, but the assessment method has to be correct and applied consistently, whatever the outcome.

Explicit accounting for CO_2 at each stage offers a further advantage, namely, it allows the dynamic modeling of emission and removal. The time dimension is crucial for systems with a long delay between removal and emission of CO_2, for example, the use of wood for buildings, furniture and wood-based materials. Such CO_2 is sequestered for decades or centuries, but eventually much or all of it will be re-emitted to the atmosphere. Different processes for the re-emission may have very different timescales. It is not appropriate to neglect such delays, even if one does not use monetary valuation and discounting in quantifying the damage costs associated with climate change.

By explicitly counting CO_2 at each stage, the analysis is consistent with the "polluter pays" principle and the Kyoto rules, which imply that each greenhouse gas contribution (positive or negative) should be allocated to the causing agent. For example, under a system of greenhouse gas taxation, the CO_2 from using wood for space heating should be taxed the same way as CO_2 from oil heating, and a credit for CO_2 removal should be paid only when and where the wood is replaced by new growth.

References

Abbey, D. E., Peterson, F., Mills, P. K. and Beeson, W. L. 1993. Long-term ambient concentrations of total suspended particulates, ozone, and sulfur dioxide and respiratory symptoms in a nonsmoking population. *Archives of Environmental Health* **48**: 33–46.

Abt 2000. *The Particulate-Related Health Benefits of Reducing Power Plant Emissions*. October 2000. Prepared for EPA by Abt Associates Inc., 4800 Montgomery Lane, Bethesda, MD 20814-5341.

Abt 2004. *Power Plant Emissions: Particulate Matter-Related Health Damages and the Benefits of Alternative Emission Reduction Scenarios*. Prepared for EPA by Abt Associates Inc. 4800 Montgomery Lane. Bethesda, MD 20814–5341.

Ball, D. J., Roberts, L. E. J. and Simpson, A. C. D. 1994. *An analysis of electricity generation health risks: a United Kingdom perspective*. Research report 20. ISBN 1 873933 60 6. Centre for Environmental and Risk Management, School of Environmental Sciences, University of East Anglia, Norwich NR4 7TJ, UK.

Brode, R. W. and Wang, J. 1992. *User's Guide for the Industrial Source Complex (ISC2) Dispersion Model*. Vols. 1–3, EPA 450/4-92-008a, EPA 450/4-92-008b, and EPA 450/4-92-008c. US Environmental Protection Agency, Research Triangle Park, NC 27711.

Burnett, R. D., Smith-Doiron, M., Steib, D., Cakmak, S. and Brook, J. 1999. Effects of particulate and gaseous air pollution on cardiorespiratory hospitalizations. *Archives of Environmental Health* **54**: 130–139.

CAFE 2005. *Damages per tonne emission of PM2.5, NH3, SO2, NOx and VOCs from each EU25 Member State (excluding Cyprus) and surrounding seas*. Report for European Commission DG Environment, by AEA Technology, Didcot, Oxon, OX11 0QJ, United Kingdom. Authors: Mike Holland (EMRC), Steve Pye, Paul Watkiss (AEA Technology), Bert Droste-Franke, Peter Bickel (IER). March 2005.

Derwent, R. G. and Nodop, K. 1986. Long range transport and deposition of acidic nitrogen species in North-West Europe. *Nature* **324**: 356–358.

ExternE 2008. With this reference we cite the methodology and results of the NEEDS (2004–2008) and CASES (2006–2008) phases of ExternE. For the damage costs per kg of pollutant and per kWh of electricity we cite the numbers of the data CD that is included in the book edited by Markandya, A., Bigano, A. and Porchia, R. in 2010: *The Social Cost of Electricity: Scenarios and Policy Implications*. Edward Elgar Publishing Ltd, Cheltenham, UK. They can also be downloaded from www.feem-project.net/cases/ (although in the latter some numbers have changed since the data CD in the book).

Heck, T., Bauer, C. and Dones, R. 2009. *Development of parameterisation methods to derive transferable life cycle inventories – Technical guideline on parameterisation of life cycle inventory data*. Report RS1a D4.1, NEEDS (New Energy Externalities Developments for Sustainability). European Commission. (www.needs-project.org/2009).

Hurley, F., Hunt, A., Cowie, H. *et al.* 2005. *Methodology for the Cost-Benefit Analysis for CAFE*: Volume 2: Health Impact Assessment. Didcot. UK: AEA

Technology Environment. Available: http://europa.eu.int/comm/environment/air/cafe/pdf/cba_methodology_vol2.pdf

Krewitt, W., Trukenmueller, A., Mayerhofer, P. and Friedrich, R. 1995. EcoSense – an Integrated Tool for Environmental Impact Analysis. pp. 192–200 in: Kremers, H., Pillmann, W. (Ed.): *Space and Time in Environmental Information Systems*. Umwelt-Informatik aktuell, Band 7. Metropolis-Verlag, Marburg 1995.

Levy, J. I., Hammitt, J. K., Yanagisawa, Y and Spengler, J. D, 1999. Development of a new damage function model for power plants: methodology and applications. *Environmental Science & Technology* **33**(24): 4364–4372.

Lindhjem, H., Navrud, S., Braathen, N. A. and Biausque, V. 2011. Valuing mortality risk reductions from environmental, transport, and health policies: A global meta-analysis of stated preference studies. *Risk Analysis* **31** (9): 1381–1407.

McDonnell, W. F., Abbey, D. E., Nishino, N. and Lebowitz, M. D. 1999. Long-term ambient ozone concentration and the incidence of asthma in non-smoking adults: The AHSMOG study. *Environ. Res.* **80**(1): 110–121.

Moolgavkar, S. H. 2000. Air pollution and hospital admissions for chronic obstructive pulmonary disease in three metropolitan areas in the United States. *Inhalation Toxicology* **12**: 75–90.

Muller, N. Z. and Mendelsohn, R. 2007. Measuring the damages of air pollution in the United States. *Journal of Environmental Economics and Management* **54**: 1–14.

NRC 2010. *Hidden Costs of Energy: Unpriced Consequences of Energy Production and Use*. National Research Council of the National Academies, Washington, DC. Available from National Academies Press. www.nap.edu/catalog.php?record_id=12794

OECD 2012. *Mortality Risk Valuation in Environment, Health and Transport Policies*, OECD Publishing. http://dx.doi.org/10.1787/9789264130807-en

ORNL/RFF 1994. *External Costs and Benefits of Fuel Cycles*. Prepared by Oak Ridge National Laboratory and Resources for the Future. Edited by Russell Lee, Oak Ridge National Laboratory, Oak Ridge, TN 37831.

Pope, C. A., Burnett, R. T. Thun, M. J. *et al*. 2002. Lung cancer, cardiopulmonary mortality, and long term exposure to fine particulate air pollution. *J. Amer. Med. Assoc.* **287**(9): 1132–1141.

Preiss, P., Friedrich, R. and Klotz, V. 2008. *Report on the procedure and data to generate averaged/aggregated data*, NEEDS project, FP6, Rs3a_D1.1 – Project no: 502687. Institut für Energiewirtschaft und Rationelle Energieanwendung (IER), Universität Stuttgart.

Rabl, A., Benoist, A., Dron, D. *et al*. 2007. How to account for CO_2 emissions from biomass in an LCA. *Int J LCA* **12** (5): 281.

Rowe, R. D., Lang CM, Chestnut, L. G. *et al*. 1995. *The New York Electricity Externality Study*. Oceana Publications, Dobbs Ferry, New York.

Schwartz J. 1995. Short term fluctuations in air pollution and hospital admissions of the elderly for respiratory disease. *Thorax* **50**(5): 531–538.

Sheppard, L., Levy, D., Norris, G., Larson, T. V. and Koenig, J. Q. 1999. Effects of ambient air pollution on nonelderly asthma hospital admissions in Seattle, Washington, 1987–1994. *Epidemiology* **10**: 23–30.

Stern, N., *et al.* 2006. *The Economics of Climate Change: The Stern Review*, Cambridge University Press, Cambridge, UK. Available at www.hm-treasury. gov.uk/stern_review_report.htm

Stieb, D. M., Burnett, R. T., Beveridge, R. C. and Brook, J. R. 1996. Association between ozone and asthma emergency department visits in St. Jon, New Brunswick, Canada. *Environmental Health Perspectives* **104**: 1354–1360.

USEPA 1999. *The Benefits and Costs of the Clean Air Act: 1990–2010.* U.S. Environmental Protection Agency. EPA Report to Congress. EPA 410-R-99-001, Office of Air and Radiation, Office of Policy, Washington, DC, 1999.

Vossiniotis, G., Arabatzis, G. and Assimacopoulos, D. (1996) *Description of ROADPOL: A Gaussian Dispersion Model for Line Sources*, Program manual, National Technical University of Athens, Greece.

Woodruff, T. J., Parker, J. D. and Schoendorf, K. C. 2006. Fine particulate matter ($PM_{2.5}$) air pollution and selected causes of postneonatal infant mortality in California. *Environ Health Perspect.* 2006 May; **114**(5): 786–790.

13   Results for power plants

Summary

This chapter discusses damage cost estimates for electric power. The damage cost of the power plant itself is straightforward, if one has the required data for the emission of pollutants (and other burdens) per kWh_e. But for many policy applications one needs to compare different technologies (e.g. coal versus nuclear or wind) that have very different life cycles. This necessitates an analysis of the entire fuel chain, including fuel extraction, plant manufacture and construction, and so on, as well as the power generation phase. For that reason we begin this chapter with the methodology of fuel chain analysis. To illustrate how the practice of fuel chain analysis has evolved over time and how the issues of concern (priority impacts) have been changing, we present a review of the main fuel chain studies that have been carried out in the last 25 years. Finally, we present current assessments of the most important power production technologies. Fossil fuels, especially coal, oil and lignite, have the largest damage costs due to greenhouse gases and health impacts of the classical air pollutants. The damage costs are low for the renewables and for the normal operation of the nuclear fuel cycle.

13.1   Scope of the analysis

13.1.1   Boundaries of the analysis

When calculating the damage costs of electric power, one has to begin by defining the boundaries of the analysis. This depends on the objective. If one wants to evaluate the benefits of different technologies for reducing the SO_2 emissions of coal fired plants, it suffices to consider the SO_2 damage costs of the plants in question (since SO_2 abatement does not appreciably alter the efficiency). If the objective is to compare different coal fired plants (e.g. pulverized coal, fluidized bed and integrated gasification combined

519

cycle), one has to evaluate the burdens for each of these plant types, and, to the extent that the efficiencies are different, one also has to account for the fact that the relative contribution of the upstream impacts is different. After all, the comparison should be made per kWh$_e$ that is produced (i.e. useful output), not per unit of fuel that is consumed.

The inclusion of upstream impacts becomes crucial for comparing technologies that use different fuels, for example coal and nuclear or coal and wind. In that case an LCA is required to examine the entire fuel chain. For fossil fuels, most of the damage cost comes from the emissions of the power plant itself, even for plants equipped with CCS (carbon capture and storage), whereas for wind or solar, the impacts of the power plant itself are negligible and damages arise only from upstream (and possibly also downstream) activities.

When evaluating fuel chains, one should also keep in mind at what time of day or year a kWh$_e$ is to be provided. Because of the difficulties and costs of storing electricity, a technology that is good for meeting base load demand (e.g. a nuclear plant) may be prohibitively expensive for peak demand (where a gas turbine is usually the most cost-effective). A simple comparison per kWh$_e$ is especially problematic for intermittent power sources such as solar and wind. We will present such simple comparisons below, but we hasten to add that they are not sufficient to address all policy questions; one also needs to take into account the value of the electricity that is produced at different times. In Section 13.7, we provide an example of the damage cost of a utility system designed to produce base load power by using wind as much as possible.

A single number for "the" damage cost of a fuel chain may capture most attention, yet it can be quite misleading. One has to be careful to clearly indicate some key assumptions, especially which technologies have been assumed. For example, it makes no sense to talk about "the biomass fuel chain" because there are radically different possibilities, from simple combustion to fermentation. The same applies to all other fuel chains. For example, for coal, what type of coal is used, what is its sulfur and ash content, what abatement technologies are fitted and to what efficiency is the plant designed to operate? The assumed technologies have to be clearly specified, and one needs to take great care in presenting the results in sufficiently clear and complete detail.

13.1.2 Priority impacts

Doing a complete fuel chain analysis from scratch is a complex and difficult undertaking (unless one is satisfied with using ready-made LCA software, without any attempt to calculate real impacts and damage costs). A major difficulty lies in the variety of possible impacts, for example when a chain has five distinct stages, each of these has its own set of impacts on public and occupational health, terrestrial and aquatic environments, and amenity. Most of them are minor or insignificant, but how does one know in advance? Coal contains a lot of potentially harmful substances, in particular toxic metals. How about emissions of Hg? What are the effects on health or ecosystems? How harmful is coal dust that gets into the air during transport and handling? Which are the priority impacts that the analysis should focus on?

Some guidance is available from earlier fuel cycle studies, particularly those discussed in this chapter. However, it must be recognized that there are fashions and there can be surprises. It is interesting to look at the history of fuel chain assessments. Initially their main focus was on accidents, mostly of workers. During the 1980s, air pollution entered as an additional concern, although mostly for environmental impacts of NO_x and SO_2, as acid rain became a hot topic in the popular press. The severity of the health impacts of air pollution was not appreciated until the 1990s, when new epidemiological studies were published and revealed substantial effects on health that had not been thought relevant, since many countries had introduced clean air legislation in the 1950s and 1960s. The first studies to find large impacts due to air pollution mortality appeared in around 1995 (ORNL/RFF, 1994, ExternE, 1995, Rowe *et al.*, 1995). Interestingly, when the ExternE and ORNL/RFF studies were launched in 1991, road damage by trucks for fuel transport was considered a priority impact. Nowadays, air pollution and greenhouse gases have become the main preoccupation. But one should always keep an open mind for possible new insights.

It is helpful to present the scope in terms of an accounting framework, a matrix of activities (or "stages") of the fuel chain and impacts of those activities to be included in the analysis. This is shown in Table 13.1 for coal and in Table 13.2 for nuclear. The scope for the other fossil fuels (oil and natural gas) is essentially a subset of that of coal. Each activity can impose a number of different burdens (e.g. emissions of SO_2 from the power generation stage for coal) and each of these burdens may cause a variety of impacts (e.g. health impacts and crop losses from SO_2). The coal fuel chain entails a wide variety of different burdens and impacts, and Table 13.1 is a rather simplified presentation with just three general categories of impacts: public health, occupational health and environment (the latter with subcategories natural environment, agricultural environment and man-made environment).

The nuclear fuel chain can also entail significant non-radiological occupational impacts. Non-radioactive emissions from the nuclear fuel chain also

Table 13.1 *Overview in matrix form of the stages and burdens of the coal fuel chain and the major impact categories (oil and gas cycles are essentially subsets, except for upstream impacts).*

		Impacts					
		Health		Environment			
Stages	Burdens	Occu-pational	Public	Natural	Agri-cultural	Man-made	Extent of impacts
Mining	Accidents	Q					L; P
	Waste water		nq	nq	nq		L; P,F
	Solid waste		nq, S	nq, S	nq, S		L; P,F
	Particles	nq	nq				L; P
	Land use			nq, S	nq, S	nq, S	L; P,F
	CH$_4$		Q	Q	Q	Q	G; F
Fuel transport	Accidents	Q	Q				L; P
	CO$_2$		Q	Q	Q	Q	G; F
	NO$_x$, SO$_2$		Q	Q	Q	Q	R; P
Construction of power plant	Land use			nq, S	nq, S	Q, S	L; P
	Accidents	Q	Q				L; P
Operation of power plant	Accidents	Q					L; P
	CO$_2$		Q	Q	Q	Q	G; F
Primary	Particles		Q			Q	R; P
air	SO$_2$		Q	Q	Q	Q	R; P
pollutants	NO$_x$		Q	Q	Q	Q	R; P
	CO		nq				R; P
	toxic metals		Q				R; P
Secondary	O$_3$ (from NO$_x$, VOC)		Q	Q	Q	Q	R; P
air	acid rain (from NO$_x$, SO$_x$)			Q	Q	Q	R; P,F
pollutants	aerosols (from NO$_x$, SO$_x$)		Q	Q	Q		R; P
	Thermal			Q, S			L; P
	Noise		Q, S			Q, S	L; P
	Waste water		nq	nq			L; P,F
	Solid waste		nq, S	nq, S	nq, S		L; P,F
Power transmission	Land use			Q, S		Q, S	L; P,F
Decommissioning							L; F

Notes: Q = quantified in at least one study
nq = not quantified (as far as we know), possibly significant
blank = not considered important
S = highly site-dependent
For extent of impacts: L = local, R = regional, G = global; P = present generation, F = future generations
Amenity includes noise, odor, impacts on buildings etc.

Table 13.2 *Overview in matrix form of the stages of the nuclear fuel chain and the major impact categories. Most of the burdens are radionuclide emissions, causing impacts of cancers and hereditary effects. Environmental impacts arise from land use.*

Stages	Occupational health	Public health	Environment
Mining and milling	Q; n, r	Q; r	nq, S
Conversion	Q; n, r	Q; r	
Enrichment	Q; n, r	Q; r	
Fuel fabrication	Q; n, r	Q; r	
Construction of reactor	Q; n	Q; n	nq, S
Electricity generation	Q; n, r	Q; r	
Decommissioning of reactor	Q; n, r	Q; n, r	
Reprocessing of spent fuel	Q; n, r	Q; r	nq, S
Low/intermediate level waste	Q; r	Q; r	nq, S
High level waste	Q; r	Q; r	nq, S
Transportation activities	Q; n, r	Q; n, r	
Reactor accident	Q; r	Q; r, S	Q, S

Q = quantified in at least one study
nq = not quantified, possibly significant
blank = not considered important
r = radiological, n = non-radiological
S = highly site-dependent

contribute some impacts but they are not significant. Environmental impacts arise mainly from land use; radiological impacts on the environment are entirely negligible, with the exception of a very large reactor accident.

13.1.3 Time and space distribution of impacts

A correct analysis requires the extending of geographic range and time horizon sufficiently far to essentially capture all of the significant impacts. For most air pollutants from fossil fuels the geographic range extends over thousands of km, and the time horizon is short, or medium in the case of chronic health impacts (but in any case limited to the present generation, less than about 50 years). For greenhouse gases the effect is global with a timescale of decades to centuries. Certain radionuclides disperse globally and may have an impact for a very long time. For ground water pathways and occupational impacts, the geographic range is local.

The types of impacts and associated uncertainty can be quite different in different time frames. This is important to keep in mind when trying to summarize the assessment results for decision makers. It is instructive to define approximate time and space categories, as indicated in Table 13.3

Table 13.3 *Possible categories for distribution of impacts in time and space. Boundaries between categories are approximate, and different choices could be made, for instance present (<50 yr) and future (>50 yr) generations.*

(a) *Coal*

Time Space	Short (immediate or <1 yr)	Medium (1 yr to 100 yr)	Long (> 100 yr)
Local (< 100 km)	occupational health public health environment	occupational health public health environment	
Regional (100 to 1000 km)	public health environment	public health environment	
Global		global warming (public health, environment)	global warming (public health, environment)

(b) *Nuclear*

Time Space	**Short** (immediate or <1 yr)	Medium (1 yr to 100 yr)	Long (> 100 yr)
Local (< 100 km)	occupational health	occupational health public health	public health
Regional (100 to 1000 km)		public health	public health
Global		public health	public health

where the major impact categories (occupational health, public health, environmental) are shown as a 3 × 3 × 3 matrix (space × time × impact). In this way decision-makers can carry out comparisons for each element of the matrix.

13.2   General issues

13.2.1   *Upstream and downstream impacts*

13.2.1.1 Upstream impacts

Upstream impacts arise from mining or drilling for fuel (accidents, land use, water pollution, etc.), fuel processing (energy use, air and water pollution, waste generation), transport (air pollution from ships, trucks or trains, accidents, etc.) and greenhouse gas emissions (e.g. some methane in coal is released to the atmosphere during mining). Fuel extraction poses a difficulty because in most cases it now takes place in a region different from where the electricity is generated. A typical example is France, where most of the coal is imported, from the USA, Australia, South Africa and so on. Accident rates per tonne of coal differ by three orders of magnitude between different countries (see Section 13.2.2). ExternE (1995), Ball *et al.* (1994) and ORNL/RFF (1994) consider only a limited source of coal (the country of the authors) and it is not clear how representative their results are of the total coal supply. Some ExternE reports have assessed upstream impacts in the countries where the fuel is imported from. Sizable quantities of air pollution are emitted during transport by ocean ships (Rabl *et al.*, 1996), but their impacts may be small because few people are affected. For the nuclear fuel chain, significant damage costs can arise from the mining of uranium if the mine tailings are left open rather than being covered.

Another type of upstream impact arises from the construction of the installations involved in a fuel chain, in particular from the pollution emitted during the production of the materials. For power plants using nuclear or fossil fuels, the impacts from the production of the materials can be neglected because they are several orders of magnitude smaller than those from the operation. For most renewable energies, on the other hand, the burdens from operation are very small or negligible, and most of the impacts arise upstream (but they tend to be small, see Section 13.4.2).

13.2.1.2 Allocation of upstream impacts for nuclear, PV and wind

LCA inventories indicate the pollution data from the industrial system of the recent past (i.e. a time for which actual data, rather than forecasts, are available). Thus an LCA of a clean technology such as PV, which requires a great deal of electric energy for its fabrication, will appear more polluting if produced in a region where the power plants are dirty. All the results for upstream impacts reported in this book are based on such LCA inventories. But is it really appropriate to penalize clean technologies in an assessment if the background emissions are high because such clean technologies are not yet used? This clearly biases towards the status quo. Should one not instead assume that the background also uses the technology one is evaluating?

For example, if the power for isotope separation for the nuclear fuel chain has been produced by coal, the conventional approach attributes a significant damage cost to nuclear due to that coal. However, isotope separation for fuel production is done exclusively for nuclear power plants and only while the plants are running (modulo short-term lags to match supply and demand); therefore the corresponding electricity input should be allocated entirely to nuclear plants. Think of it this way: LCA studies allocate the totality of emissions from coal to the coal fuel cycle (which is correct). But with an allocation of isotope separation according to the current plant mix, one assigns part of these coal emissions to nuclear in addition, a clear case of double counting. Even if the power for a particular separation plant comes from a coal plant, the produced nuclear fuel will displace pollution that would otherwise be emitted by coal.

For wind and PV, the only significant pollution damages arise from the production of the materials, e.g. steel, concrete and PV panels. Much of this pollution is emitted during the production of the required electricity, especially for the PV panels. Since industrial production of this kind runs continuously 24 hrs/day, every day of the year (except for shutdowns for maintenance and repair), the emissions correspond to base load production. Conventional studies of the damage costs of wind and PV have assumed the current mix of base load plants (with their relatively high emission of pollutants), but this is also wrong. Following the above argument for nuclear, one should really assume that such base load is supplied by wind or PV, respectively, plus NGCC (or storage). The appropriate damage costs for base load production by wind with NGCC as the cleanest fossil backup are calculated below in Section 13.7.

13.2.1.3 Wastes
Waste disposal has become an important consideration in technology choices, for instance radioactive wastes, wastes from coal, and wastes from production and disposal of PV. Impacts of solid hazardous wastes are difficult to predict to the extent that they depend on future waste management decisions. In principle, such impacts can be kept negligible by storing wastes in well managed leak proof facilities. But will the integrity of the containers and liners be maintained forever? In the case of a leak the most likely occurrence is leaching into the ground water, and the impacts tend to be limited to the local range and could be stopped or corrected, if appropriate measures are taken. Technologies for alternative methods of solid waste disposal are evolving. For example, coal waste is increasingly used as an additive in building materials. For coal, none of the studies have succeeded in quantifying physical risks from solid wastes, which have the potential of being significant at the local level.

In attempts to solve radioactive waste management problems, numerous studies have been done over the years for both hypothetical and actual sites. Disposal sites for low- and intermediate-level waste have been operating for some time, but to date no permanent long-term waste disposal of high-level radioactive waste has been implemented. Plans for only two facilities, Forsmark in Sweden and Onkalo in Finland, appear on track. Elsewhere, plans are under continual debate. Current plans in France call for permanently safeguarded and retrievable storage. The Yucca Mountain Project, long seen as the USA's solution to nuclear waste storage, was abandoned in 2010, although it may of course return to the table at some point.

There are four problems associated with long-term storage of nuclear wastes: leakage, accidental access to sites at some point in the future, use of waste materials for nuclear proliferation, and theft (e.g. for use in terrorism). Probabilities and impacts are very difficult to estimate. Estimates have been made based on scenarios such as total breach of containment after 300 years. They may or may not be representative of what will evolve. It is quite likely that these sites will be abandoned, whilst still containing large quantities of radioactive material: given that untreated high-level nuclear waste will remain hazardous for 100,000 years and more, and that recorded history to the present day covers only a few thousand years, the view that sites will remain forever secure seems optimistic at best. Even sealing off the sites does not provide an absolute guarantee of security over such a long time horizon. However, the impacts of containment failure would be limited to the local zone, because dispersion in the ground is very slow and confined and the corresponding damage cost would be small. This has been confirmed in the assessment of the French nuclear fuel cycle by Dreicer *et al.* (1995), who considered the case where the entire radioactivity of a high-level waste storage unit gets into the ground water. This assessment is discussed below in Section 13.5.1 and the results, in Fig. 13.8, show that even for zero discount rate, the damage cost due to total failure of waste containment is about two orders of magnitude smaller than the damage cost of normal operation.

This discussion highlights the role of optimism or pessimism in the assessment of long-term risks. The belief that secure waste storage can be guaranteed for the indefinite future may be as unrealistic as not recognizing the possibilities of technological progress, for example in the treatment of high-level radioactive waste. In particular, permanently safeguarded and retrievable storage has the advantage of allowing the elimination of the most troubling component of high-level waste, namely the long-lived transuranic elements, by using them as fuel in burner reactors that are now being developed. In any case, it would not be wise to reject a technology if the consequences of the alternative(s) are even worse. Are the risks of nuclear waste worse than the alternative of greenhouse gases due to fossil fuels? In Section 13.7 we show that in most regions even the cleanest alternative, wind, entails high damage costs due to climate change and air pollution, because of the requirement for fossil backup.

13.2.2 Accidents

Accidents in the energy sector can affect both workers and the general public. In the latter case their costs are obviously external costs, at least in the first instance. It is possible that they may become internalized through compensation or insurance payouts. However, even for workers, it is difficult to determine to what extent costs are already internalized. Impacts on workers are more or less internalized if workers understand their risks, have negotiated their pay accordingly and have the freedom to take employment elsewhere if they so choose. However, this may not be the case in industries without strong unions, or in areas with high unemployment rates. In any case, internalized or not, the risk of accidents should be reduced as much as can be justified by full consideration of costs and benefits. That requires an assessment of probabilities and consequences.

Historical data are a crucial source for estimating the occurrence and severity of accidents, although they are not necessarily representative of what may happen in the future. The most comprehensive database for accidents in the energy sector is the ENSAD (energy-related severe accident database) of the Paul Scherrer Institute (PSI) in Switzerland.[1] Many interesting data and studies can be found at this site. In particular, there is the report by PSI for the NEEDS project of ExternE (Burgherr *et al.*, 2008), from which we cite several results.

Figure 13.1 shows the number of fatalities for severe accidents that occurred due to natural disasters and man-made accidents in the period 1970–2005, the latter disaggregated according to energy and non-energy sectors. A breakdown by energy type can be found in Table 13.4. Chernobyl is not visible because ENSAD counts only immediate fatalities, as explained in the footnotes of Table 13.4. Likewise, none of the numbers include premature deaths due to occupational disease, such as black lung for coal miners. In Fig. 13.2, the data from Table 13.4 are normalized per energy production. For coal, oil and natural gas in OECD and EU27, the rates are around 0.1 deaths per $GW_e yr$,[2] somewhat higher for coal, somewhat lower for gas.

Example 13.1   What would the fatality rate for coal in Fig. 13.2 imply for the damage cost in OECD countries if one takes VPF = 3 M€ (the value recommended by OECD (2012), rather than what was used by ExternE)?

Solution: The fatality rate is 0.128 deaths/$GW_e yr$ = 0.0146 deaths/TWh_e. At 3 M€/death the cost is 0.0044 €cent/kWh_e, very small compared with the damage cost due to pollution of coal power plants (see Figs. 13.4 and 13.7).

---

[1] https://gabe.web.psi.ch/research/ra/   [2] 1 $GW_e yr$ = 8.766 TWh_e

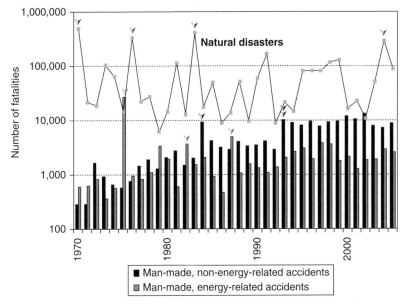

Fig. 13.1 Number of fatalities for severe (≥ 5 fatalities) accidents that occurred due to natural disasters and man-made accidents in the period 1970–2005. Years marked by an arrow indicate accidents that are discussed in Burgherr *et al.* (2008), from where this graph is copied.

It is also important to consider the contribution of each stage of the fuel chain to the total accident rates in Fig. 13.2 and Table 13.4. Such a breakdown is shown in Fig. 13.3, also from Burgherr *et al.* (2008). For coal, over 95% of the fatalities occur in coal mines, and that percentage is especially high in non-OECD countries. Actually the data from Burgherr *et al.* are not detailed enough for the case of fossil fuels produced in non-OECD countries but exported to and burned in OECD countries. Especially for coal mining, the rate of fatal accidents per tonne of coal varies a thousand-fold between different countries, and one can get very different results depending on the assumptions for the countries of origin of the coal.

Accidents in the coal mining sector have been surveyed by Holland (1996). A major factor in determining the rate is the type of mine, fatal accident rates being four to ten times lower in surface mines than in deep mines. Small mines tend to be more hazardous than large mines (this is thought to explain the very high rate shown for Turkey). For underground mines, the lowest rate is 0.1 deaths/Mt of coal in the USA and in Australia. For most countries, Holland found only average data for all mines, mostly based on ILO (1995); here, we cite a few numbers to highlight the

Table 13.4 *Summary of severe accidents with at least five immediate fatalities that occurred in fossil, hydro and nuclear energy chains in the period 1970–2005. Accident statistics are given for the categories OECD, EU 27, and non-OECD. For the coal chain, non-OECD w/o China and China alone are given separately. From Burgherr et al. (2008).*

Energy chain	OECD		EU 27		Non-OECD		World total	
	Accidents	*Fatalities*	*Accidents*	*Fatalities*	*Accidents*	*Fatalities*	*Accidents*	*Fatalities*
Coal	81	2123	41	942	144	5360	1588	31,939
					1363	24,456		
					$(818)^{(a)}$	$(11,302)^{(a)}$		
Oil	174	3338	64	1236	308	17,990	482	21,328
Natural gas	103	1204	33	337	61	1366	164	2570
LPG	59	1875	20	559	61	2610	120	4485
Hydro	1	14	1	$116^{(b)}$	12	$30,007^{(c)}$	13	30,021
Nuclear	–	–	–	–	1	$31^{(d)}$	1	31
Total	418	8554	159	3190	1950	81,820	2368	90,374

[a] First line: Coal non-OECD w/o China; second and third line: Coal China 1970–2005, and in parentheses 1994–1999. Note that only data for 1994–1999 are representative because of substantial underreporting in earlier years (Burgherr *et al.*, 2008; Hirschberg *et al.*, 2004a; Hirschberg *et al.*, 2004b).

[b] Belci dam failure (Romania, 1991)

[c] Banqiao/Shumantan dam failures (China. 1975) together caused 26,000 fatalities

[d] Only *immediate* fatalities. In the case of Chernobyl estimates for *latent* fatalities range from about 9000 for Ukraine, Russia and Belarus to about 33,000 for the whole northern hemisphere in the next 70 years (Hirschberg *et al.*, 1998). According to a recent study (Chernobyl forum, 2005) by numerous United Nations organizations (IAEA, WHO, UNDP, FAO, UNEP, UN-OCHA and UNSCEAR) up to 4000 persons could die due to radiation exposure in the most contaminated areas. This estimate is substantially lower than the upper limit of the PSI interval, which, however, was not restricted to the most contaminated areas.

variability: Poland 0.59, India 0.75, China 6.1, Pakistan 29.9, and Turkey 119 deaths/Mt of coal. Since a coal fired plant needs approximately 0.33 t of coal per MWh_e (ExternE, 1995), a rate of 1 death/Mt implies 0.33 deaths/TWh_e, and if valued at 3 M€/death, the cost would be 0.1 €cent/kWh_e. Thus, the rate for coal from the USA or Australia implies a cost of 0.01 €cent/kWh_e, about twice the result of the above example based on Burgherr *et al.* The difference is probably due to the fact that ENSAD includes only accidents with at least five immediate fatalities.

Severe accidents (more than a hundred victims) can happen with several fuel chains, e.g. a dam failure, an oil spill, an explosion of a tanker of liquefied natural gas or a nuclear power plant. The assessment of severe

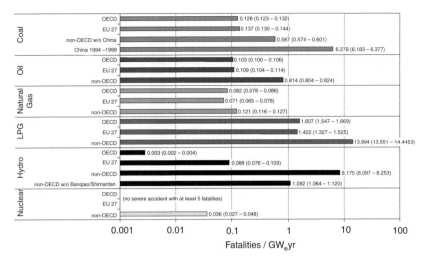

Fig. 13.2. Comparison of aggregated, normalized, energy-related fatality rates, based on historical experience of severe accidents that occurred in OECD, EU 27 and non-OECD countries in the period 1970–2005, except for coal in China, where complete data from the China Coal Industry Yearbook were only available for the years 1994–1999. Note that only immediate fatalities were considered; latent fatalities, of particular relevance for the nuclear chain, are commented on separately in the text. The exact values for each bar are shown in the figure, with values for 5% and 95% confidence intervals in parentheses. From Burgherr *et al.* (2008).

accidents raises two general issues: estimation of the probability of occurrence, and estimation of the consequences given the lack of data. The methods include complex probability safety assessments for nuclear reactor technologies, the use of complex accident consequence codes, and compilation of statistical accident data, such as for the sea transport of oil. The uncertainties are large and subjective choices are inevitable, so the results are often controversial, especially for nuclear. Many people believe that the assumed accident probabilities are too low, or that the consequences have not been adequately taken into account. Even a "worst case scenario" will not satisfy everybody, because one can always imagine something worse.

Some analysts base their assessment of a nuclear accident on Chernobyl. For power plants in countries like the USA, the European Union or Japan this is clearly not appropriate, because the technologies and the social context are totally different. But as the recent accident at the

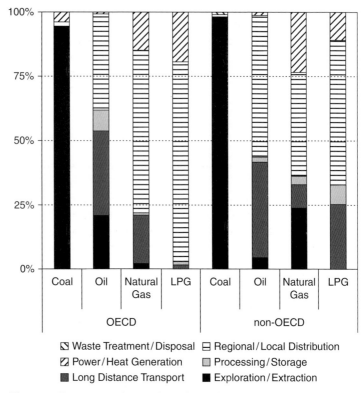

Fig. 13.3. Percentage shares of accidental fatalities in the different stages of fossil energy chains for the period 1970–2005.
From Burgherr *et al.* (2008).

Fukushima plant shows, unforeseen events may lead to catastrophic accidents, even in a country with highly developed safety standards. In Section 13.5.2 we present an estimate of the damage cost due to a catastrophic accident.

Nuclear power plants are also being planned or built in countries where coal mine accident rates are two to three orders of magnitude higher than in the EU and USA. While coal mines are not necessarily indicative of the way nuclear power would be managed, low safety standards in one sector offer no assurance that other sectors of the same country would fare better. This highlights the need to account for the manner in which a technology might be implemented at a particular site, rather than passing global judgments.

There are considerations beyond damage and cost, as the following simple example can illustrate. Suppose someone invents an energy system that will supply the world's electricity (roughly 2×10^{13} kWh/yr) at the bargain price of 1 €cent/kWh – with one little catch: there is a probability of an accident occurring once every 100 years that will kill 40,000 (the upper range of estimates of the worldwide cancer deaths due to Chernobyl). Taking a VPF of 3 million€/life and zero discount rate, the cost of such an accident has an expectation value of only,

$$4 \times 10^4 \times 3 \times 10^6 \, €/(2 \times 10^{13} \times 100\text{kWh}) = 0.006 \, €\text{cent/kWh},$$

$$(13.1)$$

less than 1% of this low electricity price. But would people accept such a deal? It would depend on how it is presented. If stated as bluntly as here, people would probably protest with a vehement "life is not for sale!". It also depends on the nature of the deaths. If all occurred at once, the deal would probably be considered unacceptable. The perception would be quite different if there were only delayed deaths occurring gradually over several decades, or if the same number of deaths occurred in many small accidents. People accept much higher mortality costs if they are gradual: road traffic accidents provide useful illustration, with over 300,000 deaths on the UK's roads between 1951 and 2006,[3] with little sign of public protest. This can be demonstrated by the results reported below in Fig. 13.4 and 13.9 for coal fired power plants after 2001, where the cost of PM_{10}, SO_2 and NO_x emissions is approximately 1 €cent/kWh, of which the mortality component is about 0.6 €cent/kWh. That is 100 times higher than Eq. (13.1). Change the numbers of this example and ask yourself how the acceptability would change. In Section 13.5.2, we present an estimate of the cost of a catastrophic nuclear accident.

13.3 Fuel chain assessments during the 1990s

Numerous fuel chain assessments have been carried out over the years, and it is instructive to look at some of them for historical perspective. The categories of concern have been changing and one may well wonder what future assessments will be like. Until the early 1990s, studies tended to focus on accidents (as expressed by the designation "comparative risk assessments") and worker accidents were an important impact category. During the 1980s there was also much concern about acid rain and its effect on forests, by now mostly forgotten as worries about dying forests and the loss of fish from rivers and lakes have subsided as a consequence of

---

[3] http://webarchive.nationalarchives.gov.uk/+/http://www.dft.gov.uk/adobepdf/162469/221412/221549/227755/rcgb2006v1.pdf. Accessed 23/3/2013.

substantial reductions in (particularly) sulfur emissions (Reis *et al.*, 2012). Impacts of the classical air pollutants on public health did not seem significant initially, but emerged during the 1990s as the dominant category in terms of costs.

A detailed review of the studies that were published during the 1990s can be found in Rabl and Dreicer (2002). The earlier studies (Hohmeyer, 1988, Ottinger *et al.*, 1991, Pearce *et al.*, 1992, Friedrich and Voss, 1993, Ball *et al.*, 1994) used a methodology called top-down, starting from national inventories of emissions, and assuming that the contribution of the power plant to the average ambient concentration of the pollutant observed in the region is the percentage of its emissions in the total emissions. Such an approach would be exact in a world where the atmosphere is uniformly mixed in the space above each region, without any transfer to or from elsewhere. That is obviously not a very realistic assumption, except for long-lived globally dispersing pollutants such as the greenhouse gases. Furthermore, there were major differences between the assumptions and the impacts included, making direct comparison of the results difficult.

In contrast, the studies since 1994 have used a bottom-up methodology, namely an impact pathway analysis. The first study of this type (ORNL/RFF, 1994) was undertaken by teams at Oak Ridge National Laboratory and Resources for the Future as part of the joint EU–US Project on the external costs of fuel chains. Detailed assessments were carried out for each of the stages of the fuel chains. Two hypothetical sites, in the southeast and in the southwest United States, were considered for the power plants. This study is fairly unique in having attempted an assessment of two non-environmental externalities (road damage and employment). Health impacts have been calculated for primary air pollutants (particles, SO_2, NO_x) as well as O_3, but not for nitrate and sulfate aerosols; the geographic range for the atmospheric dispersion calculations was 1000 miles.

The results for the nuclear fuel chain are approximately 0.010 cents/kWh at the southeast site and 0.005 cents/kWh at the southwest site. These results include two possible types of accidents, as well as the cost of waste disposal and decommissioning. The costs of a severe nuclear reactor accident were assessed using the accident consequence computer code MACCS. Two types of accidents at two sites were studied.

RCG/Hagler Bailly and Tellus Institute worked together on the New York Environmental Externalities Cost Study (Rowe *et al.*, 1995). Two sites were considered for the power plants, Upstate New York and New York City, and two stack heights. The impact pathway methodology was used, with assumptions very similar to the studies of Oak Ridge and

ExternE. Ozone damages due to NO_x emissions were estimated with a simplified model, and aerosol damages (nitrates and sulfates) were calculated by assuming that nitrate and sulfate aerosols have the same health impacts as PM_{10}. The analysis included an assessment of sensitivities and uncertainties. The atmospheric dispersion calculations cover the local and the regional range.

The ExternE Program ("external costs of energy") of the European Commission began as a collaborative project with the US Department of Energy, the US partners being Oak Ridge National Laboratory and Resources for the Future (ORNL/RFF, 1994). Whereas the US part was terminated after completion of the first phase, ExternE has been continuing with expanded scope (including transport systems and waste incineration). In the initial phase of ExternE (1995), the methodology was developed by carrying out relatively complete assessments of seven fuel chains, for installations at specific sites. An effort was made to present the results within the context of time and space to aid in the comparison process. The methodology developed in the ExternE project was then applied in a national implementation exercise, initially for France (Rabl et al., 1996) and later for all countries of the European Union (ExternE, 1998).

For the atmospheric dispersion modeling at the local scale, the ISC model (Brode and Wang, 1992) was used, a Gaussian plume recommended by the US EPA. For the regional dispersion, two models, the Harwell trajectory model (Derwent and Nodop, 1986) and results calculated by EMEP (www.emep.int), have been used independently, with good agreement (Spadaro and Rabl, 1999). ISC and the Harwell trajectory model have been integrated into the EcoSense software (Krewitt et al., 1995), which has been used for most of the calculations of ExternE.

The nuclear fuel chain was assessed to a time limit of 100,000 years for the global population. Actual sites were used in all cases except for high-level waste disposal (where an existing study was cited (EC, 1988)). For the severe reactor accident assessment, the risk was calculated for four possible scenarios, including one with a release of radioactivity comparable to Chernobyl.

The three "bottom up" studies published in 1994 and 1995 (ORNL/RFF, 1994, ExternE, 1995, Rowe et al., 1995) use essentially the same methodology and draw on the same literature of epidemiology and economic valuation. The main difference in the results comes from differences in local conditions (especially population density) and from differences in atmospheric dispersion models.

A major change occurred after the publication by Pope et al. (1995) of information on the mortality impacts of long-term exposure to air pollution ("chronic mortality" effects), which greatly increased the damage cost estimates associated with fossil fuel production via impacts of PM, NO_x and SO_2. Another significant development, this one only in Europe, was to change the valuation of mortality from VPF to VOLY. Before presenting current assessments of fuel chains, we will discuss some important aspects of specific fuel chains.

13.4   Current assessment of fuel chains

13.4.1   General remarks

Since the 1990s, several additional assessments of fuel chains have been carried out, both in the USA (Abt, 2000, Abt, 2004, Levy *et al.*, 1999, Levy *et al.*, 2009, NRC, 2010) and the EU (Rabl *et al.*, 2004, ExternE, 2008). They all use an impact pathway analysis, generally based on the same literature, although with somewhat different interpretations. In the USA, the valuation of air pollution mortality continues to be based on VPF, in the EU on VOLY. The focus in all of these studies is on greenhouse gases and the classical air pollutants. Accidents and impacts on workers have been examined in some of these (NRC, 2010, ExternE, 2008, Burgherr *et al.*, 2008) but without monetary valuation. Energy security has been examined by ExternE (2008). Issues of waste treatment, land use, and resource depletion have not been addressed. In the following, we summarize key results for USA and EU27.

To put the damage costs into perspective, we show in Table 13.5 the average selling price of electricity in the USA and in France, for industrial and for residential consumers. The data are for 2011, except those of the industrial users in France which are for 2009.

13.4.2   Damage costs of power in the EU27

The most recent assessment in the EU was carried out by ExternE (2008). Unlike earlier versions of ExternE, which used the Harwell trajectory model for the regional dispersion modeling, this assessment uses

Table 13.5 *Average selling price of electricity in the USA and in France, for industrial and for residential consumers.*

	Residential	Industrial
USA, ¢/kWh	11.0	6.7
France, €cent/kWh	11.0	6.0

source–receptor matrices of EMEP,[4] the official modeling organization
for transboundary air pollution in Europe. Source–receptor matrices are a
convenient summary of the output of a dispersion model, showing the
incremental concentration in a target grid cell due to an incremental
emission in a source grid cell. Since EMEP at that time used grid cells of
50 km × 50 km, the spatial resolution is not very fine and cannot distinguish
sources within a grid cell. For the purpose of calculating average damage
costs of power plants in Europe this is not a serious limitation.

The costs per kg of emitted pollutant by typical power plants in the
EU27 have been plotted in Fig. 12.1 of Chapter 12. The cost per tonne of
CO_2 depends on when it is emitted. For the emission by current power
plants, the CASES project assumes 21 €/tonne$_{CO2eq}$.

The emissions inventory comes from the Ecoinvent database.[5] It distinguishes four LCA stages: construction of the power plant, operation of
the power plant (for fossil plants combustion of the fuel), dismantling of
the power plant, and fuel (for fossil plants production and transport of the
fuel). No data are provided for wastes. In Fig. 13.5 we show a breakdown
of the damage costs according to operation of the power plant and the
other stages.

For the nuclear fuel cycle, ExternE (2008) estimates a damage cost of
0.21 €cents/kWh$_e$. The data CD allows us to distinguish radiological
emissions and impacts from the other burdens. The radiation dose due
to emission of radionuclides has been calculated using generic tables of
UNSCEAR (2000). These tables indicate for each radionuclide emitted
by typical sources under typical conditions, the collective dose in
person·Sv per emitted quantity in PBq (10^{15} Bq). Multiplication by the
ERFs of ICRP (2007) (see Section 3.6.11) then yields the health effects.
The monetary valuation of cancers is based on YOLL (years of life lost)
and YLD (years lived with disability), with 40,000 for the value of a
YOLL. The contribution of radionuclides to the damage costs for normal
operation is extremely small, 0.002 €cents/kWh$_e$, consistent with the
findings of the other assessments of the nuclear fuel chain.

In Fig. 13.4 and 13.5 we show selected results, with two entries for
nuclear and wind, respectively. For wind, the first entry is for a system of
wind turbine plus natural gas combined cycle to supply base load power as
do nuclear plants. The second entry is wind turbines alone, assuming
that sufficient storage is available; this entry does not include any costs for
storage. The first entry for nuclear is taken directly from the data CD of
ExternE (2008), for the second entry we have added the estimate of

---

[4] www.emep.int   [5] www.ecoinvent.ch

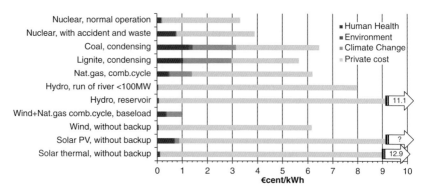

Fig. 13.4 Selected results for the costs, both external and private, of current power technologies in the EU27. The arrows indicate the estimated total cost. For the second version of nuclear, the estimate of Rabl and Rabl (2013) for accident and waste has been added. For wind, the first version is for a combination wind + natural gas combined cycle for base load (without private costs), the second version is for wind with storage (but without costs for the latter).
Data from ExternE (2008).

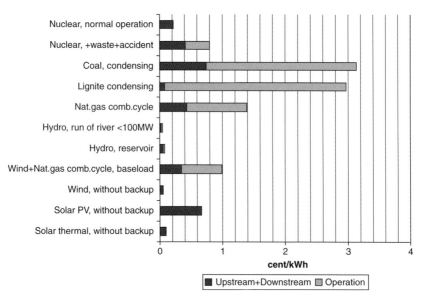

Fig. 13.5 Breakdown of damage costs of Fig. 13.4 by stages of fuel chain.
Data from ExternE (2008).

Rabl and Rabl (2013) for the cost of a catastrophic accident and waste storage (with adjustment of the numbers in Tables 13.9 and 13.12, because the costs of greenhouse gases and air pollutants are slightly different). The choice between the two estimates for nuclear power involves problematic subjective judgments: is the future accident risk for reactors under consideration as bad as the worldwide average in the past, and is the cost of waste management fully internalized by the payments to a reserve fund that are currently paid by the nuclear industry?

The arrows in Fig. 13.4 indicate the total cost where it exceeds the scale of the graph. For photovoltaics, there has been a dramatic decrease in the private cost since 2011 and we do not venture to guess. The private cost of nuclear was estimated by EdF (Electricité de France), the world's largest nuclear utility company, who in the past has been able to build nuclear power plants at impressively low cost, thanks to strong efficiencies and economies of scale and efficient centralized management. But even in France the cost of nuclear power has been increasing, driven by safety measures implemented in response to Chernobyl and Fukushima. It is interesting to compare the private cost in Fig. 13.4 with the levelized energy cost of 11 ¢/kWh_e estimated for new nuclear plants in the USA by EIA (2012).[6]

Generally the damage costs are largest for fossil fuels without CO_2 capture. Since carbon capture and sequestration (CCS) is a new technology that has not yet been tested in full scale plants, the cost estimates are extremely uncertain and we do not show them in these figures. The health and environmental costs of fossil plants with CCS can readily be estimated from Fig. 13.4. The greenhouse gas emissions are reduced by a factor of about 5 to 7. Unfortunately CCS is energy intensive and increases the fuel input by about 10 to 20%. Since the combustion process and the quantity of classical pollutants per tonne of input fuel are unchanged, CCS increases the health and non-greenhouse gas environmental impacts of the fossil plants in Fig. 13.4 by a factor 1.1 to 1.2. In Sections 13.5 to 13.7 we provide further detail on the nuclear, fossil and renewable technologies.

13.4.3   Damage costs of power in the USA

The externality study by NRC (2010) uses a methodology very similar to ExternE. The contribution of the main energy sources to the total electricity production is shown in Fig. 13.6. At the time of the study coal provided

---

[6] www.eia.gov/forecasts/aeo/electricity_generation.cfm

Table 13.6 *Emissions and damage costs for coal and gas power plants.*
The range covers the distribution from 5% to 95%; it indicates geographical
variability, not uncertainty. These data and Fig. 13.6 include only the power
plants, without upstream emissions. Data from NRC (2010).

Pollutant	g/kWh, coal mean (range)	$/kg, coal mean (range)	g/kWh, nat.gas mean (range)	$/kg, nat.gas mean (range)
SO$_2$	5.44 *(0.68–14.97)*	6.39 *(1.98–12.13)*	0.02 *(0.00–0.07)*	14.33 *(1.98–48.51)*
NO$_x$	1.86 *(0.59–4.08)*	1.76 *(0.75–3.09)*	1.04 *(0.02–2.49)*	2.43 *(0.51–5.40)*
PM$_{2.5}$	0.27 *(0.04–0.82)*	10.47 *(2.87–28.67)*	0.05 *(0.00–0.13)*	35.28 *(2.87–176.41)*
PM$_{co}$	0.06 *(0.05–0.95)*	0.51 *(0.15–1.43)*	0.00 *(0.00–0.15)*	1.87 *(0.19–8.60)*

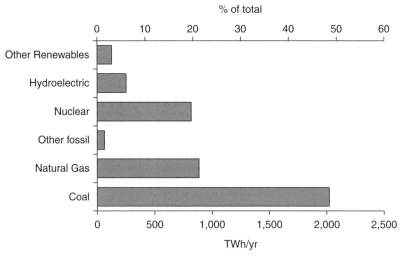

Fig. 13.6 Net electricity generation in the USA by energy. Data from
NRC (2010).

almost half of the total. However, in the meantime the growing availability
of natural gas from "fracking" (hydraulic fracturing) has spurred a rapid
expansion of natural gas combined cycle power plants.

In Table 13.6 we show the emissions and damage costs of coal and gas
fired power plants; they do not include upstream emissions. Both emission per kWh and damage cost per kg (the latter also shown in Table 12.5
of Chapter 12) vary widely, and the indicated range includes all values
from the 5th percentile to the 95th percentile (in these distributions all
plants are weighted equally, in contrast with the results shown in

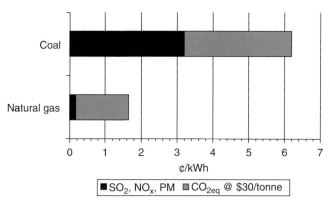

Fig. 13.7 Results of NRC (2010) for damage costs of electric power in the USA, in ¢/kWh.

For illustrative purposes, the study indicates climate damage with a hypothetical $ 30/tonne$_{CO2eq}$.

Fig. 13.7). The mean of the SO_2 and PM_{10} emissions from coal plants is an order of magnitude higher than in the EU, as shown by comparing with Table 13.10. The NO_x emissions are about three times higher. Gas is much cleaner, but even here the emissions are significantly higher than in the EU. We had some trouble interpreting the PM data in NRC (2010) because different contributors used different definitions for PM_{10}: whereas the emissions data seem to be specified with the usual definition of PM_{10}, the impacts seem to be reported according to the awkwardly unconventional definition given in the footnote on p. 313 of their Appendix C, where PM_{10} is defined as PM between 2.5 and 10 μm, a quantity designated PM_{co} in ExternE and in this book.

Finally Fig. 13.7 shows the average damage cost of coal and gas power plants in the USA, weighted according to their production. For the cost of greenhouse gases, NRC (2010) does not want to make a definitive choice but uses $ 30/t_{CO2eq}$ as a hypothetical value for the purpose of illustration (25 €/t$_{CO2eq}$ at an exchange rate of $1.3/€).[7] The damage costs in Table 13.6 and Fig. 13.7 include only the power plant emissions, not upstream or downstream impacts. As for nuclear power, NRC only reviews the studies by ORNL/RFF (1994) and ExternE (1995), without trying a

[7] The NRC report uses the same "tons" for both short tons (= 907 kg) and metric tonnes. Since the climate change literature is based on tonnes, we believe that NRC means tonnes for CO_2, in contrast with results for other pollutants which are presumably in short tons because they are calculated with the APEEP model which works with short tons (Nick Muller, personal communication 21 May 2013).

new assessment. No damage cost estimates are provided for renewable energy sources. As in Europe, the damage costs for fossil fuel power plants are very significant compared with the market price (see Table 13.5).

NRC (2010) also has some interesting data on combustion products of coal power plants. The total production of coal wastes is about 120 million tonnes/yr, of which 43% is utilized, mostly for construction materials. The quantity of combustion products is 0.06 kg/kWh of electricity from coal.

13.5   The nuclear fuel chain

13.5.1   Normal operation

Several assessments of the nuclear fuel cycle have been published since 1990, including ExternE (1995), ORNL/RFF (1994) and Ontario Hydro (1993). Some additional assessments have been carried out by ExternE (2008). In any case all of these studies are based on impact pathway analysis and LCA, and they all agree that the contribution of radionuclides from normal operation is very small, even though they make different assumptions about sites, technologies, and monetary valuation. Some studies find significant non-radiological contributions if isotope separation is done in countries where much of the electricity comes from fossil fuels, especially coal. But even those studies find that the total external costs from normal operation of the nuclear fuel chain are small compared with fossil fuels.

The only significant radiological contribution to the damage cost comes from human health, in particular cancers and hereditary effects. The reader may wonder why nothing is said about environmental impacts. Extremely high exposures, such as in the exclusion zone around Chernobyl, can indeed cause noticeable impacts on animals and plants (see e.g. IAEA, 2006). However, in terms of damage costs such impacts are negligible compared with impacts on human health. The reason lies in what we value. The health effects of nuclear radiation are an increase in the incidence of cancers and of hereditary effects. Except for extreme exposures near a catastrophic accident comparable with Chernobyl, such increases are very small compared with natural background rates.[8] When a human being gets cancer, we care a great deal, even though the effect on the survival of our species is negligible. But when an animal gets

---

[8] In NRC (2006) this is described in terms of doubling dose, i.e. the dose that would double the incidence of hereditary effects relative to the natural background rate. The doubling dose for hereditary effects is estimated to be in the order of 1 Sv per person.

cancer it is usually beyond the age of reproduction and so there is little or no effect on the survival of the species. For humans we value the individual, for ecosystems, the species. The same physical effects are valued very differently, as reflected in the high monetary valuation of cancers and hereditary effects.

Here we discuss the assessment of the French nuclear fuel chain, because it is the most thorough of all, with the most complete and available documentation (ExternE, 1995),[9] and for the radiological impacts it is still the best source of information. But we show results for a slightly more recent version (Dreicer *et al.*, 1995), because it evaluated the newer 1300 MW_e PWRs at five sites (with the same methodology), whereas ExternE (1995) considered the older 900 MW_e PWRs at only one site. These newer reactors are representative of power plants built during the 1990s in France. The spent fuel is reprocessed at a large central facility. Since France reprocesses the spent fuel of other countries as well, including Germany, the results of the French fuel chain are representative for much of the nuclear power in Europe.

Table 13.7 shows the public radiation dose for each of the main stages of the French nuclear fuel chain. Almost all of the dose is due to air emissions. The impacts are calculated by multiplying the dose by the ERFs of ICRP (1991), listed in Table 3.6. The resulting damage costs are shown in Fig. 13.8, by stage of the fuel chain, for three discount rates (0%, 3% and 10%), with a time horizon of 100,000 years.

At zero discount rate, reprocessing is the most significant contributor because of the long-lived globally dispersing C-14 and I-129 releases; but this cost drops by two orders of magnitude if one discounts instead at 3%. Discounting shrinks far future costs to insignificance. The reprocessing

Table 13.7 *Total public dose for the French nuclear fuel chain in routine operation, person·Sv/TWh.*

Mining and milling	Conversion + enrichment + Fuel fabrication	Power plant	Reprocessing	Total
1.77E-1	7.09E-5	2.16E+0	1.03E+1	1.26E+1

[9] A scanned version can be downloaded at www.externe.info/externe_d7/

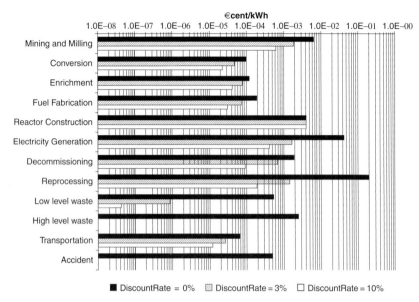

Fig. 13.8 Logarithmic plot of damage cost for the nuclear fuel chain, by stage of chain, for three discount rates: 0% (black), 3% (gray) and 10% (white), as estimated by Dreicer *et al.* (1995). Average retail price in France was 6.5 €cent/kWh in 1994.

result highlights the crucial role of the time and space boundaries of the analysis as well as the choice of technologies: other technologies capture much of the C-14 (as is done at Sellafield, UK) or do not recycle (USA, with entirely different impacts).

For another presentation, Table 13.8 shows a breakdown by impact category, in part (a), and by time and space, in part (b). It is interesting to note that the death rate among workers is about the same for cancers from radiation as for conventional accidents (e.g. a worker falling from a scaffold during construction of the reactor). The breakdown by space and time highlights the role of C-14 from reprocessing: without discounting, most of the damage cost comes from global impacts over the long term.

For nuclear, the dominant impacts from normal operation are cancers and hereditary effects. For the public, the individual risks are extremely small, but summed over the world population and over very long times, they are significant – if the dose–response function is linear without threshold, as all studies of the nuclear fuel chain have assumed.

As shown in the following paragraphs, these doses are very small compared with the background, which is generally around 2.5 mSv/yr, due to

Table 13.8 *Nuclear fuel chain impacts by time, space and impact category at 0% discount rate. Costs are based on VPF valuation; for YOLL valuation, the radiological costs would be about half as large. From Dreicer et al. (1995)*

(a) *Breakdown by impact category*

	Deaths/TWh	€cent/kWh
worker, non-radiological	0.019	0.007
worker, radiological	0.02	0.007
public, total	0.65	0.238
environment		negligible
Total		**0.252**

(b) *Breakdown by time and space.*

€cent/kWh	Local	Regional	Global
short term (< 1 yr)	0.0068	0	0
medium term (1–100 yr)	0.0084	0.006	0.019
long term (100–100,000 yr)	0.0026	0.0002	0.21

natural sources (radon, and cosmic radiation), medical x-rays, nuclear weapons tests and emissions from nuclear technologies. Epidemiological studies are unable to measure effects at doses less than about one hundred times larger than the background. Therefore, the ERFs are obtained by extrapolation from the much higher doses that were received by small case study populations, especially bomb survivors of Hiroshima and Nagasaki. This entails great uncertainty.

The total nuclear production in France was 439 TWh in 2008. Multiplying by the total public dose, 12.6 person·Sv/TWh in Table 13.7 and dividing by the population of France, 65 million, one would find a dose per person of 0.085 mSv/yr, very small compared with a typical background. The real dose per person is much smaller than this, because most of the calculated dose is global and spread over thousands of years, so one should divide by a much larger population. Therefore, it is more appropriate to estimate the dose per person by assuming a scenario of using nuclear power on a large scale and for a long time. The population of the world is about 7 billion and the electricity consumption 20,000 TWh/yr in 2010. Both will increase, especially the latter, although saturation can be expected eventually. Technologies will certainly evolve and nuclear fission reactors are unlikely to be more than a stopgap, perhaps for a century or so, until cleaner sources of energy mature: solar, fusion, or perhaps hybrid fission–fusion reactors. Since details do not matter for the following argument, we will take simple round numbers.

Let us suppose a simple "100-year nuclear scenario," where the French nuclear fuel cycle is used for 100 years to produce 30,000 TWh/yr, for a world population of 10 billion. Multiplying the total public dose in Table 13.7, 12.6 person·Sv/TWh, by 30,000 TWh/yr and dividing by a world population of 10 billion, one obtains a dose rate of 38 μSv/yr per person, if the entire dose were incurred immediately. However, only about 10% of the dose from the production of a kWh is incurred during the first 100 years (as can be seen from the global €cent/kWh numbers in Table 13.8b, since these costs are proportional to the dose). The precise time distribution of the total dose rate from the "100-year scenario" would be difficult to calculate, because of the large number of different radionuclides with different half-lives. But again, a rough order of magnitude estimate is sufficient, and we simply take 10% of the total dose rate to be imposed on the population living during this "100-year nuclear scenario,"

$$\text{total dose rate} = 3.8\ \mu\text{Sv/yr.} \qquad (13.2)$$

This can be compared with background radiation from other sources to which people are exposed. A typical value is in the range 2 to 3 mSv/yr, but that includes medical x-rays, and an average value for radon in buildings (USDOE, 1994). Probably the most meaningful comparison is with the dose rate due to cosmic radiation at sea level,

$$\text{cosmic radiation at sea level} = 260\ \mu\text{Sv/yr}, \qquad (13.3)$$

because this is the very minimum all of us are exposed to and protection from it is not practical. Thus the "100-year nuclear scenario" would increase the background exposure by, very roughly, 1.5% of the cosmic ray background at sea level. Generations living beyond those 100 years would of course also be exposed, but at a lower rate. Incidentally, the incremental dose rate of 3.8 μSv/yr is less than half the level of 10 μSv/yr that the NCRP (1993) has recommended as a "negligible dose level."

Example 13.2 Calculate the life expectancy (LE) loss for this scenario, assuming 15 YOLL/death for cancers (see Table 9.6). This has been shown in the risk comparison of Fig. 2.10.

Solution:

Dose rate due to scenario	3.8 μSv/yr	Eq. (13.2)
Lifetime exposure	300 μSv	For 80 yr
ERF	0.05 probability of death per person/Sv	Eq. (3.49)
Probability of cancer death per person/Sv	1.5E-05	Product of lines 2 and 3
YOLL/death	15 yr	
LE loss	**2.0 hr/person**	Product of lines 4 and 5, times 8766 hr/yr

13.5.2 Nuclear accidents

The cost of a large nuclear accident has been estimated both by ORNL/ RFF (1994), and by ExternE (1995) and Dreicer *et al.* (1995). ORNL/ RFF used the MELCOR accident consequence code system (MACCS) and studied two types of accidents at two sites. The results presented are 0.01 ¢/kWh for the southeast site and 0.005 ¢/kWh for the southwest site. The numbers are small despite the high cost of such an accident, because the probability is very small (as estimated by probabilistic safety assessment based on engineering fault tree analysis).

Dreicer *et al.* (1995) analyzed two types of accidents: accidents during transport of radioactive materials between sites, and accidents of the reactor. The costs due to accidents during transport turned out to be a very small fraction of the total. For a severe reactor accident with core melt, the probability was taken as 1.0E-5 per reactor·year (EdF, 1990), broadly consistent with other probabilistic assessments based on engineering fault tree analysis. Several scenarios for the source term have been evaluated. For the result reported in Fig. 13.8, the source term corresponds to a release of about 1% of the core, the same order of magnitude as the reference accident scenario used by the French national safety authorities. The COSYMA accident consequence assessment code (Ehrhardt and Jones, 1991) was used to estimate the doses, costs of countermeasures and economic losses that would be expected after an accident. The collective dose for this accident is 58,000 person·Sv (compared with about 560,000 for Chernobyl) which amounts to 0.016 person·Sv/TWh; the corresponding cost is 0.00046 €cent/kWh at zero discount rate for health costs. Again, the costs are low because of the low probability assumed for such an accident.

But is the probability of severe accidents really so low? Since they are rare occurrences, historical data do not provide much guidance. There have been two severe accidents with significant release of core material: Chernobyl and Fukushima. They are rated seven, the highest number on the scale of accident severity, which is logarithmic. The accident at Three Mile Island rates only five on this scale. Each time one learns and improves the safety. No one builds Chernobyl type reactors any more, and the protection against earthquakes and floods will be improved after Fukushima. But what unexpected risks remain lurking in the background? Fear of another catastrophic accident where operators and designers have underestimated risk is why many people are calling for an end to nuclear power, and even the shutdown of all existing nuclear plants, arguing that energy efficiency and renewables are clean and cost-effective alternatives.

It would not be wise to retire nuclear plants precipitously, if the alternatives entail total (private + damage) costs that are even higher. Rabl and Rabl (2013) have compared the damage costs of nuclear with those of the alternatives. Unlike the above assessments with complex and opaque computer calculations, they present a very simple and transparent calculation based on the actual track record of nuclear power plants. It is summarized in Table 13.9 for nuclear, and Table 13.12 for the cleanest alternative; for detailed justification of the assumptions and parameters we refer to their paper. For their central estimate they assume impacts comparable with Chernobyl and Fukushima. The release of radioactive material and the ensuing health impacts of Chernobyl were much worse than for Fukushima. The total worldwide cancer deaths due to Chernobyl have been estimated as 24,000 by Garwin (2005) and as 16,000 (95% uncertainty interval 6700 to 38,000) by Cardis *et al.* (2006). For Fukushima, Garwin (2012) estimated about 1550 cancer deaths worldwide, based on early dose assessment by IRSN (2011). A more recent assessment, with detailed modeling of dispersion and doses, by Ten Hoeve and Jacobson (2012), puts the total expected death toll at 130 (uncertainty range, 15 to 1100).

Rabl and Rabl also include an item for the cost of waste management. To account for the general tendency to underestimate future costs, Rabl and Rabl take 0.2 €cent/kWh as a damage cost of providing a permanent nuclear waste disposal site, even though much of that is already internalized because plant operators are required to put sufficient money into a waste management fund. Since the 0.2 €cent/kWh for the cost of waste management in Table 13.9 is not a damage cost, but an estimate of insufficient payment by utilities for future waste management, we label it as an external cost. In contrast, the cost of an accident and the 0.21 €cent/kWh for pollution from normal operation are damage costs.

Of course, any assessment of the external costs of nuclear is controversial, in particular with regard to accidents, proliferation, terrorism and

Table 13.9 *Assumptions and results for the external costs of nuclear power.*
From Rabl and Rabl (2013).

Cost elements	Parameter[a]	Units[10]	Central	Low	High
Fatal cancers	10,000				
Cost per cancer	5	M€/cancer			
Discount factor for cancers[b]	0.38				
Cost of cancers		**G€**	**18.8**	10	50
Lost reactors, 1GW each	6				
Cost per reactor	5	G€/reactor			
Cost of reactors		**G€**	**30**	20	40
Cost of cleanup		**G€**	**30**	20	200
Displaced persons	500,000				
Cost per displaced person	0.5	M€/person			
Cost of displaced persons		**G€**	**250**	100	1000
Area lost for agricultural production	1000	km^2			
Yield, cereals, 5 tonnes/ha per yr	500	tonnes/km^2/yr			
Price, cereals 150 €/tonne	150	€/tonne			
Loss €/km^2/yr	75,000	€/km^2/yr			
Loss duration	100	Yr			
Cost of lost agriculture		**G€**	**7.5**	5	50
Lost power production	90	TWh			
Value per kWh	0.2	€/kWh			
Cost of lost power		**G€**	**18**	10	50
Total cost of accident, if now		**G€**	**354**	165	1390
Years without accident			25	40	15
Discount rate	0.05				
Discount factor[c]			0.56	0.43	0.69
Nuclear production, 2008	2,100	TWh/yr			
Damage cost of accident, per kWh[d]		**€cent/kWh**	**0.38**	**0.08**	**2.29**
Damage cost of normal operation[e]		**€cent/kWh**	**0.21**	**0.07**	**0.63**
External cost of waste management		**€cent/kWh**	**0.20**	**0.10**	**0.30**
TOTAL EXTERNAL COST OF NUCLEAR		**€cent/kWh**	**0.79**	**0.25**	**3.22**

[a] the parameter values are used only for the central estimate
[b] to account for delay between accident and occurrence of cancer
[c] to account for an accident occurring between now and 25 yr from now
[d] cost if now × discount factor/(years without accident × TWh/yr)
[e] external cost of nuclear estimated by ExternE (2008)

waste management. Subjective choices are inevitable and any specific assumption can and will be criticized. Rabl and Rabl offer their assessment for discussion, because it is better to base decisions on an explicit analysis rather than vague impressions.

[10] T=tera, G=giga, M=mega, k=kilo

Table 13.9 shows what such assumptions imply for the central estimate of the accident cost. The cost per kWh is obtained by dividing the cost of an accident by the number of years between accidents and the annual electricity production of the countries considered here (EU, US, Canada, Japan, South Korea and Taiwan); in addition, it is multiplied by a discount factor to account for an accident occurring between now and 25 yr from now. The values listed in the parameter column are used only for the central results. The last two columns indicate lower and upper bounds; they are rough estimates of plausible ranges about the central values. For comparison, note that the accident database of Hirschberg et al. (2004b) shows an estimate of G\$ 339_{1996} (approx. 360 G€$_{2010}$) for the cost of Chernobyl, which is close to the central estimate in Table 13.9.

However, the historical accident probability assumed for Table 13.9 is about an order of magnitude higher than what has been estimated by the standard probabilistic safety assessments (PSA) with detailed fault tree modeling for the majority of reactors that have been built. This is the main reason why the accident cost in Table 13.9 is so much higher than the estimates by ORNL/RFF (1994), ExternE (1995) and Dreicer et al. (1995). Thus any estimate of the external costs of nuclear power involves problematic subjective judgments: is the future risk for reactors under consideration as bad as the worldwide average in the past, and is the cost of waste management fully internalized by the payments to a reserve fund that are currently paid by the nuclear industry? If one believes that waste management is fully internalized and that the future accident risk is as estimated by PSAs, the bottom line in Table 13.9 would be three times smaller, about 0.25 €cent/kWh.

13.6   Fossil fuel chains

For the fossil fuel chains, the lion's share of the damage costs comes from air pollutants emitted by the power plant, the main impact categories being global warming and public health. Air pollutants from upstream and downstream activities contribute less than 1% of the total, with the possible exception of wastes (impacts not yet quantified) and of greenhouse gas emissions. During the transport of coal and oil by ship, large quantities of air pollutants are emitted, but their health impact is small because the emission is far from human population.

Some greenhouse gases, especially CO_2 and CH_4, are emitted during extraction and transport of coal, oil and gas. Even though the mass of CH_4 per kWh is small, the impact can be appreciable because its GWP (global warming potential) is approximately 25 for a time horizon of 100 years. In Table 13.10, we show typical estimates of emissions from fossil plants in the

Table 13.10 *Typical emissions by coal and gas power plants in the EU, in g/kWh_e. From ExternE (2008).*

	Hard coal, condensing power plant			Natural gas combined cycle		
	upstream	plant	total	upstream	plant	total
CO_2	30.9	730.2	761.1	46.9	355.5	402.4
CH_4	2.2	0.0	2.2	1.0	0.0	1.0
CO_{2eq}[a]	84.8	730.6	815.4	71.6	355.6	427.3
$PM_{2.5}$	0.01	0.00	0.01	0.00	0.00	0.01
PM_{co}	0.02	0.03	0.04	0.00	0.00	0.00
NO_x	0.33	0.55	0.88	0.15	0.27	0.42
SO_2	0.24	0.55	0.79	0.14	0.01	0.15

[a] assuming GWP of 25 for CH_4

EU27, according to the LCA inventories[11] of ExternE (2008) for plants that respect the emission limits of the Large Combustion Plant Directive of 2001 (EC, 2001). The LCA stages in these LCA inventories in the data CD are listed as: construction, operation, dismantling, and fuel. The latter comprises the production and transport of the fuel. Wastes are not considered and the contributions of construction and dismantling to the external costs are negligible. The column "upstream" in Table 13.10 is therefore essentially for the stage called "fuel" in the data CD.

The corresponding damage costs are plotted in Fig. 13.9, with the numbers for € per kg of pollutant, also from the data CD, and indicated in the labels. To emphasize the benefit of tightened environmental regulations, we also show in Fig. 13.9 the damage costs of coal and oil plants as they were operating in France during the mid 1990s (there were no gas plants in France at the time).

To put the damage costs into perspective, note that the market price for residential customers in France was about 7 €cents/kWh during the mid 1990s, and 11 €cents/kWh in 2011. The damage costs were larger than the market price during the mid 1990s, but since then the emissions have been greatly reduced (except for greenhouse gases) and the market price has increased a great deal – probably in good part because of the improved safety of nuclear reactors. Now the damage costs of fossil power plants remain significant, mainly because of greenhouse gases.

As we explained in Section 12.2.3, we believe that the damage costs of ExternE (2008) are much too low in view of the Stern report for GHG and

[11] Based on the ecoinvent database www.ecoinvent.ch

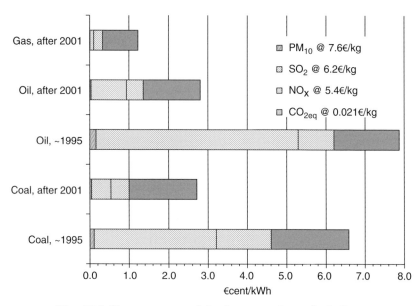

Fig. 13.9 Damage costs of fossil power plants (including upstream emissions), with damage costs per kg of pollutant and emissions for plants after 2001 according to ExternE (2008). Also shown are the external costs for coal and oil plants in France during the mid 1990s.

For comparison, the market price for residential customers in France was about 7 €cents/kWh during the mid 1990s, and 11 €cents/kWh in 2011.

the OECD (2012) report on mortality valuation. This affects most strongly the fossil fuel power plants. Therefore, we show in Fig. 13.10 how the damage cost per kWh increases if one applies the adjustment factors of Section 12.2.3. The emissions are those of Table 13.10. Only coal and gas are shown, because they are the most important fossil plants in most countries (for lignite, another important fossil fuel, the increase is similar).

A possible concern with coal and oil arises from the emissions of toxic metals. Both coal and oil contain traces of toxic metals, such as arsenic (As), lead (Pb) and mercury (Hg), and some of that escapes through the smoke stacks of power plants. The amounts of these trace metals can vary greatly with the origin and type of oil or coal. In Table 13.11, we show emissions data and damage cost for hard coal condensing power plants in the EU27, as estimated by ExternE (2008). The largest cost is due to Hg. It is interesting to note that in the USA the capture of Hg from coal plants has been proposed, after careful analysis found it to be justified in terms of

Table 13.11 *Emission (to air) and damage cost of toxic metals from coal fired power plants. Data from ExternE (2008).*

	kg/kWh	€/kg	€cent/kWh
Arsenic	7.98E-09	530	4.23E-04
Cadmium	3.92E-10	84	3.28E-06
Chromium VI	5.24E-10	66	3.47E-06
Lead	3.43E-08	278	9.54E-04
Mercury	2.54E-08	8,000	2.03E-02

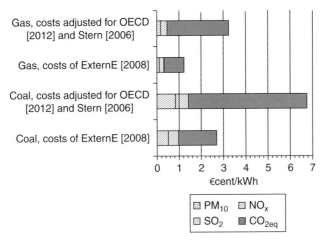

Fig. 13.10 Change of damage cost with the adjustment factors of Section 12.2.3.

social costs, but opposition by industry has prevented its implementation so far (see NRC (2010)).

13.7 Renewable energy technologies

Among renewable electricity sources there is a great variety of technologies such as hydro, wind, biomass and various forms of direct solar energy utilization, in particular photovoltaics and solar thermal power plants. Hydro, wind and solar have special appeal, being not only inexhaustible, but generally with little environmental impact or health risk. Of course, detailed studies are needed to check whether the impacts of these technologies are really benign.

For electricity from biomass, the main technologies that have been considered are combustion with steam turbine, or gasification with gas turbine. There are significant health impacts from the air pollution emitted by the power plant and by the machinery needed for the production and transport of the fuel. The net greenhouse gas emissions from the biomass itself are zero, but there are emissions from the associated machines and vehicles and any chemical inputs (fertilizers, pesticides). Combustion with steam turbine is not the most efficient way of using biomass, because temperature and pressure of the produced steam, and thus the power plant efficiency, are relatively low. In addition such plants are usually too small to maximize economies and efficiencies of scale, because the transport distances cannot be large for a fuel with low energy density. A better alternative is to add wood chips to coal in conventional coal fired power plants (up to 5% of the total fuel can be added in this form), as already practiced in some power plants. However, this option is practical only where the cost of transporting wood is not too high.

For hydro, PV and wind there are of course no emissions from power generation. There are upstream emissions from the production of the materials, although the corresponding damage costs are small. However, one must not overlook the variability of wind or insolation and the corresponding variability in the amount of electricity they supply. To achieve a reliable power supply, backup capacity must be available, especially if solar and wind provide a high fraction of the total electricity production. Of course, energy storage would be an attractive solution, but for most applications storage of the required magnitude and duration is still too expensive or the potential sites (for the most cost-effective option, pumped hydro) are too limited. Without sufficient storage, the backup capacity requirement of wind and solar implies that part of the replaced electricity will come from fossil fuels, with the attendant costs for health and environment.

Rabl and Rabl (2013) calculate the damage costs of producing base load power by wind turbines with natural gas combined cycle (NGCC) plants as the cleanest fossil backup. Their results are shown in Table 13.12, although slightly modified to make them consistent with the numbers of ExternE (2008). The fraction of base load power provided by wind is around 34%, a number for very good wind sites. Rabl and Rabl make this calculation in the context of proposals for premature retirement of existing nuclear plants and note that the damage cost of a wind plus NGCC base load plant, 1.22 €cent/kWh for their central estimate (reduced to 0.92 €cent/kWh in Table 13.12), is higher than the damage cost of nuclear with the costs of accident and waste management, 0.79 €cent/kWh for the central estimate shown in Table 13.9.

Table 13.12 *The calculations of Rabl and Rabl (2013) for the damage costs of producing base load power by wind turbines with natural gas combined cycle (NGCC) plants as cleanest fossil backup. We have slightly modified some of the numbers of Rabl and Rabl to make them consistent with ExternE (2008). GHG = greenhouse gases.*

Cost elements	Units	Central	Low	High
Cost of GHG[a]	€/tonne$_{CO2eq}$	21	7	63
Fraction of base load power provided by wind		0.34	0.38	0.27
Damage cost of wind turbines	**€cent/kWh**	**0**	**0**	**0**
GHG emissions from NGCC[a]	kg/kWh	0.43	0.43	0.43
Cost of GHG emissions from NGCC[a]	€cent/kWh	0.90	0.3	2.7
Costs of classical air pollutants from NGCC[a]	€cent/kWh	0.50	0.17	1.49
Fraction of base load power provided by NGCC		0.66	0.62	0.73
NGCC contribution due to GHG costs[a]	**€cent/kWh**	**0.59**	**0.19**	**1.97**
NGCC contribution due to classical air pollutants[a]	**€cent/kWh**	**0.33**	**0.10**	**1.09**
TOTAL DAMAGE COST[a]	**€cent/kWh**	**0.92**	**0.29**	**3.05**

[a] These numbers have been modified to be consistent with ExternE (2008)

Amenity impacts of renewables are highly dependent on local conditions, in particular the population near the site and landscape value. The impacts of hydro are so variable that specific results cannot be taken as general guidelines. The impacts can range from beneficial (for instance due to flood control or recreational facilities), to extremely harmful if large populations are displaced without compensation or if a dam breaks. The decomposition of submerged vegetation in new reservoirs can cause significant emissions of greenhouse gases. Such emissions are highly variable with site; Gagnon and van de Vate (1997) indicate 15 g$_{CO2eq}$/kWh as a typical value for cold climates, and possibly much higher values in tropical zones.

References

Abt 2000. *The Particulate-Related Health Benefits of Reducing Power Plant Emissions.* October 2000. Prepared for EPA by Abt Associates Inc., 4800 Montgomery Lane, Bethesda, MD 20814-5341.

Abt 2004. *Power Plant Emissions: Particulate Matter-Related Health Damages and the Benefits of Alternative Emission Reduction Scenarios.* Prepared for EPA by Abt Associates Inc. 4800 Montgomery Lane. Bethesda, MD 20814-5341.

Ball, D. J., Roberts, L. E. J. and Simpson, A. C. D. 1994. *An analysis of electricity generation health risks: a United Kingdom perspective.* Research report 20. ISBN 1

873933 60 6. Centre for Environmental and Risk Management, School of Environmental Sciences, University of East Anglia, Norwich NR4 7TJ, UK.

Brode, R. W, and Wang, J. 1992. *User's Guide for the Industrial Source Complex (ISC2) Dispersion Model.* Vols.1–3, EPA 450/4-92-008a, EPA 450/4-92-008b, and EPA 450/4-92-008c. US Environmental Protection Agency, Research Triangle Park, NC 27711.

Burgherr, P., Hirschberg, S. and Cazzoli, E. 2008. *Final report on quantification of risk indicators for sustainability assessment of future electricity supply options.* NEEDS Deliverable n° D7.1 – Research Stream 2b. NEEDS project "New Energy Externalities Developments for Sustainability", Brussels, Belgium.

Cardis, E., Krewski, D., Boniol, M. *et al.* (2006) *The Cancer Burden from Chernobyl in Europe.* World Health Organization, International Agency for Research on Cancer Briefing Document, April 20. www.iarc.fr/en/media-centre/pr/2006/IARCBriefingChernobyl.pdf

Derwent, R. G. and Nodop, K. 1986. Long-range transport and deposition of acidic nitrogen species in north-west Europe. *Nature* **324**: 356–358.

Dreicer, M., Tort, V. and Margerie, H. 1995. *Nuclear fuel cycle: implementation in France.* Final report for ExternE Program, contract EC DG12 JOU2-CT92-0236. CEPN, F-92263 Fontenay-aux-Roses. This report is included in Rabl *et al.* (1996). These reports by can be downloaded at www.arirabl.org

EC 1988. *Performance Assessment of Geologic Isolation Systems for Radioactive Waste,* Summary, DIR11775 EN, Brussels, Belgium.

EC 2001. Directive 2001/80/EC of the European Parliament and of the Council of 23 October 2001 on the limitation of emissions of certain pollutants into the air from large combustion plants.

EdF 1990. *Etude probabiliste de sûreté des REP de 1300 MWe.* EPS 1300. Electricité de France.

Ehrhardt, J. and Jones, J. A. 1991. An outline of COSYMA, a new program package for accident consequence assessments. *Nuclear Technology.* 94: 196–203.

EIA 2012. Levelized cost of new generation resources in the annual energy outlook 2012. US Energy Information Administration.

ExternE 1995. *ExternE: Externalities of Energy.* ISBN 92-827-5210-0. Vol.1: Summary (EUR 16520); Vol.2: Methodology (EUR 16521); Vol.3: Coal and Lignite (EUR 16522); Vol.4: Oil and Gas (EUR 16523); Vol.5: Nuclear (EUR 16524); Vol.6: Wind and Hydro Fuel Cycles (EUR 16525). Published by European Commission, Directorate-General XII, Science Research and Development. Office for Official Publications of the European Communities, L-2920 Luxembourg.

ExternE 1998. *ExternE: Externalities of Energy.* Vol.7: Methodology 1998 Update (EUR 19083); Vol.8: Global Warming (EUR 18836); Vol.9: Fuel Cycles for Emerging and End-Use Technologies, Transport and Waste (EUR 18887); Vol.10: National Implementation (EUR 18528). Published by European Commission, Directorate-General XII, Science Research and Development. Office for Official Publications of the European Communities, L-2920 Luxembourg. Results are also available at http://ExternE.jrc.es/publica.html

ExternE 2008. With this reference we cite the methodology and results of the NEEDS (2004–2008) and CASES (2006–2008) phases of ExternE. For the damage costs per kg of pollutant and per kWh of electricity we cite the numbers of the data CD that is included in the book edited by Markandya, A., Bigano, A. and Porchia, R. in 2010: *The Social Cost of Electricity: Scenarios and Policy Implications*. Edward Elgar Publishing Ltd, Cheltenham, UK. They can also be downloaded from www.feem-project.net/cases/ (although in the latter some numbers have changed since the data CD in the book).

Friedrich, R. and Voss, A. 1993. External Costs of Electricity Generation, Energy Policy **21**: 114–121.

Gagnon, L. and van de Vate, J. F. 1997. Greenhouse gas emissions from hydropower: the state of research in 1996. *Energy Policy* **25**: 7–13.

Garwin, R. L. 2005. Chernobyl's real toll. EUROPE FEATURES United Press International's "Outside View" commentaries. www.fas.org/rlg/051109-chernobyl.pdf

Garwin, R. L. 2012. *Evaluating and Managing Risk in the Nuclear Power Sector*, JSPS-USJI Risk-Management Symposium, Washington. DC, March 9, 2012 www.fas.org/rlg/Evaluating%20and%20Managing%20Risk.pd

Hirschberg, S., Heck, T., Gantner, U. *et al.* 2004a. Health and environmental impacts of China's current and future electricity supply, with associated external costs. Special Issue on China's Energy Economics and Sustainable Development in the 21st Century, *International Journal of Global Energy Issues*, Y. M. Wei, H. T. Tsai, C. H. Chen, Guest Editors, Volume **22** (2/3/4), InterScience Publishers 2004.

Hirschberg, S., Burgherr, P., Spiekerman, G. and Dones, R. 2004b. Severe accidents in the energy sector: comparative perspective. In: *Journal of Hazardous Materials* **111**: 57–65.

Hohmeyer, O. 1988. *Social Costs of Energy Consumption: External Effects of Electricity Generation in the Federal Republic of Germany*. New York: Springer.

Holland, M. 1996. *Quantifying the externalities of fuel cycle activities outside the European Union*. Maintenance Note 1 for the ExternE CORE Project. Energy Technology Support Unit, B 156 Harwell Laboratory, Didcot, Oxfordshire OX11 0RA, UK.

IAEA 2005. *Chernobyl: The True Scale of the Accident*. International Atomic Energy Agency, World Health Organization, and United Nations Development Programme. 2005. www.who.int/mediacentre/news/releases/2005/pr38/en/index.html accessed 07 Jan.2012

IAEA 2006. *Environmental Consequences of the Chernobyl Accident and Their Remediation: Twenty Years of Experience*. Report of the Chernobyl Forum Expert Group 'Environment'. International Atomic Energy Agency, Vienna, Austria.

ICRP 1991. International Commission on Radiological Protection. *Recommendations of the International Commission on Radiological Protection*, Report 60, Annals of the ICRP, Pergamon Press, UK, 1991.

ILO 1995. Recent developments in the coal mining industry. International Labor Organization, International Labor Office, Geneva.

IRSN 2011. *Assessment on the 66th day of projected external doses for populations living in the north-west fallout zone of the Fukushima nuclear accident: outcome of population evacuation measures*. Report DRPH/2011-10. Directorate of Radiological

Protection ond Human Health, Institut de Radioprotection et de Sûreté Nucléaire, France.

Krewitt, W., Trukenmueller, A., Mayerhofer, P. and Friedrich, R. 1995. EcoSense – an integrated tool for environmental impact analysis. in: Kremers, H., Pillmann, W. (Ed.): *Space and Time in Environmental Information Systems*. Umwelt-Informatik aktuell, Band 7. Metropolis-Verlag, Marburg 1995.

Levy, J. I., Hammitt, J. K. Yanagisawa, Y. and Spengler, J. D. 1999. Development of a new damage function model for power plants: methodology and applications. *Environmental Science & Technology* 33(24): 4364–4372.

Levy, J. I., Baxter, L. K. and Schwartz, J. 2009. Uncertainty and variability in health-related damages from coal-fired power plants in the United States. *Risk Anal.* 29(7): 1000–1014.

NCRP 1993. *Limitation of Exposure to Ionizing Radiation*. National Council on Radiation Protection and Measurements. NCRP report No.116. Bethesda, MD.

NRC 2006. *Health Risks from Exposure to Low Levels of Ionizing Radiation: BEIR VII – Phase 2*. Committee to Assess Health Risks from Exposure to Low Levels of Ionizing Radiation, National Research Council of the National Academies, Washington, DC. Available from National Academies Press.

NRC 2010. *Hidden Costs of Energy: Unpriced Consequences of Energy Production and Use*. National Research Council of the National Academies, Washington, DC. Available from National Academies Press. www.nap.edu/catalog.php?record_id=12794

Ontario Hydro. 1993. *Full Cost Accounting for Decision Making*. Toronto: Ontario Hydro, 700 University Ave. H18 J18 Toronto, Ontario. December 1993.

OECD 2012. *Mortality Risk Valuation in Environment*, Health and Transport Policies, OECD Publishing. http://dx.doi.org/10.1787/9789264130807-en

ORNL/RFF 1994. *External Costs and Benefits of Fuel Cycles*. Prepared by Oak Ridge National Laboratory and Resources for the Future. Edited by Russell Lee, Oak Ridge National Laboratory, Oak Ridge, TN 37831.

Ottinger, R. L. *et al.* 1991. *Environmental Costs of Electricity*. Oceana Publications, New York.

Pearce, D. W., Bann, C. and Georgiou, S. 1992. *The social costs of fuel cycles, report for the UK Department of Trade and Industry*, CSERGE, University College London.

Pope, C. A., Thun, M. J., Namboodri, M. M. *et al.* 1995. Particulate air pollution as a predictor of mortality in a prospective study of US adults. *Amer. J. of Resp. Critical Care Med* 151: 669–674.

Rabl, A. and Dreicer, M. 2002. Health and environmental impacts of energy systems. *International Journal of Global Energy Issues* 18(2/3/4): 113–150.

Rabl, A., Curtiss, P. S., Spadaro, J. V. *et al.* 1996. *Environmental Impacts and Costs: the Nuclear and the Fossil Fuel Cycles*. Report to EC, DG XII, Version 3.0 June 1996. ARMINES (Ecole des Mines), 60 boul. St.-Michel, 75272 Paris CEDEX 06.

Rabl, A., Spadaro, J., Bickel, P. *et al.* 2004. *ExternE-Pol. Externalities of Energy: Extension of accounting framework and Policy Applications*. Final Report contract N° ENG1-CT2002-00609. EC DG Research. Available at www.arirabl.org

Rabl, A. and Rabl, V. A. 2013. External costs of nuclear: Greater or less than the alternatives? *Energy Policy* 57: 575–584.

Reis, S., Grennfelt, P., Klimont, Z. *et al.* (2012) From Acid Rain to Climate Change. *Science*, 30 November 2012: 1153–1154.

Rowe, R. D., Lang, C. M., Chestnut, L. G. *et al.* 1995. *The New York Electricity Externality Study*. Oceana Publications, Dobbs Ferry, New York.

Spadaro, J. V. and Rabl, A. 1999. Estimates of real damage from air pollution: site dependence and simple impact indices for LCA. *International J. of Life Cycle Assessment* **4** (4): 229–243.

Ten Hoeve, J. E. and Jacobson, M. Z. 2012. Worldwide health effects of the Fukushima Daiichi nuclear accident. *Energy and Environmental Science* **5**: 8743–8756.

UNSCEAR 2000. United Nations Scientific Committee on the Effects of Atomic Radiation. *Sources and effects of ionizing radiation*. Report to the General Assembly, with scientific annexes, Volume **I**: Sources.[12]

USDOE 1994. *Radiation in the Environment*. Brochure published by Office of Environmental Management, US Department of Energy.

[12] http://www.unscear.org/unscear/en/publications/2000_1.html

14   Results for waste treatment

Summary

In this chapter, we evaluate the damage costs of landfill and incineration of municipal solid waste in Europe and North America, with due account for transport and for energy and materials recovery. Whilst air pollution provides some of the most significant externalities of waste management, the comparison of landfill and incineration also needs to consider potential impacts on drinking water due to leachates from landfill. A full impact pathway analysis of leachate is not possible here given that such impacts are extremely site specific. This is not to say that it could not be done for a specific site, though even this is far from straightforward given the complexity of the environmental pathways and the long time horizon of persistent pollutants. As an alternative we consider an extreme scenario, based on impact pathway thinking, to show that they are not worth worrying about if a landfill is built and managed according to regulations such as those of the EU. The damage costs due to the construction of the waste treatment facility are negligible, and so are the damage costs of waste transport, illustrated with an arbitrary choice of a 100 km round trip by a 16 tonne truck. The benefits of materials recovery make a relatively small contribution to the total damage cost. The only significant contributions come from direct emissions (from the landfill or incinerator) and from avoided emissions due to energy recovery (from an incinerator). Damage costs for incineration range from about 1.5 to 21 $€/t_{waste}$, extremely dependent on the assumed scenario for energy recovery. For landfill the cost ranges from about 11 to 14 $€/t_{waste}$; it is dominated by greenhouse gas emissions because only a fraction of the CH_4 can be captured (here assumed to be 70%). Amenity costs (odor, visual impact, noise) are highly site-specific and we only cite results from a literature survey which indicate that such costs could make a significant contribution, on the order of 1 $€/t_{waste}$.

14.1   Introduction

Management of wastes is an awkward problem, often mired in controversy. Of course, one should begin by reducing the production of waste and recycling as much as is practicable. Recycling rates are especially high for metals: among the common metals, more than 90% are recycled. Recycling is also quite successful and profitable for much of the paper, cardboard, and glass, as well as for simple plastics such as food containers and packing materials. But most specialty plastics are designed to be so durable that recycling is difficult. For organic wastes, biotreatment (composting or anaerobic digestion) is an appropriate solution and widely practiced; anaerobic digestion has the advantage of producing methane under controlled conditions to be used as fuel.

There remains a lot of waste that is not suitable for recycling or biotreatment, and it has either to be sent to landfill or incinerated (pyrolysis could be a cleaner alternative to incineration, which can produce fuels, but so far it is not widely used). In several papers we have evaluated the damage costs of landfill and of incineration of MSW (municipal solid waste) (Rabl et al., 2008, see also Rabl et al., 1998). In this chapter we present their results, updated to the current assumptions and values of ExternE (2008). The chapter progresses through consideration of the various elements that contribute to overall damage costs and then brings this information together in Section 14.6. For a broader coverage of waste management, we refer to general books, e.g. Kreith (1990), Petts and Eduljee (1994) and Hester and Harrison (2002).

Rather frequently, the performance of old waste management facilities is cited as being representative of "incineration" or "landfill." In the past, such facilities had minimal levels of emissions control. Incinerators had only some rudimentary control of particulate emissions, and landfill no controls at all. There are numerous studies in the epidemiology literature that show a link between emissions of such plant and serious health impacts. However, the current situation in the EU, North America and several other parts of the world is very different, with emission limits on discharges to air and water covering all pollutants of concern. This is not to argue that such facilities cause no harm at all (indeed, below we provide estimates of damage costs). However, what it does say is that basing waste management policies on the performance of plant built some time ago is wrong.

Further to this, there is greater control of the destiny of specific types of waste. For example, in the EU, the WEEE (waste electrical and electronic equipment) directive aims to ensure recycling of 85% of such equipment by 2016. Impacts of waste management are being further reduced by new legislation to control the use of hazardous materials in products that will eventually end up in the waste stream. A well-known example concerns controls on the mercury content of batteries. The European RoHS (restriction of hazardous substances in electrical and electronic equipment) directive restricts the use of lead, mercury, cadmium, hexavalent chromium, polybrominated biphenyls and polybrominated diphenyl ether. Under the EU's REACH regulation, further restrictions on the use of lead and mercury have been agreed in the last two years, and more substances/applications are currently under consideration for restriction. The regulation also requires users of a growing list of substances (including the phthalates of most concern) to gain authorization for continued use. The consequence of all this regulatory action is that future wastes will be inherently less hazardous than older wastes. Given the global market for affected products, it is to be anticipated that the benefits of European and US regulation will be felt across the world.

14.2   Assumptions

Since a comparison of different waste treatment options necessitates an LCA in order to properly characterize the relative performance of options, the work begins by choosing the boundaries of the analysis. For the waste that remains after sorting and recycling, the most appropriate choice is to start at the point where the waste has been collected and sorted. From here the remaining waste must be transported to the landfill or incinerator; we have included the emissions due to transport by showing a hypothetical round trip distance of 100 km, unrealistically large, but convenient for the purpose of illustration. In addition to the emission of pollutants from the landfill or incinerator, the emissions avoided by recovery of energy and materials are also taken into account, based on the LCA data from ADEME (2000). The main assumptions of the analysis are summarized in Table 14.1.

We have explicitly quantified the impacts of the stages in Table 14.1, with the exception of the construction of the landfill or incinerator. For the latter, we merely refer to the LCA of power plants carried out by ExternE, where the emissions from construction (due to materials production) were found to be about three orders of magnitude smaller than those during operation. This holds also for other combustion equipment that is used full time, in particular for waste incinerators. With landfills, the impacts of construction are also negligible compared with the utilization stage.

For the incinerator emissions, we take as starting point the limit values of the EU's Waste Incineration Directive (WID), EC (2000). In reality, if

Table 14.1 *Assumptions for the analysis of incineration and landfill of MSW. From Rabl* et al. *(2008), reprinted by Permission of SAGE.*

Stages taken into account	Construction of landfill or incinerator (negligible);
	Transport of waste (negligible);
	Emissions from landfill or incinerator;
	Avoided emissions due to energy recovery;
	Avoided emissions due to materials recovery.
Emissions from incinerator	Equal to limit values of the EU's Waste Incineration Directive (WID) EC (2000), but supplemented with current data.
Avoided emissions due to energy recovery	Equal to limit values of Large Combustion Plant Directive (EC, 2001)
Impact pathway analysis	Assumptions and damage costs per kg of pollutant of ExternE (2008).
Impacts that have been quantified	Effects of air pollutants on human health;
	Effects of air pollutants on crops;
	Effects of air pollutants on materials and buildings;
	Global warming;
	Amenity impacts (very site specific, not included in results, only order of magnitude is indicated in text, based on Walton *et al.* (2006)).
Impacts that have <u>not</u> been quantified	Effects of air pollutants on ecosystems;
	Reduction of visibility due to air pollution;
	Soil and water pollution due to leachates (but shown not to be of concern, see Section 14.3);
	Impacts from residues of incineration.

compliance is enforced the average emissions are likely to be lower to the extent that plant operators need a certain margin of safety to avoid penalties. Also, the local administrations of many areas where the ambient concentrations exceed the regulations oblige the polluters to reduce their emissions. For instance, the NO_x emissions of the incinerators of Paris for which we have been able to get data are at about one-third of the limit values, whereas for all of France the NO_x emissions are almost two-thirds of the limit.

Unfortunately, representative and usable data for the actual emissions in the EU are difficult to obtain. In theory they should be available from the European Pollutant Release and Transfer Register,[1] but at the time of finishing this book (June 2013) that website leaves much to be desired. In spite of the legal reporting requirement, emissions data are missing for most of the pollutants, at least for the incinerators that we have looked at. Furthermore, a serious shortcoming is the lack of any information on the

---

[1] http://prtr.ec.europa.eu/Home.aspx

Table 14.2 *Assumptions for the emissions from incineration of MSW. They are taken as the limit values of the flue gas concentrations, in directive EC (2000), assuming 5150 Nm^3/t_{waste}. For metals other than Hg, the directive specifies only certain sums: $0.5mg/Nm^3$ for the sum of As+Co+Cr+Cu+Mn+Ni+Pb+Sn +Sb+V, and $0.05mg/Nm^3$ for the sum of Cd+Tl; % within these sums based on ETSU (1996). The last two columns show the damage costs per kg of pollutant assumed for this chapter, and the resulting cost per t_{waste} of the direct incinerator emissions.*

Pollutant	mg/Nm^3	g/t$_{waste}$	€/kg$_{pollutant}$	€/t$_{waste}$
PM$_{10}$	10	51.5	7.6	0.39
SO$_2$	50	258	6.2	1.60
NO$_2$	200	1030	5.4	5.52
CO$_2$		807,000	0.021	16.95
As (2.8% of 0.5 mg/Nm^3)	0.014	0.072	529.6	0.04
Cd (81.2% of 0.05 mg/Nm^3)	0.0406	0.21	83.7	0.02
Cr^{VI} (6.5% of 0.2 × 0.5mg/Nm^3)^a	0.0065	0.033	66.3	0.00
Hg (0.05 mg/Nm^3)	0.05	0.26	8000	2.06
Ni (33.8% of 0.5 mg/Nm^3)	0.169	0.87	2.3	0.00
Pb (22% of 0.5 mg/Nm^3)	0.11	0.57	278.3	0.16
Dioxins	1.00E-07	5.15E-07	37,000,000	0.02

^{a} assuming that 20% of Cr from incinerators is Cr^{VI}

quantity of product or service associated with these emissions (e.g. tonnes of waste incinerated, kWh of electricity generated or tonnes of steel produced). Such information is crucial for the assessment of technology or policy options, and in most cases it is difficult to obtain except by the tedious process of contacting the companies in question (... and hoping for an answer).

We compare the results for the emission limits of directive EC (2000) with results for the actual emissions data of incinerators in France (kindly provided to us by Olivier Guichardaz of Dechets-Infos, whom we thank for the data and very helpful discussions). In Table 14.2, we list the limit values for incinerator emissions in the EU, together with the damage costs per kg of pollutant of ExternE (2008), as shown in Fig. 12.1 of Chapter 12. The last column shows the resulting contribution of each pollutant to the damage cost of the incinerator emissions. Since the limit values are fixed in terms of flue gas concentrations, one has to make an assumption for the flue gas volume per tonne of waste; we have chosen a typical value of 5150 Nm^3/t_{waste}.

Part of municipal solid waste is of biological origin and its combustion emits CO_2. In the LCA community, a special convention has been established according to which such CO_2 emission should not be counted. As we explained in Section 12.4.2, for many policy applications, that convention is inappropriate because it fails to distinguish between different options for reducing the emission of such CO_2.

To elaborate, the purpose of damage cost calculations is to help find the best solutions for environmental problems, by means of CBA and regulations. For this chapter the question is: what is the best way of treating the waste, once it has been produced? At that point the origin of the CO_2, whether biological or not, is no longer relevant: one cannot change the past. Concerning the biological component of waste, the relevant question is whether there are better treatment methods than landfill and incineration, for example separating the biomass from the rest and disposing of it by other means, such as composting. But even for such a comparison one has to count the total CO_2 emissions from each treatment facility. Of course, for a general study of waste one could and should enlarge the boundaries of the analysis even further and ask how the production of each waste component could be reduced. However, within the scope of this chapter, we take as boundary the point where the waste is to be delivered to the landfill or incinerator, and thus the damage cost has to include the total CO_2 emissions.

The principal emissions to air from landfill are CH_4 and CO_2. Figure 14.1 shows the total greenhouse gas emissions of a municipal solid waste landfill versus time. CH_4 is expressed as equivalent CO_2, using a GWP (global warming potential) of 25. Note that a modern landfill is divided into a large number of individual compartments; they are filled one after another and sealed when full. The data from ADEME (2003) are plotted in Fig. 14.1, where the time is measured from the date that a compartment is sealed. In practice it is impossible to capture all of the CH_4, and capture rates around 70% are commonly assumed (although measured data seem to be difficult to find). Here we assume a capture rate of 70% for the first 40 years, on average, after closure of a compartment; after 40 years we assume that all the remaining CH_4 escapes to the atmosphere. This last assumption reflects uncertainty in the way that sites will be managed around the time that closure is approached. One possibility is that regulators will simply sign off a site as "inert" after a certain period of time. Another is that regulators would require operators to provide evidence, such as data on emission rates, to show that a site is inert. The latter would provide a higher level of environmental protection.

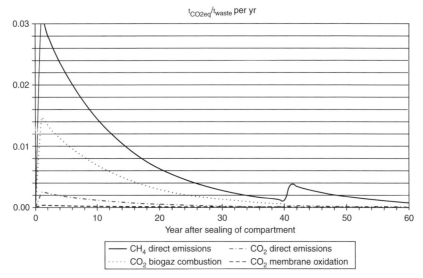

Fig. 14.1 Greenhouse gas emissions from a municipal solid waste landfill versus time, t_{CO2eq}/t_{waste} if 70% of the CH_4 is captured. Based on ADEME (2003).
From Rabl *et al.* (2008), reprinted by Permission of SAGE.

14.3   Damage cost of leachates

There are also emissions to soil and to water. Emissions to soil can occur from slag, from leaking liners under a landfill, and from the storage site of incinerator fly ash. Emissions to water arise from certain types of flue gas treatment and from the extraction of leachates under a landfill (see below for further analysis). Fly ash must be stored in specially designed sites for toxic waste, or else it must be stabilized by vitrification or by incorporation into concrete. The high level of containment ensures that resulting health impacts are entirely negligible if the management of the fly ash respects the directives of the European Union or equivalent. Slag is considered sufficiently harmless to be used as construction material, for example for roads, after it has been treated and stabilized.

Landfill emissions to soil are difficult to estimate because they depend on the integrity of the liners in the future. If the landfill is operated according to regulations and if there are no leaks or other mishaps, there are no such impacts during the foreseeable future because the operator has the obligation to maintain and safeguard the facility for 30 years after closure. In any case the impacts would remain limited to the immediate vicinity of the landfill, with the possible exception of sites with rapid ground water movement (noting that planning regulations may well prevent a landfill being located at such a site).

The quantification of damage costs due to leachates from landfills involves considerations very different from those for pollutants emitted into the air. There are so many difficulties that no complete IPA has been carried out until now. A few studies have attempted to determine damage costs of leachates, but they are not based on an IPA; instead they report the results of contingent valuations (CV), where interviewees are asked how much they are willing to pay to avoid the pollution of their drinking water by leachates (see the review by COWI, 2000). Since the interviewees had no quantitative information on the health impacts, the answers are a lump sum that has no relation to the real damage. Because of such a lack of information, CV is not an appropriate instrument for direct valuation of this type of impact (and even if one could provide such information, the variety of possible health impacts would be too detailed and complicated for a typical citizen to indicate a meaningful willingness to pay).

The problems confronting an IPA for leachates are more fundamental than the formidable technical difficulties of modeling the pathways through the environment into the drinking water. Many of the pollutants have very long lifetimes, and in particular the toxic metals stay in the environment for ever (although only a fraction may become available for intake). Even if the pathways could be analyzed in a satisfactory manner, there is no clear solution for the choice of time horizon and intergenerational discount rate (see Section 9.1.3).

In view of these difficulties we will not attempt a full IPA of leachates. Instead we offer an argument based around impact pathway thinking to show that the impacts of leachates are negligible if a landfill conforms to current regulations. The argument involves comparisons of a dose from leachates with the dose of the same pollutant from ordinary drinking water. We look at Pb, As and benzene for such a comparison because they are especially toxic.

As a source of data on pollutant concentrations in leachate we take the review by Kjeldsen $et\ al.$ (2002). Data on metal concentrations can be found in Tables 1 and 5 of that reference. Table 1 summarizes values from 14 studies, 9 of which date from before 1990; for lead, it shows a range of 0.001 to 5 mg_{Pb}/L_{leach} (per L_{leach} of leachate), and for arsenic a range of 0.01 to 1 mg_{As}/L_{leach}. Table 5 presents measurements at a large number of landfills, published since 1995; for lead, the highest value is 0.188 mg/L_{leach} and most are much lower (none are shown for As). For example, the average for 106 old Danish landfills is 0.07 mg_{Pb}/L_{leach}, and the values in 21–30-year-old German landfills range from 0.005 to 0.019 mg_{Pb}/L_{leach}.

EPA states a range of 0.008 to 1.02 mg_{Pb}/L_{leach}, with a mean estimate of 0.09 mg_{Pb}/L_{leach}; Lee and Associates (1993) report a mean value of 0.5 mg_{Pb}/L_{leach} and a range of 0.1 to 1 mg_{Pb}/L_{leach}.

To obtain an upper bound on the resulting concentration in drinking water, let us consider an extreme scenario where the landfill has no protective barriers at all and all the leachate passes directly into the water supply of the population using the landfill. For a convenient way of analyzing this case, we consider waste production and water consumption per person. We assume a waste density of 800 kg/m^3 and a production rate of MSW of 500 kg/yr per person over 30 years. If the ultimate stacking height of the waste is 10 m, the area used is 1.9 m^2/person. In a climate with an average precipitation of 3 L/m^2 per day (= 1.1 m per year) the resulting average leachate production is about 6 L_{leach}/day per person. Since the average household water consumption is around 150 L/day per person, the leachate will necessarily be diluted, by a factor of 6/150. Taking the upper limit of the Pb concentration range in Table 5 of Kjeldsen et al. (0.188 mg_{Pb}/L_{leach}), the resulting concentration is 188 × 6/150 $\mu g_{Pb}/L$ = 7.5 $\mu g_{Pb}/L$. That is lower than the limit of 10 $\mu g_{Pb}/L$ allowed by the most recent Drinking Water Directive of the EU. For arsenic, this argument implies a concentration of 40 μg/L in the water supply, compared with the regulatory limit of 10 $\mu g_{As}/L$, if one takes the upper range of 1 mg_{As}/L_{leach} of leachate concentration in Table 1 of Kjeldsen et al.

For benzene, Table 6 of Kjeldsen et al. shows a leachate concentration range of 0.0002 to 1.63 mg/L_{leach}. Average values are shown in Table 8 for American landfills; they are 0.065 mg/L_{leach} for old and 0.007 mg/L_{leach} for new landfills. The regulatory limit of the Drinking Water Directive of the EU is 1 μg/L, and that is satisfied by our extreme scenario if the leachate concentration is below 0.025 mg/L_{leach}, which is certainly the case for the average of new landfills according to Table 8 of Kjeldsen et al.

Even though the upper range of the leachate concentrations we have considered would imply some violations of the Drinking Water Directive under our extreme scenario, we emphasize that this scenario is unrealistically pessimistic. For a properly functioning landfill built to the standards required in Europe and North America, leachate would be collected and either treated on site or sent for treatment elsewhere. Water entering the public water supply would be further treated to satisfy the applicable drinking water regulations (as is done routinely, for instance to remove As in regions where natural sources cause the concentrations in the ground water to exceed limits for As). Furthermore, the concentrations observed by Kjeldsen et al. are for the low leachate production rates of landfills with barriers; with the flow-through of the extreme scenario the

Table 14.3 *Concentrations in leachate and drinking water under the extreme scenario where untreated leachate is used for water supply.*

Pollutant	Leachate, range	Leachate, EPA limit	Drinking water, extreme scenario, range	Drinking water, extreme scenario, EPA limit	Drinking water, EU directive
Units	µg/L	µg/L	µg/L	µg/L	µg/L
As	10–1000[a]	50	0.4–40	2	10
Pb	<5–188[b]	50	0.2–7.5	2	10
Benzene	<1–1630[c]	5	0.04–65	0.2	1

[a] Table 1 of Kjeldsen *et al.*
[b] Table 5 of Kjeldsen *et al.*
[c] Table 6 of Kjeldsen *et al.*

water would not have the time to absorb such a quantity of pollutants. Thus either the flow rates or the concentrations are much too high. And finally, waste that is sent to landfills is now no longer as toxic as in the past, thanks to a variety of regulations such as the prohibition of Hg in batteries and the requirement for special treatment of electric and electronic waste.

We note also that in the USA, the EPA ground water protection performance standards require the leachate concentrations of landfills to be below 0.05 mg/L_{leach} for As and Pb and below 0.005 mg/L_{leach} for benzene; at these values the Water Quality Directive would be satisfied by a wide margin. As for leachate production after final closure of a landfill, Barlaz *et al.* (2002) reviewed measured data and found leachate production rates of 0.0005 to 0.021 $L_{leach}/(m^2{\cdot}day)$, far lower than the 3 $L_{leach}/(m^2{\cdot}day)$ of our extreme scenario; within nine years of final cover installation, the rates became negligible. Thus there is no significant long-term risk of impacts from leachates. The numbers of this section are summarized in Table 14.3.

Limit values in drinking water regulations do not guarantee the total absence of possible effects, because of the uncertainties of ERFs at very low concentrations. They do, however, express the consensus of the experts that there are no significant effects below such limits. In our extreme scenario some of the highest concentrations could exceed the limits, but not by a wide margin and not if regulations such as the EPA limits on leachate concentrations are respected. In view of all the factors that reduce the concentrations far below this scenario, especially the strict requirements for protective barriers and the management of leachate, we conclude that leachate from MSW landfills is not a problem to worry about for landfills that are built and managed according to regulations.

14.4   Hg

Municipal solid waste may contain appreciable quantities of Hg, from sources such as fluorescent light bulbs, thermometers and batteries. The use of mercury in many products has been or is being phased out, but there remains a problem with contaminated articles that are still in circulation. For example, the State of Florida has estimated the amount of Hg in MSW of Florida around the year 2000 (Florida, 2002), as part of a detailed examination of data on the composition of waste, and the numbers imply a concentration of about 0.25 g_{Hg}/t_{waste}.

Calculating the damage cost of Hg emissions is not straightforward given the complexity of its environmental pathways and limited availability of data. In Section 8.3 we have described an approach taken by Spadaro and Rabl (2008b). This takes into account the worldwide damage, because Hg vapor has a long residence time in the atmosphere, about 1 to 2 yr, and its impact is therefore imposed on the entire hemisphere. Assuming a linear ERF without threshold, they find a damage cost of $3,400/$kg_{Hg}$, due to neurotoxic effects. ExternE (2008) uses 8,000 €/kg_{Hg}, because that was an earlier estimate of Spadaro and Rabl and our publication with a final estimate appeared too late to be taken into account. However, we now believe that the damage cost is higher, because we find the evidence for cardio-vascular effects of Hg increasingly convincing. Rice and Hammitt (2005), who estimated the health costs of Hg in the USA, have already included cardio-vascular effects and found that they greatly increase the damage cost. Therefore we take the value of 8,000 €/kg_{Hg} as being more correct.

Multiplying the Hg concentration in MSW by this damage cost, one finds a cost of

$$0.25 \ g_{Hg}/t_{waste} \times 8000 \ \text{€}/kg_{Hg} = 2.0 \ \text{€}/t_{waste}$$

if all the Hg escapes into the air (this emission rate is very close to the EC (2000) limit value of 0.26 g_{Hg}/t_{waste} in Table 14.2). This is the case for incineration, unless some of the Hg is removed by active carbon, a technology already used in some incinerators.

For landfills, in contrast, the release rate is quite slow, on the order of 0.1 to 1% per year during the initial operation of a landfill, and the rate seems to slow down even more after closure of the landfill (as would be expected if the integrity of the cover is reasonably secure), as indicated by the measurements of Lindberg *et al.* (2005). (In passing, we note that Lindberg *et al.* found that a significant fraction of the atmospheric emissions from landfills is in the form of methyl-Hg; however, the resulting inhalation dose is small compared with the average ingestion dose from fish (UNEP, 2002), even near the landfill.) Even if all of the Hg were to eventually get into the environment, the damage cost would be much lower than the estimate given above, because of discounting. Overall we conclude that Hg from landfills may be a significant problem that deserves attention, although it is unlikely to make a dominant contribution to the total damage cost.

14.5   Recovery of energy and materials

The damage costs and the comparison between landfill and incineration turn out to be extremely sensitive to assumptions about energy recovery. For that reason we consider a fairly large number of options, for typical installations in France, according to ADEME (2000). In the analysis in Section 14.6, we indicate the options with labels where the letters H and E refer to heat and electricity and the letters c, g, n and o to the fuel (coal, gas, nuclear and oil, respectively) displaced by energy recovery. For example, (E = c&o, H = c&o) designates a system where heat and electricity are produced, each displacing a fuel mixture of coal and oil, 50% of each. The options are for incineration:
- recovery of heat and electricity, (E=. . ., H=. . .),
- recovery of electricity only (E=. . .),
- recovery of heat only (H=. . .);

for landfill:
- no energy recovery,
- recovery of electricity, by motor (reciprocating engine) (E=. . .),
- recovery of electricity, by turbine (E=. . .),
- recovery of heat (H=. . .).

For each of these options we consider several suboptions:
- the recovered electricity displaces coal and oil fired power plants, 50% each (E = c&o),
- the recovered electricity displaces nuclear power plants (E = n),
- the recovered heat displaces gas and oil fired heating systems, 50% each (H = g&o),
- the recovered heat displaces only oil fired heating systems (H = o).

For recovery of electricity we assume a year-round demand, so all the electricity is used. Likewise for recovery of heat, we assume a year-round demand (industrial process heat loads or certain district heating systems with year-round demand, e.g. Paris and Vienna), so that little of the heat is

Table 14.4 *Assumptions for energy recovery.*
From Rabl et al. *(2008), reprinted by Permission of SAGE.*

kWh/t_{waste}	Electricity	Heat
Part load heat and electricity	202	607
Base load electricity only	270	0
Base load heat	0	1850

Table 14.5 *Assumptions for recovery rates and avoided damage costs for materials recovery from incinerators.*

	kg/t_{waste}	€/kg	€/t_{waste}
Slag	230	0	0
Iron	20.2	0.107	2.16
Aluminum	1.5	0.539	0.80

wasted. Year-round demand is essential for good recovery rates because the supply of waste tends to be fairly constant. For other load distributions, the results can be estimated by rescaling the numbers for energy recovery (see Fig. 14.3 below).

Note that for the purpose of this analysis the environmental benefit of recovered electricity is essentially zero if it displaces nuclear, because the damage costs of nuclear are very small compared with those of oil or coal; thus this option is essentially equivalent to no electricity production at all as far as damage costs are concerned. Our assumptions for energy recovery rates, to be taken through to the next section, are shown in Table 14.4. The avoided emissions are based on the Large Combustion Plant Directive (EC, 2001).

Recovery of materials is practical only for incinerators, not for landfills. Our assumptions for recovery rates and avoided damage costs are listed in Table 14.5; the avoided damage costs are based on the LCA inventory of Delucchi (2003).

14.6   Results for damage cost per tonne of waste

As pointed out already at the start, the real emissions are likely to be lower than the limit values of EC (2001). In Table 14.6, we compare the incinerator emissions and damage costs for limit values with those for the actual emissions in France, based on the data kindly provided to us by

Table 14.6 *Comparison of incinerator emissions and damage costs for limit values and actual emissions in France.*

Pollutant	Limit values EC (2000)		Actual emissions, France (2011)		Ratio actual/limit
	g/t_{waste}	$€/t_{waste}$	g/t_{waste}	$€/t_{waste}$	
PM_{10}	51.5	0.39	7.3	0.06	0.14
SO_2	257.5	1.60	54.3	0.34	0.21
NO_2	1,030	5.52	600	3.22	0.58
CO_2	807,000	16.95	863,900	18.14	1.07
As	0.072	0.04	0.013	0.01	0.18
Cd	0.209	0.02	0.007	0.00	0.03
CrVI	0.033	0.00	0.016	0.00	0.48
Hg	0.258	2.06	0.053	0.43	0.21
Ni	0.870	0.00	0.000	0.00	0.00
Pb	0.567	0.16	0.093	0.03	0.16
Dioxins	5.15E-07	0.02	8.7E-08	0.00	0.17
Total		**26.76**		**22.22**	**0.83**

Olivier Guichardaz of Dechets-Infos. For most of the pollutants the emissions have been greatly reduced. CO_2, in contrast, has increased slightly, since the composition of waste has been changing with recycling.

A summary of the total damage cost for all the options, assuming the actual emissions in France in 2011, is shown in Fig. 14.2. More detailed results for some of the options can be found in Fig. 14.3, showing the contribution of each stage and of the major pollutants (dioxins and toxic metals are shown as "Trace"). The damage costs of waste transport are negligible, as illustrated in Fig. 14.3, with an arbitrary choice of a 100 km round trip by a 16 tonne truck, for which they amount to 0.45 $€/t_{waste}$. By far the most important contributions come from direct emissions (of the landfill or incinerator) and energy recovery. Materials recovery is relevant for incineration only and brings a relatively small benefit. For incineration energy recovery can be crucial and can reduce the total damage cost to 1.5 $€/t_{waste}$.

For the direct emissions of the incinerator the total is 22.22 $€/t_{waste}$, most of that being due to NO_x and CO_2. Toxic metals and dioxins, shown in Fig. 14.3 as "Trace", contribute about 2% of the total (0.46 $€/t_{waste}$), mostly because of Hg and Pb. The contribution of dioxins is negligible, only 0.003 $€/t_{waste}$, thanks to the low emission limit of the Waste Incineration Directive (EC, 2000).

For landfill, the cost is dominated by CH_4 emissions, because only about 70% of the CH_4 can be captured. Energy recovery from a landfill

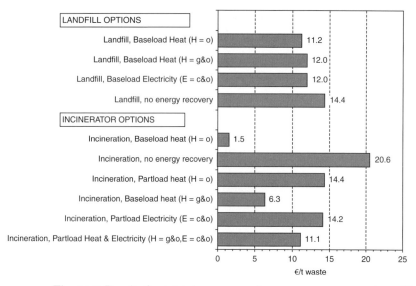

Fig. 14.2 Results for total damage cost for waste treatment options. If electricity displaces nuclear, damage costs are essentially the same as for the case without energy recovery. Amenity costs are not included; they are very site-specific and could make a contribution on the order of one €/t_{waste}.

is not very significant (and because of NO_x from electricity production, this option increases the damage cost if the electricity displaces nuclear). In contrast, energy recovery is crucial for the damage cost of incineration. Under favorable conditions (all heat produced by incinerator displaces coal and oil), the total damage cost could even be negative in some situations, i.e. a net benefit. In contrast with most other countries, in France, recovery of electricity does not bring significant environmental benefits, because it is base load power and all the base load power is produced by nuclear; the options where it displaces coal or oil are therefore not realistic (except near the border where the power can be exported), because these fuels are used only during the heating season. In any case, electricity production brings far lower benefits than heat because of the poor conversion efficiency of incinerator heat (due to its low temperature) compared with central station power plants.

Example 14.1   Estimate the LE loss due to waste incineration if the entire municipal solid waste, about 0.5 t/yr per person, is incinerated without energy and materials recovery.

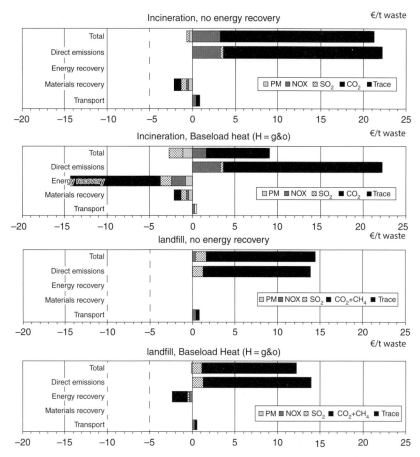

Fig. 14.3 Some detailed results, by stage and pollutant. "Trace" = dioxins and toxic metals. Benefits from energy recovery can be quite different for different countries. The direct incinerator emissions are based on actual data for France in 2011.

Solution: The LE loss in Fig. 14.3 is due to NO_x, SO_2 and PM, evaluated at 40,000 €/YOLL. The total cost of NO_x, SO_2 and PM from the incineration stage is about 7.5 €/t_{waste}, and about 2/3 of that is due to LE loss. Thus the LE loss per tonne of waste is about,

$$(2/3) \times 7.5\,(\text{€}/t_{waste})/(40000\,\text{€}/\text{YOLL}) = 0.000125\,\text{YOLL}/t_{waste}.$$

Assuming steady state and uniform distribution of the population, the total LE loss is obtained by multiplying the YOLL/t by the 40 t_{waste} of waste produced during a lifetime of 80 yr. The result is 1.8 days of life lost.

14.7   Conclusions for waste treatment

We have evaluated and compared the damage costs ("external costs") of landfill and incineration of MSW, based on the latest results of ExternE (2008) and taking into account the relevant lifecycle impacts, especially emissions avoided by recovery of energy and materials. With our assumptions, the damage cost of incineration ranges from 1.5 to 20.6 €/t$_{waste}$, depending on the type and performance of energy recovery; the benefits from energy recovery can be quite different for different countries. The damage cost of landfills, 11.2 to 14.4 €/t$_{waste}$, is mostly due to GHG, evaluated here with a unit cost of 21 €/t$_{CO2}$. In addition, there may be amenity costs on the order of 1 €/t$_{waste}$ according to a review by Walton *et al.* (2006) (highly variable with site and imposed only on the local population, thus to be internalized differently from air pollution).

The benefits of energy recovery from incinerators are largest if the heat can be used directly for process heat or district heating systems with sufficiently large base load. Electricity production brings far lower benefits than heat because of the poor conversion efficiency of incinerator heat (due to its low temperature) compared with central station power plants. Unlike incinerators, the damage cost of landfill does not vary as much with type of energy recovery because in any case the amount recovered is relatively small.

The results presented in this chapter are for typical conditions in the EU, but they can be adapted to other sites and other countries if the respective damage costs per emitted pollutant are known. Even without carrying out new calculations using the EcoSense software, one can estimate the damage costs per emitted pollutant using the UWM of Section 7.4 in Chapter 7.

The uncertainties are large and they have different effects on different policy choices. We therefore show some sensitivity analyses in Fig. 14.4; only the emissions from the incinerator and the landfill itself are included here. For the comparison between landfill and incineration the greenhouse gases (GHG) are by far the most important, especially now that the emissions of other pollutants are being reduced more and more, whereas the GHG emissions hardly change. For this comparison the assumption about the GWP of CH_4 is crucial. In Fig. 14.2 and 14.3 we have taken the GWP for the 100 yr time horizon which is 25. Because the atmospheric lifetime of CH_4 is short compared with CO_2, its GWP would be much larger for a shorter time horizon. For a 20 yr horizon it is 72. The GHG costs of incineration are not affected by that, because they are almost entirely due to CO_2. But for landfill the damage cost increases dramatically with the short time horizon.

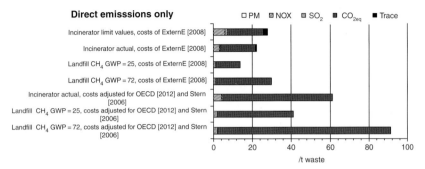

Fig. 14.4 Change in damage costs, for direct emissions from incinerator and landfill, if mortality and GHG are valued according to OECD (2012) and Stern *et al.* (2006).

What is the appropriate GWP? There is no clear answer. The high GWP expresses the damage of CH_4 in the near term, the low GWP the damage over a centurial timescale. Ideally one should calculate the damage costs expected in each future year and sum series with appropriate discounting, but practically ... (see Chapter 10). The dilemma is very general because there are numerous policy choices that involve tradeoffs between CH_4 and CO_2 emissions. Alvarez *et al.* (2012) have developed an instructive and relatively simple tool for such comparisons, called technology warming potentials. It provides a better perspective for understanding the consequences of such choices, although it does not give a definitive answer. It would be interesting to use it for the landfill versus incineration question, but here we leave it at Fig. 14.4.

As we explained in Section 12.2.3, we believe that the damage costs of ExternE (2008) are too low. In particular the mortality costs should be doubled, based on the recent OECD (2012) report, and the cost of GHGs should be increased from 21 to 65 €/t_{CO2eq}, based on Stern *et al.* (2006), to account for the appropriate intergenerational discount rate. Applying the multiplication factors of Section 12.2.3 to the direct emissions from incinerator and landfill, we find the results in Fig. 14.4. The cost of the trace metals is essentially unaffected because it is almost entirely due to neurotoxic impacts, at least according to the numbers of ExternE (2008) (although they will also have to be increased because of additional health end points for which new epidemiological evidence has been found).

In decisions about waste treatment, the full social cost (= sum of private costs and external costs) should be taken into account. We have not examined the private costs but refer to Dijkgraaf and Vollebergh (2004) whose Table 2 shows private costs in the Netherlands. The private costs of landfilling are less than half of those of incineration: for landfilling 40 €/t_{waste} without and 36 €/t_{waste} with energy recovery, versus 103 €/t_{waste} without and 79 €/t_{waste} with energy recovery for incineration. Since their estimate of the damage costs (based on CE (1996)) also indicates lower social costs for landfilling, they question the wisdom of current waste policies in the EU that favor incineration. Our damage costs are consistent with theirs.

However, we emphasize that costs (private costs + damage costs) are not the only criterion for choosing a treatment option for MSW. Above all, land use and land availability are crucial. In many regions of Europe, land is so limited that incineration is the preferred choice even if its cost is higher. Note that land availability is not simply a matter of the price of land per m^2, because for such installations one needs a sufficiently large piece of land sufficiently far from habitation to be accepted by the population.

Land taken for landfills also imposes a cost on future generations by restricting the re-use and imposing risks if the pollutants get into the environment. Such a cost has not been taken into account in our analysis.

There are almost always additional criteria, especially the preferences of the local population, that may be difficult or impossible to express in monetary terms. Such non-monetary criteria can be taken into account by means of a multicriteria analysis, preferably in consultation with the stakeholders, an approach that has been successfully tested at a stakeholder workshop in the SusTools project of the EC (Rabl et al., 2004), where earlier results of this work were discussed with policy makers, representatives of industry and environmental organizations. Demonstration of the use of the impact pathway approach and estimation of damage costs provides a mechanism for putting many of the environmental and health burdens of waste management in proper context as part of this process.

References

ADEME 2000. *Analyse environnementale de systèmes de gestion de déchets ménagers, Phase 1: analyse des parameters determinants pour les impacts environnementaux des différents modules* (Environmental system analysis of municipal waste treatment, Phase 1: analysis of the key parameters for the environmental impacts). Study by BIO Intelligence and Ecobilan for ADEME and Eco-Emballages. Agence Française de l'Environnement et de la Maîtrise de l'Energie, 27 Rue Louis Vicat, 75737 Paris CEDEX 15.

ADEME 2003. *Outil de calcul des émissions dans l'air de CH4, CO2, SOx, NOx issues des centres de stockage de déchets ménagers et assimilés.* (Tool for calculation of emissions of CH_4, CO_2, SO_x, NO_x from municipal and similar solid waste).

Agence Française de l'Environnement et de la Maîtrise de l'Energie, 27 Rue Louis Vicat, 75737 Paris CEDEX 15.

Alvarez, R. A., Pacala, S. W., James, J. *et al.* 2012. Greater focus needed on methane leakage from natural gas infrastructure. *PNAS* **109** (17): 6435–6440.

Barlaz, M. A., Rooker, A. P., Kjeldsen, P., Gabr, M. A. and Borden, R. C. 2002. Critical evaluation of factors required to terminate the postclosure monitoring period at solid waste landfills. *Environmental Science & Technology* **36** (16): 3457–3464.

CE 1996. *Financiele Waardering van de Milieu-effecten van afvalverbrandingsinstallaties in Nederland.* Centrum voor Energiebesparing, Delft. In Dutch.

COWI 2000. *A Study on the Economic Valuation of Environmental Externalities from Landfill Disposal and Incineration of Waste.* Final Main Report. European Commission, DG Environment, October 2000.

Delucchi, M. A. 2003. *A lifecycle emissions model (LEM): lifecycle emissions from transportation fuels, motor vehicles, transportation modes, electricity use, heating and cooking fuels, and materials.* UCD-ITS-RR-03-17, Institute of Transportation Studies, University of California, Davis, CA 95616.

Dijkgraaf, E. and Vollebergh, H. R. J. 2004. Burn or bury? A social cost comparison of final waste disposal methods. *Ecological Economics* **50**: 233–247.

EC 2000. Directive 2000/76/EC of the European Parliament and of the Council of 4 December 2000 on the incineration of waste.

EC 2001. Directive 2001/80/EC of the European Parliament and of the Council of 23 October 2001 on the limitation of emissions of certain pollutants into the air from large combustion plants.

ETSU 1996. *Economic Evaluation of the Draft Incineration Directive.* Report for the European Commission DG11. ETSU, Harwell Laboratory, Didcot, Oxfordshire OX11 0RA, UK.

ExternE 2008. With this reference we cite the methodology and results of the NEEDS (2004–2008) and CASES (2006–2008) phases of ExternE. For the damage costs per kg of pollutant and per kWh of electricity we cite the numbers of the data CD that is included in the book edited by Markandya, A., Bigano, A. and Porchia, R. in 2010: *The Social Cost of Electricity: Scenarios and Policy Implications.* Edward Elgar Publishing Ltd, Cheltenham, UK. They can also be downloaded from www.feem-project.net/cases/ (although in the latter some numbers have changed since the data CD in the book).

Florida 2002. *Solid Waste Management in Florida 2001–2002.* Report of Florida Department of State, downloaded 31 May 2005 from www.dep.state.fl.us/waste/categories/recycling/pages/01.htm.

Hester, R. E. and Harrison, R. M. 2002. *Environmental and Health Impact of Solid Waste Management Activities.* Royal Society of Chemistry, Cambridge, UK.

Kjeldsen, P., Barlaz, M. A., Rooker, A. P. *et al.* 2002. Present and long-term composition of MSW landfill leachate: A review. *Critical Reviews in Environmental Science and Technology* **32**(4): 297–336.

Kreith, F. 1990. *Integrated Solid Waste Management.* Genium Publishing Corp. Schenectady, NY.

Lee, G. F. and Associates. 1993. *Groundwater Pollution by Municipal Landfills: Leachate Composition, Detection and Water Quality Significance.* Sardinia 1993

IV International Landfill Symposium, Margherita di Pula, Italy, 11–15 October 1993.

Lindberg, S. E., Southworth, G., Prestbo, E. M. *et al.* 2005. Gaseous methyl- and inorganic mercury in landfill gas from landfills in Florida, Minnesota, Delaware, and California. *Atmospheric Environment* **39**: 249–258.

OECD 2012. *Mortality Risk Valuation in Environment, Health and Transport Policies.* OECD Publishing. http://dx.doi.org/10.1787/9789264130807-en

Petts, J. and Eduljee, G. 1994. *Environmental Impact Assessment for Waste Treatment and Disposal Facilities.* John Wiley & Sons Ltd. Chichester, UK.

Rabl, A., Spadaro, J. V. and McGavran, P. D. 1998. Health risks of air pollution from incinerators: a perspective. *Waste Management & Research* **16**: 365–388. This paper received the 1998 ISWA (International Solid Waste Association) Publication Award.

Rabl, A., Zoughaib, A., von Blottnitz, H. *et al.* 2004. *Tools for sustainability: Development and application of an integrated framework.* Final Technical Report for project SusTools, contract N° EVG3-CT-2002-80010. EC DG Research. Available at www.arirabl.com/sustools.htm

Rabl, A., Spadaro, J. V. and Zoughaib, A. 2008. Environmental impacts and costs of municipal solid waste: A comparison of landfill and incineration. *Waste Management & Research* **26**: 147–162.

Rice, G. and Hammitt, J. K. 2005. *Economic Valuation of Human Health Benefits of Controlling Mercury Emissions from US Coal-Fired Power Plants.* Northeast States for Coordinated Air Use Management (NESCAUM). Boston, MA. February 2005.

Spadaro, J. V. and Rabl, A. 2008b. Global health impacts and costs due to mercury emissions. *Risk Analysis* **28** (3): 603–613.

Stern, N., *et al.* 2006. *The Economics of Climate Change: The Stern Review,* Cambridge University Press, Cambridge, UK. Available at www.hm-treasury. gov.uk/stern_review_report.htm

UNEP 2002. United Nations Environment Programme. *Global Mercury Assessment.* UNEP Chemicals, Geneva, Switzerland.

Walton, H., Boyd, R., Taylor, T. and Markandya, A. 2006. *Explaining Variation in Amenity Costs of Landfill: Meta-Analysis and Benefit Transfer.* Presented at the Third World Congress of Environmental and Resource Economists, Kyoto, July 3rd–7th 2006.

15   Results for transport

Summary

In this chapter, we illustrate the use of external cost estimates for evaluating transportation options. We begin by presenting damage cost estimates in Section 15.1, with results for the EU and for the USA.

In Section 15.2 we use the damage cost estimates of ExternE to compare a hybrid passenger car with a conventional car on a lifecycle basis. In Section 15.3 we look at walking and bicycling as alternatives to commuting to work by car; here the reduction of air pollution is a significant collective benefit, but much more important is the value of the health gain for the individuals who make the switch to an active transport mode. We present sufficient detail in these two sections to show how the calculations are done.

In Section 15.4 we compare the greenhouse gas emissions of the main transport modes. In Section 15.5 we conclude the chapter with a discussion of policies that can internalize the damage costs of transport, including the low emission zones (LEZ) that have been created in many cities of Europe.

15.1   External cost estimates for transport

15.1.1   *Vehicle emissions*

In the EU the emissions of vehicles must not exceed the limits specified in the EURO standards. As an example Table 15.1 shows the standards for passenger cars. Analogous standards are in force in the USA. The regulations of China, India and Australia are based on the EURO standards, although with different implementation schedules.

These standards are to be respected in actual use, and compliance is determined by testing the vehicle with a standardized test cycle. Developing realistic test cycles is difficult because the emissions vary strongly with driving conditions (cold engine, warm engine, speed,

Table 15.1 *The EURO standards for emission limits of passenger cars, in g/km.*

Tiers	Date	CO	THC	NMHC	NO$_x$	HC+NO$_x$	PM$_{2.5}$
Diesel							
Euro 2	January 1996	1.0	–	–	–	0.7	0.08
Euro 3	January 2000	0.64	–	–	0.50	0.56	0.05
Euro 4	January 2005	0.50	–	–	0.25	0.30	0.025
Euro 5	September 2009	0.50	–	–	0.18	0.23	0.005
Euro 6 (future)	September 2014	0.50	–	–	0.08	0.17	0.0025
Petrol (Gasoline)							
Euro 2	January 1996	2.2	–	–	–	0.5	–
Euro 3	January 2000	2.3	0.20	–	0.15	–	–
Euro 4	January 2005	1.0	0.10	–	0.08	–	–
Euro 5	September 2009	1.0	0.10	0.068	0.06	–	0.005*
Euro 6 (future)	September 2014	1.0	0.10	0.068	0.06	–	0.005*

* Applies only to vehicles with direct injection engines
HC = hydrocarbons,
THC = total hydrocarbons,
NMHC = non-methane hydrocarbons

acceleration, etc.). There are always questions about how representative the tests are of typical driving conditions. The EURO standards specify the tests to be used for certifying compliance by vehicle manufacturers. The performance under other conditions can be estimated by using the COPERT 4 software of the European Environment Agency.[1]

Vehicle speed is one of the key factors that influences the emissions. In Fig. 15.1 and 15.2 we show two examples of the variation of emissions per km with speed (they are based on specific vehicle types and model years and should not be taken literally, but only as an illustration of general trends). There is a broad minimum in the range of about 50 to 70 km, below which the engine operates inefficiently at part load. At high speed, the emissions increase because more power is required to overcome wind and rolling resistance. These graphs indicate that traffic management can be a tool for reducing the emissions, for example by limiting speeds on highways or by reducing congestion. Congestion charges can thus reduce the emissions not only by limiting the number of cars but also by speeding the flow.

Furthermore, the emissions may get worse as a vehicle ages. Regular inspections, e.g. in France every two years or in the UK annually once a

---

[1] http://lat.eng.auth.gr/copert/

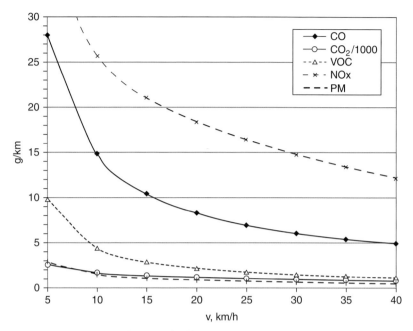

Fig. 15.1 Emissions of a diesel bus as function of speed v, based on the curve fits of Table A28 of Hickman *et al.* (1999) of Project MEET.

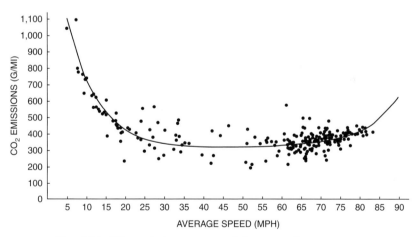

Fig. 15.2 CO_2 emissions measured for a gasoline passenger car as a function of speed. 1 mph = 1.609 km/h. From Barth and Boriboonsomsin (2009), with kind permission.

vehicle is three years old, are required to enforce the standards. In addition to these inspections, on-the-road inspections by remote sensing are a valuable check. Such a system has been developed by D. Stedman and his colleagues at the University of Denver (see Bishop and Stedman, 2008 and references therein). By analyzing the absorption of a laser beam by the exhaust of a passing car, this system can instantaneously and accurately measure the emission of all of the regulated pollutants, as confirmed by countless tests in many countries. One of the interesting findings was that even among cars of recent vintage a small percentage had very high emissions, due to a malfunctioning catalytic converter. Identifying such cars and getting them repaired is therefore a very important tool in the fight against air pollution, because dirty cars that are driven a lot are more likely to be caught before the next inspection date. Since the system is mobile, it can be placed along various busy roads or highways. It is used routinely in Colorado, for example.

In recent years much attention has also been given to non-exhaust emissions of vehicles. They are due to tire wear, brake wear, road surface wear, corrosion (of vehicle components and street furniture), and resuspension (by vehicle-generated turbulence). Based on detailed measurements, models have been developed for estimating the non-exhaust emissions of PM_{10} and $PM_{2.5}$. Data on non-exhaust emissions have been collected by EMEP as part of their emissions database. Country averaged non-exhaust PM_{10} emission factors, in g/km, for different transport modes in each country of the EU27 are presented in Table 52 of CE Delft (2011). On average, the non-exhaust PM_{10} emission factor is 0.051 g/km for passenger cars, compared with the exhaust $PM_{2.5}$ emission factor of 0.024 g/km. In terms of mass, the non-exhaust emissions are certainly very important.

However, little is known about their toxicity. A significant fraction is of crustal origin, with negligible toxicity according to several epidemiological studies (Laden et al., 2000, Pope et al., 1999, Schwartz et al., 1999). Whereas it is difficult, if not impossible to get conclusive results for the relative toxicity of all the different PM components from epidemiological studies, toxicology holds greater promise. Much toxicological research is looking at this question, but so far not enough is known.

In the meantime, some authors have made estimates based on simplistic assumptions such as equal toxicity within each size range. For example, CE Delft (2011) lists damage costs for non-exhaust emissions, in their Table 7 on which Table 15.2 is based. We do not show those numbers because they can readily be obtained by noting that all the costs per kg of PM_{10} in the CE Delft (2011) table are 0.4 times the costs per kg of $PM_{2.5}$.

Table 15.2 *Damage costs for ground level emissions, in €$_{2008}$/kg. They include impacts on human health, crop losses, and materials. Adapted from Table 7 of CE Delft (2011).*

Pollutant	PM$_{2.5}$			NO$_x$	NMVOC[b]	SO$_2$
Region type[a]	Metropolitan	Urban	Non-urban			
Austria	481.5	155.6	80.6	13.6	1.6	10.0
Belgium	495.1	159.8	106.9	8.7	2.6	10.9
Bulgaria	73.9	23.8	19	7.1	0.4	6.2
Czech Republic	381.3	122.9	94.6	10.6	1.1	9.5
Denmark	448.4	144.6	52.7	5.3	1.2	5.7
Estonia	265.1	86.2	44.7	2.8	0.6	4.5
Finland	407.9	131.4	34	2.6	0.6	3.5
France	453.8	146.1	90.8	10.5	1.4	9.9
Germany	430.5	138.8	83.9	12.7	1.4	10.9
Greece	338.4	109.1	47.6	2.7	0.6	5.8
Hungary	312.7	100.6	80.2	12.4	1.0	9.1
Ireland	535.1	172.7	55.9	4.4	1.1	5.4
Italy	426.7	137.9	77.7	9.5	1.1	8.7
Latvia	233.8	75.2	43.4	4.0	0.7	5.0
Lithuania	253.9	82.5	50.8	5.6	0.8	5.7
Luxembourg	922.4	296.9	131.4	12.7	2.4	10.3
Netherlands	495.3	159.9	96.8	8.8	2.1	12.8
Norway	393.7	126.6	38.3	3.1	0.8	3.4
Poland	234.5	75.3	70.4	7.8	1.0	8.4
Portugal	299.6	96.5	44.4	1.5	0.8	3.8
Romania	63.7	20.5	16.3	9.7	0.8	7.4
Slovakia	332.1	106.3	89.7	11.0	0.9	8.8
Slovenia	350.3	112.6	72.6	11.5	1.4	8.9
Spain	384.8	123.9	52.9	3.6	0.8	5.2
Sweden	424.4	136.5	41.3	4.1	0.8	4.2
Switzerland	475.2	152.9	78.5	19.3	1.3	13.0
UK	453.2	145.9	70.7	5.2	1.4	7.3

[a] Metropolitan = population > 5 million and Urban = population < 5 million
[b] non-methane volatile organic compounds

In view of the above, we believe that they are really upper limits, the true values being very uncertain.

15.1.2 How harmful are NO$_x$ emissions?

The standard approach taken by almost all studies that have quantified the health impacts of air pollution, in particular those by ExternE, EPA and WHO, is to use only ERFs for PM and for O$_3$. Direct effects of NO$_x$ and

SO_2 have previously been assumed to be negligible, but the secondary nitrate and sulfate aerosols created by their transformation in the atmosphere are considered as PM and their impacts are calculated by using the ERFs for PM. The reasons for this choice are that the ERFs for PM and O_3 are better established than for NO_x and SO_2, and that pathways of action within the body have been identified for primary combustion particles and for O_3, whereas it is less clear how NO_x or SO_2 could have harmful effects at the low concentrations in the environment. Even though there are questions about the toxicity of nitrate and sulfate aerosols (Reiss *et al.* 2007), the standard approach yields correct results for assessments of the total health impact of typical urban ambient concentrations, because it uses ERFs that are based on typical urban ambient PM with its mix of primary and secondary particles.

However, for the damage costs of specific sources, here vehicles, one has to evaluate something quite different, namely the contribution of a specific incremental pollution source, rather than the effect of ambient concentrations (which are due to a wide variety of sources). For the impacts of primary pollutants emitted at ground level in large cities, the regional contribution is small compared with the local contribution (see Table 15.11 below). Since the formation of nitrate and sulfate aerosols is slow and takes place over distances of tens to hundreds of km, their local contribution is negligible. The local contribution of O_3 is also small because it is a secondary pollutant created gradually in a region of tens of km from the source, and in the city the concentration is actually reduced by cars because much or most of their NO_x emission is in the form of NO, which destroys O_3 locally, before creating O_3 further away.

Thus the standard approach limits the analysis to primary pollutants, mostly $PM_{2.5}$,[2] while totally neglecting NO_x, the other pollutant emitted in large quantities by cars. This despite the fact that many experts consider NO_2 a valid indicator for the severity of automotive pollution, and there are numerous epidemiological studies that have found significant associations.

Analogy with PM suggests that by far the largest damage costs could come from chronic effects. However, in their meta-analysis of effects of chronic exposure, Chen *et al.* (2008) find nothing significant for NO_2: their RR_{10} for all-cause mortality is 1.0 (95%CI: 0.99–1.02), RR_{10} being

---

[2] There are additional automotive primary pollutants, e.g. aliphatic hydrocarbons, benzene, butadiene and formaldehyde, but their quantities and/or ERF slopes are so low that their health impacts are negligible compared with $PM_{2.5}$.

for a 10 $\mu g/m^3$ increment. For other end points they do find positive associations for NO_2 although none are statistically significant. However, the situation may be changing as more recent studies have found significant chronic effects of NO_2. The HRAPIE project (WHO, 2013) completed at the same time as this book recommends for NO_2 a RR_{10} for all-cause chronic mortality of 1.055 (95% CI 1.031 to 1.080), as well as functions for hospital admissions and bronchitis in children.

15.1.3   Damage costs of tail pipe emissions in the EU

In the EU, a first comprehensive assessment was carried out by the Transport Phase of ExternE 1999–2000 (ExternE, 2000). The dispersion of vehicle pollutants in the local zone (up to about 20 km) was calculated with the RoadPol Gaussian plume model (Vossiniotis et al., 1996). Beyond the local zone, a Lagrangian trajectory model with chemical reactions was used, covering the entire European continent.

It turns out that much of the damage is caused by primary $PM_{2.5}$, emitted in important quantities by diesel engines and also to a lesser extent by gasoline engines. In contrast with secondary pollutants, the impact of primary pollutants emitted at ground level varies strongly with the detailed relationship between the site where the emission takes place and the distribution of the population. For primary vehicle pollutants in large cities, typically around 95% of the impact is within the local zone, up to about 25 km from the source. Because of the importance of $PM_{2.5}$ for transport damage costs, one should really carry out a fairly systematic analysis of a large number of sites in order to establish a set of generalizable results.

ExternE (2000) considered only a small number of sites, one urban and one rural site in each of seven countries, and no systematic criteria were used in choosing these sites (some of the rural sites are much less urban than others). Additional calculations have been carried out since then by the HEATCO (2006) project. Unfortunately the final report is rather vague about the details of the dispersion calculations for this reassessment, and the damage costs per kg of pollutant are reported without detail about size and population density of the respective sites: the results are provided only for two types of sites, "urban" and "outside built-up areas," in each country of the EU. Thus it is a bit problematic to adapt them for specific sites.

For example, an important policy application is the cost–benefit analysis (CBA) of low emission zones (LEZ), i.e. urban areas where older vehicles are not allowed to drive. Such zones have been created in many countries in Europe, although in most cases without CBA. In fact a proper CBA would be extremely uncertain without new modeling, because HEATCO based numbers do not distinguish, e.g. Milan (population 5 million in metropolitan area) and Florence (population 370,000) from the "urban" number for Italy.

The HEATCO results have been used by the CE Delft (2011) for their assessment of Transport externalities in the EU27, reproduced here in Table 15.2. They incorporate an update for the ERFs and monetary values of the NEEDS phase of ExternE. CE Delft has also tried to distinguish "Metropolitan" (population > 5 million) and "Urban" (population < 5 million), although without explaining how.

15.1.4 Damage costs of transport in the EU

Multiplying the EURO emission factors by the costs per kg of Table 15.2, one obtains the damage cost per km.

Example 15.1   Compare the damage cost per km for diesel and gasoline vehicles of EURO3 (in force Jan. 2000 to Dec. 2004) and EURO5 (in force Sep. 2009 to Aug. 2014), in large metropolitan areas in France.

Solution: Applying the numbers of the EURO standards in Table 15.1 is not always straightforward because some of the pollutants are different. For instance, the HC are not quite the same as the NMVOC, and some of the limits are stated for the sum of NO_x and HC. Also, even though $PM_{2.5}$ emissions from gasoline cars were not regulated in the past, they are not zero. As an example we mention data of about 0.0025 g/km for EURO2 measured by CONCAWE (1998); the uncertainties are large because emissions in such a low range are difficult to measure accurately. For better emissions data one should use a model such as the COPERT 4 software of the European Environment Agency,[3] because it is based on a large set of data. Here is an attempt at combining the emissions of Table 15.1 with the damage costs of Table 15.2.

Diesel	$PM_{2.5}$	NO_x	Total
€/kg	453.8	10.5	
EURO3			
g/km	0.05	0.5	
€cent/km	2.27	0.53	2.79
EURO5			
g/km	0.005	0.18	
€cent/km	0.23	0.19	0.42

---

[3] http://lat.eng.auth.gr/copert/

Gasoline	$PM_{2.5}$	NO_x	NMVOC	Total
€/kg	453.8	10.5	1.4	
EURO3				
g/km	?	0.15	0.2	
€cent/km	?	0.16	0.03	0.19 + ?
EURO5				
g/km	0.005	0.06	0.068	
€cent/km	0.23	0.06	0.01	0.30

Comment: the damage cost of diesel cars has been greatly reduced thanks to the particulate filter. There are, however, concerns about the change in NO_x emissions because the filter converts a large fraction of the NO coming from the engine to NO_2, which is more harmful than NO (noting the WHO (2013) recommendations on ERFs for NO_2) and increases the ambient O_3 concentrations. We have put a question mark for $PM_{2.5}$ from EURO3 gasoline cars because we do not know their level. PM emissions from gasoline cars have not been regulated in the past; they were considered too low to worry about, until more recent epidemiological studies have caused concern over any $PM_{2.5}$ from combustion.

Combining the damage costs per kg of pollutant with data for transport distances, CE Delft (2011) has calculated the average external costs of passenger and freight transport in the EU27. Upstream impacts have been taken into account only for the production of the fuel or energy consumed by the vehicles, but not for the production of the vehicles. The results are shown in Tables 15.3 and 15.4. The units are in pkm (passenger km) and tkm (tonne km), assuming the occupancy or charge rates indicated under the tables. Aggregated results of this kind involve many assumptions and have to be interpreted with caution. In particular, one should keep in mind the great variability with site. The numbers are only an indication of orders of magnitude. Different assessments may well come out with different numbers. For example, in Table 15.3 the ratio of climate change damages is

$$\text{car/train/plane} = 0.3/0.03/0.8 = 1/0.1/2.7,$$

whereas in Table 15.13 it is

$$\text{car/train/plane} = 89/25/205 = 1/0.28/2.3$$

(the 205 is the average over short and long distance flights).

Table 15.3 *Average damage costs (excluding congestion) of passenger transport in the EU27, in €cent$_{2008}$/pkm. Adapted from Table 1 of CE Delft (2011).*

€cent/pkm[a]	Passenger cars[b]	Buses & coaches	Motorcycles & mopeds	Train[c]	Plane[d]
Accidents	3.23	1.23	15.66	0.06	0.05
Air pollution	0.55	0.6	1.18	0.26	0.09
Climate change[e]	0.3	0.16	0.19	0.03	0.8
Noise	0.17	0.16	1.44	0.12	0.1
Up- and downstream[e]	0.34	0.15	0.23	0.39	0.39
Nature & landscape	0.06	0.03	0.05	0.02	0.06
Biodiversity losses	0.02	0.04	0.01	0	0.01
Soil & water pollution	0.03	0.09	0.03	0.05	0
Urban effects	0.1	0.04	0.08	0.06	0
Total	**4.81**	**2.49**	**18.87**	**0.98**	**1.5**

[a] data exclude Malta and Cyprus, but include Norway and Switzerland. Data do not include congestion costs.
[b] average occupancy of passenger cars is 1.8
[c] average occupancy of passenger trains is 126
[d] only flights within EU included
[e] at 0.025 €/kg$_{CO2eq}$

15.1.5 Damage costs of transport in the USA

The external cost assessment in the USA by NRC (2010) includes results for the transport sector. The presentation is quite different from what we have shown for the EU, and we make no attempt to bring them to a common format since not enough detail is provided in that report. In any case, there are appreciable differences in the monetary values per end point, and the relation between sources and receptors is different for the EU because of the generally lower population density. A wide range of technologies, both for vehicles and for fuels/energy, has been evaluated using the GREET (greenhouse gases, regulated emissions, and energy use in transportation) model (ANL, 2004) for the LCA inventory and the APEEP model (Muller and Mendelsohn, 2007) for the calculation of the physical effects and monetary damages. Among the upstream impacts of biofuels, land use change has not been taken into account, because, as explained in NRC (2010), such impacts pass through the market and should be allocated to the new agricultural activities in order to avoid a double counting of external costs. It should, however, be noted that decisions like this on the

Table 15.4 *Average damage costs (excluding congestion) of freight transport in the EU27, in €cent₂₀₀₈/tkm.* *Adapted from Table 1 of CE Delft (2011).*

€cent/tkm[a]	LDV[b]	HDV	Train[c]	Waterborne
Accidents	5.62	1.02	0.02	0
Air pollution	1.79	0.67	0.11	0.54
Climate change[d]	0.76	0.17	0.02	0.06
Noise	0.63	0.18	0.1	0
Up- and downstream	0.84	0.17	0.24	0.08
Nature & landscape	0.09	0.07	0	0.04
Biodiversity losses	0.06	0.05	0	0.05
Soil & water pollution	0.18	0.08	0.04	0
Urban effects	0.31	0.05	0.01	0
Total	**10.28**	**2.46**	**0.53**	**0.77**

[a] data exclude Malta and Cyprus, but include Norway and Switzerland. Data do not include congestion costs.
[b] average load factor of trucks is 2.8 tonne (LDV = light duty vehicle)
[c] average load factor of freight trains is equal to 323 ton
[d] at 0.025 €/kg$_{CO2eq}$

system boundary need to be taken on a case by case basis to ensure that the analysis reflects the marginal impact of the decision under investigation.

Here we only summarize some of the results for light-duty vehicles in Table 15.5 for the classical air pollutants and Table 15.6 for GHG. They are adapted from Tables 3–18 and 3–19 of NRC (2010), which group the results in terms of impact ranges rather than showing precise numbers for each technology.

Air pollution damage in the USA is in the range 1.2 to 1.8 ¢/VMT (0.57 to 0.86 €cent/vkm), and the GHG costs range from 0.45 to over 1.8 ¢/VMT (0.28 to 1.12 €cent/vkm) if valued at \$30/tonne$_{CO2eq}$. Direct comparison with the numbers for passenger cars in the EU, in Table 15.3 above, is not meaningful because the LCA boundaries are different and so are the vehicle categories (passenger cars in Table 15.3 versus light-duty vehicles in NRC (2010), not explicitly defined but apparently including light trucks and the gas guzzling sport utility vehicles so popular in the USA).

Among biofuels the only technology in current use for ethanol production is based on fermentation of corn kernels (maize). It is not very attractive in terms of total life cycle GHG emissions, to say nothing about the resulting increase in the price of corn based food. Even though the increase in the price of food is "only" a pecuniary externality, it is problematic for its equity implications. This consideration also applies to biodiesel from soy. Technologies based on E85 from grasses or corn

Table 15.5 *Results for damage costs of light-duty vehicles in the USA due to classical air pollutants. Data from NRC (2010).*

Damage cost range, ¢/VMT			Technology[a]	Damage cost range, €cent/vkm[b]		
1.2	to	1.29	E85 from grasses E85 from corn stover Compressed natural gas HEV, grid-independent	0.57	to	0.62
1.3	to	1.39	Gasoline, conventional Gasoline, reformulated E10 Hydrogen, gaseous	0.62	to	0.66
1.4	to	1.49	Diesel, low sulfur Diesel, with 20% biodiesel from soy HEV, grid-dependent	0.67	to	0.71
1.5	to	1.59	E85 from dry corn	0.72	to	0.76
>1.6			Electric vehicle	>0.76		

[a] E10 = gasoline with 10% ethanol, E85 = gasoline with 85% ethanol, HEV = hybrid electric vehicle
[b] Conversion factors $1.3/€ and 1.609 km/VMT (vehicle mile traveled)

stover, and gaseous hydrogen, hold the promise of greatly reducing life cycle GHG emissions.

15.1.6   Do cars kill more by their front or their back?

Driving is not without risks. Motor vehicle traffic deaths account for the largest fraction of accidental deaths in the USA and EU. For example, in France in 2008, the standardized death rate due to all accidents was 26.4 per 100,000 inhabitants, and the rate due to transport accidents was 6.8 per 100,000 inhabitants. The rate of LE loss per million km due to car accidents, can be estimated from the following data (for the example of cars in France in 2008):

road deaths/yr	4443
total number of cars[a]	30,876,000
km/yr per car, average	13,000
total km/yr	4.01E+11
deaths/million km	**1.11E-02**
YOLL/death	40
YOLL/million km	**0.44**

[a] 498 cars/1000 persons in total population of 62,000,000

Table 15.6 *Results for damage costs of light-duty vehicles in the USA due to GHG. Damage cost at $30/tonne$_{CO2eq}$. Data from NRC (2010).*

Emission range, g/VMT	Damage cost range, ¢/VMT	Technology[a]	Emission range, g/vkm[a]	Damage cost range, €cent/vkm[b]
150 to 250	0.45 to 0.75	E85 from grasses E85 from corn stover	93 to 155	0.28 to 0.47
250 to 350	0.75 to 1.05	Hydrogen, gaseous	155 to 218	0.47 to 0.65
350 to 500	1.05 to 1.5	E85 from corn kernels Diesel, with 20% biodiesel from soy HEV, grid-independent HEV, grid-dependent Electric vehicle Compressed natural gas	218 to 311	0.65 to 0.93
500 to 599	1.5 to 1.8	Gasoline, conventional Gasoline, reformulated E10 Diesel, low sulfur	311 to 372	0.93 to 1.12
>600	>1.8	Tar sands	>373	>1.12

[a] E10 = gasoline with 10% ethanol, E85 = gasoline with 85% ethanol, HEV = hybrid electric vehicle
[b] Conversion factors $1.3/€ and 1.609 km/VMT (vehicle mile traveled)

For the years of life lost per accident, we assume that the average loss per death is 40 yr.

For the LE loss per million km due to pollution, let us take as an example the PM$_{2.5}$ emissions for EURO4 passenger cars, Table 15.1, assuming a mix of 50% gasoline and 50% diesel cars (in France the diesel portion has been increasing over the years and currently it is even higher). For the average damage cost in €/kg we take the Urban entry for France in Table 15.2, not very much larger than the number shown for Non-urban. Thus we obtain the numbers in this table:

PM$_{2.5}$ emission, g/km	0.0125 Table 15.1, 50% gasoline + 50% diesel, EURO4
PM$_{2.5}$ emission, kg/million km	12.5
damage cost, €/kg$_{PM2.5}$	146.1 Table 15.2, Urban, France
fraction due to mortality	0.7
mortality cost, €/kg	102.3
mortality cost, €/million km	1278
VOLY, €	40,000
YOLL/million km	**0.032**

The result, 0.032 YOLL/million km due to pollution, is more than an order of magnitude smaller than the LE loss due to accidents. But for the dirty cars of earlier vintage that was not the case. Consider diesel cars of the EURO1 standard, which was in force 1992–1995. The EURO1 and EURO4 limit values for $PM_{2.5}$ emissions of diesel cars are 0.14 and 0.025 $g_{PM2.5}$/km, respectively. As a simple approximation, let us multiply 0.032 YOLL/million km by their ratio; the result is 0.18 YOLL/million km. For those older cars driving in large cities (with damage cost of 453.8 €/$kg_{PM2.5}$) the mortality from pollution was 0.56 YOLL/million km, higher than the mortality from accidents.

In passing, we show how we calculated the life expectancy loss due to road accidents for the risk comparisons in Fig. 2.10 with data for France.

road deaths/yr	4443
total population	62,000,000
road deaths/person/yr	7.17E-05
YOLL/death, average	40 yr
YOLL/person/yr	2.87E-03
Life expectancy	80 yr
YOLL/person, life	2.29E-01
Days of life lost/person, life	8.37E+01

15.1.7   Damage costs of underground trains

In numerous underground stations and trains very high concentrations of PM have been observed, far higher than in the ambient air above ground. For example in Paris, $PM_{2.5}$ concentrations in the range 100 to 150 μg/m^3 have been measured for exposures during typical travel (Grange and Host, 2012). Naturally there are concerns about the health effects. The concentration increase is mostly due to abrasion of wheels and rails, as well as brake linings (unless the trains have electric brakes). Unfortunately not enough is known about the toxicity of these constituents. Grange and Host also review the toxicological and epidemiological evidence. Even though toxicological studies have found that these particles, with their high iron content, have a strong oxidative and inflammatory potential, no link to health effects in humans has been established. Epidemiological studies of workers in underground systems have not found any increase relative to the general population; however, the statistical power of these studies may not have been sufficient. In the following example let us suppose the same toxicity as for ordinary ambient $PM_{2.5}$.

Example 15.2 Estimate the damage cost of travel per km in the metro trains of Paris, if the passengers are exposed to 125 μg/m^3 of PM$_{2.5}$, while the average exposure above ground is 25 μg/m^3 of PM$_{2.5}$. The average speed of the trains is 25 km/h. For the exposure–response functions and monetary valuation, assume the result of ExternE (2008) in Table 12.3, namely a cost of 32.79 €/(yr. μg/m^3) per year of exposure, summed over the health end points of PM$_{2.5}$.

Solution: At this speed the exposure duration per km is 1/25 h/pkm, i.e. a fraction 0.04/(365.25 × 24) of the year. Only the exposure increment above ambient counts for the calculation. Thus the cost per pkm is 32.79 €/(yr. μg/m^3) × (125 – 25) μg/m^3 × 0.04/(365.25 × 24) yr/pkm = 1.5 €cent/pkm. This is per passenger km (pkm) because the calculation concerns the impact on an individual during a trajectory of 1 km.

Comment: This is higher than the damage cost due to air pollution of 0.55 €cent/pkm for typical passenger cars in the EU, as estimated by CE Delft (2011), see Table 15.3. However, because of the uncertainties about the toxicity these numbers are very uncertain, both in absolute magnitude and in the comparison metro/car. There is also a difference in the people affected by the pollution. In the metro, the cost is imposed only on the traveler and metro staff, whereas a car imposes its cost on the general public. In any case this result is most certainly not a reason for driving a car rather than using the underground: other considerations, including the cost of congestion, are far more important (see Fig. 15.10).

15.2   Hybrid electric versus conventional vehicles

15.2.1   Context and objectives

As an application of external costs to transport, we present a comparison of hybrid and conventional vehicles (Spadaro and Rabl, 2006). The hybrid electric vehicle (HEV) appears as an attractive option until even cleaner technologies such as all-electric vehicles are sufficiently well developed to compete in the market. Since our study in 2006, HEV sales have greatly increased and better data on their fuel consumption would be available, but we do not think that would change the basic conclusions and so we present this study without updating. Likewise, we do not try to update the pollutant costs in Table 15.7.

For a correct evaluation and comparison of different vehicle options a life cycle assessment (LCA) is required. Whereas numerous LCA studies of cars have been carried out (e.g. Delucchi, 2003, MacLeana and Lave,

Table 15.7 *Marginal damage costs in € per kg of pollutant, assumed by Spadaro and Rabl (2006) for the comparison of HEV and conventional cars.*

Pollutant	€ per kg pollutant
Particulate Matter, PM	
Tailpipe emissions ($PM_{2.5}$ urban)	750
Tailpipe emissions ($PM_{2.5}$, rural)	37.4
Tall, stationary sources (PM_{10}, urban)	44.8
Tall, stationary sources (PM_{10}, rural)	22.4
Vehicle production sources (PM_{10})	22.4
SO_2 (via sulfate aerosols, crops and material damages)	4.6
NO_x (via nitrate aerosol formation)	2.4
CO_2	0.019
NMVOC (non-methane volatile organic compounds)	1.14
CO	0.0016
Pb (includes both inhalation and ingestion)	600
Ni (inhalation dose only)	3.8

2003, Weiss *et al.*, 2000), their results for the environmental effects have been reported only in terms of the quantities of emitted pollutants. Here we evaluate the effects in terms of damage costs. Our assumptions for the cost per kg of emitted pollutant are listed in Table 15.7; they are appropriate for European conditions, as adapted from ExternE (2005).

15.2.2 Assumptions and data

The life cycle of a vehicle can be broken down into the following stages:
- Production of the materials needed for making the vehicle,
- Assembly of the materials,
- Fuel feedstock (e.g. extraction of petroleum from the ground),
- Fuel supply (refining of petroleum and transport of fuel),
- Utilization of the vehicle, and
- Disposal of the vehicle at the end of its useful lifetime.

Disposal of the vehicle at the end of its life involves as major steps, (i) dismantling the vehicle, (ii) recycling the recyclable fraction, and (iii) disposing the remaining fraction in a landfill or incinerator. The impacts due to the dismantling of the vehicle are entirely negligible, as can be seen from the fact that the energy for dismantling is about 1/3 of the energy requirement for vehicle assembly (Stodolsky *et al.*, 1995) and that the impacts of the assembly are only a very small part of the total life cycle impact. Here we have found it most convenient to include recycling in the first stage (production of the materials), because the available data for materials account for the reduction of the emissions due to recycling. The impacts from the disposal of the remaining fraction are negligible if current environmental regulations for waste disposal are respected. For these reasons we do not show explicit results for the disposal stage.

As sources of information for the life cycle stages we used:
- A large and comprehensive review paper by MacLeana and Lave (2003),
- A major LCA of many vehicle and fuel technologies by MIT (Weiss *et al.* 2000),
- The most comprehensive well-to-wheel analysis (the GREET software of ANL (2004)),
- The most comprehensive LCA of vehicle and fuel technologies (Delucchi, 2003),
- For on-the-road emissions and fuel economy, USEPA and independent test organizations.

Ideally, the HEV should be compared with a gasoline version that is identical except for the drive train. However, since the difference in damage costs is crucially dependent on the fuel consumption, we have chosen the Toyota Prius as the HEV, since it was the first to enter the market and its consumption and emissions data are more reliable than for more recent models that are offered in both versions. Unfortunately for this analysis there is no exact conventional equivalent of the Prius. The Toyota Camry comes closest, although it is slightly larger; for that reason we have also analyzed the data for the Toyota Corolla; a conventional Prius would be intermediate between Corolla and Camry. We have also considered a plug-in (grid-connected) version of the Prius (abbreviated to PHEV); at the time of that study such a version was not available. In our assessment, we assume that the batteries of the PHEV are charged overnight using base load electricity, and that there is enough energy stored in the batteries to travel 90% of the distance as an all-electric vehicle (EV), a percentage that seems reasonable for much urban driving but not for all uses of a PHEV. The remainder of the time the vehicle operates as would a hybrid car. Regarding the electricity generation mix, two scenarios were analyzed: (i) fossil fuels contribute 60% of baseload electricity (PHEV, high), and (ii) electricity supply is from nuclear power and renewable fuels (PHEV, low).

For the production of vehicles, we have taken the inventory data from Delucchi (2003), Weiss *et al.* (2000) and AMM (2003) (they are very similar) and scaled them in proportion to the actual mass of the cars under consideration. The first two of these sources provide separate inventories for conventional and for hybrid cars.

The vehicle disposal stage involves the dismantling of the vehicle and recovery of any materials that can be re-used, and the treatment of the remainder. Metals are easy to recycle, and at the present time already 90% to 95% of the metals are re-used, either in the same or in different sectors of the economy. The quantity of glass in a car is small and so are the associated impacts. Plastics are much more difficult to recycle, and it is not clear what fraction will find another use. Fluids in cars are either water or petroleum based. Most of the latter are already recycled in some form or other (including thermal recycling via incineration). In any case the recycled fractions are bound to increase under the growing pressure from governments, especially in the EU, to increase the recycling of waste. One consequence of this is standardization of plastics in vehicles, moving from the use of many types of plastic to just a few. The MIT study (Weiss *et al.*, 2000) assumes, perhaps optimistically, that 50% of plastics are recycled by 2020; Delucchi (2003) does not try to estimate a percentage. In our study we take a recycling percentage of 5% for plastics and rubber, although noting that this may well be pessimistic given action to enhance recycling.

For this analysis it does not matter whether the materials are reused in vehicles or in other sectors, because the emissions avoided by recycling are essentially independent of the sector where the production of virgin materials is avoided. We assume that an average passenger car is driven 200,000 km during its 15-year lifetime.

The assumptions for fuel consumption and tailpipe emissions are summarized in Table 15.8. Energy efficiency results by the EPA are based on standard and supplemental federal testing procedures,[4] but are not representative of real-world driving conditions, due to such factors as personal driving habits, on-road driving conditions, on-board equipment malfunction, air-conditioner use, and extra load. We have therefore looked at the results of actual driving tests by independent testing organizations such as Consumer Reports,[5] Consumer Guide,[6] and the US Department of Energy (advanced Vehicle Testing Activity)[7] and we have chosen as representative values the mean of these results. For emissions from vehicle operation, we have used, whenever available, USA published data by ACEEE, CEC, EIA, INEEL, NREL, US DOE, US DOT, US EPA and Toyota Motors. In Europe, additional test results are

---

[4] www.fueleconomy.gov   [5] www.ConsumerReports.org
[6] www.auto.consumerguide.com   [7] www.eere.energy.gov/

Table 15.8 *Vehicle data, fuel consumption and tailpipe emissions.*

Vehicle	Fuel[a] L/100km	PM$_{2.5}$[b] g/km	NO$_x$ g/km	CO g/km	NMVOC g/km	SO$_2$ g/km	CO$_{2,equiv}$ g/km
Toyota Camry 2.4L, 4-cyl (mid-size car, 1435 kg)	10.1	0.0012	0.02	0.62	0.006	0.018	240
Toyota Corolla 1.8L, 4-cyl (compact car, 1177 kg)	8.2	0.0012	0.12	2.61	0.19	0.014	195
Toyota Prius 1.5L, 4-cyl (1311 kg)	5.5	0.001	0.02	0.13	0.006	0.01	131
Plug-in Prius (PHEV) (HEV driving mode, 1538 kg)	6.2	0.001	0.02	0.13	0.006	0.011	147
Plug-in Prius (PHEV) (EV driving mode)	2.6 (L of gasoline equivalent)	0	0	0	0	0	0

[a] expressed as gasoline equivalent for "real-world" driving conditions; vehicle distance traveled 200,000 km over lifetime; useful vehicle lifetime is 15 years.
[b] brake and tire wear particulate emissions (PM$_{10}$) are 0.013 g/km, but we have not included them here because of uncertainty about their toxicity.

available from the ARTEMIS project of the EU.[8] In the absence of data, we assume the regulatory limit values of the national Tier 2 emission standards of the USA. It turns out that perfect knowledge of actual emissions is not critical, because the external costs are dominated by CO$_2$ releases, which depend only on fuel consumption. We emphasize that the fuel consumption of HEVs is extremely variable with driving conditions: for the typical stop-and-go traffic of urban driving the HEV can be far more fuel efficient than conventional cars (especially if a large percentage of the driving is powered by electricity charged at night), but on highways the HEV offers no advantage. This is why firm data are difficult to establish and different studies can reach quite different conclusions.

---

[8] www.trl.co.uk/artemis/

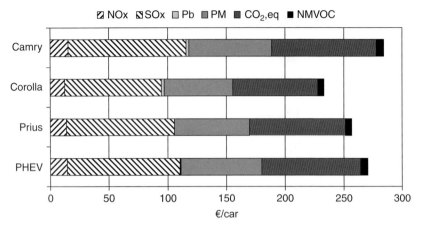

Fig. 15.3 Damage cost due to the vehicle production stage. PHEV = plug-in version of Prius.

15.2.3   Results

The damage cost of the vehicle production stage is shown in Fig. 15.3, with breakdown by pollutant. It represents very roughly one percent of the price of the car. Greater detail for the contribution of the individual materials and of vehicle assembly is provided in Fig. 15.4 for the Camry. The corresponding numbers for the Prius are very similar, slightly smaller in proportion to the vehicle mass; if the rest of the car were identical the only difference would come from the battery (Pb for the Camry, a larger quantity of the much less toxic Ni for the Prius). Compared with the total cost of this stage the contribution of vehicle assembly is only about a quarter. NMVOCs are relatively more important for assembly than for the individual materials because of emissions during painting and coating (Toyota has switched to water-based paints, which should reduce emissions, and consequently damage costs, by a factor of two). In terms of damage costs, the most important materials are iron, steel, aluminum and plastics.

For the well-to-tank and tank-to-wheel stages we used the GREET model (version 1.5a, dated April 2001). This software was developed by ANL (2004). Since its first release in 1996, GREET has been used extensively by industry, government and academia. Although the default input data and assumptions are for the United States (e.g. electricity mix for feedstock production, fuel specifications, upstream boiler emission

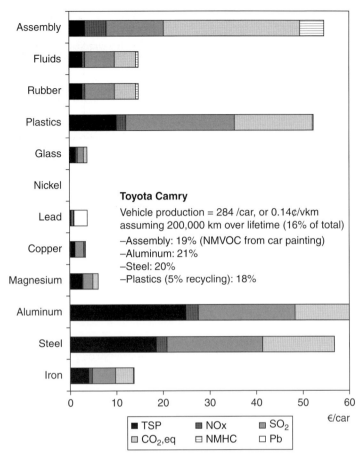

Fig. 15.4 Contribution of materials and vehicle assembly to damage cost of vehicle production.

factors, etc.), the technologies are sufficiently similar in the EU. Figure 15.5 shows the breakdown by pollutant for the stages of the well-to-wheel analysis and vehicle production. We do not show the damage cost of CO because it is so small that its contribution to the total would not even be visible in the graph. Only the Camry is shown, because these impacts are proportional to fuel consumption.

The results for the total damage cost, in €cents/vkm, are shown in Fig. 15.6. The hybrid vehicles cause by far the lowest damage. Vehicle utilization accounts for the largest contribution to the total damage cost in the case of the Camry, Corolla and Prius, but its share of the total drops

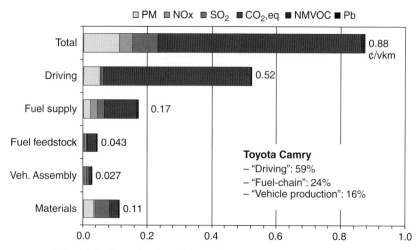

Fig. 15.5 Contribution of the pollutants to the well-to-wheel and vehicle production stages.
Here the ¢ are €cents.

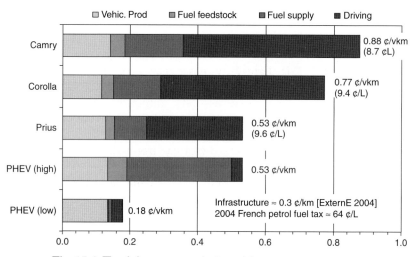

Fig. 15.6 Total damage cost, in €cent/vkm.
PHEV = plug-in version of Prius.

drastically to between 10% and 20% for the plug-in hybrid. It is quite evident that the well-to-tank damage cost of the plug-in Prius is sensitive to the assumption of the share of fossil fuels in the electricity supply mix, a factor of three separates the PHEV low and high estimates. Damages from

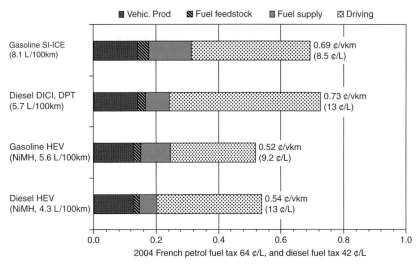

Fig. 15.7 Damage costs of gasoline and diesel passenger cars, in €cents/vkm.

vehicle production are essentially proportional to vehicle mass since the relative composition does not vary much.

Finally, Fig. 15.7 compares the damage costs of gasoline and diesel (direct injection, DI) light-duty passenger cars in around 2010. Although diesel vehicles have lower $CO_{2,eq}$ emissions per km compared with gasoline engines, particulate and NO_x tailpipe emissions are higher (of particular concern is the potential impact on health from exposure to increasing emissions of NO_2 from diesel cars equipped with a particulate filter). The net effect is that the diesel car has a slightly higher total damage cost than the gasoline car on the basis of €cent/km. Hybrid (NiMH battery) options, of course, have significantly lower damage costs.

To conclude, the HEV has much lower damage costs than the conventional alternative, the difference being proportional to the fuel savings (e.g. 0.53 ¢/km for the Prius versus 0.88¢/km for the Camry, or a 40% damage cost reduction). A further advantage can be gained from the plug-in option (charging the HEV with baseload electricity at night); this option is strongly dependent on the driving pattern and share of fossil fuels in the electricity generation mix.

Example 15.3 Estimate the benefit of an HEV compared to a conventional car, using the above data for Prius, Corolla and Camry, assuming that a conventional equivalent of the Prius would be intermediate between Corolla and Camry. Use the damage costs per km in Fig. 15.6, and assume that the cars last $N = 15$ yr and are driven 13,333 km/yr. Take a discount rate of $r_{dis} = 5\%$.

Solution: The average damage cost of Corolla and Camry is 0.825 €cent/km, that for the Prius is 0.53 €cent/km. Multiplying the difference by the km/yr we find an annual damage cost saving of 39.33 €/yr. To find the present value of the annual savings, one needs to divide the annual saving by the capital recovery factor $(A/P,N,r_{dis})$, Eq. (9.15). Here it is equal to $(A/P,15yr,5\%) = 0.0963$. Hence the present value of the damage cost savings is,

$$(39.33€/yr)/(0.0963/yr)€ = 408€.$$

Of course, there are also the fuel savings. With the data from Table 15.8 we find that the Prius consumes 5.5 L/100 km, compared with 9.15 L/100 km for the conventional equivalent. The annual gasoline saving with the Prius is 487 L/yr. In France in 2011 the price of gasoline is about 1.50 €/L, and the present value of the fuel saving is,

$$(487 L/yr \times 1.50€/L)/(0.0963/yr) = 7580€.$$

However, this number cannot simply be added to the avoided damage cost because it includes a hefty fuel tax. The 7580 € is relevant for the cost–benefit analysis of a private car buyer, but not for social cost analysis. Without tax, the price of gasoline is 0.85 €/L and the present value of the fuel saving is 3285 €. Thus the benefit for a social cost–benefit analysis is,

$$408€ + 3285€ = 3693€.$$

We do not know the extra cost of a hybrid car. Manufacturers tend to keep such information confidential. Market prices are not a reliable indicator for new technologies because manufacturers set their prices strategically with a view to long-term prospects. Price data, in 2011, for the Honda Civic, which is available in both hybrid and conventional versions, suggest that the difference is around $5000 in the USA. At $5000/($1.3/€) = 3846 € an HEV would already be nearly justified today, both from the private and the social perspective. In any case, costs are declining as the technology is still evolving and the production volume still increasing rapidly, so the HEV appears to be a good solution.

15.3   Costs and benefits of active transport

15.3.1   *Context and objectives*

As another illustration of the use of external costs, let us look at the costs and benefits of active transport: "get out of your car and walk or bicycle instead."[9] A shift from driving to active transport can bring about significant benefits for our health and environment. To help policy makers, urban planners and local administrators make the appropriate choices, it is necessary to quantify all the significant impacts of such a change. There are countless possible effects, some of which are extremely difficult to evaluate, for instance impacts on the social fabric of a community, on the sense of well-being of the population, even on the crime rate. But health impacts of the physical activity (PA) and of air pollution are especially important, and at least the benefit of reduced mortality can be evaluated quite reliably.

It is convenient to calculate cost and benefits per individual driver who switches to active transport, because such information can be used to evaluate a wide variety of policy options. We evaluate in detail four effects when people change their transportation mode from driving to bicycling or walking:
- the health benefit of the physical activity,
- the health benefit for the general population due to reduced pollution,
- the change in air pollution impacts for the individuals who make the change,
- changes in accidents.

There is a wide variety of possible health impacts, but here we focus on mortality, because the exposure–response functions and accident data for this end point have the lowest uncertainty. In monetary terms, the mortality impacts are especially large, and they also tend to weigh heavily in public perception. But we also indicate how the conclusions might change if other health impacts are included. We also cite estimates of the benefits of reduced noise and congestion in Section 15.3.9.

15.3.2   *Outline of the methodology*

Table 15.9 outlines the key assumptions. The following subsections present more detail. We begin by choosing the scenarios, namely a change in the transport mode for commuting to and from work. For the assessment of

---

[9] This section is based on Rabl and de Nazelle (2011), to which we refer for detailed explanations. As with the other case studies we cite, we do not update the details (ERFs, monetary values, etc.), because the differences would not be very significant in view of the uncertainties.

Table 15.9 *Key assumptions for analysis of active transport.*
ERF = exposure–response function, LE = life expectancy, PA = physical activity,
RR = relative risk, VOLY = value of life year, VPF = value of prevented fatality.

---

(1) Scenarios
(a) Use bicycle instead of car for commuting to work 5 days/week, 46 weeks/yr
   trajectory 5 km one way, 2300 km/yr
   by car: average speed 20 km/hr, duration of one-way trip 0.25 hr
   by bicycle: average speed 17 km/hr, duration of one-way trip 0.33 hr
(b) Walk instead of driving for commuting to work 5 days/week, 46 weeks/yr
   trajectory 2.5 km one way, 1150 km/yr
   by car: average speed 20 km/hr, duration of one-way trip 0.125 hr
   on foot: average speed 5 km/hr, duration of one-way trip 0.5 hr

(2) Benefit of PA
Life table calculation of LE change, with the following RR
(a) for bicycling: based on Andersen *et al.* (2000) and applying a correction for the difference in bicycling duration compared with our scenario, assume RR = 0.709 for age-specific mortality from age 25 to age 65, as result of bicycling from age 20 to age 60.
(b) for walking: based on WHO (2010) and applying corrections for our scenario, assume RR = 0.735 for age-specific mortality from age 25 to age 65, as result of walking from age 20 to age 60.

(3) Health impacts of air pollution
ERF for mortality due to $PM_{2.5}$ is linear without threshold and is expressed as LE loss, with slope S_{ERF} = 6.51E-04 years of life lost per person per year per $\mu g/m^3$ of $PM_{2.5}$ (Eq. (3.12)). Impact change of individuals is proportional to duration of exposure/dose change.

(4) Public benefit from reduced pollution
(a) Avoided emissions: 0.031 $g_{PM2.5}$/km, based on COPERT 4 software.
(b) Calculation of avoided air pollution mortality: based on results of the Transport phase of ExternE (2000), but updated to current best values for ERF and monetary valuation.

(5) Effect of exposure change from car to bicycle and from car to walking
Based on measured concentration data in representative busy streets of eight cities in the EU (EEA, 2008), assume 23 $\mu g/m^3$ of $PM_{2.5}$ and 57 $\mu g/m^3$ of NO_2 at side of street.
Modifying factors for exposure (due to increased concentration) and dose (due to increased inhalation) during different transport modes: 1.5 for cars, 2 for pedestrians, 3 for bicyclists.

(6) Accidents
Accident statistics for Paris, Belgium and the Netherlands.
Cost of nonfatal bicycle accidents based on Belgian data from Aertsens *et al.* (2010)

(7) Monetary valuation
Monetary valuation of fatal accidents based on VPF = 1.6 M€
Monetary valuation of PA and air pollution based on VOLY = 40,000 € but adjusted for inflation to 43,801 €
Cost of CO_2 emissions based on 25 €/tonne$_{CO2}$

---

bicycling, we consider an individual who switches from car to bicycle for a trajectory of 5 km one way. The assumptions for trip duration and average speed are typical of bicycling. For cars they are realistic for the congestion in large cities; for smaller cities or rural sites, the speed would be higher and the emission of pollutants lower. For a switch from car to walking, the typical distance would be much shorter and we take 2.5 km, commuting time being a crucial determinant for the choice of transportation mode.

15.3.3 Benefits of physical activity

That physical activity brings large health benefits has been established beyond any doubt, by countless epidemiological studies in many countries all over the world; see for example the review by the US Department of Health and Human Services (US DHHS, 2008). It presents explicit exposure–response functions (ERF) for several end points, in particular the reduction of all-cause mortality. We use this review as the basis, because it is the most comprehensive we have found. In particular, we use their ERF for all-cause mortality, shown here as a solid line in

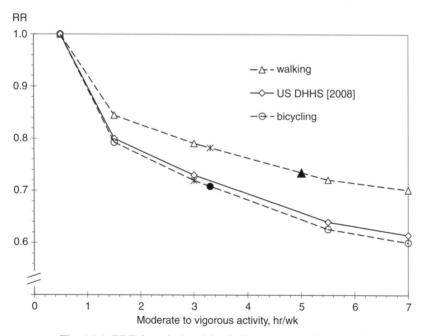

Fig. 15.8 ERF for relative risk of all-cause mortality, as a function of hours/week of physical activity. Solid line: data from US DHHS (2008). Dashed lines are obtained by scaling (1-RR) in proportion to the (1-RR) of WHO (2010) and Andersen et al. (2000) at the points indicated by the stars. The solid black points on the dashed lines indicate the RR chosen for our scenarios. For details, see Rabl and de Nazelle (2011), from which this graph is taken, with permission of Elsevier.

Fig. 15.8, drawn as linear interpolation of the data points (the other lines in this figure will be explained below). It has been obtained as a median of the ERFs of 12 studies that are sufficiently comparable to be summarized in such a manner. The general pattern is typical of the various health benefits of activity: it is nonlinear, the incremental benefit being greatest at low levels of activity.

In addition to mortality, activity also reduces the incidence of a wide range of morbidity end points, especially coronary heart disease, stroke, hypertension, and type 2 diabetes; activity is also associated with significantly lower rates of colon and breast cancer, as well as improved mental health (US DHHS, 2008). The range of morbidity benefits is much wider than for air pollution, where morbidity involves mostly cardio-pulmonary effects. In monetary terms the ratio of morbidity over mortality benefits may thus be significantly larger than the ratio 0.5 that ExternE finds for air pollution; but further research is needed to examine this question.

For the health benefits of bicycling, we invoke WHO (2008).[10] The authors of this report carried out a thorough review of health benefits of bicycling and concluded that it would be best to consider only mortality, using as a basis a large epidemiological study of cyclists in Copenhagen (Andersen et al., 2000). That study is a prospective cohort study of the effects of PA on all-cause mortality, involving 30,896 men and women, with mean follow-up of 14.5 yr. The bicycling results are based on the subset of 6954 individuals who bicycle to work. Such a large sample and follow-up was possible because Copenhagen is one of the cities with the highest percentage of bicycling to work, more than 35%. After adjustment for age, sex, educational level, leisure time, physical activity, body mass index, blood lipid levels, smoking, and blood pressure, the relative risk was RR = 0.72 (95% CI, 0.57–0.91) for individuals who bicycle to work (average 3 hr/wk) compared with those who do not. Since our scenario involves a bicycling time of 3.3 hr/wk, slightly different from the 3 hr/wk of Andersen et al., we have adjusted that RR to obtain RR = 0.709, as indicated by the solid circle in Fig. 15.8.

The World Health Organization has also evaluated the benefits of walking (WHO, 2010). The recommended relative risk for the reduction of mortality is RR = 0.78 (95% confidence interval: 0.64–0.98) for a walking exposure of 29 minutes seven days a week = 3.38 hr/wk. After adjustment for the walking duration of our scenario, we obtain RR = 0.735, as indicated by the solid triangle in Fig. 15.8.

---

[10] This report is a user guide for a software tool, HEAT, that evaluates the benefits of bicycling scenarios, based on avoided deaths and VPF. We do not use it because LE change and VOLY is more appropriate for small changes in LE.

We consider a bicycling cohort of age 20 to 60 yr and assume a time delay of five years for the full attainment of benefit. Thus we assume that the age-specific mortality is reduced by a factor of 0.709 from age 25 to 65. We carried out life table calculations, using data for age-specific mortality for a wide range of countries, in particular for the EU in 2007 from Eurostat.[11] The LE gain is 1.20 yr for EU25. It is not very different within the EU, varying by less than about 0.1 yr. For the USA, the gain is 1.32 yr. The gains tend to be larger in countries with lower LE because lower LE is due to higher age-specific mortality, generally at all ages; thus a reduction of RR between 25 and 65 has a larger effect. In Romania where LE is only 73 yr, the LE gain from bicycling is 1.69 yr, and for Russia the corresponding numbers are LE = 67.5 yr and LE gain = 2.67 yr.

Since these LE gains are the result of bicycling from age 20 to 60, but we want an equivalent annual benefit, we multiply the LE gain by VOLY and divide by the 40 years from age 20 to 60. Such allocation per year, without discounting, is appropriate because discounting is already implicit in the VOLY of Desaigues et al. (2011). Multiplying the LE gain of 1.20 yr by VOLY we find that the average annual benefit of our bicycling scenario in the EU25 is 1310 €/yr. In like manner and assuming RR = 0.735 for our walking scenario, we find that the average LE gain in the EU25 is 1.09 yr, worth 1192 €/yr.

One of the uncertainties of this calculation, and one that may not be correctly taken into account by the indicated confidence intervals, arises from the nonlinearity of the ERF. The incremental benefit of 1 hr of activity for a couch potato is several times higher than for someone who is already very active. The RR of Andersen et al. (2000) includes corrections for the level of physical activity and is appropriate for populations with a mix of activity levels comparable with the population of Copenhagen. But it may significantly overestimate the marginal benefit of more active populations or individuals.

15.3.4 Car emissions

To estimate the emissions of a car, we use the COPERT 4 software, version 8.0, of the European Environment Agency.[12] The user specifies the vehicle types, as well as the percentage of each of three main driving conditions (urban, rural and highway) and the corresponding average speeds. Vehicle types are specified in terms of EURO standards. We consider passenger cars conforming with the EURO4 and EURO5

---

[11] http://epp.eurostat.ec.europa.eu/portal/page/portal/eurostat/home
[12] http://lat.eng.auth.gr/copert/

Table 15.10 *Passenger car emissions for urban driving, as calculated by COPERT 4. CO_2 is the same for EURO4 and EURO5. Here we have chosen the values in bold face. From Rabl and de Nazelle (2011), with permission of Elsevier.*

g/km	CO_2 at 20 km/h	CO_2 at 50 km/h	$PM_{2.5}$, EURO4 at 20 km/h	$PM_{2.5}$, EURO4 at 50 km/h	$PM_{2.5}$, EURO5 at 20 km/h
Gasoline cars	306.7	198.7	0.012	0.011	0.012
Diesel cars	250.0	177.0	0.050	0.039	0.013
50% gas+50% diesel	**278.3**	187.8	**0.031**	0.025	0.013

standards, under conditions of urban driving. EURO4 has been in force since January 2005, and EURO5, fully in force since January 2011.

Results are shown in Table 15.10. COPERT distinguishes between different cylinder sizes, but we show only simple averages over the respective cylinder sizes, because the $PM_{2.5}$ emissions per km are the same while the CO_2 emissions (which increase somewhat with cylinder size) are not our main focus here. We assume a rather low speed of 20 km/hr because of congestion in large cities; for instance the measured average speed in Paris is approximately 20 km/hr (EQT, 2004). Since a vehicle mix equivalent to 50% gasoline 50% diesel of EURO4 is fairly representative of the current situation in the EU, we take 0.031 g/km $PM_{2.5}$ and 278.3 g/km CO_2 for the present calculations. For short urban trips at 20 km/hr, the $PM_{2.5}$ emissions are about twice as large as the respective EURO limits in Table 15.1. For higher speeds or cars of more recent vintage the emissions would be lower, and the reader can readily scale the public health impact in proportion to the emissions of Table 15.10. Rural emissions are lower, but we do not bother to indicate these because their public health impact is so small as to be negligible, as shown at the end of Section 15.3.6 below.

15.3.5 Change in exposure for individuals who switch from car to bicycle or to walking

Several studies have measured the exposures of drivers and bicyclists on selected trajectories, for example in Paris, in Toulouse (France), in Arnhem (The Netherlands) (Zuurbier et al., 2010) and in Brussels, Louvain-la-Neuve and Mol (Belgium) (Int Panis et al., 2010). The data show that the change in exposure of individuals who leave their car to bicycle or to walk is extremely variable from one case to another.

However, as our calculations will show, this does not matter since the health impact of such changes is entirely negligible compared with the overall benefits of the physical activity.

As a starting point we take the concentrations that have been measured in streets of large cities. For European cities such data have been reported in Fig. 5.2 of EEA (2008). That figure shows annual average concentrations for monitoring stations along busy roads in major European cities: Vienna, Prague, Paris, Berlin, Athens, Krakow, Bratislava, Stockholm and London for NO_2, and Prague, Copenhagen, Berlin, Reykjavik, Rome, Bratislava, Stockholm and London for PM_{10}. Numbers for NO_2 are shown for each of the years 1999 to 2005; they vary slightly around 57 $\mu g/m^3$, without any clear long-term trend and significantly above the 40 $\mu g/m^3$ specified as upper limit by the air quality guidelines of the WHO (2005). Unfortunately the EEA report has no data for $PM_{2.5}$. Numbers for PM_{10} are shown for each of the years 2002 to 2005; they vary between 40 and 37 $\mu g/m^3$, with a slight declining trend. To estimate the corresponding values for $PM_{2.5}$, we multiply 38 $\mu g/m^3$ by a typical ratio of $PM_{2.5}/PM_{10} = 0.6$ to obtain 23 $\mu g/m^3$. This, too, is well above the WHO guideline of 10 $\mu g/m^3$.

The exposures encountered by the commuters depend on the detailed conditions of each trip. Concentrations inside a car tend to be higher than roadside concentrations, but in newer cars with good air filters the exposure can be much lower. A cyclist in the middle of a busy street is exposed to concentrations higher than at the side of the road, but on a separate bike path the exposure could be up to two times lower. Here we assume that the concentrations of $PM_{2.5}$ and NO_2 inside a car are 50% higher than the roadside concentrations measured by EEA, whereas the bicyclist is exposed to the roadside concentration. We also take the roadside concentration for pedestrians.

Whatever the exposure, one also has to account for the fact that the pollutant dose increases with the inhalation rate. Both the number of breaths per minute and the volume per breath increase (Int Panis et al., 2010). We assume that the dose is proportional to the total air intake, and that the latter is proportional to the metabolic rate. Metabolic rates are expressed as MET (metabolic equivalent), one MET being defined as 1 kcal/kg/hour, which is roughly equal to the energy cost of sitting quietly. Metabolic rates for different activities have been measured systematically, see e.g. Ainsworth et al. (2000). A detailed catalog of MET values[13] shows the following:

---

[13] http://prevention.sph.sc.edu/tools/docs/documents_compendium.pdf

transportation riding a car or truck:	1.0 MET
transportation automobile or light truck driving:	2.0 MET
walking, 2.5 mph (miles/hr), firm surface:	3.0 MET
walking, 2.0 mph, level, slow pace, firm surface:	2.5 MET
bicycling, <10 mph, leisure, to work:	4.0 MET
bicycling, 10–11.9 mph, leisure, slow, light effort:	6.0 MET

To determine the modifying factor for the DRF, we assume that the MET rate for driving is the same as the 24 hour population average that is implicit in the epidemiological studies of air pollution mortality. Based on these considerations we choose the following modifying factors to account for exposure (due to increased concentration) and dose (due to increased inhalation) during different transport modes. For cars, we assume that the concentrations are 50% higher than what is reported by the measuring stations of EEA (2008) because the latter are at the curb side and at about 2 m above street level, whereas drivers in busy streets are much closer to the exhausts of other cars. Such levels have been observed by measurements in cars by e.g. AIRPARIF (2009). For pedestrians, we assume the curb side data of EEA, together with a MET rate that is about twice the 24 hour population average. For bicyclists, we assume the curb side data of EEA, together with a MET rate that is about three times the 24 hour population average. Thus our modifying factors are: 1.5 for cars, 2 for pedestrians, and 3 for bicyclists. For the change of the health impact we assume proportionality with the exposure duration. This choice of modifying factors is somewhat arbitrary, but for any reasonable choice the effect turns out to be negligible compared with the health benefit of the physical activity.

15.3.6   Impact of car emissions on the general public

To estimate the mortality impact for the general population, we use ExternE (2000) rather than HEATCO (2006). Specifically we take the results for the seven urban sites in Table 13.26 of ExternE (2000), shown here in columns two to four of Table 15.11. They include all health end points.

In ExternE (2000), mortality contributed 71% of the total. As explained in Section 5.7.3 of that report, it was calculated with an ERF of 1.57E-04 YOLL/person/year per $\mu g/m^3$ of PM_{10}, which we divide by 0.6 to convert to 2.61E-04 YOLL/person/year per $\mu g/m^3$ of $PM_{2.5}$, and it was valued at VOLY of 96,500 ϵ_{2000}. Here we take only mortality, and we use the ERF and monetary valuation of ExternE (2008), namely an ERF of 6.51E-04

Table 15.11 *Results for the damage cost in* ϵ_{2000}/kg *(columns 2 to 4), of* $PM_{2.5}$ *emitted by cars in 7 countries of the EU, as calculated by ExternE (2000). The last column shows the cost of mortality in large cities, obtained by multiplying column 4 by the adjustment factor of Eq. 15.1.*

Site[a]	Local ϵ_{2000}/kg	Regional[b] ϵ_{2000}/kg	Total ϵ_{2000}/kg	Local/Total	Mortality ϵ_{2010}/kg
Brussels (1.0, 1.8)	388.5	30.1	418.6	0.93	336.1
Helsinki (0.6, 1.3)	170.5	4.3	174.8	0.98	140.4
Paris (2.2, 11.8)			1170.0		939.4
Stuttgart (0.6, 5.3)	193.0	29.7	222.7	0.87	178.8
Athens (0.7, 3.1)	916.8	10.0	926.8	0.99	744.1
Amsterdam (1.4, 6.7)	361.9	22.1	384.0	0.94	308.3
London (7.6, 13)	675.0	30.1	705.1	0.96	566.1
Average					**459.0**

[a] the numbers next to the city name indicate the population in million, of the city and the metropolitan area, mostly based on Wikipedia (the definitions of city and metropolitan area are not uniform).
[b] in Table 13.26 of ExternE (2000) the sum of Local and Regional is slightly larger than Total because of overlap of population grids; since the numbers for Total are correct, we have slightly reduced the ones for Regional to eliminate this overlap.

YOLL/person/year per $\mu g/m^3$ of $PM_{2.5}$ and, after adjustment for inflation, a VOLY of 43,801 ϵ_{2010}. Thus we multiply the damage costs of column 4 of Table 15.11 by an adjustment factor,

$$\text{Adjustment factor} = 0.71 \times (6.51E{-}04/2.61E{-}04)$$
$$\times (43{,}801/96{,}500), \tag{15.1}$$

to obtain the last column. In the following we take the mean for large cities, 459 €/kg of $PM_{2.5}$. For the rural data of ExternE (2000) (not shown here) we find a mean of 28.3 €/kg of $PM_{2.5}$.

15.3.7 Accidents

Changes in accidents are difficult to estimate, because they are extremely dependent on the specifics of the change: even though bicyclists are more vulnerable than drivers, their accident risk can become very small if bike paths are provided or if bicycling is as widely adopted as in the Netherlands or Denmark (in Amsterdam and Copenhagen more than a third of the commuters use the bicycle). In these countries bicycle accidents are much lower than elsewhere because traffic management is better adapted to bicycling and both drivers and bicyclists have learned to coexist – there is safety in numbers. Referring for details to the paper by Rabl and de Nazelle, we take the fatal accident rates of Amsterdam and Paris as lower and upper bounds, as shown in Table 15.12.

15.3.8. Results

The steps of the calculations for an individual who switches from car to bicycle are shown in Table 15.12 and the results are plotted in Fig. 15.9. The calculation for drivers who switch to walking are similar.

For our walking scenario the benefit of PA is 1192 €/yr. The public benefit is only 16.5 €/yr, because the trip is half as long as for bicycling. The change in pollution exposure and intake implies a cost of 15 €/yr for the individual. We have not evaluated a possible change in accident risk for walking.

The error bars in Fig. 15.9 indicate one standard deviation confidence intervals. For the gain from PA they are calculated by repeating the life table calculation with the lower and upper bounds (0.094 and 0.447) of (1-RR) of the ERF for bicycling. For pollution, we estimate the confidence intervals according to Section 11.3.1. For fatal accidents, the error bars indicate the range between the values for Amsterdam and Paris. We do not include the uncertainty of the monetary valuation in these error bars because it affects the costs in the same manner (although for accidents there is an additional uncertainty due to the ratio VPF/VOLY). The reader can readily scale the graph for a different valuation of mortality. For the uncertainty of the latter we estimate that the valuation could be a factor of two higher or lower.

15.3.9 Discussion

In spite of the uncertainties, and whatever one assumes about the scenarios and the impacts of car emissions, the key conclusions about the health impacts are not affected: by far the most important item is the health benefit due to physical activity. The benefit for the general population due to reduced air pollution is much smaller, and in large cities it is larger than the cost due to changed exposure for a driver who switches from car to bicycle; in small cities or rural zones the public benefit is small or negligible. The exposure change for the individuals who switch implies a loss with our assumptions, but could be a gain if the bicycle can travel on a path with lower pollution. The concern about pollution exposure of bicyclists, often evoked in the context of bicycling in cities, is unfounded when compared with the benefits of the cycling activity; of course, such

Table 15.12 *Calculations and results for mortality impacts of a switch from car to bicycle.*

Item	Value	Unit	Explanation
Health gain from PA			
RR	0.709		Health gain of individual due to physical activity
LE gain	1.20	Yr	Solid circle in Fig. 15.8
Lifetime benefit	52418	€	Life table calculation for EU25
Benefit per year	**1310**	**€/yr**	LE gain × VOLY
			Lifetime benefit/40 yr
Public health gain			
PM$_{2.5}$ emission/km	0.031	g/km	Due to reduced emission of pollution
Length of trip	5	km	Table 15.10, average diesel and gasoline EURO4
Number of trips/yr	460	/yr	One way
PM$_{2.5}$ emission/yr	71.8	g PM$_{2.5}$/yr	2 × 5 trips/week, 52–6 weeks/yr
Avoided damage cost	459.0	€/kg of PM$_{2.5}$	Avoided emissions due to shift to bicycling
Benefit per year	**33**	**€/yr**	Table 15.11, average large cities
Change of individual dose[a]			
Concentration	23	μm^{-3}	Due to change in exposure and intake
ERF	0.000651	YOLL/(pers.yr.μg/m^3)	Concentration of PM$_{2.5}$ in street
Duration – car	0.25	hr/trip	Slope of ERF for mortality due to PM$_{2.5}$
Modifying factor – car	1.5		Duration of car trip
			For exposure and inhalation of driver, relative to ERF of general population
Cost – car	4.30[c]	€/yr	Avoided cost, relative to general population
Duration – bicycle	0.33	hr/trip	Duration of bicycle trip
Modifying factor – bicycle	3		For exposure and inhalation of bicyclist, relative to ERF of general population
Cost – bicycle	22.9[c]	€/yr	Cost increase relative to general population
Benefit per year	**–18.6**	**€/yr**	Negative, i.e. cost, of exposure change car – bicycle

Table 15.12 (cont.)

Item	Value	Unit	Explanation
Fatal accidents[b]			
Accident rate	6.6E-05	accidents/yr per bicyclist	Increased mortality due to accidents
Accident rate	2.5E-05	accidents/yr per bicyclist	Paris
Cost/accident	1.6	M€$_{2010}$	Amsterdam
Benefit per year	**-72.3**	**€/yr**	VPF
			Average of -39, Amsterdam, and -105, Paris
			Negative, i.e. cost, of risk change car – bicycle

[a] highly dependent on details of trajectory, could even have opposite sign

[b] highly dependent on details of trajectory and behavior of drivers and bicyclists in the city

[c] the ratio cost-bicycle/cost-car is proportional to $[(3-1) \times 0.33]/[(1.5-1) \times 0.25]$, not $[3 \times 0.33]/[1.5 \times 0.25]$, because these are changes relative to the background mortality risk

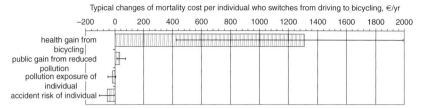

Fig. 15.9 Results for mortality costs and benefits per individual who switches from car to bicycle for commuting to work (2 × 5km roundtrip, 5 × 46 weeks/yr) in large cities of EU.

Error bars indicate one standard deviation confidence intervals.

In view of the meta-analysis by OECD (2012) (see Section 9.3.1) the monetary amounts should be doubled. From Rabl and de Nazelle (2011), with permission of Elsevier.

exposure should be minimized as far as is practical. Accidents can be a more serious problem, however, and more should be done to reduce the risks.

The conclusions about the relative magnitude of the effects also hold for individuals who switch from driving to walking. Incidentally, the role of physical activity (walking to the station, standing, climbing stairs to the subway) is not negligible when people switch from driving to public transportation and the associated benefits may well outweigh the increased exposure to PM that has been observed in subways and many buses.

Our results for the effects of pollution are entirely consistent with the site specific calculations of de Hartog *et al.* (2010) and Woodcock *et al.* (2009), but they are more general because we have considered many sites. Our estimate of the LE gain due to bicycling is about twice as large as that of de Hartog *et al.* because our life table calculation considers the full steady-state benefit, attained when the entire 20 to 60 yr cohort has been bicycling. In the near term the benefit is smaller because the risk reduction is applied only for a limited number of years.

We have considered only mortality. Had we included morbidity end points, the numbers for public and individual air pollution impacts would be about 50% larger according to the ERFs and monetary values of ExternE (2008). Since the health benefits of physical activity span a wider variety of important end points than for air pollution, the value of the benefit may be increased by more than 50%, but we have no specifics to support this possibility. The cost of bicycle accidents would be very much larger than our numbers, as demonstrated by a detailed investigation of nonfatal bicycle accidents in Belgium by Aertsens *et al.* (2010). These authors find that the average cost of such accidents is 0.125 € per km bicycled. Applied to our scenario this implies a cost of 286 €/yr for the individuals who switch to bicycling.

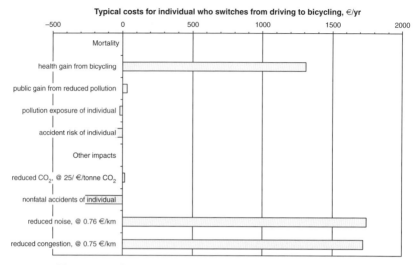

Fig. 15.10 Comparison of mortality costs and benefits with other impacts, for our bicycling scenario.

In view of the meta-analysis by OECD (2012) (see Section 6.3.1) the mortality benefits should be doubled. From Rabl and de Nazelle (2011), with permission of Elsevier.

In addition to health, such a switch can bring several other important benefits, especially reduced congestion and reduced street noise. For these we refer to estimates in CE Delft (2008), as shown there in Tables 6.2 and 6.3. For our bicycling scenario, we choose a noise cost of 0.76 €/km and a congestion cost of 0.75 €/km.

In Fig. 15.10 we show what these numbers imply for our bicycling scenario. Typical average benefits from reduced congestion and noise may well be even larger than the health gain from physical activity. In this figure we have also added the benefit of reduced greenhouse gas emissions, assuming 25 €/t_{CO2}. But compared with all the other items it is negligible, because the distances are small compared with the total distance driven by cars during the year.

15.4   Greenhouse gas emissions by car, plane and train

Numerous websites offer to calculate the carbon footprint of travel by car, plane or train. Of course, such calculations make various assumptions, in particular the number of passengers per vehicle. The fuel consumption of cars varies strongly with size of car. The emissions by train depend on

Table 15.13 *Greenhouse gas emissions of different travel modes for long distance travel, calculated by LIPASTO.*[14] *Essentially all is due to CO_2, the contribution of CH_4 and N_2O being less than 1% in all cases.*

	CO_{2eq}, g/vkm	CO_{2eq}, g/pkm
Car[a]	170	89[b]
Train[c]	3029	25[d]
Plane: Europe, short-distance		260[e,f]
Plane: Europe, long-distance		149[e,g]

[a] gasoline cars, EURO2 and EURO3, also fleet average in 2009, highway driving, fuel consumption 7.1 L/100 km.
[b] 1.9 passengers/car.
[c] electric fast train; tilting train technology (for greater speeds on conventional tracks); 12.5 kWh_e/vkm.
[d] 309 seat/train, 40% load factor
[e] CO_2 emission factor: 3.169 kg/(kg fuel)
[f] short-distance < 463 km, long-distance > 463 km
[g] 155 passengers for scheduled flights, 84 for charter flights

whether the train is diesel or electric and how the electricity is produced. It does not come as a surprise that the results from different websites can differ by more than a factor of two. Here we have chosen results from the LIPASTO-calculation system, created by the VTT Technical Research Centre of Finland, because we find the results the most plausible and the assumptions are clearly presented.

LIPASTO provides quite a bit of detail, including the emission of other air pollutants. Here we only look at greenhouse gas emissions, because we would not know what to do with the other emissions of airplanes. Especially important are their NO_x emissions and their costs would have to be calculated as a function of the emission height above ground, for the most part far above the mixing layer height where the usual atmospheric models are no longer applicable. Such calculations have not been done. So we show only a small subset of the LIPASTO numbers, in Table 15.13. To account for occupancy, the cost is also stated per passenger km (pkm), in addition to vehicle km (vkm).

Example 15.4   Compare travel from Paris to Frankfurt, by car, plane and train.

---

[14] www.nap.edu/catalog.php?record_id=12794

Solution: By car the distance is 572 km and the travel time is estimated as 6 hr, as per Google Maps. For travel by a single person the emissions are 572×170 kgCO$_{2eq}$ = 97.2 kgCO$_{2eq}$.

By train the travel time is 3:49 hr by high speed train, and the emissions are 572×25 kgCO$_{2eq}$ = 14.3 kgCO$_{2eq}$.

By plane the flight time is about 1:10 hr, but one has to add the travel to and from the airport as well as the requirement to arrive well before departure. Since the flight distance, 471 km, is just around the separation between short and long distance in Table 15.13, we take the average and estimate the emissions as $471 \times (260+149)/2$ kgCO$_{2eq}$ = 96.3 kgCO$_{2eq}$.

Comment: of course in practice CO$_2$ emissions are not the primary criterion for most people. Usually cost, convenience and the value of time matter the most. The marginal cost of travel by car is at least the 572 km $\times$ 7.1 L/100 km $\times$ 1.40 €/L = 56.86 € for gasoline. The price of a train or plane ticket depends on various factors, especially whether it is bought in advance or not. For the train we found a range of 39 € to 110 €. For many destinations served by low cost carriers we found that travel by plane can be the cheapest if one is flexible for the date. If the cost of CO$_2$ were internalized by a tax of 25 €/t$_{CO2}$, the price of a plane ticket would increase by 96.3 kg$_{CO2eq}$ $\times$ 25 €/t$_{CO2}$ = 2.41 €, in any case a very small percentage of the ticket price.

15.5   Internalization of damage costs of transport

For greenhouse gases the simplest internalization would be a tax per tonne of CO$_{2eq}$. For the other air pollutants the difficulty lies in the site dependence of the damages. Ideally one should know where how much of each pollutant is emitted. This is not practical with current technologies. As a simple compromise one could calculate the tax for each motor vehicle on the basis of the emission tests that are (or should be) required and the km driven since the last test, assuming that it has been driven according to the typical driving pattern, for the location where the car is registered.

As an alternative, one could impose a single tax rate in an entire country or region and internalize the rest of the damage in urban areas by specific city charges that also include charges for noise and congestion. The numbers in Tables 6.2 and 6.3 and Fig. 15.10 suggest that externalities due to noise and congestion in large cities are much larger than those due to pollution.

At the present time (2013) congestion charges are in effect for the central areas of a few large cities, in particular London, Singapore and

Stockholm. The numbers in Tables 6.2 and 6.3 indicate that noise externalities in large cities can be around 0.76 €/vkm during the day and 1.39 €/
vkm at night. The marginal cost of congestion could be around 2.00 €/
vkm in the central zones of large cities, according to Table 6.3. The
amount of the charge in London is £10 per day, an amount that seems
quite appropriate in view of these numbers. In cities with such charges the
traffic in the respective zones has been reduced by 10 to 30%, and air
pollution levels have decreased.[15]

In spite of all the good reasons for such charges, in most cities analogous
proposals have been rejected. Unfortunately most voters perceive only the
cost, not the benefit. There is in fact a serious equity problem: a congestion charge affects rich and poor very differently. The rich can easily
pay it and continue driving, while the poor feel the pressure to find
alternatives. However, in view of the benefits of congestion charging,
more effort should be made to design charging schemes that are accepted
by voters and that minimize equity problems.

It is interesting that many cities in Europe have created low emission
zones (LEZ) where older more polluting vehicles are not allowed to
circulate. The decision whether or not to create such zones is rarely
based on scientific analysis. Political considerations seem to carry the
day. For example, there are numerous LEZ in Germany, but the Swiss,
very much aware of what is going on in Germany and why, decided that it
would be too costly.

Only for two cities, London and Antwerp, has a CBA been carried out.
For Antwerp the conclusion was that the proposed scheme would bring
significant net benefits. The study for London (2003) analyzed three
effects: health, administrative costs, and costs for the owners of vehicles
that are excluded. The administrative costs are negligible compared with
the other two categories. At the time of the study only the exclusion of
trucks and buses was under consideration. For these the study found that
the benefit is comparable to the cost. However, in view of the magnitude
of noise costs in Table 6.2 it seems likely that the benefit could be much
larger. The UK assumptions on air pollution damage are also conservative
compared to those that are adopted elsewhere (e.g. in analyses for the
European Commission and the USEPA).

In France the creation of LEZ is currently under discussion, because
several cities do not meet the air quality guidelines of the European
Commission and risk having to pay penalties. One of the authors (A. R.)
has a contract to carry out a CBA, but has no definitive results yet. Several

---

[15] http://en.wikipedia.org/wiki/Congestion_pricing.

options for the definition of the LEZ (geographic extent and class of vehicles to be excluded) are on the table. Generally the restrictions would affect only diesel vehicles, but that is more than half of the passenger cars and light-duty vehicles because the government has for many years already encouraged the use of diesel by a lower fuel tax compared to the tax on gasoline.

LEZs are an example of policy options for which a simple CBA is not sufficient because of the very unequal distribution of costs and benefits. The public at large benefits but the cost is borne by the owners of old vehicles. Even if they can sell, the price is lower than in the absence of LEZs because of reduced demand for old vehicles. A system of adequate compensation for the losers by the winners would be difficult to devise. For this reason LEZs are a rather problematic option for reducing air pollution in cities, in contrast with emission controls for power plants and waste incinerators where the costs are shared more or less equally because everybody uses electricity and produces waste.

As another example of internalization, there are proposals to tax imported goods for the pollution emitted in their countries of origin, if these countries have not internalized the damage costs of their pollution. These ideas often go under the subject of "embodied pollution." Such pollution should really be taxed in order to avoid the incentive for companies to move their production to countries with lesser environmental regulation.

Example 15.5   How much would the price of a car increase if the CO_2 emitted during its fabrication is taxed at 30 €/t$_{CO2}$?

Solution: The required information can be found in Fig. 15.4, which shows that CO_2 contributes about 80 €/car to the external cost of the production of mid-sized cars. Since for this graph a carbon cost of 19 €/t$_{CO2}$ was assumed, as per Table 15.8, we multiply by 30/19 to obtain a carbon cost of 126 €/car.

References

Aertsens, J., de Geus, B., Vandenbulcke, G. et al. 2010. Commuting by bike in Belgium, the costs of minor accidents. *Accident Analysis and Prevention* **42**: 2149–2157.

Ainsworth, B. E., Haskell, W. L., Whitt, M. C. et al. 2000. Compendium of Physical Activities: An update of activity codes and MET intensities. *Medicine and Science in Sports and Exercise* 2000;**32** (Suppl):S498–S516. http://prevention.sph.sc.edu/tools/docs/documents_compendium.pdf

AIRPARIF 2009. AIRPARIF Actualité N° 32, February 2009. Paris.

AMM 2003. American Metal Market. www.amm.com/index2.htm?/ref/carmat98.htm.

Andersen, L. B., Schnohr, P., Schroll, M. and Hein, H. O. 2000. All-cause mortality associated with physical activity during leisure time, work, sports and cycling to work. *Archives of Internal Medicine* **160**(11): 1621–1628.

ANL 2004. Well-to-wheel analysis. Argonne National Laboratory, Center for Transportation Research. Available at http://transtech.anl.gov/v2n2/well-to-wheel.html.

Barth, M. and Boriboonsomsin, K. 2009. Traffic congestion and greenhouse gases. *ACCESS*: the Magazine of University of California Transportation Center, number **35**: 2–9.

Bishop, G. A. and Stedman D. H. 2008. A decade of on-road emissions measurements. *Environ. Sci. Technol.* **42**: 1651–1656.

CE Delft. 2008. *Handbook on estimation of external costs in the transport Sector*. Produced within the study Internalisation Measures and Policies for All external Cost of Transport (IMPACT), Version 1.1. CE Delft, February, 2008. http://en.wikipedia.org/wiki/Congestion_pricing

CE Delft 2011. *External Costs of Transport in Europe: Update Study for 2008*. CE Delft, Oude Delft 180, 2611 HH Delft, The Netherlands. www.cedelft.eu/publicatie/deliverables_of_impact_%28internalisation_measures_and_policies_for_all_external_cost_of_transport%29/702

Chen, H., Goldberg, M. S. and Villeneuve, P. J. 2008. A systematic review of the relation between long-term exposure to ambient air pollution and chronic diseases. *Reviews On Environmental Health* **23** (4): 243–297.

CONCAWE 1998. *A Study of the Number, Size & Mass of Exhaust Particles Emitted from European Diesel and Gasoline Vehicles under Steady State and European Driving Cycle Conditions*, CONCAWE Report #98/51, Madouplein 1, B-1210 Brussels, Belgium.

de Hartog, J. J., Boogaard, H., Nijland, H. and Hoek, G. 2010. Do the health benefits of cycling outweigh the risks? *Environ Health Perspect* **118** (8): 1109–1116.

Delucchi, M. A. 2003. *A lifecycle emissions model (LEM): lifecycle emissions from transportation fuels, motor vehicles, transportation modes, electricity use, heating and cooking fuels, and materials*. UCD-ITS-RR-03–17, Institute of Transportation Studies, University of California, Davis, CA 95616.

Desaigues, B., Ami, D., Bartczak, A. *et al.* 2011. Economic valuation of air pollution mortality: a 9 – country contingent valuation survey of value of a life year (VOLY). *Ecological Indicators* **11**(3): 902–910.

EEA 2008. *Climate for a transport change. TERM 2007: indicators tracking transport and environment in the European Union*. EEA Report No 1/2008. European Environment Agency.

EQT 2004. *Les déplacements des franciliens en 2001–2002*. Enquête globale des transports. Plan de Déplacements Urbains. Direction Régionale de l'Equipement Ile-de-France.

ExternE 2000. External Costs of Energy Conversion – Improvement of the Externe Methodology And Assessment Of Energy-Related Transport Externalities. Final Report for Contract JOS3-CT97–0015, published as *Environmental External Costs of Transport*. R. Friedrich & P. Bickel, editors. Springer Verlag Heidelberg 2001.

ExternE 2005. *ExternE – Externalities of Energy: Methodology 2005 Update.* www. cedelft.eu/publicatie/external_costs_of_transport_in_europe/1258

ExternE 2008. With this reference we cite the methodology and results of the NEEDS (2004–2008) and CASES (2006–2008) phases of ExternE. For the damage costs per kg of pollutant and per kWh of electricity we cite the numbers of the data CD that is included in the book edited by Markandya, A., Bigano, A. and Porchia, R. in 2010: *The Social Cost of Electricity: Scenarios and Policy Implications.* Edward Elgar Publishing Ltd, Cheltenham, UK. They can also be downloaded from http://www.feem-project.net/cases/ (although in the latter some numbers have changed since the data CD in the book).

Grange, D. and Host, S. 2012. *Pollution de l'air dans les enceintes souterraines de transport ferroviaire et santé.* Observatoire Régional de Santé Île-de-France, 43, Rue Beaubourg 75003 Paris. June 2012.[16]

HEATCO 2006. *Developing Harmonised European Approaches for Transport Costing and Project Assessment (HEATCO),* Deliverable D5: Proposal for Harmonised Guidelines. EC Contract No. FP6-2002-SSP-1/502481. P. Bickel et al. Stuttgart : IER, Germany, Stuttgart, 2006.

Hickman J., Hassel, D., Joumard, R., Samaras Z. and Sorenson, S. 1999. *Methodology for calculating transport emissions and energy consumption for the Project MEET* Report for Contract ST-96-SC.204 European Commission / DG VII. Transport Research Laboratory, Old Wokingham Road, Crowthorne, RG45 6AU, United Kingdom.

Int Panis, L., de Geus, B., Vandenbulcke, G. *et al.* 2010. Exposure to particulate matter in traffic: A comparison of cyclists and car passengers. *Atmospheric Environment* **44**: 2263–2270.

Laden F., Neas, L. M., Dockery, D. W. and Schwartz, J. 2000. Association of fine particulate matter from different sources with daily mortality in six U.S. cities. *Environmental Health Perspectives – New Series* **108** – issue 10: 941–948.

London 2003. The London Low Emission Zone Feasibility Study A Summary of the Phase 2 Report to the London Low Emission Zone Steering Group 2003. www.ors-idf.org/dmdocuments/2012/Synthese_pollution_enceintes_souterrai nes_web.pdf

MacLeana, H.L. and Lave, L.B. 2003. Evaluating automobile fuel/ propulsion system technologies. *Progress in Energy and Combustion Science* **29**: 1–69.

Muller, N. Z. and Mendelsohn, R. 2007. Measuring the damages of air pollution in the United States. *Journal of Environmental Economics and Management* **54**: 1–14.

NRC 2010. *Hidden Costs of Energy: Unpriced Consequences of Energy Production and Use.* National Research Council of the National Academies, Washington, DC. Available from National Academies Press. www.tfl.gov.uk/assets/downloads/ roadusers/lez/phase-2-feasibility-summary.pdf

OECD 2012. *Mortality Risk Valuation in Environment, Health and Transport Policies.* OECD Publishing. http://dx.doi.org/10.1787/9789264130807-en

Pope, C. A., Hill, R. W. and Villegas, G. M. 1999. Particulate air pollution and daily mortality on Utah's Wasatch Front. *Environmental Health Perspectives* **107** (7): 567–573.

---

[16] http://www.externe.info

Rabl, A. and de Nazelle, A. 2011. Benefits of shift from car to active transport. *Transport Policy* **19**: 121–131.

Reiss, R., Anderson, E. L., Cross, C. E. *et al.* 2007. Evidence of health impacts of sulfate- and nitrate-containing particles in ambient air. *Inhalation Toxicology* **19**: 419–449.

Schwartz, J., Norris, G., Larson, T. *et al.* 1999. Episodes of high coarse particle concentrations are not associated with increased mortality. *Environmental Health Perspectives* **107**(5): 339–342.

Spadaro, J. V. and Rabl, A. 2006. *Hybrid Electric vs Conventional Vehicle: Life Cycle Assessment and External Costs.* 2nd conference 'Environment & Transport' including 15th conference 'Transport and Air Pollution'. Reims, France, 12–14 June 2006.

Stodolsky, F., Vyas, A., Cuenca, R. and Gaines, L. 1995. *Life Cycle Energy Savings Potential from Aluminum-Intensive Vehicles* Conference Paper presented at the 1995 Total Life Cycle Conference & Exposition, Vienna, Austria, October 16–19, 1995.

US DHHS 2008. *Physical Activity Guidelines Advisory Committee Report, 2008.* Physical Activity Guidelines Advisory Committee. Office of Public Health and Science, U.S. Department of Health and Human Services. Washington, DC 20201.

Vossiniotis, G., Arabatzis, G. and Assimacopoulos, D. 1996. *Description of ROADPOL: A Gaussian Dispersion Model for Line Sources,* program manual, National Technical University of Athens, Greece.

Weiss, M. A., Heywood, J. B., Drake, E. M., Schafer, A. and AuYeung, F. F. 2000. *On the road in 2020: a life-cycle analysis of new automobile technologies.* Energy Laboratory Report # MIT EL 00–003, Energy Laboratory, Massachusetts Institute of Technology, Cambridge, Massachusetts 02139-4307.

WHO 2005. Air Quality Guidelines for Europe. www.who.int/mediacentre/fact sheets/fs313/en/index.html (accessed 21 June 2010).

WHO 2008. Methodological guidance on the economic appraisal of health effects related to walking and cycling. Health Economic Assessment Tool for Cycling (HEAT for Cycling). User Guide, version 2. WHO Regional Office for Europe, Coperhagen, Denmark.

WHO 2010. *Development of Guidance and a Practical Tool for Economic Assessment of Health Effects from Walking.* Consensus Workshop, 1–2 July 2010, Oxford, UK. World Health Organization, Europe.

WHO 2013. HRAPIE: Health risks of air pollution in Europe. Recommendations for concentration–response functions for cost–benefit analysis of particulate matter, ozone and nitrogen dioxide. WHO Regional Office for Europe, Bonn, Germany.

Woodcock, J., Edwards, P., Tonne, C. *et al.* 2009. Public health benefits of strategies to reduce greenhouse-gas emissions: urban land transport. *Lancet* **374**(9705): 1930–1943.

Zuurbier, M., Hoek, G., Oldenwening, M. *et al.* 2010. Commuters' exposure to particulate matter air pollution is affected by mode of transport, fuel type and route. *Environ Health Perspectives* **118**: 783–789.

16   Lessons for policy makers

Summary

This chapter provides an overview of issues linked to damage cost assessments, drawing on the material provided in earlier chapters. It also provides a list of applications of external costs analysis to demonstrate that the approaches outlined here are part of the policy toolkit for many authorities. Much of the focus is on applications within Europe, the area with which the authors are most familiar, although applications in other parts of the world are also discussed. The examples provided demonstrate a great breadth in policy applications, covering not only environmental quality standards but also the energy, industry, waste, transport, chemicals and domestic sectors.

16.1   Choice of method

There is a range of methods for sustainability appraisal of projects and strategies, as noted in Chapter 2 of this book. In addition to the impact pathway approach (IPA) to the quantification of externalities and associated cost–benefit analysis (CBA) they include life cycle analysis (LCA), cost-effectiveness analysis (CEA) and multi-criteria decision analysis (MCDA). Work performed for the Sustools project (Rabl *et al.*, 2004) reviewed these methods and came to the view that each has a role to play: individually they all have their limitations, but used together they can provide a thorough overview of issues and present information in a form that is directly relevant to the decision making process. This is an important lesson for policy makers, that no matter how convinced an analyst is of the superiority of his or her own method, for most policy applications a single tool is unlikely to provide all of the answers.

A second lesson is that it is important to stand back before commencing analysis and develop an overview of the possible consequences of actions. An important example concerns current plans for climate controls, as these have a very wide range of consequences for society in the form of co-benefits and tradeoffs (i.e. impacts that will not show up in a simple cost–benefit analysis that equates only the change in climate impacts with the direct costs of planned emission controls. There is a huge diversity of these co-benefits and tradeoffs (see Smith, 2013). Effects of reduced fossil fuel use, for example, include the following:

- Reduced emissions of the regional pollutants that are the primary focus of this book
- Change in occupational health risks
- Changes in major accident risks
- Changes in demand for land, e.g. to grow biofuels
- Impacts on landscape through the large-scale development of wind farms and other renewable technologies
- Changes in demand for rare metals, e.g. for photovoltaic technologies
- Changes in energy security at a national level
- Changes in energy poverty (considered when energy costs exceed a certain level of household income)
- And so on.

Returning to the first point made above: In this case, LCA will provide valuable information on emissions and resource use, but nothing on accident risks, landscape effects, etc. The IPA takes the analysis of impacts further to quantify specific health impacts of pollution and accident risks. MCDA can assist with the integration of otherwise unquantified effects, and so on. Each method plays its part and the combined analysis is far more comprehensive than would be possible if only single methods were used.

It might be thought that the issues just raised are peculiar to climate policy assessment, given its known complexities. However, this is not the case. Drawing on the lessons of the Sustools project, Holland et al. (2009) applied a variety of methods in the context of further regulation of non-methane volatile organic compounds (NMVOCs) for the UK government. The first driver for this research was recognition that European policies directed at ground level ozone control could be made more efficient if they were targeted at those VOCs that were emitted in significant quantity and had a high photochemical oxidant creation potential (POCP), in other words, those most responsible for generation of ozone from anthropogenic emissions rather than the simple mass of NMVOC emitted. The need to account for this was strengthened by the large fall in emissions of NMVOCs that had been achieved in the UK; a 64% fall between 1990 and 2007. This implied that most of the lowest cost measures for abatement of NMVOCs collectively were already in place and hence that additional measures would need closer targeting to pass a benefit:cost test. It was also noted by policy makers that these pollutants were relevant to a number of different policy areas each of which considered at best a subset of impacts, for example:

- Direct chemical effects on human health (including cancers), potentially affecting both workers and the general public depending on exposure routes
- Other occupational risks from use of VOCs (e.g. fire hazard)
- Direct chemical effects on ecosystems
- Global warming effects
- Ground level ozone formation
- Stratospheric ozone depletion
- Formation of secondary organic aerosols with associated health impacts
- Life cycle burdens generated by VOC production, use and disposal (e.g. energy use and release of pollutants to air, land and water).

The failure to assess all of the impacts simultaneously creates the potential for inconsistency in policy, with different policies having quite different priorities for VOC control. It also generates potential inefficiency for industry and additional burdens on society and the environment.

Climate policies, and policies in many other areas, cannot therefore be seen as one-dimensional as far as impacts are concerned. The co-benefits and tradeoffs need to be factored into decision making, and the use of external costs analysis is a useful approach for doing that (Forster et al., 2013). It cannot answer all of the questions, but it can answer some of them in a policy-relevant form (economic value). By providing a partial quantification it is possible to bring other effects into the equation, considering which are likely to be significant and which are likely to be insignificant relative to those that have been quantified. Gradually, a more complete picture of effects emerges. Some effects will defy quantification for the foreseeable future and some of them, like effects of air pollutants on ecosystems and landscape impacts of renewable technologies may well be important. However, impact pathway *thinking*, even with little or no quantification can help to express them in a way that is more policy relevant. Impact pathway thinking may also prompt analysts to provide at least some data that better describe the impact. This thinking may take the following form (taking the example of the impacts of wind farms on landscape/visual amenity):

1. How many wind turbines would be needed to meet the objectives of the policy in question? And from that, how much land would be required for the turbines and associated infrastructure including additional transmission lines?

2. Where would the wind industry like to locate the turbines? This could be extended to a classification of sites according to the preference of the industry (accounting for windiness, accessibility and so on).
3. What is preferable – a few very large wind farms, many smaller wind farms, or something in between? This question also links with issues over the size of each turbine – one may prefer to adopt smaller turbines though this would be at the expense of geographically more extensive wind farms and higher cost.
4. Over what distance would wind farms be visible?
5. How visible would the wind farms be from areas with high visitor numbers or areas with a high existence value relative to their unspoilt beauty?
6. How long would wind farms last? Are they likely to be a permanent feature of the landscape, or, like the windmills of old, a stop-gap technology that will be phased out when more energy-dense and less variable alternatives such as wave or nuclear fusion become available?

None of this approaches valuation, but together it provides a structure for understanding the nature of the impact under investigation and its effects on society.

Further to this, the simple identification of co-benefits and tradeoffs develops opportunity to enhance the former and mitigate the latter. Without this knowledge, some climate policies will present something of a "hostage to fortune" situation, which is clearly not a driver for optimal policy development. Policies that are seen to generate unanticipated negative impacts inevitably attract a great deal of criticism and negative publicity for those who have developed them.

Returning to the question of the choice of method for the analysis, the examples shown here, and others presented in this book (e.g. the analysis of externalities for electric, hybrid and standard gasoline cars in Chapter 15) highlight the benefits that can be gained by bringing together different tools and using them in combination. It is of course not necessary to use the full toolbox in every case. A preliminary screening phase can be useful in identifying where priorities lie for the assessment and the method used can be tailored accordingly.

16.2 Trends in the use of valuation over time

Whilst there has been interest in valuation of the impacts of pollutants for many years, the use of valuation in the decision making process is still somewhat limited. Whilst it is very pronounced in some countries (e.g. the UK and the USA), in others, alternative approaches to the justification of environmental legislation have dominated in the decision making process.

Within Europe the use of valuation is growing, not only in relation to the issues addressed specifically in this book (air quality including its links to green accounting, energy, transport, etc.). One specific area of growth concerns valuation in the context of the REACH (registration, evaluation, authorisation and restriction of chemicals) regulation. This is a particularly challenging sphere given the diversity of chemicals that need to be addressed and the lack of information in many areas. When considering lessons for policy makers it is instructive to consider how valuation has been used over time.

As with many things associated with air pollution, the systematic use of economics in the field can be traced back to the Great London Smog of 1952. A thorough account is provided by the Office of the Mayor of London (2002), reflecting on the events around the smog and the response to it. London's problems had persisted for centuries. Whilst to many, the stories of 60 years past may seem nothing more than a historic footnote, we believe that important lessons were learned that are still relevant today, relating not just to the specific policies enacted but also to the way in which evidence was gathered, opinions formed and decisions reached.

Amongst the earliest legislation in the UK against pollution was a ban on the burning of "sea coale"' in London dating to 1257, though it appears to have had no effect. In 1819, the poet Shelley had written that *"Hell is a city much like London, a populous and smoky city"*. Hence, in the eyes of the government, London had always had smog and always would. Davis (2002) sums up the attitude of the authorities at the time of the great London smog by stating that the government *"sprang into inaction"*, setting up a committee (the Beaver Committee on Air Pollution, under the chairmanship of Sir Hugh Beaver) to investigate the problem, although considering it unlikely that its recommendations would have any effect. The government's irritation when pressed on the issue was clear, with Ian Macleod, the Minister for Health, reported as saying *"Really you know, anyone would think fog had only started in London since I became a Minister."* Suspicions were cast that the 4000 extra deaths identified in the week of the smog were a result of a flu epidemic, though this was quite unfounded (Bell *et al.*, 2004). Bell also reports that the number of deaths directly linked to the smog should be increased to about 12,000.

Some may think that we know enough about the world now that there are no more "surprise" findings of this nature in wait. However, this is not the case. During the 1990s it became clear from the work of Arden Pope and his colleagues that long-term (chronic) exposure to fine particles had a substantial effect on health, and indeed, this effect underpins much air pollution CBA work and policy development to this day. At the time of writing there is debate in Europe also about effects of long-term exposure to ozone and NO_2, with the possibility that an additional effect similar in magnitude to that of $PM_{2.5}$ on mortality should be factored into future analysis.[1] A clear lesson is that the value of research should not be underestimated, and similarly the value of projects that seek to convert research into a form that can be applied to inform decision makers. Experience with the REVIHAAP and HRAPIE[2] studies led by WHO-Europe to provide input to the EU's review of its Thematic Strategy on Air Pollution demonstrates the value of such exercises, and also their complexity.

The Beaver Committee proved far more persuasive than many, probably most, thought likely. In the context of this book it is particularly interesting to see that the Beaver report contains a very early example of environmental economic analysis (noting that the valuations described are as shown in the original 1954 source and have not been updated):

The Committee estimate that the smoke, grit, dust and noxious gases, emitted into the air from domestic dwellings and industrial plant, cause damage to property and other harmful effects to the tune of about £250 million a year. To this must be added the value of the heat wasted through excessive smoke, which is assessed by the Committee at between £25 million and £50 million a year. These figures take no account of injury to health and loss of life, as for example, the 4,000 deaths caused by smoke-laden fog in London two years ago.

The issue of efficiency was central to the conclusions reached by Clinch (1955), who considered that . . .*the average citizen will believe in the possibility of smoke abatement only after it has been proved to him beyond all reasonable doubt that he will benefit financially by preventing smoke. . . .*

The Committee's recommendations were far-reaching:
(1) *That, subject to certain exceptions, the emission of dark smoke should be prohibited by law.*
(2) *That industries, when installing new plant, should be required to take all practical steps to prevent the emission of grit and dust.*
(3) *That, subject to confirmation by the Government, local authorities should be empowered to designate "Smokeless Zones" and "Smoke Control Areas."*
(4) *That the duty of inspection and enforcement should be placed upon local authorities, except in the case of certain industrial processes, which should be supervised by Government Inspectors.*

---

[1] There is not a similar debate in Europe about the effects of SO_2 on health, as urban concentrations of SO_2 are now far lower than they used to be. However, this should not be taken as an indication that effects of SO_2 on health should be ignored elsewhere. Indeed, there may well be some European cities where it is still a significant problem.

[2] REVIHAAP: Review of Evidence of Health Aspects of Air Pollutants. HRAPIE: Health Risks of Air Pollution in Europe.

(5) *That householders in smoke restricted districts should be required to burn only smokeless fuel, and that the cost of converting domestic fireplaces for this purpose should be met, to a large extent, by grants from the Exchequer and the local authorities.*

In spite of the evidence on health impacts, the partial quantification of externalities demonstrating substantial economic effect to materials and the recommendations of the Beaver Committee, the Government was still reluctant to act. Hansard (1955) provides an interesting exchange between the Minister responsible for local government and a number of MPs who were keen to see legislation introduced. Indeed, given Government inaction, one MP, Gerald Nabarro, brought forward a Private Members Bill, though this was dropped once the Government gave assurance that it would introduce its own legislation. In spite of much further legislation on air quality in the years since, the basic framework described by the Beaver Report describes much of the structure of current UK legislation in the area to this day.

The air quality control measures identified by Beaver, large-scale relocation of industry and setting controls on the fuels that people could use in their own homes have a draconian quality to them that politicians would now probably struggle with even more than their counterparts of 60 years ago. However, the following points are worth noting:

1. It is very difficult to criticise the Beaver report. In spite of limited knowledge at the time compared to the present day there is little different that would be done. We now place more emphasis on abatement technologies, though most of the options of this type that are available now were unavailable in the 1950s.

2. The public accepted the new legislation, recognising the severity of the problem and accepting the vision that something could be done about it. This is an important lesson both for those cities in the world that remain subject to very high particle and SO_2 levels linked to coal and oil burning, and to policy makers concerned with convincing the public that climate change needs to be addressed as a matter of urgency.

3. Improved knowledge of health impacts of air pollution means that current estimates of impacts in countries like the UK are considerably higher than the 4000 deaths that prompted the Beaver report and subsequent legislation.

4. Targets for reducing greenhouse gas emissions are of a similar magnitude to the improvements that have been achieved in air quality over the last 60 years.

5. One area that Beaver did not address concerned the increasing importance of diesel powered vehicles, with diesel buses, for example,

replacing electric trams. The lesson to be learned is that apparently minor sources of pollution can become important once other things are addressed. There is also a clear parallel here with the emphasis on the use of biomass to mitigate greenhouse gas emissions.

With the passage of clean air legislation in the UK and other countries between the 1950s and 1970s there was a perception that air pollution problems had been consigned to history. Interest in the valuation of externalities was for many years of little more than academic interest.

Air pollution became prominent again with the acid rain debate that commenced in the 1970s. Some of the concerns were linked to issues similar to those considered by Beaver: extremely high concentrations of air pollutants, albeit over relatively small areas, best illustrated by the destruction of forests in an area where Poland, East Germany and Czechoslovakia met which became known as the Black Triangle. Other concerns, however, were raised in relation to the effects of pollution in regions far removed from major sources of pollution, especially in Scandinavia. Odén (1967) reported accelerated levels of acidification that could only be explained by the long-range transport of pollution over hundreds of kilometres. In the years that followed, impacts of acidification were observed on both terrestrial and aquatic ecosystems, most notably perhaps the loss of freshwater salmonid species (salmon and trout). Through observations on the remains of different species of diatom in sediments, Renberg and Battarbee (1990) demonstrated that the onset of acidification corresponded to the dawn of the Industrial Revolution. Concern linked to these findings led to the development of the Convention on Long Range Transboundary Air Pollution (the LRTAP Convention) through the UNECE (United Nations Economic Commission for Europe, which includes representation for all European countries and the USA and Canada).

The finding that significant levels of acidification had taken place and were causing damage to ecosystems was sufficient for a number of countries to sign up to the first sulfur protocol to the LRTAP Convention, requiring countries to reduce emissions of SO_2 by 30%. Notably, the UK, one of Europe's largest emitters of SO_2 at the time, refused to sign the Protocol, arguing that its effects would be disastrous for the UK economy. This refusal led to the UK being branded "the dirty man of Europe". Ironically, the same government was already taking measures to liberalise the energy sector that led to what became known as the "dash for gas". This led to a massive reduction in coal burning that meant that the country achieved the 30% sulfur reduction target without the need for specific emission control measures. The lesson to be learned is that significant opportunity for emission reductions may be available through unexpected means: the engineering solution to pollution problems may be the most obvious solution, but it is not necessarily either the cheapest or best. Structures should therefore be in place to ensure that decision makers are aware of the full range of options available to them.

The earliest analysis of health impacts in the ExternE Project was limited to a zone within close proximity of power plants (at most, 50 km). However, the interdisciplinary nature of ExternE highlighted inconsistency in approach between ecosystems analysis and health impact assessment. For ecosystems it was accepted that modeling to identify areas subject to exceedance of critical loads and critical levels (thresholds for damage from deposition and air concentrations of pollutants) should consider the whole European domain. For health, in contrast, there was then, as now, no evidence of a threshold for effects of exposure to most air pollutants. So why restrict the range of assessment for one effect with no threshold, whilst not restricting it for an effect that does have a threshold? Expansion of the range of health impact assessment led to a major increase in damage estimates, not least as it gave time for the formation of secondary aerosols. The decision not to limit the range to a set distance has been adopted widely ever since (though in practice, range is almost always limited in assessments as a consequence of the limited geographic domains adopted for dispersion models). There are two lessons here:

- The first, obviously, is that analysis should not be artificially constrained, either with respect to the distance or set of impacts considered.
- The second lesson is that much insight can be gained from interdisciplinary working (Reis *et al.*, 2012), and hence that this should be encouraged through Research Councils and other funding bodies.

Some countries have been quick to adopt cost–benefit analysis in the environmental field as an input to the decision-making process, the UK and USA being obvious examples. Guidance on economic appraisal and impact assessment is available from various sources.[3] A lesson here, for countries that currently lack such guidance, is that tested examples already exist, as shown in the next section, so the process of drafting guidance should be relatively straightforward.

---

[3] Examples include: USEPA – Guidelines for preparing economic analysis, http://yosemite.epa.gov/ee/epa/eed.nsf/pages/guidelines.html; European Commission – Impact Assessment, http://ec.europa.eu/governance/impact/index_en.htm; UK – Impact Assessment Guidance, www.hm-treasury.gov.uk/data_greenbook_impact_assessments.htm).

A number of other countries have taken a different approach, considering that the demonstration that pollution control techniques are available ("availability" here can have a broad definition, including cost as well as technical dimensions) is sufficient to warrant their adoption (Germany being a good example). The European Commission has taken something of a middle route until recently, basing air quality legislation primarily on cost-effectiveness considerations, particularly using the RAINS[4] model of IIASA, and its successor, GAINS.[5] These models have been mainly focused on the following pollutants and impacts (though the list of pollutants and effects continues to expand, now including short-lived climate forcers and mercury):

	NH$_3$	NOx	PM$_{2.5}$	SO$_2$	VOCs
Ecosystem acidification	✓	✓		✓	
Ecosystem eutrophication	✓	✓			
Ecosystems and ozone		✓			✓
Health and particles	✓	✓	✓	✓	✓
Health and ozone		✓			✓

An obvious lesson from this table is that it is appropriate to consider the pollutants listed together. It would not be possible to identify the most cost-effective route for reducing health impacts from particle exposure, for example, by focusing on individual pollutants. There is a very strong likelihood that such a piecemeal approach would be extremely inefficient. This was clearly recognized during the 1990s when Protocols to the UNECE Convention on Long Range Transboundary Air Pollutants (LRTAP) switched from dealing with individual pollutants to a more integrated approach. The same of course applies to policies in a number of other fields, obviously climate, but also chemicals, for example in relation to endocrine disrupters.

The RAINS/GAINS analysis proceeds through the following stages:

1. **Define scenarios for the target year** (including demand for energy, transport, etc.). This part of the analysis includes close collaboration with Member States. However, some significant differences in expectation between Member States and projections made by consultants advising

---

[4] Regional Air pollution INformation and Simulation model (http://webarchive.iiasa.ac.at/Research/TAP/rains_europe/intro.html)

[5] Greenhouse Gas and Air Pollution Interactions and Synergies Model (http://gains.iiasa.ac.at/models/)

the European Commission arise in the modeling, linked to different assumptions, for example linked to growth and energy costs.

2. **Estimate emissions** for the target year based on these scenarios assuming full implementation of current legislation (CLE). This part of the analysis is based on extensive research on emission factors over many years, estimates being based on literature, industry and regulatory (e.g. via emissions reporting) sources.

3. **Estimate the extent to which emissions can be reduced** using the measures included in the RAINS/GAINS model. This gives a range from current legislation to what is often described as the "maximum technically feasible reduction scenario" (MTFR). It should be recognized that the MTFR does not represent the true maximum that emissions could be reduced. There are inevitably additional measures that could be introduced, including:

   a. Technical measures, for example actions that are currently under research but would be available by the target year

   b. Fuel switching (this can be introduced to the analysis, though generally as a sensitivity analysis)

   c. Some behavioral measures, for example encouraging reduced meat consumption or modal shift for transport

   The MTFR therefore represents a pragmatic estimate of potential abatement. However, a lesson to be learned is that estimates of total abatement potential for pollutants arising from many sources are likely to be conservative. This is not simply a function of the RAINS/GAINS models, but a consequence of the complexity of the modeling. This has consequences for the subsequent economic analysis, which will be discussed below.

4. **Generate the cost curve** covering the gap between the CLE and MTFR scenarios. At this stage this is assessed against the change in emission for each pollutant rather than the change in impact.

5. **Quantify key indicators of impact** for the CLE and MTFR scenarios. RAINS/GAINS includes a subset of human health and ecosystem effects, as follows:

   a. Effects of ozone on mortality

   b. Effects of $PM_{2.5}$ (including both primary and secondary particles) on mortality

   c. Exceedance of the critical load for acidification

   d. Exceedance of the critical load for eutrophication (nitrogen deposition)

   e. Exceedance of the critical level for ozone exposure

6. By this stage we have estimates of emissions, potential emission reductions costs, and some of the benefits associated with the reductions.

The model can then be used to **optimize emission controls** in each country to achieve specific levels of improvement between the two extreme scenarios at least cost.

The European policy debate has tended to focus on the information provided to this stage. There has been some acceptance of the view that an appropriate target would be around the point of inflection of the cost curve, beyond which costs per unit abatement start to increase rapidly. However, there is argument as to where this position actually is for each country and there is potential for bias in discussion, with only those countries that feel unduly challenged by targets engaging with the debate.

The first application of externalities data for the European Commission concerned the EU's Acidification Strategy (Holland and Krewitt, 1996), with results indicating that the benefits of the strategy would exceed costs. However, like much of the CBA work that followed, the analysis was used as little more than a rubber-stamping exercise for policies that had to a large extent been defined by the time that the analysis was performed. Most importantly, this work succeeded in bringing health impacts to the fore in the international air pollution debate in Europe for the first time. The Acidification Strategy was focused on environmental impacts, whilst most of what was quantified in the CBA addressed health. By the time that the Clean Air For Europe Programme commenced in the early 2000s the health impacts were central to the optimization analysis for emission reductions as well as the CBA. However, this raised the question as to why the Acidification Strategy was focused on ecosystem damage when the largest benefits related to health; the logic being that a policy targeted directly at health would have even greater benefits.

Until recently, the quantification of benefits and resulting CBA has been used only as a final check that benefits exceed costs, ensuring conformity with the impact assessment requirements (e.g. IIASA, 2010; Holland *et al.*, 2011). Current negotiations on the review of the EU's "Thematic Strategy on Air Pollution," however, have taken a different approach with (an albeit conservative assessment of) marginal benefits being compared with marginal costs and the ambition level being defined as the point where the two curves cross (e.g. IIASA, 2013; Holland, 2013).

It was noted (paragraph 3 in the list above) that it is difficult for models like RAINS/GAINS to include all possible measures within reason (e.g. non-technical as well as technical). This has several effects on the analysis:

- The true extent of potential abatement is underestimated
- Costs are exaggerated for reaching any level of abatement beyond the point at which an omitted measure would be introduced
- This in turn biases the CBA against finding a net benefit for such an ambition level.

Further to this, it is not possible to include all impacts in the benefits assessment in a monetized form (e.g. the benefits from reduced deposition of acidifying or eutrophying pollutants to ecosystems). Again, the effect of omission is to bias the CBA against finding a net benefit. MCDA could be applied, but in practice there has so far been no clear demand for it.

There are several lessons from this. Decision makers need a good level of understanding of the limitations of the results presented to them. Then, an assumption that costs and benefits both have their uncertainties and that these broadly cancel against each other may be wrong: as shown, the consequences of omission of abatement measures and omission of impacts combine to make it harder to demonstrate that benefits exceed costs.

Another lesson to be drawn from the review of the EU's Thematic Strategy concerns the availability of low cost abatement measures. Given the large volume of air quality regulation already in place it might be assumed that all low cost measures have already been adopted. However, IIASA (2013) shows that this is not the case, with marginal benefits exceeding marginal costs for at least 75% of the set of measures included in the MTFR.

It should be remembered that cost-curves are one-dimensional with respect to impacts related to measures. For example, consider the consequences of reducing GHG emissions from a power plant by fitting some form of carbon capture and storage (CCS) technology. As a first step, cost-effectiveness may be assessed by comparing the reduction in GHGs with the cost, and such data are all that is needed for entry to a cost curve. We may assume that the cost assessment accounts for the use of electricity for CCS and hence a reduction in the overall efficiency of the power plant. However, consider the following that are illustrative of the further consequences that are not factored into the cost curves:

- Increased emission of regional pollutants (SO_2, NO_x, etc.) per unit of power sent out from the power plant
- Increased demand for coal, with associated risks to workers and burdens linked to processing and transport
- Increased abstraction of water for cooling
- Burdens associated with CCS technology (e.g. reagent manufacture).

This leads back to an issue raised above, the potential for new policies to have unforeseen negative impacts. This can be avoided through the use of an extended analytical framework that accounts for all effects of the measures applied.

In the coming years one of the most active areas of application of externalities research and analysis is likely to be in the field of chemicals control under the EU's REACH (registration, evaluation and authorisation of chemicals) regulation. Hitherto there has been little interest in quantification of the benefits of chemicals regulation, with analysis firmly focused on somewhat traditional chemical risk assessments. However, under REACH this is supplemented by socio economic analysis (SEA), which is in essence the same as CBA. Risk assessment and SEA have a similar standing under the regulation, both having their own expert committees required to assess submitted information and make recommendations to the European Commission on the desirability of the proposed action.

16.3 Examples of application

This section identifies a number of applications of externalities analysis that have been used for policy making. Much of the work, though extensively debated and reviewed, has only been published in the gray literature (e.g. contractor's reports to policy making bodies), rather than through the academic literature. We do not attempt to describe these applications in detail. However, simply by listing them across the fields of air quality policy, climate, energy, waste, chemicals, industry and transport we demonstrate a number of important points:

1. That the methods described in this book have a wide applicability.
2. That methods are already available "off the shelf." They will need review and updating over time, but there is no need to reinvent the wheel in terms of the overall structure of the analysis.
3. The concept of economic valuation of environmental goods and the general CBA framework is very widely accepted.
4. That expertise in bringing research into policy is also available.

Air quality at local and national levels

- The four daughter directives to the EU's Framework Directive on Ambient Air Quality (IVM, 1997; AEA, 1998; AEA, 1999, AEA, 2001, Entec, 2001)
- The UK's National Air Quality Strategy (IGCB, late 1990s to the present day)
- USEPA's CBA studies of the Clean Air Act (USEPA, 1997; 1999; 2011).

Regional air pollution

- The EU's Acidification Strategy and Large Combustion Plant Directive (Holland and Krewitt, 1996)

- The EU's Clean Air For Europe Programme and the Thematic Strategy on Air Pollution (AEA, 2002–2006)
- The EU Directive on the sulfur content of liquid fuels (Bosch *et al.*, 2009)
- Revision of the National Emission Ceilings Directive (IIASA, 2010; AEA, 2010)
- The Gothenburg Protocol to the UNECE/CLRTAP (IIASA, 1999; Holland *et al.*, 1999; IIASA, 2011; Holland *et al.*, 2011; Miller *et al.*, 2011)
- Review of the EU's Thematic Strategy on Air Pollution (IIASA, 2013; Holland, 2013).

Climate co-benefits

- Assessment of the co-benefits for air quality of greenhouse gas control policies for the European Commission (AEA, 2006a; EMRC, 2008; Holland *et al.*, 2012)
- Review of the impacts of carbon budget measures on human health and the environment for the UK's Climate Change Committee (Forster *et al.*, 2013).

Energy

- European renewable energy policy (based on work in ExternE, 1998)
- Appraisal of power generation options for developing countries by the International Atomic Energy Agency (IAEA, 2003)
- The costs and benefits of emission controls on Large Combustion Plant (Pye and Holland, 2007).

Waste

- The EU's draft directive on incineration of municipal waste (ETSU *et al.*, 1997)
- Economic Analysis of the Options for Managing the Biodegradable Fraction of Municipal Waste undertaken for DG Environment (Eunomia *et al.*, 2001)
- Analysis of the external damage costs associated with residual waste treatments carried out as part of an international review of waste management policy undertaken for the Republic of Ireland (Eunomia, 2009).

Chemicals

- Various socio-economic assessment studies linked to proposals for restriction of chemicals under the EU's REACH legislation (the latest proposals are available at www.echa.europa.eu/restrictions-under-consideration).

Industry

- Assessment by the European Environment Agency of the impacts of emission sources included in the EPER and EPRTR databases (EEA, 2011)
- Development of a methodological convention for estimating environmental externalities by the German Federal Environment Agency (UBA, 2008)
- Use of Market Based Instruments for emission control (Entec, 2010).

Transport

- Assessment of various transport policies in London (AEA, 2006b)
- Assessment in relation to a possible Nitrogen Emission Control Area (NECA) in the North Sea (Hammingh *et al.*, 2012)
- Setting road user charges for vehicles (EEA, 2013).

16.4   Final thoughts

We end with a series of final thoughts, designed both to underline key messages and to suggest how analysis may proceed in the coming years.

First, externalities analysis enhances the transparency of decision making. Without it, there will still be valuation of health and environmental impacts whenever, for example, governments decide how to allocate their budgets. However, no one outside of the decision making group will understand how different effects were weighted against one another.

The "willingness to pay" paradigm is not an artificial construct of environmental economics but the fundamental requirement for any economic transaction (alongside a "willingness to accept" on behalf of the seller).

What's unquantified may be as important as what is quantified. Examples include the effects of regional air pollutants on ecosystems with great concern now focused on nitrogen deposition, the "difficult" climate impacts such as conflict and the effects of extreme events, and the long-term risks from nuclear power (will society in the future be in a position to manage the wastes that we leave them to manage?). The inability to quantify everything is not a reason for dismissing externalities analysis or CBA (there is a mistaken understanding in some circles that a complete quantification is essential for the latter). If the full quantification cannot be done using externalities assessment techniques it cannot be done using others either, but the externalities work at least takes what can be quantified through to a policy-relevant indicator. Discussion can then turn to omitted impacts – which of these are likely to be as or more important than those that have been assessed? Can we develop a sense of scale for these effects, for example using multi-criteria decision analysis?

Uncertainty can be accounted for. In many cases the uncertainties in externalities quantification have very little impact on the results of cost–benefit analysis as shown by Rabl *et al.* (2005).

No single tool solves all problems. However, different tools used in combination can provide far more complete and reliable answers. The problem may be in finding analysts who understand the strengths of different tools and who have the expertise to apply them, emphasizing the multi-disciplinary nature of the problem.

The importance of different effects will vary in different parts of the world. For example, the results presented here have shown that air pollution damage to materials is unimportant relative to health impacts, at least in Europe and North America. However, in other parts of the world there are a number of reasons to believe that effects on materials will be more important, for example, because SO_2 levels are higher and standards for building and material quality may be lower. With respect to water quality, a focus in Europe on (e.g.) endocrine disrupting chemicals seems unlikely to be of any significant relevance in countries where few have access to water that is free of microbial contamination.

We acknowledge that there are problems in the valuation of impacts in different parts of the world. This caused problems for the 1995 IPCC report, and as noted earlier, it has led to a reluctance to monetize impacts by IPCC ever since. The general principle is clear, that people should be free to exercise their own value systems. There are very good and obvious reasons why the priorities of someone in a very poor country will be different to those of someone in a richer part of the world. Willingness to pay for clean air will not be high for people struggling to provide food and good water for their family. However, this principle gets into difficulties when dealing with impacts imposed on the people of one country by those elsewhere with the obvious example relating to climate change. In such a case it is appropriate to adjust values for example using equity weighting. Hence, even in this difficult situation, there is a remedy within the economics literature.

Finally, and leading on from this, there is nothing immoral about the use of economic valuation. Indeed, it can be said that a failure to apply valuation is the more immoral position as it implies that one does not care about the efficiency of resource allocation, and that one refuses to look at all the important consequences of a decision.

References

AEA 1998. *Economic Evaluation of Air Quality Targets for Tropospheric Ozone* Part C: Economic Benefit Assessment. Report to European Commission DG XI. http://ec.europa.eu/environment/enveco/air/index.htm#_Toc240787025

AEA 1999. *Economic Evaluation of Air Quality Targets for CO and Benzene.* Report to European Commission DG XI. http://ec.europa.eu/environment/enveco/air/index.htm#_Toc240787024

AEA 2001. *Economic Evaluation of Air Quality Targets for PAHs.* Report to European Commission DG XI. http://ec.europa.eu/environment/enveco/air/index.htm#pahs

AEA 2002–2006. Various reports relating to the development and application of methods for cost-benefit analysis of European air pollution policies, available at: http://cafe-cba.org/,

AEA 2006a. *Assessing the air pollution benefits of further climate measures in the EU up to 2020.* Report to European Commission DG Environment. www.cafe-cba.org/assets/further_climate_measures_benefits.pdf

AEA 2006b. London Low Emission Zone Health Impact Assessment Final Report. Report to Transport for London. http://www.tfl.gov.uk/assets/downloads/roadusers/LEZ/health_impact_assessment.pdf

AEA 2010. Cost Benefit Analysis for the Revision of the National Emission Ceilings Directive Interim Report. Report to European Commission DG Environment. http://ec.europa.eu/environment/air/pollutants/pdf/necd_cba.pdf

Bell, M. L., Davis, D. L. and Fletcher, T. (2004) A retrospective assessment of mortality from the London Smog Episode of 1952: The role of influenza and pollution. *Environmental Health Perspectives* **112**: 6–8.

Bosch, P., Coenen, P., Fridell, E. *et al.* 2009. Cost Benefit Analysis to Support the Impact Assessment accompanying the revision of Directive 1999/32/EC on the Sulphur Content of Certain Liquid Fuels, for European Commission DG Environment.

Clinch, H. G. 1955. *Atmospheric Pollution in London and the Home Counties: A Report on Known Facts.* London and Home Counties Smoke Abatement Advisory Council.

Davis, D. 2002. The Great Smog. *History Today,* 52. www.historytoday.com/devra-davis/great-smog.

EEA. 2011. *Revealing the costs of air pollution from industrial facilities in Europe.* European Environment Agency, Copenhagen, Denmark. www.eea.europa.eu/publications/cost-of-air-pollution.

EEA 2013. *Road user charges for heavy goods vehicles (HGV).* European Environment Agency, Technical report No 1/2013. www.eea.europa.eu/publications/road-user-charges-for-vehicles.

EMRC (2008) The co-benefits for health of strong climate change policy. Report to the Health and Environment Alliance (HEAL). www.climnet.org/index.php?option=com_docman&task=doc_download&gid=682&Itemid=55

Entec 2001. *Economic Evaluation of Air Quality Targets for Heavy Metals*. Report to European Commission DG XI. http://ec.europa.eu/environment/enveco/air/index.htm#heavy_metals

Entec 2010. *Assessment of the Possible Development of an EU-wide NOx and SO₂ Trading Scheme for IPPC Installations*. Report to European Commission DG Environment.

ETSU 1997. *Cost-benefit analysis of the draft directive on the incineration of non-hazardous waste*. Report to European Commission DG XI.

Eunomia 2001. *Economic Analysis of the Options for Managing the Biodegradable Fraction of Municipal Waste*, Report to European Commission DG Environment.

Eunomia 2009. *International Review of Waste Management Policy*, Report for Department of Environment Heritage and Local Government (Republic of Ireland), September 2009.

ExternE 1998. Volume 10, National Implementation, pp. 150–154. www.externe.info/externe_d7/?q=node/41.

Forster, D., Korkeala, O., Warmington, J., Holland, M. and Smith, A. 2013. *Review of the impacts of carbon budget measures on human health and the environment*. Report for the UK's Climate Change Committee.

Hammingh, P, Holland, M. R., Gellenkirchen, G. P., Jonson, J. E. and Maas, R. J. M. 2012. *Assessment of the environmental impacts and health benefits of a Nitrogen Emission Control Area in the North Sea*. Report to PBL / Netherlands Environmental Protection Agency. www.pbl.nl/en/publications/2012/assessment-of-the-environmental-impacts-and-health-benefits-of-a-nitrogen-emission-control-area-in-the-north-s.

Hansard. 1955. Parliamentary debate on the Air Pollution Committee's Report. http://hansard.millbanksystems.com/commons/1955/jan/25/air-pollution-committees-report.

Holland, M. 2013. *Cost-benefit Analysis of Policy Scenarios for the Revision of the Thematic Strategy on Air Pollution*: Version 1 Corresponding to IIASA TSAP Report #10, Version 1. March 2013. Contract report to European Commission DG Environment.

Holland, M. and Krewitt, W. 1996. Benefits of an Acidification Strategy for the European Union and CBA of Pollutant Abatement Options for Large Combustion Plant. In *ExternE: Externalities of Energy* Volume 10, National Implementation, p. 150–154: www.externe.info/externe_d7/?q=node/41

Holland, M., Spadaro, J., Derwent, R., Jenkin, M. and Murrells, T. 2009. *Costs, Benefits and Trade-Offs: Volatile Organic Compounds*. Contract report for Department for Environment, Food and Rural Affairs, London, UK.

Holland, M., King, K. and Forster, D. 1999. Cost-Benefit Analysis for the Protocol to Abate Acidification, Eutrophication and Ground Level Ozone in Europe www.unece.org/env/lrtap/TaskForce/tfeaas/4meeting/aea2_final.pdf.

Holland, M., Wagner, A., Hurley, F., Miller, B. and Hunt, A. 2011. Cost Benefit Analysis for the Revision of the National Emission Ceilings Directive: Policy Options for revisions to the Gothenburg Protocol to the

UNECE Convention on Long- Range Transboundary Air Pollution http://ec. europa.eu/environment/air/pollutants/pdf/Gothenburg%20CBA1%20final% 202011.pdf.

Holland, M., Amann, M., Heyes, C. *et al.* 2012. *ClimateCost study*. Technical Policy Briefing Note number 6: Ancillary Air Quality benefits, www.climate cost.cc/images/Policy_Brief_master_REV_WEB_medium_.pdf.

IAEA 2003. Workshop on the use of the Simpacts Model for estimating Human Health and Environmental Damages from Electricity Generation. Materials from a workshop in Trieste, Italy, 12–23 May 2003. http://cdsagenda5.ictp. trieste.it/full_display.php?ida=a0266.

IIASA. 1999. *Integrated Assessment Modelling for the Protocol to Abate Acidification, Eutrophication and Ground-level Ozone in Europe*. Air & Energy 132, Ministry of Housing, Spatial Planning and the Environment, Directorate Air and Energy, The Hague, The Netherlands.

IIASA. 2010. *Baseline Emission Projections and Further Cost-effective Reductions of Air Pollution Impacts in Europe – A 2010 Perspective*. International Institute for Applied Systems Analysis, Laxenburg Austria. http://ec.europa.eu/environ ment/air/pollutants/pdf/nec7.pdf.

IIASA. 2013. Policy Scenarios for the Revision of the Thematic Strategy on Air Pollution. TSAP Report #10, Version 1.2 International Institute for Applied Systems Analysis, Laxenburg Austria. Report for European Commission DG Environment. http://ec.europa.eu/environment/air/pdf/TSAP-Report-10.pdf.

IVM (1997) *Economic Evaluation of Air Quality Targets for sulphur dioxide, nitrogen dioxide, fine and suspended particulate matter and lead*. Contract report for European Commission DG XI.

Mayor of London. 2002. 50 years on: The struggle for air quality in London since the great smog of December 1952. Greater London Authority. http://legacy. london.gov.uk/mayor/environment/air_quality/docs/50_years_on.pdf.

Odén, S. 1967. The acidification of precipitation. *Dagens Nyheter*, 27/10/1967.

Pye, S. and Holland, M. 2007. Evaluation of the costs and benefits of the imple- mentation of the IPPC Directive on Large Combustion Plant. www.cafe-cba. org/assets/ippc_ec_lcplant.pdf

Rabl, A., Zoughaib, A., von Blottnitz, H. *et al.* 2004. Tools for sustainability: Development and application of an integrated framework. Final Technical Report for project SusTools, contract N° EVG3-CT-2002–80010. EC DG Research. www.arirabl.org/Research_files/Rabl+04%20SusTools.pdf.

Rabl, A., Spadaro, J. V. and van der Zwaan, B. 2005. Uncertainty of pollution damage cost estimates: to what extent does it matter? *Environmental Science & Technology* **39**(2): 399–408.

Reis, S., Grennfelt, P., Klimont, Z. *et al.* 2012. From acid rain to climate change. *Science* **338**.

Renberg, I. and Battarbee, R. W. (1990). The SWAP Palaeolimnology Programme: a synthesis. In: *The Surface Waters Acidification Programme*. Edited by B. J. Mason. Cambridge University Press, Cambridge, UK.

UBA 2008 *Economic Valuation of Environmental Damage: Methodological convention for estimates of environmental externalities*. Umweltbundesamt: http://www. umweltdaten.de/publikationen/fpdf-l/3482.pdf

USEPA. 1997. *Retrospective Study: The Benefits and Costs of the Clean Air Act, 1970 to 1990*. United States Environmental Protection Agency. www.epa.gov/cleanairactbenefits/retro.html

USEPA. 1999. *Prospective Study: The Benefits and Costs of the Clean Air Act, 1990 to 2010*. United States Environmental Protection Agency. www.epa.gov/cleanairactbenefits/prospective1.html

USEPA. 2011. *Prospective Study: The Benefits and Costs of the Clean Air Act, 1990 to 2020*. United States Environmental Protection Agency. www.epa.gov/cleanairactbenefits/prospective2.html

Appendix A: Nomenclature, symbols, units and conversion factors

Symbols

Our notation is somewhat different from many reports of the EPA and other organizations because we follow the custom of physics and engineering textbooks where a single letter is used for the "family name" of a variable, with subscripts to distinguish different variants. We choose subscripts that are fairly explicit and in most cases self-explanatory.

It is helpful to distinguish different substances by adding subscripts to some units: for instance m_{wat}^3 for a m^3 of water. Likewise we sometimes add a subscript to the mass for clarity, e.g. kg_{soil} for a kg of soil.

To minimize the risk of confusion about units for items that can be stated as quantities or as rates (i.e. quantity per time), we indicate rates by dots over the respective symbol, the usual notation for time derivatives; for example if m = mass of emitted pollutant (kg), $\dot{m}$ = emission rate (e.g. kg/yr).

For certain variables we sometimes add the location **x** as argument to indicate a possible dependence on the location where they are evaluated; when **x** is not shown, the average over the entire region is understood, for example k_{dep} = average of $k_{dep}(\mathbf{x})$ over all locations **x** and S_{ERF} = population-weighted average of $S_{ERF}(\mathbf{x})$.

A = area (m^2)

B = Transfer factor from medium 1 to medium 2, with subscripts $[(kg/kg_{media2})/(kg/kg_{media1})]$

B_R = breathing rate (m^3/day)

c = concentration

$c(\mathbf{x}, \dot{m})$ = Concentration increment at a location **x** due to emission rate $\dot{m}$, with subscripts for media ($\mu g/m^3$ for air, $\mu g/kg_{media}$ for other media)

$c_{air}(\mathbf{x}, \dot{m})$ = increase in concentration ($\mu g/m^3$) at a location **x** = (x,y) due to the emission rate $\dot{m}$,

C = cost (€, $, ...)

D = damage cost (€, $, ...)

D_u = unit damage cost (€/kg)

DR = pollutant dilution rate (m^2/s) = ground wind velocity × characteristic mixing depth for ground-level sources

$\dot{d}(\mathbf{x}, \dot{m})$ = Dose rate due to emission rate $\dot{m}$, for an individual living at $\mathbf{x}$ (kg/(pers·yr))

$\dot{d}$ = Dose rate (kg/(pers·yr))

F_{dep} = deposition flux (μg/(m^2·s))

h_e = effective source height (m)

h_{mix} = mixing layer height (m)

h_s = source height, stack height (m)

I = impact in physical units (damage units/kg of pollutant)

I_u = unit impact (damage units/kg)

$\dot{I}(\dot{m})$ = Impact rate (damage units/yr) due to emission rate $\dot{m}$

$\dot{I}$ = impact rate in physical units (damage units/yr)

iF = intake fraction

k_{dep} = depletion velocity (m/s)

k_{ct} = transformation velocity (primary to secondary pollutant) (m/s)

LE = life expectancy

$\dot{m}$ = Emission rate of pollutant (kg/yr)

$\dot{m}_{j,in}$ = Inflow rate of pollutant into the jth compartment (kg/s)

m = quantity of pollutant

$M(\dot{m}, \bar{r})$ pollutant removal flux (kg/m^2/s)

P = unit cost ("price") (€/case)

q_R = pollutant dilution rate (or ventilation index) for elevated sources (m^2/s) = wind speed times h_{mix}

$Q_{food,j}$ = Quantity of food product j consumed per pers per year (kg_{food}/(pers·yr))

$\mathbf{r}$ = (r,ϕ) = location of a receptor, with cylindrical coordinates

r_{dis} = discount rate = rate at which one is neutral between a payment P_0 today and a payment $P_n = P_0 (1+r_{dis})^{-n}$ in n years from now

R = radius (m)

S(x,z) = plume reflection term for Gaussian plume model

S_{ct} = modification factor of UWM for chemical transformations

S_{sh} = modification factor of UWM for site and stack height

S_{ERF} = slope of ERF (cases/(receptor·yr·μg/m^3))

$S_{ERF}(\mathbf{x})$ = slope of ERF at $\mathbf{x}$ ((cases/yr)/(receptor·(μg/m^3))).

$S_{ERF,ingest}$ = slope of ERF for ingestion (cases/kg)

$S_{ERF,inhal}$ = slope of ERF for inhalation (cases/(pers·yr·(μg/m^3)))

t = time

t_{cut} = cutoff time for analysis (yr)

T = temperature

T_a = ambient air temperature

u = wind speed (m/s)

v_{dry} = dry deposition velocity (m/s)

v_{wet} = wet deposition velocity (m/s)

x = age

$\mathbf{x}$ = (x,y) = location of a receptor, with Cartesian coordinates

$X = c_{food}/c_{air}$ = Transfer factor from air to food (with subscript for food type) $((kg/kg_{food})/(kg/m^3))$

z_0 = surface roughness (m)

α_p, α_s = parameter of dispersion model, subscript for primary and secondary pollutant

β_p, β_s = parameter of dispersion model, subscript for primary and secondary pollutant

$\phi = \ln(\sigma_g)$

κ = rate constant (inverse of time constant) (yr^{-1})

μ = mean

μ_g = geometric mean

$\mu(x)$ = age-specific mortality = probability of members of the cohort of ages from x to x+Δx to die during 1 year

ρ = density of receptors (population, buildings, crops, ...) (receptors/m^2)

ρ_{eff} = effective density of receptors = weighted average of local and background densities

$\rho(\mathbf{x})$ = density of receptors (population, buildings, crops, ...) (receptors/m^2) at $\mathbf{x}$

σ = standard deviation

σ_g = geometric standard deviation

σ_y = dispersion parameter of Gaussian plume model, horizontal direction

σ_z = dispersion parameter of Gaussian plume model, vertical direction

τ = time constant (yr)

$\xi = \ln(\mu_g)$

Units and conversion factors

In this field the conversion of units is a ubiquitous chore and a frequent source of errors. We have seen quite a few instances of errors by a factor of 1000. A convenient trick for converting units is to multiply the quantity in question by a ratio of units that is equal to unity, the ratio chosen to cancel the undesirable unit. For example if the fuel consumption of a car is stated as 25 mpg (miles per gallon) the conversion to km/L is

$$\frac{25 miles}{1 gallon} \times \frac{1 gallon}{3.785L} \times \frac{1.609 km}{1 mile} = \frac{10.627 km}{L}$$

Since many of the variables and parameters are commonly stated in units that are not SI, we find it convenient to arrange the spreadsheets for our calculations in the following manner: listing variables and parameters sequentially in rows, we input the parameter values in a column reserved for conventional units; in a separate column we state the values in units of m, kg and s, and we carry out all the calculations in these latter units. For results to be reported, we transform back to conventional units.

ppb conversions at 20°C:

1 ppb O_3 = 2.00 µg/m^3 of O_3
1 ppb NO_2 = 1.91 µg/m^3 of NO_2
1 ppb SO_2 = 2.66 µg/m^3 of SO_2
1 ppm CO = 1.16 mg/m^3 of CO

1 barrel = 42 gal = 0.159 m^3
1 mile = 1.609 km
1 ft (US foot) = 12 in (US inch) = 0.3048 m
1 gallon (US) = 3.785 L
1 kWh = 3.6 MJ
1 ton (US short ton) = 2000 lb (US pound) = 907 kg
1 tonne = 1000 kg
1 lb (US pound) = 0.45356 kg
1 Btu (US British thermal unit) = 1.055 kJ
1 Ci (US curie) = 3.7×10^{10} Bq (1 Bq = 1 decay per second)
1 rem = 0.01 Sv

Decimal multiples

Factor	Prefix	Symbol
10^{12}	Tera	T
10^{9}	Giga	G
10^{6}	Mega	M
10^{3}	Kilo	k
10^{2}	Hecto	h
10^{1}	Deca	da
10^{-1}	Deci	d
10^{-2}	Centi	c
10^{-3}	Milli	m
10^{-6}	micro	µ
10^{-9}	Nano	n
10^{-12}	Pico	p

Abbreviations

ADEME = Agence de l'Environnement et de la Maîtrise de l'Energie (French agency for environment and energy efficiency)

As = arsenic

BAF (BCF) = Bioaccumulation factor (Bioconcentration factor)

BS = black smoke

CB = chronic bronchitis

CBA = cost–benefit analysis

CEA = cost-effectiveness analysis

CFC = chlorofluorocarbon

CHA = cardiovascular hospital admission

CI = confidence interval

CO = carbon monoxide

COPD = chronic obstructive pulmonary disease

CRF = concentration–response function

Cr-VI = chromium in oxidation state 6

CV = contingent valuation

DR = pollutant dilution rate (m^2/s) = ground wind velocity × characteristic mixing depth for ground-level sources

DRF = dose–response function (also known as exposure–response function ERF or concentration–response function CRF)

EC = European Commission

EU = European Union (followed by a number to indicate number of countries included)

ECU = European currency unit (before 1999) = Euro (since 1999)

EEA = European Environment Agency

EMEP = organization charged with modeling transboundary air pollution in Europe.

EPA = Environmental Protection Agency of USA

ERF = exposure–response function

GWP = global warming potential (kg of substance with same radiative forcing as 1 kg of CO_2)

HA = hospital admission

Hg = mercury

Hg_p = particulate mercury

IPA = impact pathway analysis

IPCC = intergovernmental panel on climate change

kWh_e = kWh of electricity

kWh_t = kWh of heat

LCA = life cycle assessment

LE = life expectancy

LLE = loss of life expectancy

LOAEL = Lowest Observed Adverse Effects Level

LRS = lower respiratory symptoms

Marginal = incremental

MeHg = methyl-Hg

Morbidity impacts = illness

Mortality impacts = increased number of deaths

mRAD = minor restricted activity day

MSW = municipal solid waste

N = nitrogen

Ni = nickel

NMVOC = non-methane volatile organic compounds

NOAEL = No Observed Adverse Effects Level

NO_x = unspecified mixture of NO and NO_2

O_3 = ozone

OR = odds ratio = output of case-control studies

PA = physical activity

PAH = polycyclic aromatic hydrocarbons

Pb = lead

pkm = passenger km

PM_{co} = coarse fraction of particulate matter = particles with aerodynamic diameter between 2.5 and 10 μm

PM_d = particulate matter with aerodynamic diameter smaller than d μm.

ppb = parts per billion

PPP = purchasing power parity

PRTP = pure rate of time preference (one of the two components of the social discount rate, the other component being economic growth)

RAD = restricted activity day

RGM = reactive gaseous mercury Hg^{++}

RHA = respiratory hospital admission

RR = relative risk

S = sulfur

TEQ = toxic equivalence 2,3,7,8-tetrachlorodibenzo-p-dioxin (TCDD)

URF = unit risk factor = probability that a person of standard weight of 70 kg will develop cancer due to exposure (by inhalation) to a concentration of 1 μg/m^3 of a pollutant over a 70-year lifetime.

URS = upper respiratory symptoms

UWM = uniform world model for simplified approximate calculation of typical impacts and damage costs

vkm = vehicle km

VMT vehicle miles traveled

VOC = volatile organic compounds

VOLY = value of a life year

VPF = value of prevented fatality = VSL

VSL = "value of statistical life"

WLD = work loss days

WHO = World Health Organization

WTP = willingness to pay

YOLL = years of life lost

Key concepts

Age-specific mortality = probability of dying at a specified age (typically stated as number of deaths per year per 100,000 population between ages x and x + 1)

Burden = pollutant or nuisance (e.g. noise, odor) emitted by a source

Characterization factor (in LCA) = a substance-specific factor calculated with a characterization model for expressing the impact from the particular elementary flow in terms of the common unit of the category indicator.

Concentration–response function, see Exposure–response function

Cost–benefit analysis (CBA) = comparison of costs and benefits of a choice

Cost-effectiveness analysis (CEA) = ranking of choices in the order of increasing ratio improvement/cost (Section 1.3)

Criteria air pollutants = group of 6 air pollutants (CO, NO_x, SO_x, O_3, PM, Pb) = first set of air pollutants identified by US EPA to require national regulations.

DALY = Disability Adjusted Life Years

Damage function = function that describes the damage as function of the quantity of a burden

Discount rate = rate that specifies how the value of a given amount of money changes with time (see Section 6.1)

Dose–response function, see Exposure–response function

Exposure–response function (ERF), often called dose–response function or concentration–response function = relation between exposure to a substance or other causative agent and the resulting effect

External cost = cost imposed on non-participants in a transaction.

Impacts = impacts in physical units

Levelized energy cost = cost per kWh, taking into account all cost contributions over the life of a power plant, with discounting

Life table = table of mortality data by age group (age-specific mortality)

Odds ratio OR = measure of effect observed in case-control studies (see Section 3.3.5).

Primary pollutant = a pollutant that is harmful in the form in which it is emitted

QALY = Quality Adjusted Life Years

Receptor = anyone or anything that is affected by the pollution emitted by a source

Relative risk RR = change in the occurrence of a health end point due to exposure to a causative agent (see Section 3.3.5)

SDI = sustainable development indicators

Secondary pollutant = a pollutant that is created from a primary pollutant by chemical transformation in the environment

Source–receptor matrices = concise and very convenient presentation of the results of atmospheric dispersion models (indicating how much the concentration in receptor grid cell z is increased by an incremental emission in source cell y)

VOLY = value of a life year

Appendix B: Description of the RiskPoll software

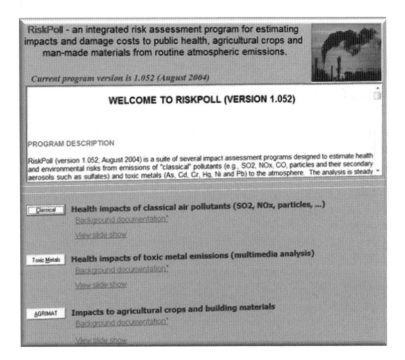

RiskPoll is an integrated risk assessment program for calculating health and environmental impacts and damage costs (also known as external costs) associated with routine atmospheric pollutant emissions from stationary sources. RiskPoll includes several routines for assessing inhalation and ingestion doses from exposure to particulate matter (PM_{10}), sulfur dioxide (SO_2), oxides of nitrogen (NO_x), carbon monoxide (CO), secondary aerosols (nitrates and sulfates), and toxic metals (arsenic, cadmium, chromium, mercury, nickel and lead).

Health impacts of air pollution are calculated using an impact pathway analysis, which traces the fate of a pollutant in the environment from point of emission into the air, to environmental dispersion and eventual receptor uptake. Health impacts include loss of life expectancy (mortality) and various morbidity diseases (e.g., asthma attacks, chronic bronchitis and hospital visits). Quantification of the impacts and damage costs follows the methodology developed by the ExternE Project of the European Commission. The multimedia impact assessment is based on the transfer factors published by the US Environmental Protection Agency. In addition to health effects, RiskPoll also computes impacts and costs to agricultural crops and building materials (based on the work by ExternE).

In contrast with existing environmental impact assessment models, the RiskPoll methodology is more transparent, easier to implement and requires fewer user input data, this is an especially important advantage for applications in developing countries where lack of data oftentimes excludes the use of very detailed programs. The RiskPoll user can clearly identify relationships between input parameters that matter the most in the impact analysis and then do sensitivity studies to investigate how the final results might change. RiskPoll can also serve as a pedagogic tool for both beginners and experts, because very detailed documentation, with explanations of the concepts and links to internet resources, is directly integrated into the software.

RiskPoll requires a minimum or Core input data set. For particulate health risks, as an example, data requirements include pollutant emission and depletion rate, population density for a circular area centered at the source with radius of 500 to 1000 km (depending on near source urban development), and dose–response functions for assessing health impacts. If more input data are known (e.g. meteorological statistics and stack variables), improved results are calculated. In spite of the simplicity of the RiskPoll approach, the model user is able to carry out calculations capable of providing robust guidance in decision taking at the local, industry or government level.

Monetizing physical impacts is optional. The monetization process requires country specific unit costs ($ per impact case) for each impact quantified. In the absence of such cost values, a procedure is described in the RiskPoll documentation for transferring known costs from one country to another location. Default pollutant depletion rates for various regions around the world, and lists of dose–response functions and monetary unit costs are included in RiskPoll.

Appendix C: Equations for multimedia model of Chapter 8

Table C.1 *Equations and input data for concentrations in soil*

Parameter	Definition and units	Equation	Default values for this chapter		
c_{soil}	Average soil concentration [kg/kg$_{soil}$]	$\dfrac{c_{soil}}{c_{air}} = \dfrac{k_{dep}}{h_{soil}\rho_{soil}\kappa_{soil}}$			
c_{air}	Ambient air concentration [kg/m^3 $_{air}$]				
k_{dep}	Total (dry + wet) deposition velocity [m/s]		0.0049		
h_{soil}	Soil depth [m]		Ingestion: 0.01 Animal feed: 0.10 Food crops: 0.20		
ρ_{soil}	Soil bulk density [kg$_{soil}$/m^3 $_{soil}$]		1500		
κ_{soil}	Total soil loss constant [1/s]	$\kappa_{soil} = \kappa_{soil,leach} + \kappa_{soil,ro} + \kappa_{soil,er} = \dfrac{1}{\tau_{soil}}$ τ_{soil} = soil time constant [s]			
$\kappa_{soil,leach}$	Loss constant for leaching [1/s]	$\kappa_{soil,leach} = \dfrac{k_{prec} + k_{irr} - k_{ro} - k_{evap}}{h_{soil}} \cdot \dfrac{1}{\theta_{sw} + K_{sw}\rho_{soil}}$			
k_{prec}	Mean precipitation rate [mm/yr]		750		
k_{irr}	Mean irrigation rate [mm/yr]		110		
k_{ro}	Mean surface run-off from pervious area [mm/yr]		100		
k_{evap}	Mean evaporation rate [mm/yr]		300		
K_{sw}	Soil–water partition coefficient [(kg/kg$_{soil}$)/(kg/m^3 $_{wat}$)]	$K_{sw} = \dfrac{c_{soil}}{c_{wat}}$ c_{wat} = water concentration [kg/m^3 $_{wat}$]	at pH As Cd Cr Hg Ni Pb	6.8 (default) 0.029 0.075 0.019 7.0 0.065 0.9	4.9 0.025 0.015 0.031 0.016	8.0 0.031 4.3 0.014 1.9
$\theta_{s,wat}$	Soil volumetric water content [m^3 $_{wat}$/m^3 $_{soil}$]		0.2		

Symbol	Description	Equation	
$\kappa_{soil,ro}$	Soil loss constant for surface runoff [1/s]	$\kappa_{soil,ro} = \dfrac{k_{ro}}{h_{soil}} \dfrac{1}{\theta_{sw} + K_{sw}\rho_{soil}}$	
$\kappa_{soil,er}$	Soil loss constant for erosion [1/s]	$\kappa_{soil,er} = \dfrac{L_{soil} R_{sed} R_{en}}{h_{soil}} \dfrac{K_{sw}}{\theta_{sw} + K_{sw}\rho_{soil}}$	
L_{soil}	Unit soil loss [$kg_{soil}/(m^2.s)$]	$L_{soil} = RF \times K \times LS \times C \times PF \times \dfrac{907.18\ kg/ton}{4047 m^2/acre}$	RF = Erosivity factor = 122.5 [1/yr] K = Erodibility factor = 0.39 ton/acre LS = Length slope factor = 1.5 C = Cover management factor = 0.1 PF = Supporting practice factor = 1.0 $L_{soil} = 1.61$ [$kg_{soil}/(m^2.s)$]
R_{sed}	Sedimentary delivery ratio [–]		0.1065
R_{en}	Soil enrichment ratio [–]		1.0
$X_{soil,ing}$	Air-to-soil transfer factor for ingestion [$(kg/kg_{soil})/(kg/m^3_{air})$]	$X_{soil,ing} = \dfrac{c_{soil,ing}}{c_{air}}$	

Table C.2 *Equations and input data for concentrations in water*

Parameter	Term definition and units	Equation	Default values for this chapter
$c_{wat,tot}$	Total water body concentration [kg/m³_wat] (water column + upper benthic sediment layer)	$c_{wat,tot} = \dfrac{\dot{m}_{wat}}{\bar{V}_{wat}\,f_{wc} + \kappa_{wat}\,A_{wat}(h_{wc}+h_{bs})}$	
$\dot{m}_{wat}$	Total pollutant flux to water [kg/yr]	$\dot{m}_{wat} = A_{wat}\,c_{air}\,k_{dep} + \dot{m}_{ro,i} + \dot{m}_{ro,p} + \dot{m}_{er}$ $A_{wat}c_{air}k_{dep}$ = direct deposition to water area A_{wat}	A_{wat} = 2% of A_{land} A_{land} = watershed surface area A_{imp} = 10% of A_{land}
$\dot{m}_{ro,i}$	Runoff from impervious area to water [kg/yr]	$\dot{m}_{ro,i} = A_{imp}\,c_{air}\,k_{dep}$ A_{imp} = impervious area of watershed	
$\dot{m}_{ro,p}$	Runoff from pervious area to water [kg/yr]	$\dot{m}_{ro,p} = k_{ro}\left(A_{land} - A_{wat} - A_{imp}\right)\dfrac{c_{soil}\rho_{soil}}{\theta_{s,wat}+K_{sw}\rho_{soil}}$	
$\dot{m}_{er}$	Erosion from pervious area to water [kg/yr]	$\dot{m}_{er} = L_{soil}\left(A_{land}-A_{imp}\right)R_{sed}\,R_{en}\,\dfrac{c_{soil}K_{sw}\rho_{soil}}{\theta_{s,wat}+K_{sw}\rho_{soil}}$	
$\dot{V}_{wat}$	Water flow rate through watershed [m³_wat/yr]	$\dot{V}_{wat} = A_{land}\left(k_{prec}-k_{evap}\right)$	
f_{wc}	Fraction of $c_{wat,tot}$ that is in water column [–]	$f_{wc} = \dfrac{(1+K_{sw}c_{sussol})h_{wc}}{(1+K_{sw}c_{sussol})h_{wc}+(\theta_{bs}+K_{sw}c_{bs})h_{bs}}$	
c_{sussol}	Total suspended solids concentration [kg_sed/m³_wat]		0.01
h_{wc}	Water column depth [m]		1.0
θ_{bs}	Bed sediment porosity [m³_wat/m³_sed]		0.6
c_{bs}	Bed sediment concentration [kg_sed/m³_wat]		100
h_{bs}	Upper benthic sediment layer [m]		0.03
κ_{wat}	Overall water body loss constant [1/s]	$\kappa_{wat} = f_{bs}\,k_b$ $f_{bs} = 1 - f_{wc};\; k_b = \dfrac{L_{soil}A_{land}R_{sed} - \dot{V}_{wat}c_{sussol}}{A_{wat}c_{bs}h_{bs}}$ k_b = Benthic rate burial constant	
$c_{wc,tot}$	Water column concentration [kg/m³_wat] (dissolved phase + suspended sediment)	$c_{wc,tot} = f_{cs}\,c_{wat,tot}\,\dfrac{(h_{wc}+h_{bs})}{h_{wc}}$	
$c_{wat,d}$	Dissolved phase water concentration [kg/m³_wat]	$c_{wat,d} = \dfrac{c_{wc,tot}}{1+K_{sw}c_{sussol}}$	
X_{wat}	Air-to-water transfer factor [m³_air/m³_wat]	$X_{wat} = \dfrac{c_{wat,d}}{c_{air}}$	

Table C.3 *Equations and input data for concentrations in vegetation*

Parameter	Term definition and units	Equation	Default values for this chapter
$c_{plant,foliar}$	Plant concentration from foliar absorption [kg/kg$_{DW}$]	$c_{plant,foliar}$ $= \dfrac{[k_{dep,dry}c_{air} + f_{wet}(k_{dep,wet}c_{air}+f_{irr}k_{irr}c_{wc,tot})]f_{int}[1-\exp(-\kappa_p t_p)]}{Y_p\kappa_p}$	
$k_{dep,dry}$	Dry deposition velocity [m/s]	$k_{dep,dry} = k_{dep} - k_{dep,wet}$ with $k_{dep,wet} =$ wet deposition velocity [m/s] $= \dfrac{k_{dep}}{2} \times \left(\dfrac{k_{prec}}{750mm/yr}\right)$	60%
f_{wet}	Fraction of wet deposition that adheres to plant [–]		12%
f_{irr}	Fraction of irrigated cropland [–]		Food crops: 39% Forage: 50% Silage: 46%
f_{int}	Fraction of deposition intercepted by edible portion of plant [–]		
k_{irr}	Irrigation rate, averaged over time and all cropland [mm/yr]	Typically, around 10 kg$_{wat}$ / (m^2.day) during irrigation days; 12% of cropland in EU is irrigated 3 months of the year	110
κ_p	Plant surface loss constant [yr^{-1}]		18
t_p	Exposure time of edible portion of plant per harvest [yr]		Food crops: 0.16 Forage: 0.12 Silage: 0.16
Y_p	Yield per planted area [kg$_{DW}$/m^2]		Food crops: 2.24 Forage: 0.24 Silage: 0.8

Table C.3 (*cont.*)

Parameter	Term definition and units	Equation	Default values for this chapter
$c_{plant,root}$	Plant concentration from root uptake [kg/kg$_{DW}$]	$c_{plant,root} = c_{soil} \times B_{root}$	
B_{root}	Plant–soil transfer factor [(kg/kg$_{DW}$)/(kg/kg$_{soil}$)]		**Green vegetables** As: 0.00633 Cd: 0.125 Cr: 0.00488 Hg: 0.0294 Ni: 0.00931 Pb: 0.0136 **Root vegetables** As: 0.008 Cd: 0.064 Cr: 0.0045 Hg: 0.099 Ni: 0.008 Pb: 0.009 **Grains** As: 0.004 Cd: 0.062 Cr: 0.0045 Hg: 0.3 Ni: 0.006 Pb: 0.009 **Animal feed** As: 0.036 Cd: 0.364 Cr: 0.0075 Hg: 0.3 Ni: 0.032 Pb: 0.045
$c_{plant,tot}$	Total pollutant concentration in plant [kg/kg$_{DW}$]	$c_{plant,tot} = c_{plant,foliar} + c_{plant,root}$	
X_{veg}	Air-to-vegetables transfer factors [m$^3_{air}$/kg$_{DW}$]	$\text{Green}: X_{gv} = \dfrac{c_{plant.tot.gv}}{c_{air}} ; \text{Root}: X_{rv} = \dfrac{c_{plant.tot.rv}}{c_{air}}$	
X_{grains}	Air-to-grains transfer factor [m$^3_{air}$/kg$_{DW}$]	$X_{grains} = \dfrac{c_{plant.tot.grains}}{c_{air}}$	
X_{feed}	Air-to-animal feed transfer factors [m$^3_{air}$/kg$_{DW}$]	$\text{Forage}: X_{forage} = \dfrac{c_{plant.tot.forage}}{c_{air}} ; \text{Silage}: X_{silage} = \dfrac{c_{plant.tot.silage}}{c_{air}}$	

Table C.4 *Equations and input data for concentrations in meat, milk and fish*

Parameter	Term definition and units	Equation	Default values for this chapter
c_m	Pollutant concentration in milk or meat from cattle [kg/kg$_{milk}$ or kg/kg$_{beef,FW}$]	$c_m = B_m \sum_i Q_{feed,i} \times c_{feed,i}$ $Q_{feed,i}$ = quantity of feed intake of type i, including water, forage, silage, grains and ingestion of surface soil $c_{feed,i}$ = concentration in feed of type i (for water $c_{wc,tot}$)	$Q_{feed,i}$ beef cattle milk cattle water [m³$_{wat}$/day]: 0.04 0.06 forage [kg$_{DW}$/day]: 8.8 13.2 silage [kg$_{DW}$/day]: 2.5 4.1 grains [kg$_{DW}$/day]: 0.47 3 soil [kg$_{soil}$/day]: 0.5 0.4
B_m	Biotransfer factor for milk or meat from cattle [day/kg$_{milk}$ or day/kg$_{meat,FW}$]		For meat For milk As: 0.002 0.006 Cd: 0.00012 0.0000065 Cr: 0.0055 0.0015 Hg: 0.00078 0.000338 Ni: 0.006 0.001 Pb: 0.0003 0.00025
X_m	Transfer factor air to milk or meat [m³$_{air}$/kg$_{milk}$ or m³$_{air}$/kg$_{beef,FW}$] $X_m = c_m/c_{air}$	$c_{fish} = c_{wc,tot} \times B_{fish}$	
c_{fish}	Pollutant concentration in fish [kg/kg$_{fish,FW}$]		
B_{fish}	Bioconcentration factor water→fish for freshwater fish [m³$_{wat}$/kg$_{fish,FW}$]		As: 0.1 Cd: 0.2 Cr: 0.01 Hg: 100 Ni: 0.2 Pb: 0.05
X_{fish}	Air-to-freshwater fish transfer factor [m³$_{air}$/kg$_{fish,FW}$]	$X_{fish} = \frac{c_{fish}}{c_{air}}$	

Table C.5 *Sensitivity of intake fractions to changes of input parameters (normalized by base case with $t_{cut} = 100$ yr).*

Parameters changed in base case	As	Cd	Cr	Ni	Pb
Base case (values in Chapter 8)	1.000	1.000	1.000	1.000	1.000
$k_{dry}=k_{wet}=0.335$ cm/s	0.996	0.995	0.994	0.995	0.974
$k_{dry}=k_{wet}=1$ cm/s	0.990	0.987	0.984	0.985	0.963
$k_{dry} = 3$ x v_{wet}	**1.107**	1.048	**1.108**	1.092	1.071
$k_{dry} = 0.333$ x v_{wet}	**0.893**	0.952	**0.892**	0.908	0.929
$h_{soil,ingestion} = 3$ cm (3 x base case)	1.000	1.000	1.000	1.000	0.979
$k_{irr} = 33$ cm/s (3 x base case)	0.979	0.868	0.984	0.956	0.966
Soil–water partition factors for pH=4.9 (no values for Pb, use pH=6.8)	0.995	**0.638**	1.012	0.939	1.000
Soil–water partition factors for pH=8.0 (no values for Pb, use pH=6.8)	1.003	**1.734**	0.995	**1.466**	1.000
$h_{wc} = 3$ m (3 x base case)	0.998	0.997	0.998	0.997	0.998
$c_{bs} = 1$ kg/L (10 times base case)	1.000	1.000	1.000	1.000	1.000
$c_{sussol} = 0.1$ kg$_{soil}$/m$^3_{wat}$ (10 times base case)	1.000	1.000	1.000	1.000	1.000
Yield per planted area $Y_p = 3$ x base case	**0.427**	**0.745**	**0.424**	**0.507**	**0.618**
Soil–plant transfer factors = 0.333 x base case	0.983	**0.692**	0.992	0.960	**0.852**
Soil–plant transfer factors = 3 x base case	1.051	**1.923**	1.023	**1.120**	**1.445**
Transfer factors for meat = 0.333 x base case	0.968	0.996	**0.868**	**0.856**	0.983
Transfer factors for meat = 3 x base case	1.096	1.012	**1.397**	**1.431**	1.050
Transfer factors for milk = 0.333 x base case	**0.639**	0.999	**0.863**	0.911	0.957
Transfer factors for milk = 3 x base case	**2.083**	1.002	**1.411**	**1.267**	**1.130**
Transfer factors for fish = 0.333 x base case	0.982	0.954	0.997	0.947	0.982
Transfer factors for fish = 3 x base case	1.055	**1.139**	1.009	**1.158**	1.054
$t_{cut} = 30$ yr	0.998	0.806	0.999	0.980	**0.826**
$t_{cut} = \infty$	1.000	1.039	1.000	1.003	**1.706**

Index